Springer Water

More information about this series at http://www.springer.com/series/13419

Sangam Shrestha · Anil K. Anal
P. Abdul Salam · Michael van der Valk
Editors

Managing Water Resources under Climate Uncertainty

Examples from Asia, Europe, Latin America, and Australia

Springer

Editors
Sangam Shrestha
Water Engineering and Management
Asian Institute of Technology
Klong Luang, Pathum Thani
Thailand

Anil K. Anal
Food Engineering and Bioprocess Technology
Asian Institute of Technology
Klong Luang
Thailand

P. Abdul Salam
School of Environment, Resources and Development
Asian Institute of Technology
Klong Luang
Thailand

Michael van der Valk
The Netherlands National Committee IHP-HWRP
Amsterdam
The Netherlands

ISBN 978-3-319-10466-9 ISBN 978-3-319-10467-6 (eBook)
DOI 10.1007/978-3-319-10467-6

Library of Congress Control Number: 2014951154

Springer Cham Heidelberg New York Dordrecht London

Printed on acid-free paper

Springer is part of Springer Science+Business Media (www.springer.com)

Foreword by WMO

The Fifth Assessment Report (2013) by the Intergovernmental Panel on Climate Change (IPCC), adopted by 110 governments, provides conclusive new scientific evidence that human activities are causing unprecedented changes in the Earth's climate. The report confirms that it is extremely likely (95–100 % probability) that most of the warming since 1950 has been due to human influence.

The new report further states that greenhouse gas emissions at or above current rates would induce changes in the oceans, ice caps, glaciers, the biosphere and other components of the climate system. Some of these changes would very likely be unprecedented over decades to thousands of years. Limiting climate change would require substantial and sustained reductions in emissions of carbon dioxide (CO2) and other greenhouse gases.

In a changing climate, our valuable water resources will be one of those areas most impacted. For example, there is very high confidence that glaciers have continued to shrink and lose mass worldwide, with very few exceptions. By 2100, global glacial volume could, under one scenario, decline further by as much as 35–85 %. Meanwhile, the extent of Northern Hemisphere snow cover has decreased since the mid-twentieth century, especially in spring, and this decline, too, will continue.

It is likely that human influences have affected the global water cycle and its patterns since 1960. For example, in recent decades, precipitation has increased in the mid-latitude land areas of the Northern Hemisphere.

The UN-wide Global Framework on Climate Services (GFCS) led by the World Meteorological Organization (WMO), with a wide range of partners, assists governments to produce and use climate information and predictions for adapting to and mitigating climate change while transitioning to a green economy. Climate services can empower decision-makers, making water resources management decisions more climate resilient. The Integrated Flood Management and Integrated Drought Management approaches, adopted by the WMO in partnership with the Global Water Partnership are just two risk-based, resilience building methodologies that will improve the coordination and collaboration between the climate and water communities as part of the GFCS User Interface Platform.

This book contains a collection of individually authored chapters, which provide increased knowledge of the impacts of climate change on the water cycle and identify practices and procedures that can assist in adaptation to a changing and variable climate. I commend the authors of these chapters for their contributions and the Editors for bringing the material together and urge readers to critically examine, review and make use of the material in the most relevant and practical manner.

Michel Jarraud
Secretary-General of WMO

Foreword by UNESCO

Water—the basic ingredient of life and a fundamental human right—holds the key to global sustainability. The UN International Year of Water Cooperation, 2013, emphasised that cooperation *around* water, *for* water and *through* water must happen everywhere—between states and within states. While we talk about water, we are really talking about human rights, about the sustainable development of our societies, about sustainability of ecosystems. This publication constitutes a joint effort between scientists and other experts from around the world, and is testament to the spirit of the UN International Year of Water Cooperation, during which the preparations for the book started.

UNESCO's International Hydrological Programme (IHP) is the only intergovernmental water science programme of the United Nations. Over the past 12 years, The Netherlands, through the Secretariat of its National IHP Committee, has been one of the most active countries worldwide contributing to the Programme. The Secretary personally has also been instrumental in supporting many water professionals from developing countries and countries in transition. In addition, the Asian Institute of Technology (AIT) has indeed a long history of working together with the United Nations, in research, education and capacity-building, at a high level.

The IHP is an intergovernmental programme that is implemented in phases. IHP operates in accordance with the needs of its Member States, and thrives thanks to their support and contributions. In 2014, the eighth phase of IHP began, focused on six themes along three axes:

- Mobilizing international cooperation to improve knowledge and innovation to address water security challenges;
- Strengthening the science–policy interface to reach water security at local, national regional and global levels;
- Developing institutional and human capacities for water security and sustainability.

Despite the intergovernmental nature of IHP, the essential contributions to the Programme have always been the work of dedicated individuals with their hearts in the right place. They are the ones who deliver the substance and tangible results that advance humanity. This book is an excellent contribution to the Programme that highlights these achievements. East and West come together: water cooperation and science diplomacy in the true meaning.

It is also due to my personal involvement in the Intergovernmental Panel on Climate Change (IPCC) that I am delighted to see this publication '*Managing Water Resources under Climate Uncertainty*', addressing one of the most important current issues globally in water resources management. I sincerely hope that it will raise awareness of sensitive and urgent questions related to water resources management and climate uncertainty, in order to ensure that ecological principles, including hydrology, are at the heart of economic development and decision-making. In this regard, I would like to thank the editors of the publication for their excellent work. It reflects the rich expertise of participants from various geographical and cultural backgrounds, and thus the true spirit of water cooperation!

Blanca Elena Jiménez Cisneros
Director, Division of Water Sciences
Secretary, International Hydrological Programme, UNESCO

United Nations Educational, Scientific and Cultural Organization
International Hydrological Programme

Foreword by SEA-EU-NET

The availability of safe water is a major global challenge for the future owing to a rapidly growing population and unsustainable consumption pattern, increasingly urbanised populations, rapid shifts in land use and climate change. Global water demand has tripled in the past 50 years.[1] and just 2.5 % of the world's water resources are freshwater of which only 0.4 % are available and accessible for use. Water is intrinsically linked to the most pressing challenges we face today, including food security and safety, health, climate change, economic growth and poverty alleviation.

The United Nations projects that by 2025, half of all countries worldwide will face water stress or outright shortages. By 2050, three out of four people around the globe could be affected by water scarcity. Water problems in Asia today are severe—one out of five people (700 million) does not have access to safe drinking water and half of the region's population (1.8 billion people) lacks access to basic sanitation. Although Asia is home to more than half of the world's population, it has less freshwater, i.e. 3,920 cubic meters per person per year, than any other continent. As population growth and urbanization rates in the region rise, the stress on Asia's water resources is rapidly intensifying. Climate change is expected to worsen the situation. According to the Intergovernmental Panel on Climate Change (IPCC), by 2050, more than one billion people in Asia alone are projected to experience negative impacts on water resources as a result of climate change. Experts agree that reduced access to freshwater will lead to a cascading set of consequences, including impaired food production, the loss of livelihood security, large-scale migration within and across borders, and increased economic and geopolitical tensions and instabilities.

Within ASEAN, overall water demand is expected to increase by one-third by 2015.[2] Although most Southeast Asian countries do not experience physical water

---

[1] UNEP – A Tale of Two Trends: providing information and knowledge for decision-making in water-scare regions through water assessments – http://www.unwater.org/downloads/www.Singh.pdf.

[2] ASEAN (2005) ASEAN Strategic Plan of Action on Water Resources Management. Accessed http://environment.asean.org/files/ASEAN%20Strategic%20Plan%20of%20Action%20on%20Water%20Resources%20Management.pdf 27 May 2011.

scarcity, seasonal water scarcity can be an issue, e.g. in Cambodia and Vietnam. High rates of development put pressure on the sustainable water supply and sanitation, and increase competition for water resources. Some ASEAN member states are unlikely to meet the Millennium Development Goals relating to drinking water and sanitation. The key water challenges for the ASEAN region have already been set out in the ASEAN Strategic Plan of Action of Water Resources and Management.[3] They plan includes aspects such as collecting and maintaining high quality data, mitigating the effects of extreme events on water resources (especially to subsistence farmers and the poor), sustaining and improving water quality, improving governance systems and acquiring financing for the development of new water infrastructure.[4]

To address these challenges, we have initiated the project 'SEA-EU-NET 2—EU-ASEAN S&T cooperation to jointly tackle societal challenges'. The SEA-EU-NET 2 project aims at strengthening the bi-regional dialogue on international S&T cooperation between Europe and Southeast Asia, particularly by tackling societal challenges, creating direct linkages to the policy dialogue, development of additional funding sources and improved dissemination of project results to the interested public. The SEA-EU-NET 2 project is working within the framework of the official EU-SEA cooperation in Science, Technology and Innovation. Cooperation between EU and ASEAN, which has been ongoing for 30 years, has gained significant momentum over the last decade.

One of the primary aims of the project is to stimulate deeper and more productive cooperation in three global societal challenges: health, food and water. The rationale for the selection of the three societal challenges was recognition that these are areas in which the EU and Southeast Asia have strong and complementary interests. In health, Southeast Asia is increasingly coming to resemble Europe, with non-communicable diseases burdening health systems and taking over from infectious disease as the leading cause of death. Yet the region still suffers from high incidences of infectious diseases which Europe—though climate change and global connectedness—is also exposed to. Southeast Asia is a major exporter of food to Europe, providing a strong rationale to work with the region to ensure the security and safety or Europe's food supply. Disruption caused by flooding in Southeast Asia affects the production facilities of European companies and disrupts the plans of holidaymakers, and tensions over transboundary water resources threatens the stability of the region. These challenges are also interlinked; extreme weather events could threaten food supplies while also spreading waterborne diseases. These societal challenges also reflect the areas in which much EU-ASEAN collaboration already takes place.

---

[3] ASEAN (2005) ASEAN Strategic Plan of Action on Water Resources Management. Accessed http://environment.asean.org/files/ASEAN%20Strategic%20Plan%20of%20Action%20on%20Water%20Resources%20Management.pdf 27 May 2011.

[4] ASEAN (2005) ASEAN Strategic Plan of Action on Water Resources Management. Accessed http://environment.asean.org/files/ASEAN%20Strategic%20Plan%20of%20Action%20on%20Water%20Resources%20Management.pdf 27 May 2011.

International exchange and collaboration is necessary to tackle these complex and interrelated issues. This book is proving that there is much international knowledge and expertise which should help develop innovative solutions.

Christoph Elineau
SEA-EU-NET Coordinator
International Bureau of the German Federal Ministry
of Education and Research (DLR)

Acknowledgments

A number of individuals have contributed to the preparation of this book, and to whom we extend our deepest gratitude, but because of space constraints it is not possible to mention all the names here. However, it would be an injustice if we failed to mention a few individuals whose contributions are particularly significant. Our sincere thanks to all contributing authors who prepared the chapters despite their busy schedules, and who were always supportive despite constant and frequent reminders. We would also like to thank all the reviewers for their valuable feedback.

Special thanks to Smriti Malla and Kshitij Parajuli for their tireless efforts in communicating with the contributing authors and compiling and formatting all the chapters throughout the publication process.

Contents

Part I Understanding the Impacts

Impact of Climate Change on the Water Cycle 3
Velma I. Grover

Uncertainty Assessment of Climate Change Impacts on Hydrology: A Case Study for the Central Highlands of Vietnam 31
Dao Nguyen Khoi and Phan Thi Thanh Hang

Assessment of the Impact of Climate Change on Water Availability in the Citarum River Basin, Indonesia: The Use of Statistical Downscaling and Water Planning Tools 45
I. Putu Santikayasa, Mukand S. Babel and Sangam Shrestha

Impact of the Uncertainty of Future Climates on Discharge in the Nam Ou River Basin, Lao PDR 65
Manisha Maharjan and Mukand S. Babel

Integrated Modelling of Climate Change and Urban Drainage. 89
Ashish Shrestha, Mukand Singh Babel and Sutat Weesakul

Estimating the Impacts and Uncertainty of Climate Change on the Hydrology and Water Resources of the Koshi River Basin 105
Anshul Agarwal, Mukand S. Babel and Shreedhar Maskey

Uncertainty Analysis of Statistically and Dynamically Downscaled Precipitation Data: A Study of the Chaliyar River Basin, Kerala, India. 127
Santosh G. Thampi and N.R. Chithra

Assessment of the Impact of Projected Climate Change on Streamflow and Groundwater Recharge in a River Basin 143
Santosh G. Thampi and K.Y. Raneesh

Assessing the Impacts of Climate Change on Water Resources: Experiences From the Mediterranean Region 177
Ivan Portoghese, Michele Vurro and Antonio Lopez

Effect of Climate Change in Wastewater Treatment Plants: Reviewing the Problems and Solutions 197
Anastasios Zouboulis and Athanasia Tolkou

Part II Adaptation

Managing Hydropower Under Climate Change in the Mekong Tributaries 223
Thanapon Piman, Tom A. Cochrane, Mauricio E. Arias, Nguyen D. Dat and Ornanong Vonnarart

Managing Water Resources Under Climate Uncertainty: Opportunities and Challenges for Cambodia 249
Thun Vathana

Transboundary Water Resources Management in the Context of Global Environmental Change: The Case of Bhutan Himalaya 269
Om N. Katel, Dietrich Schmidt-Vogt and Ngawang Dendup

Addressing Climate Change Impacts Through Community-Level Water Management: Case Studies from the South-Eastern Coastal Region of India 291
Ronjon Chakrabarti, Ni Luh Made Ashanapuri Hertz, Sibylle Kabisch, Alice Reil and Till Wolf

Climate Adaptation and Governance and Small-Scale Farmers' Vulnerability Through Artificial Glacier Technology: Experiences from the Cold Desert of Leh in North-West Himalaya, India 319
Falendra Kumar Sudan and Jennifer McKay

Governing Disaster Risk Reduction in Nepal 343
Dallas Blaney

Role of Sustainability Policy Entrepreneurs in Building Water-Sensitive Cities to Respond to Climate Change: A Case Study in Adelaide, Australia 359
Ganesh Keremane

Managing Water Resources in Mexico in the Context of Climate Change 377
Úrsula Oswald Spring

Rainwater Harvesting as an Effective Climate Change Adaptation Strategy in Rural and Urban Settings 405
Ezgi Akpinar Ferrand

Coevolving Water Infrastructures for Adaptation to Climate Change 421
Tse-Hui Teh

Abbreviations

ADB	Asian Development Bank
AGloCAP	Adaptation to Global Changes in Agricultural Practices
ANN	Artificial Neural Networks
ANOVA	Analysis of Variance
AOGCM	Atmosphere-Ocean General Circulation Model
APHRODITE	Asian Precipitation Highly Resolved Observational Data Integration Towards Evaluation of Water Resources
AR4	Fourth Assessment Report
ASR	Aquifer Storage and Recharge
BEA	Bhutan Electricity Authority
BOD	Biochemical Oxygen Demand
BPL	Below Poverty Line
BRO	Border Roads Organisation
CA	Christian Aid
CBOs	Community-based Organisations
CC	Climate Change
CCS	Climate Change Scenarios
CER	Certified Emission Reduction
CH_4	Methane Gases
CIRCE	Climate Change and Impact Research
CNR	National Research Council of Italy
CNRS-L	National Council for Scientific Research of Lebanon
CO_2	Carbon Dioxide
COD	Chemical Oxygen Demand
CPRC	Chronic Poverty Research Centre
CSO	Combined Sewer Overflows
DCA	DanChurchAid
DD	Dynamical Downscaling
DEFRA	Department for Environment, Food and Rural Affairs
DEM	Digital Elevation Model

DHM	Department of Hydrology and Meteorology
DWT	Depth to Water table
EDA	Exploratory Data Analysis
EI	Efficiency Index
ENSO	El Nino Southern Oscillation
EPA	Environmental Protection Agency
ESD	Ecologically Sustainable Development
ESS	Ecosystem Services
EWEs	Extreme Weather Events
FAO	Food and Agriculture Organization
GCM	General Circulation Model
GDP	Gross Domestic Product
GEC	Global Environmental Change
GHF	Global Heritage Fund
GHG	Greenhouse Gases
GHGES	Gas Emission Scenarios
GHR	Greater Himalayan Region
GIS	Geographic Information System
GLOF	Glacial Lake Outburst Floods
GoI	Government of India
GWL	Global Water Law
HadCM3	Hadley Centre Coupled Model version 3
HFCs	Hydrofluorocarbons
HRUs	Hydrological Response Unit
ICIMOD	International Centre for Integrated Mountain Development
IDFs	Intensity Duration Frequency Curves
IITM	Indian Institute of Tropical Meteorology
IMD	Indian Meteorological Department
INCCA	Indian Network for Climate Change Assessment
INEGI	Instituto Nacional de Estadística, Geografía e Informática
IPCC	Intergovernmental Panel on Climate Change
IRI	International Research Institute for Climate Prediction
IRIN	Integrated Regional Information Networks
IUWM	Integrated Urban Water Management
IWMI	International Water Management Institute
K–S test	Kolmogórov–Smirnov test
LAF	Leaf Area Index
LARS-WG	Long Ashton Research Station—Weather Generator
LDOF	Landslide Dam Outburst Floods
LGA	Local Government Authority
LULC	Land Use and/or Land Cover
MB	Mean Bias
MEoF	Ministry of Environment and Forests
MONRE	Ministry of Natural Resources and Environment
MRC	Mekong River Commission

MSL	Mean Sea Level
MVMC	Metropolitan Valley of Mexico City
N_2O	Nitrous Oxide
NAFOSTED	National Foundation for Science and Technology Development
NAFTA	North American Free Trade Agreement
NCEP	National Center for Environmental Prediction
NCRMP	National Cyclone Risk Mitigation Project
NDN	Nitrification/Denitrification
NDTV	New Delhi Television Limited
NGOs	Non-Governmental Organisations
NIW	Nahr Ibrahim Watershed
NMHS	National Meteorological and Hydrological Service
NMSE	Normalised Mean Square Error
NRRC	Nepal Risk Reduction Consortium
NSE	Nash-Sutcliffe Efficiency
NWC	National Water Commission of Mexico
PBIAS	Percent Bias
PDFs	Probability Density Functions
PDS	Public Distribution System
PFCs	Perfluorocarbons
PRECIS	Providing Regional Climate for Impacts Studies
PWD	Public Works Department
R&D	Research and Development
R^2	Coefficient of Determination
RACCM	Regional Assessment of Climate Change in the Mediterranean
RCM	Regional Climate Model
RCOF	Regional Climate Outlook Forum
RGOB	Royal Government of Bhutan
RMA	Royal Monetary Authority
RMSE	Root Mean Square Error
SD	Statistical Downscaling
SDSM	Statistical Downscaling Model
SI	Statistical Interpolation
SRES	Special Report on Emission Scenarios
SRTM	Shuttle Radar Topography Mission
SSO	Sanitary Sewer Overflow
STI	Science Technology and Innovation
SWAT	Soil and Water Assessment Tool
SYB	Statistical Year Book of Bhutan
Tmax	Maximum Temperature
Tmin	Minimum Temperature
ToRs	Terms of Reference
TSS	Total Suspended Solids
UNDP	United Nations Development Program
UNEP	United Nations Environment Program

UNESCO	United Nations Educational, Scientific and Cultural Organization
UNFCCC	United Nations Framework Convention on Climate Change
UNICEF	United Nations International Children's Fund
UNISDR	United Nations International Strategy for Disaster
UPaRF	UNESCO-IHE Partnership Research Fund
URE	Urban Rainfall Effect
USACE	US Army Corps of Engineers
USDA	United States Department of Agriculture
UW CIG	University of Washington's Climate Impacts Group
VE	Volume Error
VNU-HCM	Vietnam National University Ho Chi Minh City
VOCs	Volatile Organic Compounds
WEAP	Water Evaluation and Planning
WHO	World Health Organization
WF	Water Footprint
WMO	World Meteorological Organization
WWT	Wastewater Treatment
WWTPs	Waste Water Treatment Plants

About the Editors

Dr. Anil Kumar Anal is an Associate Professor in Food Engineering and Bioprocess Technology at the Asian Institute of Technology (AIT), and coordinator of Food Agriculture and Biosystems cluster. His background expertise is in the valorization as well as bioprocessing of agro-industrial waste and its application in functional foods, neutraceuticals, cosmetics and pharmaceuticals as well as the formulation and delivery of cells and bioactivity for human and veterinary applications, controlled release technologies, and particulate systems. He also has interests in physicochemical characterization, interactions and applications of biopolymers and bioactive compounds in the functionality and delivery of cells and other bioactive compounds, detection and control of food pathogens using biosensor/nanotechnology; and biopackaging. His recent research focusses on the valorization of waste from plant and animal sources including seafood and other marine waste, bioprocess technology, post-harvest technology towards green growth, food safety and food security issues in developing countries, micro-/nanoencapsulation technology of marine omega-3 fatty acids, cells, probiotics, immunoglobulins, peptides, enzymes, vitamins, and antioxidants for gastrointestinal targeted delivery to enhance the stability and bioavailability for optimising health benefits. He has published various articles in peer-reviewed and internationally referred Life Science Journals, books, and conferences. One of his recent works on the encapsulation of probiotics has been granted a U.S. Patent and World Patent. His two books on Food Waste Valorisation and Utilisation and Functional Foods published by Wiley, are under preparation and due to be published soon. He is a guest author in the book series of the Pharmaceutical Manufacturing Handbook and Pharmaceutical Sciences Encyclopedia: and similarly for the chapters in various books. He is currently serving as an Editorial Board Member in some of the relevant International and Regional Journals. Dr. Anal received his Ph.D. in Bioprocess Technology from the AIT, Thailand. He has held previous academic and research positions in industries and academia including Otago University and Massey University, New Zealand.

Dr. P. Abdul Salam received his B.Sc. (Eng.) (Honours) in Mechanical Engineering (1991) from the University of Peradeniya, Sri Lanka; Master of Engineering (1994) in Renewable Energy and Doctor of Engineering (2005) in Energy Technology from the AIT. He is currently an assistant professor at the Energy Programme of School of Environment, Resources and Development, AIT. He has more than 15 years of teaching, research and consultancy experience in the area of renewable energy, rational use of energy, climate change mitigation, clean coal technologies, carbon capture and sequestration, low carbon cities, etc. He has carried out, so far, 17 sponsored research projects and organised 20 seminars and workshops at AIT. Dr. Salam has published 4 books/monographs; 35 internationally refereed journal articles and conference papers; 15 non-refereed development reports and articles. He has also had 3 years' experience working as the General Manager and Chief Design Engineer of an energy service company.

Dr. Sangam Shrestha is an Assistant Professor of Water Engineering and Management at the Asian Institute of Technology (AIT), Thailand. He is also a Visiting Faculty Member of the University of Yamanashi, Japan, and Research Fellow of the Institute for Global Environmental Strategies (IGES), Japan. His research interests are within the field of hydrology and water resources including, climate change impact assessment and adaptation on the water sector, water footprint assessment, and groundwater assessment and management.

After completing his Ph.D., Dr. Shrestha continued his postdoctoral research in the GCOE project at the University of Yamanashi in Japan until 2007, where he was involved in the development and application of a material circulation model and groundwater research in the Kathmandu Valley. He then worked as a policy researcher at the IGES, where he was actively involved in research and outreach activities related to water and climate change adaptation and groundwater management in Asian cities. Dr. Shrestha has published more than three dozen peer-reviewed international journal articles and presented more than three dozen conference papers ranging from hydrological modelling to climate change adaptation in the water sector. His recent publications include 'Kathmandu Valley Groundwater Outlook' and 'Climate Change and Water Resources'.

His present work responsibilities at AIT include delivering lectures at the postgraduate and undergraduate levels, supervising research to postgraduate students and providing consulting services on water related issues to government and donor agencies, and research institutions. He has conducted several projects relating to water resources management, climate change impacts and adaptation, with awards from International organisations such as APN, CIDA, EU, FAO, IFS, IGES, UNEP and UNESCO.

Dr. Michael R. van der Valk was trained as a hydrologist with degrees in physical geography and geographical hydrology from VU University, Amsterdam, The Netherlands. Since its initiation in 1993, he has been final editor of Stromingen, the professional magazine of The Netherlands Hydrological Society, where he currently is the board member for international relations. From 2002 to 2008, he served as the water expert in the International Relations and Strategy department of the Royal Netherlands Meteorological Institute (KNMI). From 2008 onwards, he has been coordinator of the Communication and Information portfolio of the Cooperative Programme on Water and Climate (CPWC), which included organising sessions on climate change adaptation during the fifth and sixth World Water Forums, coining the Water and Climate Days during the World Water Week in Stockholm, and a high-level panel on groundwater and climate change in Africa during the UNFCCC COP-15 in Copenhagen. He is currently the Scientific Secretary of The Netherlands National Committee IHP-HWRP (the scientific water programmes of UNESCO and WMO), an advisory commission for the Government of The Netherlands. He is a board member of The Netherlands' chapter of the International Association for Hydrogeologists (IAH) and director of CrossVision Communications.

Michael van der Valk has contributed to the UNECE Guidance on Water and Adaptation to Climate Change, and he was author of the UN Commission's report on Transboundary Flood Risk Management. Over the years he has served on several commissions and working groups of the International Hydrological Programme (IHP) of UNESCO and as a member of the Commission for Hydrology of the World Meteorological Organization (WMO). He is editor of about 25 publications on hydrology, water resources and climate change (adaptation) including Climate Change Adaptation in the Water Sector.

About the Authors

Mr. Anshul Agarwal is a Doctoral Researcher at Asian Institute of Technology (AIT), Thailand. Mr. Agarwal has a Master's degree in Water Engineering and Management from AIT and a Bachelor's degree in Agriculture Engineering from MPUAT, Udaipur, India. He has recently completed his Ph.D. research regarding Assessment of climate and land use change impacts on hydrology and water resources of the Koshi river basin, Nepal. His recent work involves statistical downscaling of future climate projections, uncertainty analysis, GIS-based analysis, hydrological modelling, crop modelling etc. He has been dealing with the statistical downscaling tools: SDSM, LARS-WG, ANN; hydrological models: SWAT, HEC-HMS, MIKE-11; crop models: DSSAT, CROPWAT and remote sensing and GIS tools: ERDAS, ENVI, ArcGIS, Quantum-GIS. Mr. Agarwal has published many peer-reviewed international journal papers and conference papers ranging from crop modelling, hydrological modelling to climate change impacts and adaptation on the agriculture and water sector.

Dallas Blaney is an Assistant Professor of Political Science in the Department of Public and Environmental Affairs at the University of Wisconsin-Green Bay. Mr. Blaney earned his Ph.D. at Colorado State University, where his research explored the role of non-government organisations in the global governance of freshwater resources. Mr. Blaney has published on the topic of global water resource governance, generally, and the governance of environmental flows, specifically. His current research projects explore the role of non-governmental organisations in the governance of water-induced disaster risk reduction in Nepal, examine the linkages between groundwater resource management and environmental flows, and evaluate the post-2015 development agenda. Mr. Blaney teaches courses on Global Environmental Politics, the Politics of Developing Areas and International Relations.

Ronjon Chakrabarti is Scientist and Project Manager at Adelphi. His research and work focusses on integrated water resource management, water supply and innovative water treatment technologies, climate change and renewable energies, technology cooperation and participatory approaches. Currently, he is collaboration manager and leading scientist for the Indo European water treatment research collaboration project 'ECO-India' aiming at developing energy efficient community-based water supplies. He also works as a technical expert for climate change adaptation measures in the water sector and their participatory development with public and private stakeholders.

He studied Environmental Engineering and Philosophy at the Technical University in Berlin, where he focussed on water quality control and integrated water resource management. In addition to his employment at Adelphi, he is writing and pursuing his applied research on IWRM and surface water treatment solutions in rural areas with the Technical University, Berlin and the Jadavpur University, Kolkata.

Ezgi Akpinar Ferrand is an Assistant Professor of Geography at Southern Connecticut State University. She is a physical geographer with research interests in water resources, past-to-present human-environment interactions, climate change and sustainable development.

She has authored and co-authored papers in these areas for Environmental Science and Policy, Geography Compass, Ancient Mesoamerica, Global Water Forum, and Encyclopedias of Global Warming and Climate Change and Energy. She is one of UC Irvine's Empowering Sustainability Fellows. Akpinar Ferrand holds a Ph.D. in Physical Geography from the University of Cincinnati.

Dr. Velma I. Grover is currently a Bryant Drake Guest Professor at Kobe College, Japan. After completing her Ph.D. (University of London, U.K.), Dr. Grover has worked in the area of environment and international development with Climate and Development Knowledge Network, UNU, UNESCO, the Economic Research Forum and various additional non-governmental organisations. She teaches at the McMaster and York Universities, and has consulted in the areas of water and waste management. Her work involves capacity building in integrated water resources management; impacts of climate change on the water cycle (health and food security); trans-boundary lake basin management; integrating supply of safe drinking water, sanitation, health and the reduction of poverty; and introduction of good governance practices in Asia, Africa and North America. Dr. Grover has been a visiting research fellow at Kalmar University, the Smith School of Enterprise and Environment at Oxford, and the School of International Relations and Public Affairs, Fudan University, Shanghai China.

Dr. Om Katel is a Bhutanese citizen and has completed his Ph.D. from Natural resources management field of studies, under School of Environment Resources and Development, Asian Institute of Technology, Thailand. Currently, he is working as a lecturer at the Department of Forestry, College of Natural Resources, Royal University of Bhutan. Dr. Katel teaches Integrated Watershed Management, Climate change Adaptation, Environmental Governance, Natural Resource Economics and Applied Conservation Science. Since the year 20013, he has been involved in three small scale research projects; (1) The cost of land use change on improvement of ecosystem services at Toebesa, Bhutan, (2) Farmer's vulnerability to climate variability in Punakha-wangdue valley and (3) Human wildlife conflict and species conservation in biological corridor in southern Bhutan. Dr. Katel's research interest is broad and falls within the theme of Biodiversity conservation, Water resources management, Adaptation to climate change, Ecosystem services, Land use and land cover change.

Dr. Ganesh Keremane is a research fellow at the School of Law, University of South Australia. He completed his Ph.D. at the University of South Australia for which he received the 2008 CRC Irrigation Futures Director's Award. He also has a Master's degree in Agricultural Economics from the University of Agricultural Sciences, India. After completing his Ph.D. Dr. Keremane continued at the CCWPL and is currently working with his colleagues on different research projects at the Centre including institutional analysis of implementing desalination projects and an integrated urban water management strategy. Dr. Keremane's research interests lie in the field of natural resources management with an emphasis on institutional and policy analysis of surface and groundwater management. He is also interested in assessing community attitudes and perceptions towards alternative water sources and has published several peer-reviewed international journal papers and book chapters related to these topics.

Mr. Dao Nguyen Khoi got a Ph.D.'s degree in integrated river basin management from University of Yamanashi, Japan in March 2013. He then joined the University of Science, a member of Vietnam National University Ho Chi Minh City (VNU-HCM), since April 2013, as a lecturer of Faculty of Environmental Science. His work responsibilities in the university are giving lectures and supervising research to graduate and undergraduate students. Besides that, he has also worked as Head of Department of Research and Development at Center of Water Management and Climate Change (WACC), which belongs to VNU-HCM. He has been conducting several projects related to climate change impacts on eco- and social-hydrology being funded by VNU-HCM and NAFOSTED.

Manisha Maharjan is a research associate at Asian Institute of Technology (AIT), Thailand. She received her B.Eng. in Civil Engineering from Tribhuvan University (Institute of Engineering, Nepal) in 2010. She then received her M.Eng. in Water Engineering and Management from AIT and Masters of Science Hydroprotech (Dual Degree) from University of Nice, France in 2012. She is currently working as a postgraduate researcher for Postdoctoral Research Programme on Adaptation to Climate Change

(PRoACC2), the project under UNESCO-IHE, The Netherland. Her research interest includes climate change and water resources, river hydrology, water modelling and sediment management.

Dr. Piman is a Technical Advisor under Climate Change and Adaptation Initiative, Mekong River Commission (MRC), Lao PDR and Research Associate of University of Canterbury, New Zealand. He graduated his Master's degree in Water Resources Engineering from Chulalongkorn University and doctoral degree in Water Resource Engineering and Management at Asian Institute of Technology, Thailand. He was a postdoctoral fellow in water resources under the Department of Civil and Natural Resources Engineering, University of Canterbury from 2011–2012 under the 'River at Risk' project. Dr. Piman has published a number of peer-reviewed international journal papers in climate change, hydrology, and water resources development and management. His present work at MRC is supporting Lower Mekong Countries (Cambodia, Lao PDR, Thailand and Vietnam) in climate change adaptation planning. His main responsibilities include developing database, monitoring and reporting system on climate change and adaptation, climate change and hydrological modelling, supporting basin-wide vulnerability assessment and formulating Mekong Adaptation Strategy.

Ivan Portoghese is Research Scientist at the National Research Council of Italy—Water Research Institute (IRSA) since 2005. His research is addressed to the hydrological characterisation of Mediterranean catchments and the development of methods for water resources assessment. At the IRSA, he is responsible for the research programme on the evaluation of climate change impacts on water resources management. Current activities concern the assessment of hydrological response under climate change, with particular emphasis on groundwater, and the development of water resources scenarios and related implications on water management.

He is author of more than 40 papers published in international journals, book chapters and conference proceedings, and reviewer of several science journals in the field of hydrology. He was lecturer in post-graduate courses in the field of hydrology and water resources.

I. Putu Santikayasa currently, is a Ph.D. candidate in Water Engineering and Management at Asian Institute of Technology, Thailand. He received his M.Sc. on the Information Technology for Natural Resources Management (IT for NRM) from Bogor Agricultural University, Indonesia. Previously, he was worked in the various areas related to hydrology, water resources management and climate change, analysing the data using the tool such as GIS, remote sensing, climate analysis and downscaling tools. He is also one of the teaching staff at Geophysics and Meteorology field of study, Bogor Agricultural University, Indonesia since 2005 to date. As part of his Ph.D. research, he is working on the hydro-economic model to analyse agriculture water use sustainability. The model combines the hydro-economic model to evaluate the scenario impacts on the water use sustainability under climate change. On the economic assessment, the model uses the optimization approach to maximize the expected utility in the farm level.

Mr. Ashish Shrestha is a Research Associate at Asian Institute of Technology (AIT), Thailand. He completed his joint degree masters programme on Urban Water Engineering and Management from AIT, Thailand and UNESCO-IHE Institute for Water Education, The Netherlands, in 2013. His research interests are focussed on climate change, urban hydroinformatics, hydrology, urban floods, urban water/waste water systems, environmental engineering, environmental assessment and water-energy-carbon nexus. His recent research works involve climate modelling to project future climate variables at high temporal and spatial resolution using statistical downscaling tools like SDSM, LARS WG and rainfall disaggregation tools; hydrodynamics 1D and 2D modelling of urban floods using tools like PCSWMM, MIKE MOUSE, MIKE URBAN, MIKE FLOOD, MIKE 21 and GIS tools.

Úrsula Oswald Spring (Mexico), full time Professor/ Researcher at the National Autonomous University of Mexico (UNAM) in the Regional Center for Multidisciplinary Research (CRIM) and lead author of IPCC. She was national coordinator of water research for the National Council of Science and Technology, first Chair on Social Vulnerability at the United National University Institute for Environment and Human Security; founding Secretary-General of El Colegio de Tlaxcala; General Attorney of Ecology (1992–1994), first Minister of Ecological Development in the State of Morelos (1994–1998). She was President of the International Peace Research Association and General Secretary of the Latin-American Council for Peace Research. She has studied medicine, clinical psychology, anthropology, ecology, and classical and modern languages. She has a Ph.D. from the University of ZÏrich. She received the following prices: Sor Juana Inés de la Cruz, Environmental Merit, Tlaxcala, UN Development Prize, Academic Women, from UNAM and Women of the Year 2000.

Dr. Falendra Kumar Sudan is currently working as Professor at Department of Economics, University of Jammu, Jammu and Kashmir, India with specialisation in Environmental and Natural Resource Economics. He is Adjunct Research Fellow, Centre for Comparative Water Policies and Laws, University of South Australia, Adelaide. Dr. Sudan has been awarded various interdisciplinary postdoctoral research projects and national and international fellowships in multidisciplinary perspectives. He has been awarded Senior Professional Research Fellowship for the 'International Training and Research Program 2006–2007' on 'Groundwater Governance in Theory and Practice' by International Water Management Institute, Sri Lanka. He has served as Editor of International Journal of Environment and Development for the period June 2004 to June 2008 and Member, Editorial Board of Applied Economics and Policy Analysis (An International Journal) and Journal of Social and Economic Policy. He has extensively visited foreign universities for academic and research collaborations and participation.

Dr. Thampi is a Professor in the Department of Civil Engineering at the National Institute of Technology Calicut (NITC) at Kozhikode, Kerala, India. After a stint of about two and a half years in the industry, he joined the then Calicut Regional Engineering College (CREC) in 1991 and continued with NITC since the transformation of CREC to NITC. He served as Visiting Professor in the Water Engineering and Management FoS in the School of Engineering and Technology, Asian Institute of Technology, Bangkok for one term during the period September 2013 to December 2013 on secondment from the Government of India. His present work responsibilities at NITC include delivering lectures for undergraduate, postgraduate and research students, supervising research and providing consultancy services to various government departments and organisations. At present, he is also discharging the responsibility of Registrar of NITC. He has been very active in teaching, research, and consultancy in areas related to water and waste management. He has supervised four Ph.D. and 40 Masters students, and is presently supervising nine Ph.D. and five Masters students. He has conducted short courses and delivered invited lectures in many short courses and conferences. He has successfully completed few sponsored R&D projects and presently one project is in progress. He has published extensively in international/ national journals and conferences and is a reviewer for many reputed journals in the water and environmental sectors.

Dr. Thun Vathana graduated from the Royal University of Agriculture in Cambodia, receiving his bachelor's degree in 1996. From 1999 to 2005, he studied in Japan, receiving his master's degree from Tokyo University of Agriculture and Technology and his doctoral degree from Nagoya University.

Dr. Vathana has a significant experience in development industry, government and private sector. He has been involved directly or indirectly in many research projects. So far, he has conducted a wide range of research on agriculture and environment, economy and trade, poverty and micro-finance.

Dr. Vathana is currently the Executive Director of Prek Leap National College of Agriculture. Before becoming the director, he had worked as a teacher at the Royal University of Agriculture from 1996 to 2014. He used to work from 2008 to 2012 as the Research Head and Adviser at Angkor Mikroheranhvatho Kampuchea Co. Ltd, Cambodia's leading micro-finance institution. He also used to work as a Research Manager at Cambodia Development Resource Institute, Cambodia's leading policy research institute.

Dr. Anastasios Zouboulis (Ph.D. Chemist), born in 1959, is a Professor of Chemical and Environmental Technology in Aristotle University of Thessaloniki (A.U.Th.), Department of Chemistry and Division of Chemical Technology and from 2005 is the Director of the Laboratory of General and Inorganic Chemical Technology, Department of Chemistry of A.U.Th. He is author (and co-author) of several articles in refereed journals, review chapters, co-author and co-editor of several books, conference proceedings, etc., guest editor in special issues of scientific journals (Separation Science and Technology, Desalination, etc.) and has more than 200 publications in international scientific journals. He participated in more than 45 National and European R&D Projects, dealing with water or wastewater treatment, hazardous solid waste management and biotechnological applications, and has been the administrative coordinator of several projects.

Introduction

The Intergovernmental Panel on Climate Change (IPCC)'s Fifth Assessment Report (AR5) provides an overview of what is known about climate change. The global average surface temperature for 2081–2100 is likely to be 0.3 °C in the most optimistic scenario and 4.8 °C in the most pessimistic, above the average of 1986–2005. Precipitation changes will vary from region to region. Under RCP8.5, mid-latitude wet regions are likely to see increases in precipitation, while many mid-latitude and sub-tropical dry regions are likely to experience decreases. Similarly, the IPCC reported that the loss in glacier volume outside Greenland and Antarctica since the 1960s is equivalent to an average of 0.76 mm/year sea level rise during 1993–2009 and even 0.83 mm/year for the period 2005–2009 (IPCC 2013).

Climate change will have a significant impact on the hydrological cycle. Shorter wet seasons with more intense rainfall and extended dry seasons will affect water availability, water distribution, and agricultural planning and management (Seiller and Anctil 2013; Kang and Khan 2009). Delta areas, where the ecosystem is highly threatened, will not only be affected by climatic factors, but also by the indirect implications of the sea level rise, upstream flow of sea water and salt-water intrusion. Climate change will also affect agriculture through higher temperatures and more variable rainfall, with substantial reductions in precipitation likely in the mid-latitudes, where agriculture is already precarious and often dependent on irrigation. Historical observations of agricultural productivity on a global scale reflect higher productivity in higher latitudes and a relatively lower yield for regions closer to the equator. This difference is expected to grow with climate change progression (Ray et al. 2012).

Several research studies have been carried out to forecast climate change on a global scale. The sign and magnitude of projected changes in future climates vary between different GCMs, leading to substantial uncertainty in precipitation or temperature projections (IPCC 2008). Projections of future precipitation changes are more robust for some regions than for others. As precipitation is the key driver of the hydrological cycle, the uncertainty in precipitation projection leads to an accumulation of uncertainties during impact studies on a local or regional scale.

The climate model studies suggest that the observed trends in mean precipitation are due to changes in radiative forcing. The uncertainties in predicting extreme events such as floods and droughts are a major concern for planners and policy makers.

As climate change is unavoidable, there is no alternative other than to follow the proper adaptive measures. It is the only available solution to the problems we are going to face in the coming decades (Dessai and Sluijs 2007). Implementation of adaptation policies, however, needs to be combined with major investment, and is unsustainable for developing countries and some countries in transition. For the poorer countries facing problems with hydrological variability, it will be even more difficult to achieve water security.

Documentation and dissemination of the impacts of climate change on water resources and water use sectors is a very important step towards managing the water resources in a sustainable manner. Therefore, this book aims to provide information to water managers and decision makers about climate change and its impacts on water resources and selected water use sectors and how to adapt to these changes, taking examples from various countries in Asia and Europe. The chapters are arranged in two parts.

The first part, with 10 chapters, discusses the impact of climate change on water resources and different water sectors. The potential impacts of climate change on hydrology and water resources are illustrated, including urban drainage and waste water treatment plants. Uncertainties associated with climate change projections derived from different GCMs downscaled with different techniques are also discussed.

The second part depicts adaptation strategies to reduce or offset the negative impacts of climate change on water resources and water use sectors. It illustrates how different water use sectors, such as hydropower, water supply, and agriculture can adapt to climate change in an attempt to reduce its impacts. A review on transboundary water and disaster risk reduction is also presented with special reference to countries in Asia.

'Impact of Climate Change on the Water Cycle' focusses on the link between climate and the water cycle and how changes in climate are impacting on the water cycle creating a converse impact on water quality and availability, health, agriculture (food security), biodiversity and water security.

In 'Uncertainty Assessment of Climate Change Impacts on Hydrology: A Case Study for the Central Highlands of Vietnam', the uncertainty of the impact of climate change on the streamflow was analysed in the Srepok watershed in Vietnam through the application of climate scenarios to the SWAT hydrological model. The results from seven CMIP3/IPCC-AR4 GCMs and four emission scenarios in the study show an increase in mean temperature for the future. It also states that the GCM structure is the key to uncertainty in the impact of climate change on streamflow. Therefore, the multi-model is suggested to evaluate climate change impacts on streamflow.

'Assessment of the Impact of Climate Change on Water Availability in the Citarum River Basin, Indonesia: The Use of Statistical Downscaling and Water Planning Tools' examines the impact of future climate change on water availability using a combination of the downscaling technique and water planning tools in the Citarum River Basin in Indonesia. The streamflow and evapotranspiration were projected for three future periods: 2010–2039, 2040–2069 and 2070–2099. The results show an increase in precipitation and run-off, as well as potential evapotranspiration losses in the sub-catchment of the study area.

'Impact of the Uncertainty of Future Climates on Discharge in the Nam Ou River Basin, Lao PDR' investigates the uncertainty of climate change impact on discharge in the Nam Ou River Basin in Lao PDR. LARS-WG was used to project the future climate under two GHGES A1B and A2 scenarios for the 2020s, 2055s, and 2090s. Furthermore, the Soil and Assessment Water Tool (SWAT) model was used to simulate present and future changes in discharge in the river basin. The results show that both increases and decreases in discharge can be expected in future periods.

The significance of the integrated modelling approach was emphasised in 'Integrated Modelling of Climate Change and Urban Drainage' to assess the direct implications of climate change on hydraulic performance in urban drainage and the need to update the urban drainage design criteria. Adaptation measures are required together with a reassessment of the storm water management design and planning process.

The impacts of climate change on hydrology of the Koshi River Basin in Nepal using a multi-model, multi-scenario approach is presented in 'Estimating the Impacts and Uncertainty of Climate Change on the Hydrology and Water Resources of the Koshi River Basin'. Multiple GCM projections for each of the three SRES scenarios (B1, A1B, and A2) were used. A statistical downscaling model (LARS-WG) was selected to downscale the global scale projections to basin scale. Furthermore, the SWAT was used to analyse the impacts of climate change on hydrology. In future, the Koshi River Basin may become warmer as projected by all GCMs under three SRES scenarios.

'Uncertainty Analysis of Statistically and Dynamically Downscaled Precipitation Data: A Study of the Chaliyar River Basin, Kerala, India' examines the assessment of uncertainty analysis on statistically and dynamically downscaled monthly precipitation data in the Chaliyar River Basin, in Kerala, India. Several tests were performed such as the Wilcoxon signed rank test, Levene's test, Brown-Forsythe test, and the non-parametric Levene's to evaluate errors in the mean data. Results shows that the error is not significant in the case of the statistically downscaled data using predictors generated from the reanalysis data. It also highlights the necessity for performing an uncertainty analysis of the downscaled data.

In 'Assessment of the Impact of Projected Climate Change on Streamflow and Groundwater Recharge in a River Basin', the impact of climate change on streamflow and groundwater recharge in the Chaliyar River Basin was investigated. The SWAT hydrological model was used to evaluate the impact of projected climate change on streamflow in the river. Results show an average increase in temperature (2 °C) and decrease in rainfall (11.5 %) in the southwest monsoon period for the A2 scenario, whereas, there is an average increase in temperature of 1 °C and a decrease in rainfall of 8.79 % for the B2 scenario. In the A2 scenario, the SWAT analysis shows an increase in potential evapotranspiration and a decrease in streamflow. In the B2 scenario, SWAT predictions show an increase in potential evapotranspiration of 1.12 % and a decrease in streamflow of 4.62 %. Similar trends are predicted for the northeast monsoon period also.

'Assessment of the Impact of Projected Climate Change on Streamflow and Groundwater Recharge in a River Basin' discusses the link between climatic change on Mediterranean society and the environment through two case studies in Southern Italy and Lebanon. This chapter also highlights the snow-dominated hydrological system, the combined effect of changes in temperature, and precipitation. It emphasises the suitability of the methodological approach used in the Mediterranean area for regions in Southeast Asia in order to undertake similar impact studies.

A comprehensive review of climate change, wastewater treatment and the solutions is presented in 'Effect of Climate Change in Wastewater Treatment Plants: Reviewing the Problems and Solutions'. This chapter emphasises the need to develope future adaptation strategies and knowledge to manage emissions, together with a vulnerability climate assessment to interact with the adaptive responses to address emission sources.

'Managing Hydropower Under Climate Change in the Mekong Tributaries' examines how projected climate change scenarios affect the flow regimes in the 3S basin compared to flow alterations induced by hydropower development. It further investigates the cumulative impacts on flows within the 3S basin by using the combined scenarios of climate change and hydropower development.

'Managing Water Resources Under Climate Uncertainty: Opportunities and Challenges for Cambodia' provides a review of the link between climate and water issues in relation to the hydrological cycle and water resource, as well as the effects of climate change and potential ways of managing water resources.

Based on their research in Bhutan for the improvement of policies and management, Katel et al. ('Transboundary Water Resources Management in the Context of Global Environmental Change: The Case of Bhutan Himalaya') emphasise the need for a multidisciplinary approach when dealing with transboundary water issues, the need for a better framework for scientific cooperation on adaptation, and the need for models with a higher spatial and temporal resolution than that currently available in the Himalayas.

'Addressing Climate Change Impacts through Community-Level Water Management: Case Studies from the South-Eastern Coastal Region of India' presents an assessment of climate change impacts on the east of India followed by a methodology based on the community participation processes to address the aforementioned key impacts manifesting in the area. An example is provided as part of the project 'Strengthening Adaptation Capacities and Minimizing Risks of Vulnerable Coastal Communities in India'.

'Climate Adaptation and Governance and Small-Scale Farmers' Vulnerability through Artificial Glacier Technology: Experiences from the Cold Desert of Leh in North-West Himalaya, India' examines the vulnerability and livelihood interactions, use of the artificial glacier technology to adapt to climate variability by small farmers in Leh (Ladakh), and the need for a multidisciplinary approach to improve the climate adaptations and governance among different stakeholders.

'Governing Disaster Risk Reduction in Nepal' emphasises the expansion of private sector participation in disaster risk governance, summarises the endogenous and exogenous drivers of water-induced disaster risk and recent international mitigation efforts in the case of Nepal.

'Role of Sustainability Policy Entrepreneurs in Building Water-sensitive Cities to Respond to Climate Change: A Case Study in Adelaide, Australia' describes the challenges and strategies used by policy entrepreneurs to encourage sustainable water management by designing innovative water management initiatives, and further illustrates the key role of policy entrepreneurs in shaping policy outcomes.

'Managing Water Resources in Mexico in the Context of Climate Change' addresses problems such as how countries severely affected by global environment and climate change can cope with the present unequal access of water without destroying the already precarious water and food security and how small-scale farmers in rain fed agriculture can contribute to food security, thus also improving the livelihood of extremely marginal people.

A case study on rainwater harvesting applications as effective climate change adaptation strategies in rural and urban settings to increase human resilience is presented in 'Rainwater Harvesting as an Effective Climate Change Adaptation Strategy in Rural and Urban Settings'.

Finally, 'Coevolving Water Infrastructures for Adaptation to Climate Change' presents a coevolution theoretical approach to understand how in London people can adapt their water infrastructures and practices in extreme water conditions as anticipated by climate change, with suggestions for more sustainable water resources in the future.

References

Dessai S, Sluijs JP (2007) Uncertainty and climate change adaptation—a scoping study, report NWS-E-2007-198, Department of Science Technology and Society, Copernicus Institute, Utrecht University, pp 95

IPCC (2008) Climate change mitigation measures and water. pp 115–125 In: Bates B, Kundzewicz ZW, Wu S, Palutikof J (eds). Climate change and water. Technical paper of the intergovernmental panel on climate change. IPCC Secretariat, Geneva

IPCC (2013) Climate change 2013: the physical science basis. In: Stocker TF, Qin D, Plattner G-K, Tignor M, Allen SK, Boschung J, Nauels A, Xia Y, Bex V, Midgley PM (eds) Contribution of working group I to the fifth assessment report of the intergovernmental panel on climate change. Cambridge University Press, Cambridge, UK and New York, NY, USA (in press)

Kang Y, Khan S (2009) Climate change impacts on crop yield, crop water productivity and food security—A review. Prog Nat Sci 19(12):1665–1674

Ray DK, Ramankutty N, Mueller ND, West PC, Foley A (2012) Recent patterns of crop yield growth and stagnation. Nat Commun. doi:10.1038/ncomms2296

Seiller G, Anctil F (2014) Climate change impacts on the hydrologic regime of a Canadian river: comparing uncertainties arising from climate natural variability and lumped hydrological model structures. Hydrol Earth Syst Sci 18:2033–2047. doi:10.5194/hess-18-2033-2014

Part I
Understanding the Impacts

Impact of Climate Change on the Water Cycle

Velma I. Grover

Abstract Global warming has accelerated in recent years with an increase of about 0.75 °C during the past 100 years. The rate of temperature increase in the past 25 years has been over 0.18 °C per decade. Global warming has been observed more over land than over the ocean. This rise in temperature is leading to a rise in sea levels, glacier melt, and changes in precipitation patterns. In addition to urbanisation where roads and buildings impact on the amount of groundwater percolation, large infrastructures such as dams are impacting on the microclimate cycle, which changes the evapotranspiration rate in the region leading to a change in the amount of precipitation. The focus of this chapter is on the impact of climate change on the water cycle, particularly in relation to freshwater, including how a change in the climate cycle is impacting on the water cycle, followed by the impacts of change on water quality and availability, health, agriculture (food security), biodiversity, and water security.

Keywords Climate change · Water cycle · Health · Malaria · Great lakes

1 Background

Climate change is happening. No one is debating that fact anymore, even though the reasons for the change in climate (i.e. natural or anthropogenic) might still be debatable. Natural hazards are increasing as can be observed from the increased intensity and frequency of floods, typhoons, hurricanes, and increased famine (especially in Africa) (Dasgupta et al. 2009). Water, which is so important for

---

This chapter is based on the book, Grover (2012) "Impact of climate change on water cycle and health".

V.I. Grover (✉)
Kobe College, Nishinomiya, Japan
e-mail: velmaigrover@yahoo.com

S. Shrestha et al. (eds.), *Managing Water Resources under Climate Uncertainty*,
Springer Water, DOI 10.1007/978-3-319-10467-6_1

human survival, can also cause a lot of damage. For example, the beginning of 2011 was marked by climate-related disasters with serious implications for human well-being: In Queensland, Australia, floods surged through the region, setting Brisbane, Australia's third largest city, under water, and killing at least 19 people. In Brazil, more than 500 people perished when mudslides caused by heavy rain covered and destroyed homes, making it the worst natural disaster for several decades (Webersik 2012). In 2011, Thailand witnessed the worst flooding in at least five decades, displacing millions of residents and killing 884 people. Out of 77 provinces, 65 were affected and the World Bank estimated the damage to be about U.S. $45.7 billion making it one of the five costliest natural disasters in modern history (Aon Benfield 2011). IPCC reports predict that there will be an 80 % chance of increased mortality and morbidity due to climate change-related extreme weather events (EWEs) (Parry et al. 2007). In 2007 alone, 95 % of the 16,000 global fatalities from EWEs can be directly attributable to climate change. Because of the change in temperature and precipitation patterns, EWEs have changed in frequency and intensity. This is not only causing loss of property, but is also increasing diseases (in the affected areas) and adding an economic burden.

Some of the recent EWEs have crossed all the historical records. Heavy rainfall within 24 h, glacial melt, cloud burst, and lake burst caused havoc in the state of Uttarakand in India in June 2013, stranding thousands of pilgrims, killing thousands, and causing severe damage to the infrastructures, e.g. destroying about 150 bridges, washing out most of the roads, and wiping out houses and hotels in some regions (with an estimated cost of U.S. $867 million in infrastructure damage alone, not including losses due to non-electricity generation, loss in tourist revenue, and manufacturing).[1] Typhoon Haiyan, in the Philippines in November 2013, has been recorded as the strongest storm on land so far and unofficially is also recorded as the strongest typhoon in terms of wind speed.[2] This was by far one of the deadliest typhoons in the Philippines affecting 7 million people, killing about 6,155 persons, leaving 600,000 homeless[3] and causing agricultural and fisheries a loss of about U. S. $225 million[4] and U.S. $1.5 billion in non-life insurance payments.[5]

Floods in Toronto, Canada (caused by 126 mm of rain in 2 h), in summer 2013, left many homes with property damage, flights cancelled, and thousands stranded on commuter trains submerged in water; it has been one of most expensive natural disasters in the history of Ontario with about $850 million paid out in insurance as at August 2013[6]). The snowstorm in December 2013 again stranded a lot of air

[1] http://www.downtoearth.org.in/content/heavens-rage and http://www.downtoearth.org.in/content/what-really-happened-uttarakhand.

[2] Fischetti, Mark (November 27, 2013). "Was Typhoon Haiyan a Record Storm?". *Scientific American.*

[3] http://www.businessinsider.com/typhoon-haiyan-damage-is-worse-than-hell-2013-11.

[4] http://blogs.wsj.com/economics/2013/11/19/typhoon-haiyan-caused-225-million-in-agricultural-damage/.

[5] http://www.businessinsurance.com/article/20131226/NEWS09/131229953.

[6] http://www.cbc.ca/news/business/toronto-s-july-storm-cost-insurers-850m-1.1363051.

travellers, left many people in darkness, and caused a lot of property and infrastructure damage (where the post-storm crisis has caused CAN $8–10 million damage to the city so far, [7] and it will be a while before the total damage is known).

Global warming has accelerated in recent years. While the past 100 years saw an increase of about 0.75 °C, the rate of temperature increase in the past 25 years has been over 0.18 °C per decade. Global warming has been observed more over land than ocean. This rise in temperature is leading to a rise in sea levels, glacier melt, and changes in precipitation patterns (for example, from 1900 to 2005, precipitation has increased in eastern parts of North America, Northern Europe, Northern and Central Asia, and Southern Europe while it has declined in Southern Africa, parts of Southern Asia, the Sahel, and the Mediterranean) (WHO 2008).

Melting ice and thermal expansion of oceans are causing sea levels to rise. In addition to exposing coastlines, where the majority of the human population lives, greater erosion and flooding pressures, together with rising sea levels, might also lead to salt water contamination of groundwater supplies. The quality and quantity of climate change and water resources threaten freshwater access to a large percentage of the population. According to Water Aid 2007, a one metre rise in sea level will displace 80 % of the population of Guyana.

Climate change is not only impacting on sea level rise and availability of water but also both freshwater resources and oceans in terms of acidification and coral reef bleaching. A change in ocean acidity is likely to reduce the ocean's capacity to absorb CO_2 from the atmosphere, thus compounding the effects of climate change, and will affect the entire marine food chain. Also, large-scale, irreversible system disruption and the destabilisation of the Antarctic ice sheets are serious risks: changes to polar ice, glaciers, and rainfall regimes have already occurred.

The focus of this chapter is mainly on the impact of climate change on the water cycle, mainly freshwater. Freshwater is important because of its related challenges: too much water, too little water, or the quality of water (pollution) are all exacerbated by climate change (Bates et al. 2008a, b). The chapter provides a global picture and includes some country or community level case studies. As explained by Gupta (Grover 2012), "… the problem of climate change is about the economy, our production, and consumption system. Climate change is about society; our lifestyles, our jobs, our food, our recreation. Climate change is about the environment, about how land use changes affect the climate, and how climate change affects species and ecosystems. Climate change is about so many issues and can be defined in so many different ways, that we often forget that climate change is also about water; water that makes our planet quite unique; water that makes life possible; water that makes the economy flourish. The link between climate and water is very critical". The next section of the chapter looks at the climate and water cycle and the impact of climate change on water quality and availability, health,

[7] http://www.thestar.com/news/gta/2013/12/29/ice_storm_7400_in_toronto_still_without_power.html.

agriculture (food security), biodiversity, and water security followed by a concluding section. This chapter relies heavily on secondary sources and is a synthesis of the data therefrom.

2 Climate Change and the Water Cycle

The climate system and the water/hydrological cycle are driven by solar radiation. As described in the IPCC report (2007), the climate system is a complex interactive system that consists of the atmosphere, land, water (in the form of oceans, lakes, rivers, snow, and ice etc.), and living things (including humans). The climate system is affected by changes in greenhouse gas compositions, some natural phenomenon (i.e. volcanic eruptions), and anthropogenic factors (e.g. change in land use etc.). All these changes have an impact on the absorption of solar radiation and their reflections in the atmosphere, causing changes in the earth's temperature thus impacting on the climate.

In the natural water cycle, solar energy enters the earth's system leading to the process of evaporation, formation of clouds, and then precipitation in the form of rain, hail, or snow. Once precipitation has occurred, water evaporates or transpires back into the atmosphere, or it can be a run-off (in the form of a river, etc., going into lakes or oceans). It can also penetrate the surface to become part of groundwater. This means that water remains within the water cycle and is considered as a renewable resource. However, in some locations, water takes a much longer time to cycle, virtually making it a non-renewable resource.[8]

The water cycle is changing due to a few reasons: firstly, because of unlimited withdrawal of water for drinking, agriculture, industries, extracting groundwater, and in certain places at a rate more than it can replenish itself. Secondly, in other places, we are building dams or diverting river water; this essentially means that the amount of water present in a particular location has changed (increasing where a dam is built and decreasing where water is diverted), thus altering both the local water cycle (changing evapotranspiration rate and precipitation) and the microclimate of the place. This also alters the water cycle at deltas because the amount of water present there is less than before since it is being used for other purposes or stored in dams (see Box 1.1). Run-off from agriculture, industrial, and municipal waste deteriorates the quality of water.

Box 1 Impact of urbanisation and large infrastructures on the water cycle Urbanisation has led to changes in land cover, concrete houses, buildings, asphalt roads, etc., which has led to changes in groundwater percolation and evaporation. Studies have shown that urban mega centres have

[8] https://www.ec.gc.ca/eau-water/default.asp?lang=En&n=1C100657-1.

an impact on local precipitation patterns (Shepherd 2005). For example, studies of precipitation in major urban centres such as Kolkata, India (Mitra et al. 2011), the Pearl River Delta in China (Kaufmann et al. 2007), and Atlanta, USA (Shem and Shepherd 2009), have all noted a significant increase in precipitation. Shem and Shepherd (2009) estimate the increase to be as much as 10–13 % for areas less than 50 km downwind of the urban centre. Though there remains some controversy over the size and even direction of the urban rainfall effect (URE), it is generally accepted in the literature that for large urban centres, the URE is identifiable (Hossain et al. 2012). Large infrastructures such as dams also change the water cycle and contribute to local climate change. Research by Degu et al. (2011), using 30 years of North American reanalysis data of 92 large dams in various climate regions of the USA, has identified three primary and relevant parameters impacting mesoscale precipitation patterns: convective available potential energy, specific humidity, and surface evaporation to elucidate climate and precipitation patterns. Their findings also show that the increased frequency and intensity of rainfall in mesoscale is due to increased evaporation, changes in land use, and/or land cover (LULC). Increased deforestation for dams can impose significant changes to the local climate and weather. It may also lead to increased saline soil resulting in decreased agriculture productivity (IPCC 2008). Similar observations have been made by Degu et al. (2011) for Mediterranean and semi-arid regions. Dams, particularly shallow dams, are believed to be emitters of GHGs due to evaporation and decomposition of aquatic organisms, and UNFCCC has struck major hydro projects from its Clean Development Mechanism list (IPCC 2008).

Thirdly, changes in the climate also affect the hydrologic cycle (World Bank 2009). As discussed in the IPCC report, water is part of all components of the climate system, i.e. atmosphere, hydrosphere, cryosphere, land surface, and biosphere. Essentially, this means that any change in the climate impacts on the water (cycle) through different means. Climate change has been associated with changes in the hydrological systems such as change in precipitation patterns, melting of snow and ice, increased evaporation, increased atmospheric water vapour, and changes in soil moisture and run-off. (Bates et al. 2008a, b). As the temperature increases, saturation of vapour pressure in air increases, and it is expected that with a warming climate (increasing temperature), the amount of water vapour suspended in the air will increase (IPCC 2013). Increase in the water vapour has also been noted in the IPCC Fifth Assessment Report, "Observations from surface stations, radiosondes, global positioning systems, and satellite measurements indicate increases in tropospheric water vapour at large spatial scales. It is very likely that tropospheric specific humidity has increased since the 1970s. The magnitude of the observed global change in water vapour of about 3.5 % in the past 40 years is consistent with the observed temperature change of about 0.5 °C during the same period, and the relative humidity has stayed approximately constant" (IPCC 2013, p. 13).

As discussed in the IPCC report (2013), it is not always easy to establish a direct link between precipitation and evaporation, but some trends can be concluded based on the observed oceanic surface salinity, which is strongly dependent on the difference between evaporation and precipitation. "The spatial patterns of the salinity trends since 1950, the mean salinity, and the mean distribution of evaporation and precipitation are similar to each other: regions of high salinity where evaporation dominates have become more saline, while regions of low salinity where rainfall dominates have become fresher. This provides indirect evidence that the pattern of evaporation and precipitation over the oceans has been enhanced since the 1950s (*medium confidence*). The inferred changes in evaporation minus precipitation are consistent with the observed increased water vapour content of the warmer air" (IPCC 2013).

Whenever there is a discussion on different greenhouse gases, the greenhouse gases that are often mentioned (and discussed) are carbon dioxide and methane. Water vapour is hardly ever discussed, and the role of water vapour in both its natural and anthropogenic aspects remains unmentioned (Gupta 2012). Yet water vapour not only holds the pole position concerning the natural greenhouse effect, but also participates in the additional absorption of heat in the atmosphere.

The main concern with water vapour is because of "secondary effects". Essentially, it means that if the average temperature of the atmospheric layers near to the ground is rising (either naturally or as a consequence of anthropogenic CO_2 and methane emissions), then the evaporation of water increases. Hence, more water vapours will get into the atmosphere, and this additional increase of water vapour will also absorb more heat. Since most of the solar energy received by the earth is used by the hydrological cycle, higher levels of solar energy trapped in the atmosphere will lead to an intensification of this cycle, resulting in changes in precipitation patterns. These changes will result in increased floods and drought, which will have a significant impact on the availability of freshwater. These impacts on freshwater will be further compounded by rising sea levels and melting glaciers (Gupta 2012).

"Observational records and climate projections provide abundant evidence that freshwater resources are vulnerable and have the potential to be strongly impacted by climate change, with wide ranging consequences on human societies and ecosystems" (Bates et al. 2008a, b, p. 4). "However, observed changes in the hydrologic cycle at the sub-continental scale have consistently been associated with climate warming over the past several decades. These changes are often referred to as intensification and acceleration of the hydrologic cycle. These changes include increasing atmospheric water vapour content; changing precipitation patterns, intensity, and extreme events; reduced snow cover and faster and widespread melting of ice; and changes in soil moisture and run-off. It is projected that many of these phenomena will become more pronounced with climate change" (World Bank 2009).

Essentially, warmer average global temperatures will result in higher evaporation because a warmer atmosphere will be able to hold more moisture aloft which will come back as precipitation, more moisture leading to more precipitation thus increasing the potential for flooding. However, in drier regions, even a slight rise in temperature leads to a greater loss of moisture, exacerbating drought and desertification. Drought will not only lead to decreased water availability but also poor water quality in water-scarce regions,[9] such as Southern and Northern Africa, Central America, and Central Asia. In sub-Saharan Africa, for example, there have been reoccurring incidences of long periods of drought which are predicted to become more widespread. While some farmers have been able to survive these long periods of drought by selecting seed varieties for the changing conditions, it has been difficult for poorer farmers to adapt (Water Aid 2007).

3 Global Predictions for Changes in the Hydrological Cycle

As discussed below, the IPCC Fourth Assessment Report findings (as quoted in the World Bank, 2009) report some changes in the hydrological cycle because of climate change (some of these findings of the IPCC fourth report are also compared with the results in the IPCC Fifth Assessment Report 2013), for example, in the points 1.f and 2.e. In some cases, these findings are also followed by an example:

1. *Precipitation (including extreme events) and water vapour:*

   a. The average atmospheric "water vapour" content has increased since the 1980s.
   b. Over the twentieth century, on the one hand, the mean precipitation has increased over land in high northern latitudes between 1901 and 2005, while on the other hand, since the 1970s, mean precipitation has decreased between 10° S and 30° N.
   c. There has been an increase in heavy precipitation events in the mid-latitude regions.
   d. Soil moisture has decreased and droughts are more intense and longer, especially in the tropics and sub-tropics.
   e. Intense tropical cyclone activity has increased in some regions.
   f. According to the IPCC Fifth Assessment Report (2013), "Confidence in precipitation change averaged over global land areas is low prior to 1950 and medium afterwards because of insufficient data, particularly in the earlier part of the record … Northern hemisphere mid-latitude land areas do show a

[9] Low precipitation will exacerbate many types of water pollution problems such as increased sediments, nutrients, dissolved organic carbon, pathogens, pesticides, salt, and thermal pollution (because concentration of these elements will increase) leading to higher algal blooms, increase of bacterial and fungal content and maybe reduced oxygen concentrations. This will obviously have an impact on human health, ecosystems, and on operating costs of water systems. (Bates, et al. 2008).

likely overall increase in precipitation (medium confidence[10] prior to 1950, but high confidence afterwards) …: (page 11)."[11]

g. Case Study. Changes in temperature and precipitation in Bangladesh: A preliminary investigation on the impacts of climate change on small isolated wetlands (ponds) in the Satkhira district (located in Southern Bangladesh) in late 2010 was reported by Rabbani et al. (2012). The district is surrounded by a complex river network consisting of Kobadak, Sonai, Kholpatua, Morischap, Raimangal, Hariabhanga, Ichamati, Betrabati, and Kalindi-Jamuna. Satkhira is one of the most vulnerable coastal districts of the country and the hot spot for any type of climate-induced hazards. The most recent cyclone events, e.g. Cyclone Sidr and Cyclone Aila, hit most parts of the district. The long-term trend in average maximum temperature shows a decline between 1976 and 2005. It has, on average, decreased by 0.009 °C per annum over the period. Similarly, the average annual minimum temperature has also declined, on average, by 0.001 °C over the same time period. In contrast, the annual rainfall in the region has increased by 9.5 mm over the period 1990–2005. The pattern of total rainfall for different years of the last decade was quite irregular: while pre-monsoon rainfall followed a decreasing pattern (sharp and gradual) from 1997 to 2005, the monsoon of 2002 received the highest (1271 mm) rainfall as compared to other years in the last decade. A gradually decreasing pattern of pre-monsoon rainfall has been observed from 1997 to 2005, while the total post-monsoon rainfall shows an increasing pattern from 2002 to 2005. Apart from the observed data, a survey of the people in the region also confirms the changes observed: on average, about 70 % responded that winters are getting shorter while summers are getting longer, rainfall is late and at times there is a lack of rainfall. More than 50 % of people mentioned that the frequency and intensity of cyclone and storm surge have increased in the last decade. It revealed that 95 % of people think that the temperature has increased, especially during the pre-monsoon period.

As noted by the IPCC (2013, p. 14), "Global-scale precipitation is projected to gradually increase in the twenty-first century. It is *virtually certain*, that

[10] "Confidence in the validity of a finding, based on the type, amount, quality, consistency of evidence (e.g., mechanistic understanding, theory, data, models, expert judgment), and the degree of agreement. Confidence is expressed qualitatively." (Mastrandrea 2010, p. 2). "A level of *confidence* is expressed using five qualifiers: "very low," "low," "medium," "high," and "very high." It synthesises the author teams' judgments about the validity of findings as determined through evaluation of evidence and agreement." (Mastrandrea 2010, p. 4).

[11] There are uncertainties in projections of the impact of climate change on water resources because of internal variability of the climate system, uncertainty in future emissions, translation of these emissions into climate change by global climate models, its mismatch with hydrological models, and other factors such as estimating impacts on ground water recharge, water quality, or flooding/drought as translation of climate into response is less well understood. (World Bank 2009).

precipitation increase will be much smaller, approximately 2 % K^{-1}, than the rate of lower tropospheric water vapour increase (~7 % K^{-1}), due to global energetic constraints. It is virtually certain that changes of average precipitation in a much warmer world will not be uniform, with some regions experiencing increases, and others with decreases or not much change at all. The high latitudes are likely to experience greater amounts of precipitation due to the additional water carrying capacity of the warmer troposphere. Many mid-latitude arid and semi-arid regions are likely to experience less precipitation. The largest precipitation changes over Northern Eurasia and North America are projected to occur during the winter". Table 1 compares the observed trends and projects for the twenty-first century for the changes in the hydrological cycle.

Table 1 Changes in key hydrologic variables (*Source* World Bank 2009)

Key variables	Observed trends	Projections for twenty-first century
Total precipitation	Trend is unclear and generally increases in precipitation over land from 30° N to 85° N. Notable decreases from 10° S to 30° N	Increase (about 2 %/°C) in total precipitation. High-latitude areas generally projected to increase. High/low to mid-latitude areas projected to decrease. Changes at the regional scale vary
Atmospheric water vapour content	Increasing in lower atmosphere (lower troposphere; a tout 1 %/decade) in specific humidity; little change in relative humidity	Increasing
Intensity of precipitation	Disproportionate increase in volume of precipitation in heavy or extreme precipitation events	Increasing (about 7 %/°C)
Droughts	Drought, as measured by the Palmer Drought Severity Index, increased in the twentieth century, although some areas became wetter	Increasing in many areas, particularly lower latitudes. Decreasing in many high-latitude areas. Patterns are complex
Tropical cyclones	Increases in intensity, particularly in North Pacific, Indian Ocean, and south-west Pacific	Increase in intensity. Changes in frequency and track are uncertain
Glaciers and snow cover	Decrease in mass of glaciers, but not in all regions. Decrease in snow cover in regions in the Northern Hemisphere. Earlier peak run-off from glacier and snowmelt	Continued decrease in glacial mass and snow cover
Sea level	Increased about 0.2 m over the twentieth century. A rise equivalent to 0.3 m per century was recorded since the early 1990s, but it is not clear if this is an acceleration of long-term sea level rise	IPCC projects 0.2–0.6 m by 2100, but upper end could be much higher

2. *Snow and land ice:* Just as the precipitation (rainfall) pattern is changing, changes in snowfall events have also been observed. Due to an increase in the winter temperatures, a significant reduction in snow cover in the Northern Hemisphere over the past 90 years has been observed (most reduction occurred in 1980s). Due to earlier spring snowmelt, the duration of snow season in the Northern Hemisphere has declined by 5.3 days per decade since 1972/73 winter (IPCC 2013).

   a. There has been a considerable loss of ice/snow on the majority of ice caps and glaciers.
   b. Snow cover has decreased in most of the regions, especially in spring and summer seasons.
   c. Degradation of permafrost and seasonally frozen ground has occurred in many areas.
   d. Freeze-up and break-up dates for river and lake ice have been delayed and taken place earlier, respectively, especially in the Northern Hemisphere where data are available (World Bank 2009. Although the Great Lakes in the Northern Hemisphere have experienced a heavy snow fall between December 2013 and March 2014 (setting a record for most ice cover on Lake Michigan in 41 years[12]), there had been a significant loss of ice cover between 1973 and 2012 (The Great Lakes have lost about 71 % of ice cover since 1973[13]). The loss in ice cover, mainly due to increasing air and water temperatures, causes more water to evaporate in the winter season leading to lower lake levels and heavier snow in some areas along the lake shore. In addition, the Great Lakes are also thawing a little earlier each spring. Studies have shown that the trend of earlier thawing has been observed since 1846, but the rate of change has become three times greater since 1975 (Tip of the Mitt Watershed Council 2014).
   e. According to the IPCC Fifth Assessment Report (2013), the arctic sea ice has also decreased over the period 1979–2012 (with an annual decrease between 3.5 and 4.1 % per decade). According to the report, for most of the regions, overall glaciers have continued to shrink (shown by time series measurements of glacier length, area, volume, and mass). Also during the last decade, about 80 % of ice loss can be attributed to the loss of ice from glaciers in Alaska, the Canadian Arctic, the periphery of the Greenland ice sheet, the Southern Andes, and the Asian mountains. "There is high confidence that current glacier extents are out of balance with current climatic conditions, indicating that glaciers will continue to shrink in the future even without further temperature increase" (IPCC 2013, p. 12).

---

[12] http://www.washingtonpost.com/blogs/capital-weather-gang/wp/2014/03/10/lake-michigan-sets-41-year-record-for-most-ice-cover/.

[13] http://www.watershedcouncil.org/water%20resources/great%20lakes/threats-to-the-great-lakes/great-lakes-water-levels/other-opinions-on-low-water-levels/.

3. *River flows:*
a. Milly et al. (2005) showed the total (global) annual river run-off is projected to increase, although there is a considerable variability across regions with a significant decrease in mid-latitudes and some parts of the dry tropics and a significant increase in high latitudes and wet tropics.
b. In areas fed by glacier melt, in the short term, there will be an increase in river flow which will decline as glaciers recede.
c. Box 2 describes the impact observed in the Kumaon region in India.

Box 2 Impact of climate change on the Kumaon region in India India is a vast country and its water cycle is impacted by climate change in addition to other drivers, such as population increase and rise in demands from the agriculture and industrial sectors. This section of the chapter is based on a smaller case study area (as discussed by Tiwari and Joshi 2012): Kosi is one of the major rivers of the West Ramganga system of Kumaon Himalaya which ultimately drains into the Ganges system. The headwater of the Kosi River (upstream Someshwer), which encompasses an area of 39.9 km^2 (3,990 ha) and lies between 1,405 and 2,720 m altitude above mean sea level in the Kumaon Lesser Himalaya, is one of the critical rain-fed headwaters identified for priority conservation of water and other natural resources in Kumaon Lesser Himalaya (Tiwari 2008).

Rainfall pattern is governed by the south-west monsoon, and nearly 80 % of the total annual rainfall occurs during the monsoon season, normally between 15 June and 15 September. The maximum proportion of rainfall flows out from the headwater through overland flow. It has been observed that rainfall is gradually decreasing due to continuously changing climatic conditions. The interpretation of hydro-meteorological data shows that the numbers of annual rainy days have declined from an average of 60 days to nearly 50 days with a few exceptions. During 2003 to 2009, the number of rainy days fluctuated between 58 and 59, whereas in 2010, the number of rainy days increased abruptly to as many as 65 and devastated the entire region with extreme weather conditions, such as several cloud bursts and flash floods during the peak monsoon month of September 2010. The continued decline in the number of rainy days has also adversely affected the availability of annual rainfall and its distribution over time, particularly affecting the precipitation pattern during the winter and summer months, creating drought conditions. Consequently, the amount of annual rainfall has decreased from an average 135 cm to approximately 112 cm during the last 20 years with the exception of the years 1999 and 2010 when the region recorded an average annual rainfall amounting, respectively, to 138 and 140 cm which was comparatively very high.

These above-mentioned observed changes in rainfall pattern over the last 20 years have also disrupted the hydrological system of the watershed and contributed significantly towards depletion of water resources in the headwater region. The hydrological imbalances have been observed in the forms of (i) decline in groundwater reserve, (ii) drying of natural springs, (iii) decrease in the water discharge in streams and springs, and (iv) drying of stream heads. Recent hydrological investigations carried out in Central Himalaya reported that the average groundwater storage level in the region was nearly 12 % (Rawat 2009) as against the recommended norm of minimum 31 % (Hewelet 1985). The decreasing rainfall has played a very important role in drastically reducing the recharge of groundwater in the region. The present study revealed that out of a total of 107 springs in the Kosi headwater, nearly 39 have completely dried up and more than 20 % have become seasonal during the last 20 years. As mentioned above, besides changes in rainfall pattern and erratic rainfall, several other factors, particularly the land use changes, are also responsible for the drying of springs in the region (Sharma et al. 2007).

Hydrological investigations carried out in other parts of the catchment also revealed that perennial streams are disappearing at a rate of 4.5 km per year (Rawat 2009). It was observed that the discharge of the Kosi River in its headwater region declined from 550 m^3 per second in 2001 to as low as 220 m^3 per second in 2007 thus bringing a total decline of 60 % during a short period of 7 years. However, the water discharge of the river recorded an increasing trend during 2007–2009, but this was mainly due to erratic high-intensity rainfall during the peak monsoon time in the entire Kumaon Himalaya. Further, strikingly, the lean period discharge of the Kosi River reduced by as many as 9 times between 1998 and 2009, and the river was found to be completely dry at several places during the dry summer months since 2003 (Rawat 2009).

These hydrological changes and resultant decrease in the flow of water in streams and springs have considerably reduced the availability of water for domestic purposes and irrigation, undermining the food and health of rural communities in the entire region

4. *Groundwater*: Impacts of climate change on groundwater have not been studied much, but a few imminent threats to groundwater because of climate change have been observed:
    a. Reduced groundwater recharge: for example, studies suggest that by 2050, there may be a significant decrease in recharge (up to 70 % less) in North-Eastern Brazil, Western Southern Africa, and along the southern rim of the Mediterranean Sea (references).
    b. Increased use of groundwater in some regions, especially if there is a decrease in surface water because of climate change.
    c. Contamination of coastal aquifers due to salt water intrusion as sea water level rises in addition to contamination by storm surges.

5. *Sea level:* As discussed by Briggs (2012), "Changes in sea level will expose infrastructure similar to Fukushima to storm surges, flooding and other natural events. Although the IPCC Fourth Assessment Report (2007) only predicted 18–59 cm rise in sea level by 2100, other scientists have criticised their approach as too conservative and only taking account of thermal expansion and not ice sheet melt. Other studies say that the upper bounds of sea level will rise and approach 2–3 m by 2100, which can flood large areas of coastal regions worldwide (Hansen 2007). This would flood transport networks, energy systems (reactors and refineries are always placed near water), and force millions of people from their homes. The indirect health effects of such displacement are highly significant, including decreased health form refugee status, decreased economy output from flooded infrastructure, and food security risks from flooded agriculture lands (Renaud et al. 2007). Low lying areas in Bangladesh and Vietnam are at obvious risk, but modern industrial societies may also be unable to withstand rapid changes, especially if combined with increasing severity of tropical storms".

4 Regional Impacts: Examples of North America and Africa

4.1 Impact in North America

Usually, predictions in temperature rises are global averages; essentially, the temperature will increase in some regions more than in others. Even a one degree temperature rise will have an impact on water systems, small glaciers, ecosystems, and crops (affecting food security). Regional impacts of climate change at places will be more than the global average.[14] For example, some models of climate

[14] There are some uncertainties inherent in modelling. However, in this case, trying to fit hydrological projections within climate change models also leads to further uncertainties (since the spatial scale for both global climate models and hydrological models varies). However, to begin with, there are uncertainties in projected changes in the hydrological system itself because of internal variability of the climate system, uncertainty in future greenhouse gas and aerosol emissions, the translation of these emissions into climate change by global climate models, and hydrological model uncertainty. Another source of uncertainty in hydrological projects is due to the structure of current climate models, since they generally exclude things such as feedbacks from vegetation change to climate change and anthropogenic changes in land cover. Despite these uncertainties, some robust results of change in precipitation with change in temperature are available (Bates et al. 2008).

change predict that in the US annual mean temperatures might rise by 2–3 °C over the next 100 years. However, northern regions will experience a greater increase of up to 5 °C, and some places such as Northern Alaska might also see an increase of up to 10 °C. In relation to this increase in temperature, other changes are also expected, for example, in the USA precipitation is predicted to continue to increase overall. Some GCMs predict a 20 % increase for the northernmost part of North America, a 15 % increase in the winter precipitation for the north-western regions, and a general increase in winter precipitation for the central and eastern regions. Models predict a 20 % decrease in summer precipitation, especially for the south-western regions, and a general decrease in the summer precipitation is projected for the southern areas. Although there are predictions that precipitation will increase in most regions of the USA, there will be a net decrease in water availability in those areas due to offsetting increases in evaporation (Adams and Peck 2008).

According to Angel and Kunkel (2009), temperature and precipitation in the Great Lakes region will increase. Merely for the Great Lakes region, 1.5–4 °C increase above the 1970–1999 average is predicted by the end of the twenty-first century. The change values for the A1B emission scenario (slightly larger) range from 2–5 °C as compared to the A2 high emission scenario where the range is from 3.5 to 7 °C, by the end of the twenty-first century. The annual temperature in these GCM simulations is sensitive to the level of emissions. Although the three scenarios yield different results for temperatures, they have similar results for precipitation; the majority of cases showed wetter conditions. Results based on the A2 emission scenario yielded a slightly larger range in precipitation changes, from a 5 cm drop to a 20 cm increase in annual precipitation. Although there is an uncertainty about the Great Lake levels over a period of time, the studies indicate a drop in those levels.

4.2 How Is Africa's Climate Changing and How Does This Impact on the Water Cycle?

As discussed by Cruickshank and Grover (2012): according to the IPCC's Fourth Assessment Report, Africa's warming trend will be 1.5 times the global mean (Eriksen et al. 2008). Climate models project an increase in global temperatures of between 1.4 and 5.8 °C by 2100 and between 3 and 4 °C for Africa compared to 1980–1999 (IPCC 2007). Evidence of Africa's changing climate already exists; during the last century, the continent as a whole warmed by 0.7 °C and the maximum temperature experienced in some regions increased by as much as 3.5 °C (Magrath 2006). Temperature increases can also be seen through melting glaciers and shifting ecosystems.

Global warming will alter the water cycle by increasing rates of evaporation and causing fluctuating precipitation levels and patterns, and changes in run-off. The number and severity of extreme events will also rise (Bates et al. 2008a, b). At its most basic, predictions state that "*areas that already get a lot of rainfall will get more and areas that get little will get less*" (Fields 2005). For Africa, the IPCC (2007) predicts that the Northern Sahara, Mediterranean Africa, and Southern Africa will experience a decrease in rainfall, whereas East Africa is likely to have an increase. The global models do not consider vegetation and dust aerosol feedbacks, nor is El Nino Southern Oscillation (ENSO), deemed one of the key controlling factors for rainfall in Africa, adequately represented (Hulme et al. 2001). These discrepancies create a level of uncertainty, and in the cases of the Sahel, the Guinean Coast, and Southern Sahara, the models have produced particularly conflicting results. Climatic changes are not uniform and are strongly influenced by localised variables, climatic zones, and elevation across the continent.

Overall, the impact of climate change on freshwater resources across Africa is expected to be negative. More areas will be under water stress due to net reductions in rainfall and higher temperatures causing melting glaciers and increased evaporation of surface water. By the 2020s and 2050s, an additional 75–250 million and 350–600 million Africans, respectively, are expected to live in areas of water stress (Arnell 2004 as quoted in Bates et al. 2008a, b). Even areas experiencing increased rainfall will contend with higher evaporation rates, more intense events, and an increased variability in the timing of onset and distribution of rains. The drier ground and high evaporation mean that less run-off will reach rivers, and surface water will have little time to replenish aquifers. In Southern Africa, climate models predict a 30 % reduction in run-off for a 2 °C increase in global temperatures and 40–50 % reduction for a 4 °C increase (Stern 2007). At the same time, warmer ocean temperatures will result in more intense rainfall events. In areas where soil becomes quickly saturated, run-off will create flash floods. While the net increase would likely improve water security in developed countries, most African countries lack the necessary infrastructure to harvest and store water from these events for use during the dry periods.

Incidences of extreme events are not limited to the increased frequency of floods and droughts. Sea level rises will be accompanied by increased storm surges, salt water intrusion, coastal erosion, and flooding (Dasgupta et al. 2009). The IPCC (2007) predicts that storms and cyclones will change their paths and become more intense with higher wind speed and heavier rainfall. Coastal ecosystems, such as coral reefs and mangroves, which are vital for coastal protection and fisheries, are being impacted by salt water intrusion, sedimentation, and increased wave action. The IPCC (1998) estimated that 60 % of mangrove areas in Senegal have been lost due to a combination of factors including increased water and soil salinity. According to the Stern Report (2007), inundation and coastal erosion are expected to place millions at risk around the African coastline.

5 Impact of Changes in the Hydrological Cycle

Changes in the climate and hydrological cycle will impact on water quality, water availability, health, movement of people (increasing climate change refugees[15]), agriculture (food security), energy, and industries (because of variation in availability of water). The direct supply side effects of climate change outlined previously, including increased water scarcity, flooding, accelerated glacial melting, and rising sea levels, have the potential to accelerate human migration.

5.1 Impact on Water Quality and Availability

The following changes will be observed on the water quality because of changes in the water cycle (World Bank 2009):

- Increase in the water temperatures in lakes and reservoirs will have an impact on certain properties such as oxygen solubility, stratification, and mixing, in addition to other biochemical processes.
- Increased water temperatures will also impact on the assimilative capacities of rivers to breakdown organic waste.
- Increased precipitation intensity on the one hand will increase dilution, but on the other hand, it can also increase the erosion, thus potentially increasing sediment loads, nutrients, pathogens, and toxins transported to downstream water bodies.
- Decrease in rainfall or longer periods of low flows can reduce the dilution capacity of the water bodies. It will also reduce dissolved oxygen, increase algal blooms, and magnify the impact of water pollution affecting human health, ecosystems, and water supplies.
- Reduced lake levels may also lead to resuspension of bottom sediments and nutrient cycling.
- Rise in the sea level can increase salt water intrusion in estuaries and coastal aquifers. It may also interfere with storm water drainage and sewage disposal.

5.2 Impact on Availability of Water

The impact of climate change on water availability will vary from region to region. Some areas will see a decrease in precipitation and increase in water stress, while some other areas face a risk of submergence due to floods or a rise in sea level.

[15] Studies have shown that climate change refugees will range from 250 million to about 1 billion between now and 2050.

Taken together, the potential changes in water availability and use may aggravate global water stresses. Most studies have found that the levels of water stress will increase, although there are significant differences in estimates across studies (Arnell 2004). Arnell (2004), who accounts for population growth and the impact of climate change, has noted that the number of people projected to experience an increase in water stress is between 0.4 and 1.7 billion in the 2020s and between 1 and 2.7 billion in the 2050s (using the A2 population scenario for the 2050s). When environmental flow needs are incorporated, that is, the amount of water required for sustaining a functioning ecosystem, the degree of water stress is projected by some to increase further (Smakhtin et al. 2003). Based on these and other studies, the IPCC concluded with a high degree of confidence that, globally, the negative impacts of future climate change on freshwater systems and ecosystems are expected to outweigh the benefits (World Bank, 2009). Box 3 discusses the change in water availability in Nepal.

Box 3 Emerging narratives of climate change impacts in Nepal So far, the chapter has given scientific facts based on literature establishing linkages between climate change and the hydrological cycle, and Dixit et al. (2012) also looks at print media. Some conclusions can be drawn based on what the local population in Nepal is observing and saying about water availability in the regions. For example, popular writer Jagadish Ghimire [as quoted in Dixit et al. (2012)] captures the plight of residents of the Ramechhaap District in the mid-eastern hills, where he helped build drinking water systems. He recounts the difficulties faced by women now that those systems yield much less water than they once did:

"The Southern Ramechhaap District is experiencing the worst drought it has seen in the last century. No rain has fallen for the last seven years. As a result, neither streams nor springs have received any run-off water, nor have there been any floods. The people in the village of Bhangeri, which is 2 kosh [about 3 km] away from the district headquarters Manthali, faces severe water scarcity".

Another example from the East about water shortages in Sindhuli DIstrict, as discussed by Jagadish Chandra Baral and Shanker Adhikary (2010), and quoted in Dixit et al. (2012):

"For the past three decades, the area in and around the village of Ghokshila has experienced erratic monsoon precipitation: sometimes there is rain during the pre-monsoon season while other times there is none and the monsoon is often late. According to the villagers, major flood events have damaged large sections of agricultural land. Since 2060 B.S. [2002 A.D.], drinking water has been scarce, agriculture yields have declined, and biodiversity has been adversely impacted, all due to insufficient rainfall…"

As discussed further by Dixit et al. (2012), a study conducted in 2007 suggests that while an erratic climate is one reason behind source depletion, there are other factors in play, e.g. social and institutional factors. Also, a study examined 50 springs in Nepal's mid-hills, where people did remark in general that the discharge of sources has declined, but the degree of decline and indeed whether or not there was a decline at all depended on local characteristics of rainfall. The interdependence of springs, rural drinking water systems, and climate change is clear, though its exact nature cannot be precisely defined.

6 Impact on Health

Climate change not only affects the availability of water and its quality, but also poses a direct challenge to human health: waterborne, water wash, and vector-related diseases are spreading to wider geographical areas posing challenges to the existing health systems. People living in the southern countries and vulnerable (also poor) people in the industrialised countries have been the main victims, suffering the most from both the climate-induced physical impacts (temperature and sea level rise, precipitation change, increase in the number and intensity of natural hazards such as drought, heat waves, storms, and floods) and the societal effects (famine, food protests, diseases, migration) (UNESCO-WWAP 2009). However, the greatest health impact of climate change will be borne by the poor (mainly in developing countries, but also in developed countries) who are already facing a host of health-related problems due to their socio-economic conditions. **Kofi Annan** (President, Global Humanitarian Forum) introduced the Human Impact Report *Anatomy of a Silent Crisis* (GHF 2009) by noting that, "**The first hit and worst affected by climate change are the world's poorest groups. 99 % of all casualties occur in developing countries. A stark contrast to the 1 % of global emissions attributable to some 50 of the least developed nations**".

As shown in Fig. 1, climate change will impact on us in different ways: it will change the regional weather and create more extreme weather conditions, i.e. too much or too little rainfall will lead to floods and droughts, higher temperatures will cause heat waves resulting in illness and death because of the heat (especially in vulnerable people), smog will worsen air quality, and the incidence of water and food borne diseases will increase.

The IPCC report shows that the impact of climate change on health issues will be mainly negative (Fig. 2).

With the change in temperature and precipitation pattern, the spread of malaria will vary (since the spread of disease is directly related to the amount of precipitation and temperature for mosquitoes to thrive). In some areas, the geographical range of the disease will contract, while in the other areas, it will expand and even the season when it is transmitted might change. There is a direct co-relation between higher minimum temperatures and incidences of malaria outbreaks, which has been shown in case of Ethiopia. Malaria is expected to spread to more regions

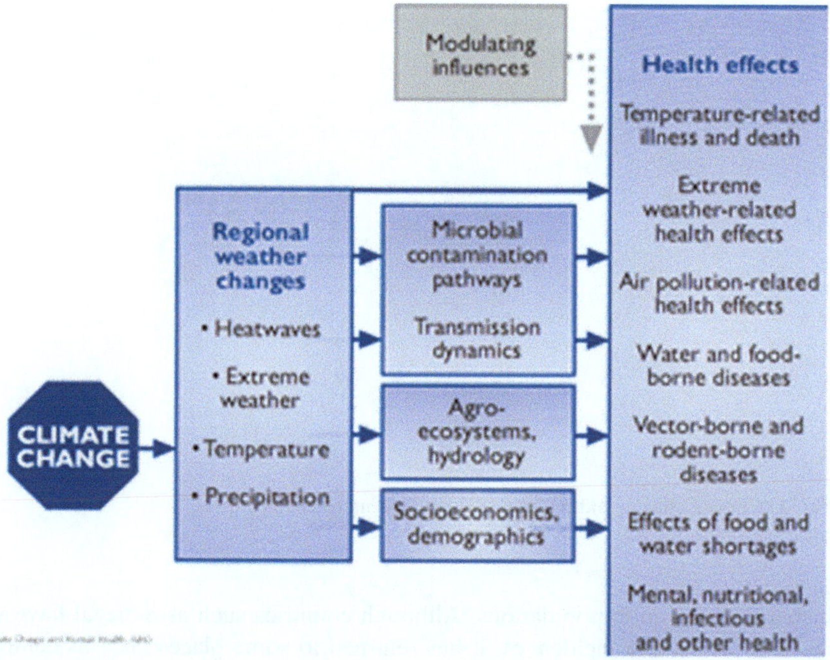

Fig. 1 Impacts of climate change on health (*Source* McMichael et al. 2003)

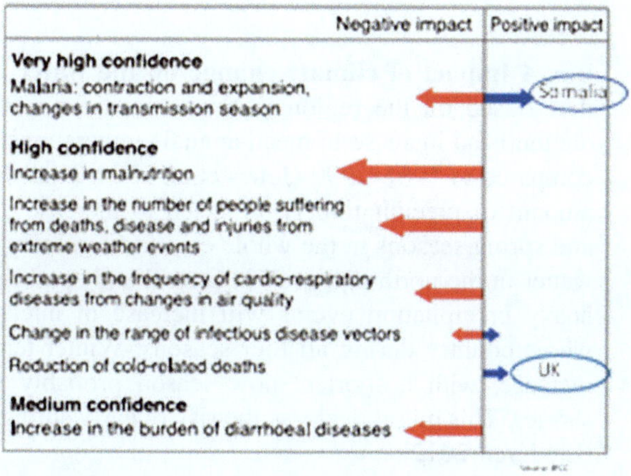

Fig. 2 Direction and magnitude of change on selected health impacts of climate change (*Source* Parry et al. 2007)

including higher altitudes because of favourable conditions due to change in climate (also shown in Fig. 3). However, countries such as Senegal have seen a 60 % drop in malaria cases in the past 30 years because of reduced precipitation. However, in areas where temperatures are rising, in the traditionally cool climates and higher latitudes, these areas will become more suitable reproductive habitats for the mosquitoes spreading malaria. In other places, already warm zones may also see an

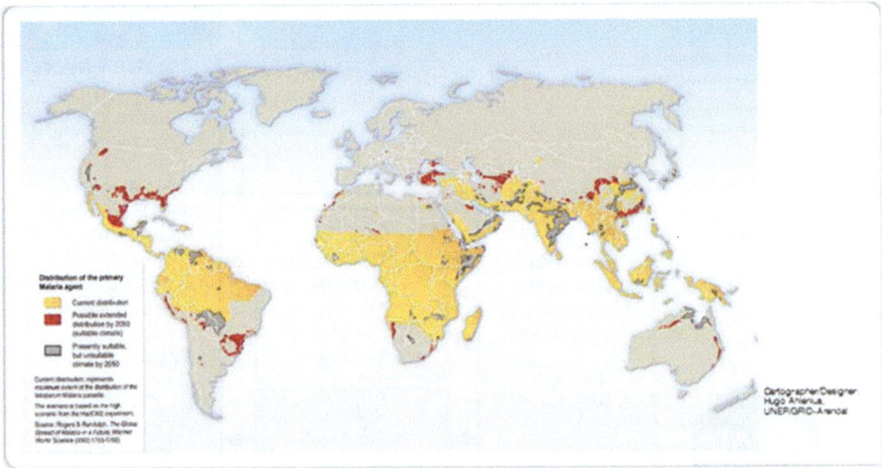

Fig. 3 Climate change and malaria (*Source* Accenture 2011)

increase in mosquito populations. Although countries such as Senegal have seen a decrease in malaria incidences, it has returned to some places such as central and northern South America, much of Asia, some Mediterranean countries, and much of the former USSR. Study estimates show that by 2080, approximately 260–320 million more people are likely to be affected by malaria (Water Aid 2007).

Box 4 Impact of climate change on the water cycle and health in Sweden Based on the regional climate scenario models for Sweden, the prediction is an increase in mean annual temperatures with 35 °C by 2071–2100 compared to 1961–1990 (Jones et al. 2004; Kjellström et al. 2005). Also, the amount of precipitation is projected to increase during the autumn, winter, and spring seasons in the whole country. Summer seasons are expected to be wetter in the north and drier in central and southern parts. It is expected that heavy precipitation events will increase in intensity and frequency in the whole country during all four seasons. Winter temperatures are expected to increase, with a shorter snow season probably making winters markedly shorter. This might decrease the risk of early spring floods from snow melting (Lindgren 2012).

According to the models developed by the Rossby Research Centre at the Swedish Meteorological and Hydrological Institute, run-off will increase in most parts of the country, and the so-called 100-year flows will occur more frequently in the western regions of Sweden. Flood and landslide events are projected to become more common in risk prone areas. As the precipitation and heavy rainfall events increase in combination with an increase in water temperature, at times, it also leads to reduced water quality. Trends of increasing humus concentrations in surface water sources have been observed in Sweden. Increased turbidity often occurs after heavy rain with increased run-off and has in some studies been shown to be correlated with outbreaks of gastroenteritis. Increased turbidity and changes in water colour make people drink less tap water, which can have implications for risk groups like the elderly during heat waves. Increased temperature and heavy rains will also lead to contamination of recreational water. Higher sea water temperatures will stimulate growth of pathogens such as Vibrio vulnificus and in some cases also lead to algal bloom (Lindgren 2012).

Also, as discussed by Lindgren (2012), still-standing flood waters are also acting as breeding sites for mosquitoes and an emerging problem in Sweden. Essentially, milder winters, increased spring precipitation, and floods have contributed to standing still water leading to extreme numbers of mosquitoes. "These insects are not transmitting disease but due to their extreme abundance they are becoming an environmental hazard. Extreme invasion of the floodwater mosquito species *Ochlerotatus sticticus* after heavy rain is an emerging threat, in particular, in the lowland near the River Dalälvenin, Central Sweden (Schäfer et al. 2008). The mosquitoes are so abundant that people and cattle are forced to stay indoors" (Lindgren 2012).

Box 4 shows the impact of climate change in Sweden. Clearly, the health implications of changes to water supply are far-reaching; in some cases, it is disease-causing mosquitoes, and in other situations as discussed in the case of Sweden, just the number of mosquitoes can cause nuisance. Currently, more than 3 million people die each year from avoidable water-related diseases, and most of whom are in developing countries. The effects of climate change on water will exacerbate the existing implications of water shortages on human health as follows:

- Waterborne diseases: resulting from the contamination of water, chances of which are more likely to occur during periods of flood and therefore intensify with the projected increases in natural disasters under climate change. Diseases are transmitted directly with water consumption or in food preparation.
- Water-washed diseases: resulting from inadequate personal hygiene because of scarcity or inaccessibility of clean safe water.
- Water-dispersed diseases: infections for which the agents proliferate in fresh water and enter the human body through the respiratory tract (e.g. legionella). (Water Aid 2007)

7 Impact on Agriculture

Agriculture is by far the largest consumer of freshwater. Globally, about 70 % of freshwater is used in irrigated farming, and far greater volumes of water are used in rain-fed agriculture (references Accenture 2011). Although average temperatures are predicted to increase more dramatically in the northern hemisphere, the changes in temperature dependent agriculture will be felt more significantly in the developing regions because of their heavy reliance on small-scale farming, dependence on rain-fed agriculture, a fragile infrastructure, and a limited capacity to respond to emergencies. Many African communities will be at risk, particularly subsistence farmers with low incomes in sub-Saharan Africa. This will essentially impact on the income of these small farmers, as increased droughts may exacerbate poverty levels and increase the vulnerability of these people. The United Nations' scientists warned in 2005 that one in six countries are facing food shortages because of severe droughts that could become semi-permanent. National communications report that climate change will cause a general decline in many subsistence crops, for example, sorghum in Sudan, Ethiopia, Eritrea and Zambia; maize in Ghana; millet in Sudan; and groundnuts in Gambia. Africa already accounts for a large proportion of the total additional people at risk of hunger as a result of climate change; by the 2080s, it may account for the majority. As discussed by Webersik (2012), rising mean temperatures will also have socio-economic consequences. The impact of higher temperatures on agricultural yields is well documented, as discussed above. With rising temperatures, overall productivity of agricultural crops will decline (Battisti and Naylor 2009), for example, Maize in the sub-Saharan Africa (since it is also less receptive to carbon fertilisation when CO_2 concentration increases). The same effect is, however, less clear in Asia where rice is the main staple diet (Xiangzheng et al. 2010). Higher temperatures also foster evaporation of water before it can reach soil, posing a challenge to rain-fed agriculture, the dominant form of agriculture in the sub-Saharan Africa. Still, there are uncertainties in predicting changes in growing season temperatures and precipitation (Lobell and Burke 2008).

As can be seen in Fig. 4, as the temperature rises, the impact gets more severe. For example, in the case of food, a slight or one degree rise might raise yield in some places at higher altitudes while any further rise will actually lead to failed crops in many areas with a more drastic impact in developing countries.

Also, changing temperature, precipitation, humidity, rainfall, and extreme weather-related incidents will make food security more complex. For example, studies in countries such as Mali and Nepal suggest that by 2050, 72 % of the population could face serious food shortages (Accenture 2011). Current climate variability already presents serious challenges for food security in many developing countries. Rural-based populations in countries that rely on rain-fed agriculture and primarily depend on subsistence farming systems are especially vulnerable. In general, water scarcity can limit food production and supply, putting pressure on food prices and increasing countries' dependence on food imports.

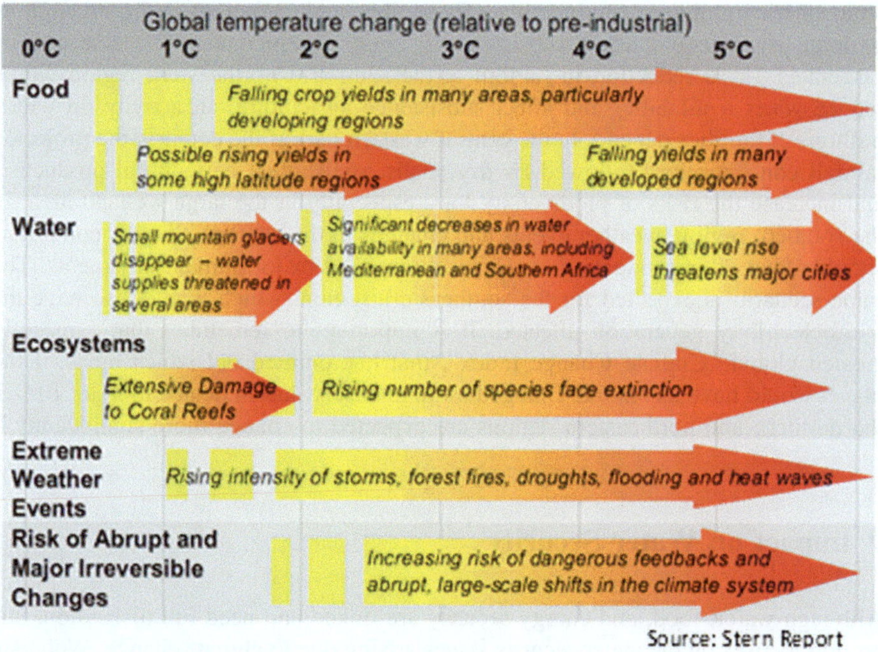

Fig. 4 Projected impacts of climate change (*Source* Stern Report)

IPCC report (Parry et al 2007) also predicts that malnutrition will be negatively impacted by the change in climate (Fig. 2), which is expected if agriculture is affected by a change in precipitation thus threatening food security. Even today, malnutrition is one of the most serious global health problems (FAO and WHO); about 178 million children globally are stunted and 1.5 million die annually from wasting, both important indicators of malnutrition. According to the IPCC report (Parry et al. 2007), there is an 80 % chance that climate change will increase malnutrition and consequent disorders (Accenture 2011).

8 Impact on Biodiversity: A Case Study from Brazil

The regional (downscaled) climate scenarios for Brazil indicate that, during the current century, most regions of the country will experience an increase in temperature ranging from 2 to 8 °C (Ulisses et al. 2012). Also, average precipitation is projected to decrease in most regions, but there will be an increase in the frequency and intensity of extreme weather events (especially storms) in the southern and south-eastern regions. As discussed by Ulisses (in Grover 2012), "In the Amazon, a temperature increase and a projected 20 % reduction in precipitation could change the forest cover, which would be transformed into a savanna. A decrease in atmospheric humidity would facilitate forest fires, and reductions in the stream flow of large rivers could affect subsistence fishing. A temperature increase and an associated 15 % reduction in rainfall at the semi-arid northeastern region would reduce water availability and affect subsistence agriculture, an activity on which millions of families depend. In the central-western part of the country, the projected rainfall concentration, followed by dry spells, shall affect agricultural production and increase the risks of biodiversity loss, especially in the large wetland area. In the southern region, weather events and changes in the average climatic conditions should affect grain production, water resources, and major human settlements. The same scenario is expected for the southeastern region, which would also have the hydroelectricity generation affected. It is important to remember that a recently created global "Climate Change Index", that has pointed out which areas of the planet would have its climate changed more intensely, has indicated that, for Brazil, the northern and northeastern regions are expected to change more significantly".

9 Impact on Water Security

Although water, food, and energy security are linked and need full discussion, this section focuses on the water security issues arising due to climate change. Webersik (in Grover 2012) discusses the impact of climate change on water and human health and its relevance for human security. There is little evidence that water stress or health risks will lead to armed confrontation or "water wars", but there is compelling evidence that water stress and health risks accelerated by climate change will affect human security. By focusing on human security, the threats of a warmer and more variable climate become more tangible. With growing populations, the challenge will be the provision of safe and clean drinking water while avoiding water pollution. Sustainable water management is key here. As the US President Barack Obama stated in his Nobel Lecture, "… the world must come together to confront climate change. There is little scientific dispute that if we do nothing, we will face more drought, more famine, more mass displacement; all of which will fuel more conflict for decades. For this reason, it is not merely scientists and environmental activists who call for swift and forceful action; it's military leaders in my own country and others who understand our common security hangs in the balance" (Obama 2009).

10 Conclusion

Understanding any global environmental change necessitates identifying and measuring the impact of human activity in the environment. Since water is so central to climate and environmental change, both as a catalyst and as a vulnerable resource, much emphasis needs to be placed on how humans interact with the water cycle. The discussion above shows that climate change, urbanisation, etc., are changing the water cycle.

There are several human–environment interactions that affect water, weather, and climate. These include dams, diversions, irrigation, deforestation, wetlands drainage, and the production of impermeable surfaces such as concrete. Though humans have been "pushing water" around since early civilisation, the twentieth century has seen a rapid acceleration in the scope, scale, and intensity of these interventions (Conca 2006). Before heavy industrialisation, the greatest anthropogenic transformation of natural ecosystems was due to agricultural expansion. During the twentieth century, the balance shifted and urban expansion has played an increasingly important role. As discussed above, urbanisation is leading to urban heat islands as well as entailing significant changes to the hydrological cycle. The growth of impermeable surfaces changes run-off, erosion, and soil moisture patterns, as well as degrading water quality when pollutants from road surfaces are carried to lakes and streams. Recent studies have also shown that urban mega centres have an impact on local precipitation patterns as well (Shepherd 2005). Shem and Shepherd (2009) estimate the increase to be as much as 10–13 % for areas less than 50 km downwind of the urban centre. Dams and large infrastructures are also changing the microclimate and the local hydrological cycle at the locations where water is stored and also in the delta region. As discussed in the chapter above, studies in the USA have shown that in the Mediterranean and semi-arid regions, large dams contribute to increased frequency and intensity of rainfall in the mesoscale.

The change in climate change and the water cycle is also leading to extreme weather events (i.e. heavy floods or persistent droughts) that are impacting on the availability of safe drinking water, agriculture (and food security), health, biodiversity, and increasing water stress in many regions.

Changes in the water cycle will mean adapting policies for better irrigation systems and better health systems and will also require changes in infrastructure. (For example, capacity of dams can be a challenge, waste water or rainwater infrastructure might need to be changed if the 100-year event becomes more frequent, and pipes to drain rain water might need to be changed, etc.). Climate change is usually referred to as the "supply driver" since it will determine the amount of water available in different places/regions. This essentially also means there is a need to manage interactions between human needs (demand-side drivers such as economic, social, and demographic pressures) of water consumption/supply and the changes in climate patterns (Source: Water in a Changing World, UNWWAP project). All these demand drivers will affect climate change, so we need better policies for water management at municipal, national, and federal levels. There needs to be a paradigm shift towards a more integrated and holistic approach to balance all the competing needs of water.

References

Accenture (2011) Climate change and health: framing the issue. Accessed from http://nstore.accenture.com/acn_com/PDF/Accenture_Climate_Change_and_Health.pdf

Adams RM, Peck DE. (2008) Effects of climate change on water resources, choices, 23 (1) accessed from http://www.choicesmagazine.org/2008-1/theme/2008-1-04.htm

Angel JR, Kunkel KE (2009) The response of Great Lakes water levels to future climate scenarios with an emphasis on Lake Michigan-Huron. J Great Lakes Res 36(2010):51–58

Aon Benfield (2011) Annual global climate and catastrophe report, impact forecasting-2011. Accessed from http://thoughtleadership.aonbenfield.com/Documents/20120110_if_annual_global_climate_cat_report.pdf

Arnell NW (2004) Climate change and global water resources: SRES emissions and socio economic scenarios. Glob Environ Change 14:31–52

Bates BC, Kundzewicz ZW, Wu S, Palutikof JP (eds) (2008) Climate change and water. Technical Paper of the Intergovernmental Panel on Climate Change, IPCC Secretariat, Geneva

Bates B, Kundzewicz ZW, Wu S, Palutikof J (2008) Climate change mitigation measures and water. In: Climate change and water. Technical Paper of the Intergovernmental Panel on Climate Change, IPCC Secretariat, Geneva, pp 115–25

Battisti DS, Naylor RL (2009) Historical warnings of future food insecurity with unprecendented seasonal heat. Science 323: 240–244

Briggs C (2012) Environmental change and water-related health risks: an arctic security approach. In: Grover V (ed) Impact of climate change on water and health. CRC Press Christian Aid. 2007. Human Tide: The Real Migration Crisis. A Christian Aid Report. Christian Aid, London

Conca K (2006) Pushing rivers around: the cumulative toll on the world's watersheds and freshwater ecosystems. In: Governing water: contentious transnational politics and global institution building. MIT Press, Cambridge, pp 73–93

Cruickshank A, Grover V (2012) Impact of climate change in Africa. In: Grover V (ed) Impact of climate change on water and health. CRC Press, Boca Raton

Dasgupta S, Laplante B, Murray S, Wheeler D (2009) Sea-level rise and storm surges [Policy Research Working Paper 4901]. The World Bank- Development Research Group- Environment and Energy Team, Washington DC

Degu AM, Hossain F (2011) Investigating the mesoscale impact of artificial reservoirs on frequency of rain. Water Resources Research

Degu AM, Hossain F, Niyogi D, Pielke R, Shepherd JM, Voisin N, Chronis T (2011) The influence of large dams on surrounding climate and precipitation patterns. Geophys Res Lett 38 (4)

Dixit A, Gyawali D, Pandey U (2012) Improved access to safe drinking water, sanitation and health: foundations for adapting to climate change impacts in Nepal. In Grover V (ed) Impact of climate change on water and health. CRC Press, Boca Raton

Eriksen S, O'Brien K, Rosentrater L (2008) Climate change in Eastern and Southern Africa: impacts, vulnerability and adaptation. Global Environmental Change and Human Security Project, Olso, Norway

Fields S (2005) Continental divide: why Africa's climate change Burden is greater. Environ Health Perspect 113:A534–A537

GHF 2009 [Global Humanitarian Forum] Human impact report-climate change: the anatomy of a silent crisis. Global Humanitarian Forum, Geneva, Switzerland

Grover VI (ed) (2012) Impact of climate change on water and health. CRC Press, Boca Raton

Gupta J (2012) Climate change and water law: the regimes compared. In: Grover V (ed) Impact of climate change on water and health. CRC Press, Boca Raton

Hansen J (2007) Scientific reticence and sea level rise. Environ Res Lett 2:024002. doi:10.1088/1748-9326/2/2/024002

Hewelet JD (1985) Principles of hydrology. The University of Georgia Press, Athens

Hossain F, Degu A, Woldemichael AT, Yigzaw W, Mitra C, Shepherd JM, Siddique-E-Akbor AHM (2012) Analysis of rainfall patterns in the context of rapid urbanization and water resources vulnerability of an Asian Mega City. Water Resources Research

Hulme M, Doherty R, Ngara T, New M, Lister D (2001) African climate change: 1900–2100. Clim Res 17:145–168

IPCC (1998) The regional impacts of climate change—an assessment of vulnerability. A special report of Intergovernmental Panel on Climate Change WGII. Cambridge University Press, Cambridge, UK

IPCC (2007) Climate change 2007: synthesis report. Contribution of Working Groups I, II and III to the Fourth Assessment Report of the Intergovernmental Panel on Climate Change. Core Writing Team, Pachauri RK, Reisinger A (eds) IPCC, Geneva, Switzerland, http://www.ipcc.ch/publications_and_data/publications_ipcc_fourth_assessment_report_synthesis_report.htm

IPCC (2008) Climate change mitigation measures and water. pp 115–125 In: Bates B, Kundzewicz ZW, Wu S, Palutikof J (eds). Climate change and water. Technical paper of the intergovernmental panel on climate change. IPCC Secretariat, Geneva

IPCC (2013) Climate change 2013: the physical science basis. In: Stocker TF, Qin D, Plattner G-K, Tignor M, Allen SK, Boschung J, Nauels A, Xia Y, Bex V, Midgley PM (eds) Contribution of working group I to the fifth assessment report of the intergovernmental panel on climate change. Cambridge University Press, Cambridge, UK and New York, NY, USA (in press)

Jones CG, Willén U, Ullerstig A, Hansson U (2004) The rossby centre regional atmospheric climate model part I: model climatology and performance for the present climate over europe. Ambio 33:(4–5) 199–210

Kaufmann RK, Seto KC, Schneider A, Liu Z, Zhou L, Wang W (2007) Climate response to rapid urban growth: evidence of a human-induced precipitation deficit. J Clim 20(10):2299–2306

Kjellström E, Döscher R, Meier HEM (2005) Atmospheric response to different sea surface temperatures in the Baltic Sea: coupled versus uncoupled regional climate model experiments. Nordic Hydrology 36:(4–5) 397–409

Lindgren E (2012) Water-related health effects of climate change in Sweden. In: Grover V (ed) Impact of climate change on water and health. CRC Press, Boca Raton

Lobell DB, Burke MB (2008) Why are agricultural impacts of climate change so uncertain? the importance of temperature relative to precipitation. Environ Res Lett 3

Magrath J (2006) Africa: up in smoke 2. New Economics Foundation and International Institute for Environment and Development, London

Mastrandrea MD, Field CB, Stocker TF, Edenhofer O, Ebi KL, Frame DJ, Held H, Kriegler E, Mach KJ, Matschoss PR, Plattner G-K, Yohe GW, Zwiers FW (2010) Guidance note for lead authors of the IPCC fifth assessment report on consistent treatment of uncertainties. Intergovernmental Panel on Climate Change (IPCC). Available at http://www.ipcc.ch

McMichael AJ, Campbell-Lendrum DH, Corvalan CF, Ebi KL, Githeko AK, Scheraga JD, Woodward A (2003) Climate change and human health: risks and responses, WHO. Accessed from http://www.who.int/globalchange/publications/climchange.pdf

Milly PCD, Dunne KA, Vecchia AVE (2005) Global pattern of trends in streamflow and water availability in a changing climate. Nature 438(17):347–350

Mitra C, Shepherd JM, Jordan T (2011) On the relationship between the premonsoonal rainfall climatology and urban land cover dynamics in Kolkata city, India. Int J Climatol, 7 June 2011. doi:10.1002/joc.2366. Accessed 24 Sept 2011

Obama BH (2009) A just and lasting peace. In: Nobel Lecture, Oslo

Parry ML, Canziani OF, Palutikof JP, Van der Linden PJ, Hanson CE (eds) (2007) Climate change 2007: impacts, adaptation and vulnerability. In: Contribution of working group II to the fourth assessment report of the intergovernmental panel on climate change, Cambridge University Press, Cambridge, UK

Rabbani G, Huq S, Rahman SH (2012) Impacts of climate change on water resources and human health: empirical evidences from a coastal district (Satkhira) in Bangladesh. In: Grover V (ed) Impact of climate change on water and health. CRC Press, Boca Raton

Rawat JS (2009) Saving Himalayan Rivers: developing spring sanctuaries in headwater regions. In: Shah BL (ed) Natural resource conservation in Uttarakhand, Ankit Prakashan, Haldwani, India

Renaud F, Bogardi JJ, Dun O, Warner K (2007) Control, adapt or flee how to face environmental migration?, vol 7. Intersections UNU, Bonn

Schäfer ML, Lundström JO, Petersson, E (2008) Comparison of mosquito (Diptera: Culicidae) populations by wetland type and year in the lower river dalälven region, Central Sweden. J Vector Ecol 33:(1) 150–157

Sharma E, Bhuchar S, Xing M, Kothyari BP (2007) Land use change and its impact on hydro-ecological linkages in Himalayan watersheds. Trop Ecol 48(2):151–161

Shem W, Shepherd M (2009) On the impact of urbanization on summertime thunderstorms in Atlanta: two numerical model case studies. Atmos Res 92(1):172–189

Shepherd JM (2005) A review of current investigations of urban-induced rainfall and recommendations for the future. Earth Interact 9:1–12

Smakhtin V, Revenge C, Doll P, Tharme R (2003) Putting the water requirements of into the global picture of water resources assessment. IWMI, Colombo

Stern N (2007) The economics of climate change. Cambridge University Press, Cambridge

Tip of the Mitt Watershed Council (2014) Influences on Great Lakes water levels. Accessed from http://www.watershedcouncil.org/water%20resources/great%20lakes/threats-to-the-great-lakes/great-lakes-water-levels/other-opinions-on-low-water-levels/

Tiwari PC (2008) Management of common pool wasteland for rural livelihood improvement in the Himalayas, sustainable solutions: a spotlight on South Asian research. In: Proceedings of 10th international sustainable development conference, Dec 2007. Sustainable Development Policy Institute, Islamabad, Pakistan, pp 75–101

Tiwari PC, Joshi B (2012) Rainfall variability and its impacts on water resources and rural health in Kumaon Himalaya, India. In: Grover V (ed) Impact of climate change on water and health. CRC Press, Boca Raton

Trenberth KE, Fasullo JT. Kiehl J (2009) Earth's global energy budget. Bull Am Meteorol Soc 90 (3):311–323 (29 July 2009). doi:10.1175/2008BAMS2634.1

Confalonieri UEC, Ferreira IB, Barros BF (2012) Climate change and health impacts in Brazil: a discussion on "hotspots". In: Grover V (ed) Impact of climate change on water and health. CRC Press, Boca Raton

UNESCO-WWAP (2009) Climate change and water: an overview from the world water development report 3: water in a changing world, A United Nations World Water Assessment Programme Special Report: WaterAid, Climate Change and water Resources (May 2007). Accessed from http://www.wateraid.org/documents/climate_change_and_water_resources_1.pdf

Water Aid, May (2007) Climate change and water resources. Accessed from http://www.wateraid.org/documents/climate_change_and_water_resources_1.pdf

Webersik C (2012) Climate change, water and health: current impacts and future challenges for human security. In: Grover V (ed) Impact of climate change on water and health. CRC Press, Boca Raton

[World Bank 2009] Alavian VHA, Qaddumi HH, Dickson E, Diez SM, Danilenko AV, Hirji RF, Puz G, Pizarro C, Jacobsen M, Blankespoor B (2009) Water and climate change: understanding the risks and making climate-smart investment decisions. The World Bank, Water Sector Board, Bank Netherlands Water Partnership Program and Norwegian Ministry of Foreign Affairs

WHO (2008) Protecting health from climate change—World Health Day 2008. World Resources Institute, United Nations Development Programme, United Nations Environment Programme, and World Bank. World Resources, 2000–2001, 2000: People and Ecosystems, the Fraying Web of Life. World Resources Institute, Washington, DC

World Water Assessment Programme (2009) The United Nations World Water Development Report 3: Water in a Changing World. UNESCO, Paris and Earthscan, London

Xiangzheng D, Jikun H, Fangbin Q, Naylor RL, Falcon WP, Burke M, Rozelle S, Battisti D (2010) Impacts of el nino-southern oscillation events on china's rice production. J Geogr Sci 20: 3–16

Uncertainty Assessment of Climate Change Impacts on Hydrology: A Case Study for the Central Highlands of Vietnam

Dao Nguyen Khoi and Phan Thi Thanh Hang

Abstract This paper focuses on quantifying the uncertainty in climate change and its impacts on hydrology in the Srepok watershed in the Central Highlands of Vietnam. The uncertainty associated with the general circulation model (GCM) structure from a subset of CMIP3 (CCCMA CGCM3.1, CSIRO Mk3.0, IPSL CM4, MPI ECHAM5, NCAR CCSM30, UKMO HadGEM1, and UKMO HadCM3), SRES emission scenarios (A1B, A2, B1, and B2), and prescribed increases in global mean temperature (0.5–6 °C) using the soil and water assessment tool (SWAT) was investigated. For prescribed warming scenarios using HadCM3, linear decreases in mean annual streamflow ranged from 2.0 to 9.8 %. Differences in projected annual streamflow between SRES emission scenarios using HadCM3 were small (−3.8 to −3.3 %). Under the A1B scenario and 2 °C increase in global mean temperature using seven GCMs, there was substantial disparity, of −3.7 to 21.0 % and −6.0 to 16.1 %, respectively. It was concluded that, in the case of the Srepok watershed, the most important source of uncertainty comes from the GCM structure rather than from the emission scenarios and climate sensitivity.

Keywords Climate change · Srepok watershed · Streamflow · SWAT model · Uncertainty

D.N. Khoi (✉)
Faculty of Environmental Science, University of Science,
Vietnam National University, Ho Chi Minh City,
227 Nguyen van Cu Street, District 5, Ho Chi Minh City, Vietnam
e-mail: dnkhoi86@gmail.com

D.N. Khoi
Center of Water Management and Climate Change,
Vietnam National University, Ho Chi Minh City,
IT Park, Linh Trung Ward, Thu Duc District, Ho Chi Minh City, Vietnam

P.T.T. Hang
Institute of Geography, Vietnam Academy of Science and Technology,
18 Hoang Quoc Viet Street, Cau Giay, Hanoi, Vietnam
e-mail: hangphanvn@yahoo.com

S. Shrestha et al. (eds.), *Managing Water Resources under Climate Uncertainty*,
Springer Water, DOI 10.1007/978-3-319-10467-6_2

1 Introduction

The Intergovernmental Panel on Climate Change (IPCC) Fourth Assessment Report (AR4) reaffirmed that global warming is occurring (IPCC 2007). It is widely acknowledged that climate change can affect the spatial and temporal distribution of water resources as well as the intensity and frequency of extreme hydrological events (Bae et al. 2011). Therefore, studies of climate change impacts on hydrology have recently become a hot topic. Evaluating the hydrological impacts of climate change is most commonly based on the use of a hydrological model with climate change scenarios derived from the general circulation model (GCM) forced with emission scenarios (Thompson et al. 2013). However, these results are rarely used by decision-makers and managers in managing and planning water resources because of the existence of uncertainties in assessments of climate change impacts on hydrology and the difficulty of quantifying these uncertainties (Bae et al. 2011).

Vietnam has experienced changes in climate that includes rising air temperatures and more variable precipitation. In the period 1958–2007, the annual average temperature increased by about 0.5–0.7 °C. The annual precipitation decreased in northern areas while increasing in the southern regions. On an average for the whole country, the rainfall over the past 50 years (1958–2007) decreased by approximately 2 % (MONRE 2009). These changes have impacted significantly on the availability of water resources in Vietnam. The studies on impacts of climate change on hydrology have also gained the attention of Vietnamese scientists, and most analyses of these studies in Vietnam have, to date, mainly focused on climate projection forced by individual GCMs or an ensemble of GCMs. For example, Kawasaki et al. (2010) used the HEC-HMS model feeding climate projections from the Japanese Meteorological Agency GCM for the SRES A1B scenario to assess the climate change impacts on water resources in the Srepok watershed. Thai and Thuc (2011) employed MIKE 11/NAM hydrological model and climate change scenarios of the Vietnam Ministry of Natural Resources and Environment (2009) downscaled by MAGICC/SCENGEN and PRECIS model from GCMs to evaluate the impacts of climate change on the flow in Hong–Thai Binh and Dong Nai River Basins. Khoi and Suetsugi (2012a, b) used the SWAT hydrological model and projected climate from an ensemble of four GCMs (CGCM3.1 (T63), GFDL CM2.0, GFDL CM2.1, and UKMO HadCM3) to estimate the projected river discharge in the Be River Catchment. In these studies, the uncertainty associated with the GCM structure has not been investigated yet. In fact, the projection of future climate (especially precipitation) from different GCMs often disagrees even in the direction of change (Kingston et al. 2011).

The main objective of this study is to estimate the uncertainty in projection of climate change on hydrology through the application of a range of climate scenarios[1] (obtained from the QUEST-GSI project; Todd et al. 2011) to the SWAT hydrological model; a case study for the Srepok watershed in the Central Highlands

[1] http://www.cru.uea.ac.uk/~timo/climgen/data/questgsi/, accessed June, 2011.

of Vietnam. The climate scenarios were generated from different GCMs, emission scenarios (A1B, A2, B1, and B2), and prescribed warming in global mean air temperature (from 0.5 to 6 °C), including a 2 °C threshold of "dangerous" climate change. Baseline climate was obtained from CRU-TS 3.0 dataset (Mitchell and Jones 2005) for the period 1971 to 2000.

The paper first presents a brief description of the watershed. A detailed description of the SWAT hydrological model and climate change scenarios is then presented. The paper concludes with a presentation of results and a conclusion.

2 Study Area

The Srepok watershed, a sub-basin of the Mekong River basin, is located in the Central Highlands of Vietnam and lies between latitudes 11°45′–13°15′ N and longitudes 107°30′–108°45′ E (Fig. 1). The Srepok River is formed by two main tributaries: the Krong No and Krong Ana Rivers. The total area of this watershed is approximately 12,000 km^2 with a population of 2.2 million (2009). The average altitude of the watershed varies from 100 m in the northwest to 2,400 m in the

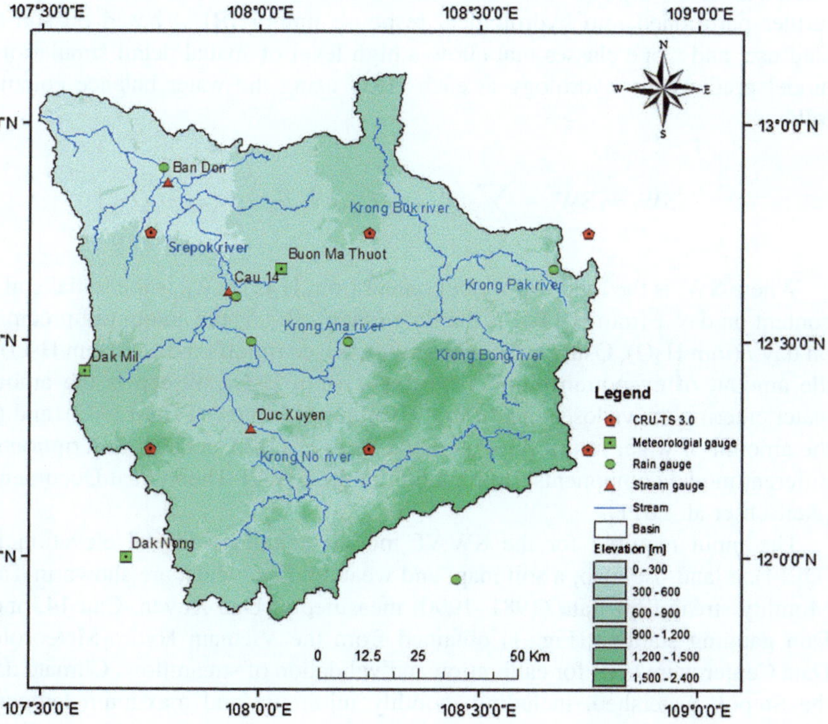

Fig. 1 Location of weather stations and CRU-TS3.0 grid points in the Srepok watershed

southeast. The climate in the area is very humid (78–83 % annual average humidity) with annual rainfall varying from 1,700 to 2,300 mm and features a distinct wet and dry season. The wet season lasts from May to October (with peak floods often in September and October) and accounts for over 75–95 % of the annual precipitation. The mean annual temperature is 23 °C.

In this watershed, there are two dominant types of soils: grey soils and red-brown basaltic soils. These soils are highly fertile and consistent with the agricultural development. Agriculture is the main economic activity in this watershed of which coffee and rubber production are predominant.

3 Methodology

3.1 The SWAT Hydrological Model

The SWAT model is a physically based distribution model designed to predict the impact of land management practices on water, sediment, and agricultural chemical yields in large complex watersheds with varying soil, land use, and management conditions over long periods of time (Neitsch et al. 2011). With this model, a catchment is divided into a number of sub-watersheds or sub-basins. Sub-basins are further partitioned into hydrological response units (HRUs) based on soil types, land use, and slope classes that allow a high level of spatial detail simulation. The model predicts the hydrology at each HRU using the water balance equation as follows:

$$\mathrm{SW}_t = \mathrm{SW}_0 + \sum_{i=1}^{t} \left(R_{\mathrm{day}} - Q_{\mathrm{surf}} - E_{\mathrm{a}} - w_{\mathrm{seep}} - Q_{\mathrm{gw}}\right) \tag{1}$$

Where SW_t is the final soil water content (mm H_2O), SW_0 is the initial soil water content on day 1 (mm H_2O), t is the time (days), R_{day} is the amount of precipitation on day i (mm H_2O), Qsurf is the amount of surface runoff on day 1 (mm H_2O), E_a is the amount of evapotranspiration on day 1 (mm H_2O), Wseep is the amount of water entering the vadose zone from the soil profile on day 1 (mm H_2O), and Q_{gw} is the amount of water return flow on day 1 (mm H_2O). A detailed description of the different model components can be found in the SWAT Theoretical Documentation (Neitsch et al. 2011).

The input required for the SWAT model includes a digital elevation model (DEM), a land-use map, a soil map, and weather data, which are shown in Table 1. Monthly streamflow data (1981–1990) measured at Duc Xuyen, Cau 14, and Ban Don gauging stations (Fig. 1) obtained from the Vietnam Hydro-Meteorological Data Center were used for calibration and validation of streamflow. Climate data for the Srepok watershed, including monthly minimum and maximum temperature, precipitation, and number of wet days, were obtained from the gridded (0.5° × 0.5°)

Table 1 Input data used in the SWAT model for the Srepok watershed

Data type	Description	Resolution	Source
Topography map	Digital elevation map (DEM)	90 m	SRTM
Land-use map	Land-use classification	1 km	GLCC
Soil map	Soil types	10 km	FAO
Weather	Monthly precipitation, minimum and maximum temperature	0.5°×0.5°	CRU-TS 3.0 Dataset

CRU-TS 3.0 observational dataset (Mitchell and Jones 2005). Monthly data for 6 grid cells (Fig. 1) that cover the watershed were disaggregated to daily data using a weather generator (MODAWEC model, Liu et al. 2009). In the MODAWEC model, daily precipitation was generated using a first-order Markov chain and exponential distribution based on the monthly precipitation and monthly wet days. Daily temperature was determined using a multivariate normal distribution based on monthly means of maximum and minimum temperatures and their standard deviations.

3.2 Model Calibration and Validation

In this study, the Nash–Sutcliffe efficiency (NSE) and percentage bias (PBIAS) were used to assess the model performance in flow simulation. The NSE determines the relative magnitude of the residual variance compared with the measured data variance, and the PBIAS measures the average tendency of the simulated value to be larger or smaller than their observed counterparts. The NSE value is defined by

$$\mathrm{NSE} = 1 - \frac{\sum_{i=1}^{N} (O_i - S_i)^2}{\sum_{i=!}^{N} \left(O_i - \overline{O}\right)^2} \tag{2}$$

and the PBIAS value is defined by

$$\mathrm{PBIAS} = \left[\frac{\sum_{i=1}^{N} (O_i - S_i) \times 100}{\sum_{i=1}^{N} (O_i)} \right] \tag{3}$$

Where 'O' is the observed discharge, S is the simulated discharge, and N is the amount of observed discharge data. According to Moriasi et al. (2007), the values of NSE greater than 0.5 and the PBIAS values of less than 25 % indicate satisfactory model performance for flow simulation.

Table 2 Hydrological model runs

Model	Scenario	Period	Description
HadCM3	A1B	2006–2100	Hadley centre model
HadCM3	A2	2006–2100	
HadCM3	B1	2006–2100	
HadCM3	B2	2006–2100	
HadCM3	+0.5 °C	2040–2069	0.5–6 °C increase in average global temperature
HadCM3	+1.0 °C	2040–2069	
HadCM3	+1.5 °C	2040–2069	
HadCM3	+2.0 °C	2040–2069	
HadCM3	+2.5 °C	2040–2069	
HadCM3	+3.0 °C	2040–2069	
HadCM3	+4.0 °C	2040–2069	
HadCM3	+5.0 °C	2040–2069	
HadCM3	+6.0 °C	2040–2069	
CCCMA CGCM31	A1B	2006–2100	
CSIRO Mk30	A1B	2006–2100	
ISPL CM4	A1B	2006–2100	
MPI ECHAM5	A1B	2006–2100	
NCAR CCSM 30	A1B	2006–2100	
UKMO HadGEM1	A1B	2006–2100	
CCCMA CGCM31	+2.0 °C	2040–2069	2 °C increase in average global temperature
CSIRO Mk30	+2.0 °C	2040–2069	
ISPL CM4	+2.0 °C	2040–2069	
MPI ECHAM5	+2.0 °C	2040–2069	
NCAR CCSM 30	+2.0 °C	2040–2069	
UKMO HadGEM1	+2.0 °C	2040–2069	
CRU dataset	Baseline	1970–2000	Control run

The SWAT flow predictions were calibrated against monthly flow from 1981–1985 and validated from 1986–1990 at Duc Xuyen, Cau 14, and Ban Don gauging stations. The simulated monthly flow based on station-based data matched well with the observed data for both calibration and validation periods with NSE and PBIAS varying from 0.70 to 0.90 and −8.0 to 3.0 %. In the of CRU dataset, the model performance over the calibration and validation periods for all gauging stations is satisfactory as indicated by the acceptable values of the NSE and PBIAS ranging from 0.52 to 0.66 and −14.5 to 8.3 %. In general, the agreement between the observed and simulated streamflow was calculated using CRU data as input is not as good as that obtained using station-based data. However, the simulated streamflow using CRU data can be considered reasonable. It can be summarised that the calibrated SWAT model can be used to simulate the impact of climate change scenarios.

3.3 Climate Change Scenarios

Future climate scenarios for temperature and precipitation were generated at a monthly scale using the ClimGen pattern-scaling technique described in Osborn (2009) and Todd et al. (2011). Scenarios were generated for (1) greenhouse gas (A1B, A2, B1, and B2) and (2) a prescribed warming of global mean temperature of 0.5, 1, 1.5, 2, 2.5, 3, 4, 5, and 6 °C using HadCM3 GCM as well as (3) A1B emission scenario and (4) prescribed warming of 2 °C ("dangerous" climate change) using six additional GCMs: CCCMA CGCM3.1, CSIRO Mk3.0, IPSL CM4, MPI ECHAM5, NCAR CCSM3.0, and UKMO HadGEM1. These models were chosen following the analyses described by Todd et al. (2011) to span a range of "plausible" different modelled global climate futures (e.g. Indian monsoon weakening/strengthening and magnitude of Amazon dieback). Table 2 summarises the model runs that were evaluated.

4 Results and Discussion

4.1 Uncertainty in Greenhouse Gas Emission Using HadCM3

Changes in mean climate associated with different SRES emission scenarios (A1B, B1, B2, and A2) using HadCM3 GCM are shown in Fig. 2a. Increases in mean annual temperature range from approximately 1.6 to 2.4 °C. Projected precipitation decreases by approximately 3 % with small variation between emission scenarios. In the case of the most severe emission scenario, A2, the temperature increase is highest (2.4 °C) and the precipitation decrease is largest (3.5 %) compared with the other scenarios (Fig. 2a). Figure 2b shows the projected changes in annual river discharge projected by HadCM3 for each of the four SRES scenarios. A decrease in annual river flow compared with the baseline is projected under four scenarios. The magnitude of decreases for annual river discharge ranges from 3.3 to 3.8 %. The projected monthly discharge under all four scenarios mostly decreases in the dry season and increases in the rainy season (Fig. 2c).

4.2 Uncertainty in Prescribed Warming Using HadCM3

Figure 3a indicates that the changes in air temperature are projected by HadCM3 to be linear with a rise in global mean temperature. Temperatures increase from 0.34 to 6.93 °C with an increasing global mean temperature of 0.2–6.6 °C. Annual precipitation is projected to decrease, relative to baseline, at a near-linear rate by 1.7–10.8 %.

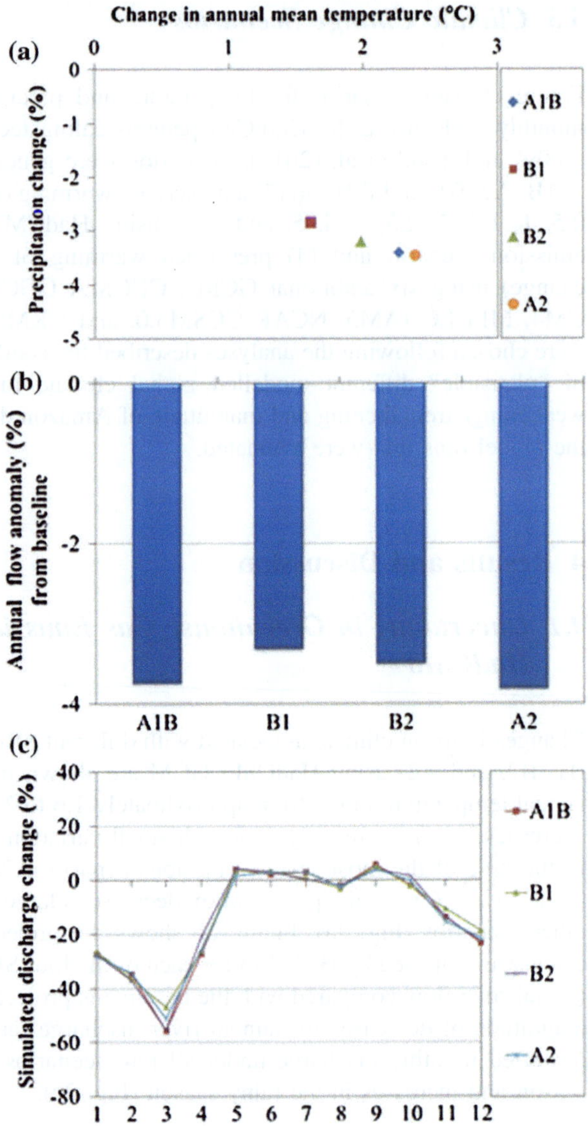

Fig. 2 Projected changes (%) in **a** annual climate, **b** annual discharge, and **c** monthly discharge for HadCM3 GCM with different emission scenarios

Mean annual river discharge is estimated to decrease under the scenarios of prescribed increases in global mean temperature from 0.5 to 6 °C using HadCM3 (Fig. 3b). Decreases in mean annual river flow with increasing global temperature are nearly linear by 2.0–9.8 %. Figure 3c summarises the changes in monthly discharge for all nine scenarios. The monthly river discharge in the dry season (May–October) decreases dramatically from 1.3 % in October for the 0.5 °C scenario to 97.0 % in March for the 6 °C scenario, and monthly discharge in the wet season (November–April) also changes significantly from −8.4 % in September for

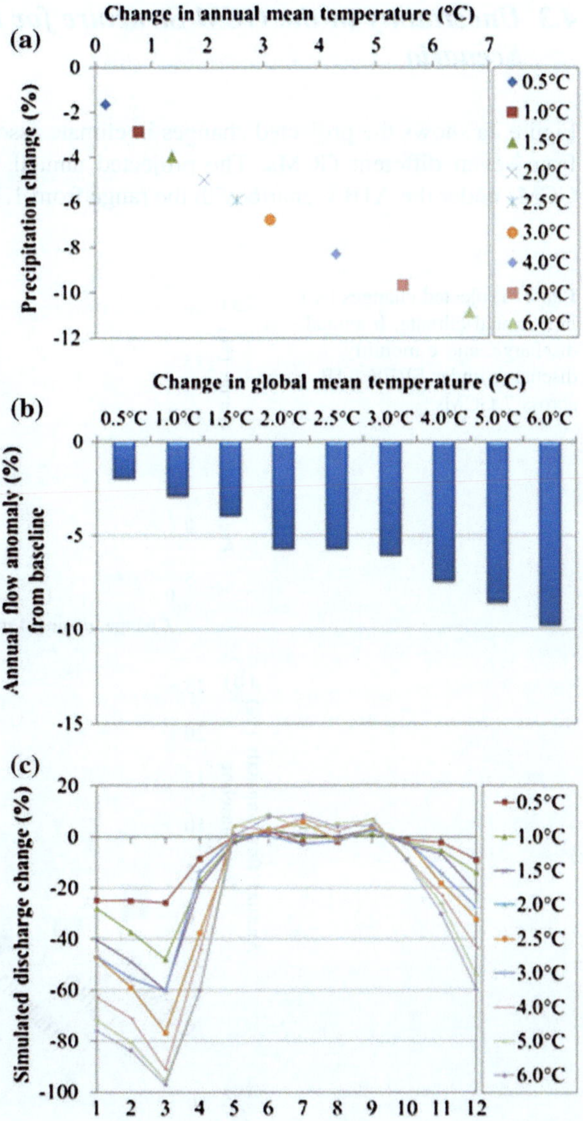

Fig. 3 Projected changes (%) in **a** annual climate, **b** annual discharge, and **c** monthly discharge for HadCM3 prescribed warming scenarios

the 6 °C scenario to 8.7 % in July for the 5 °C scenario. Uncertainty in predicted monthly river discharge increases with a rise in global mean temperature from the range of −26.0 to 2.0 % for the 0.5 °C scenario to the range of −97.0–4.3 % for the 6 °C scenario.

4.3 Uncertainty in the GCM Structure for the A1B Emission Scenario

Figure 4a shows the projected changes in climate associated with the A1B scenario from seven different GCMs. The projected annual temperature increases for all GCMs under the A1B scenario with the range from 1.1 to 2.3 °C. Projected changes

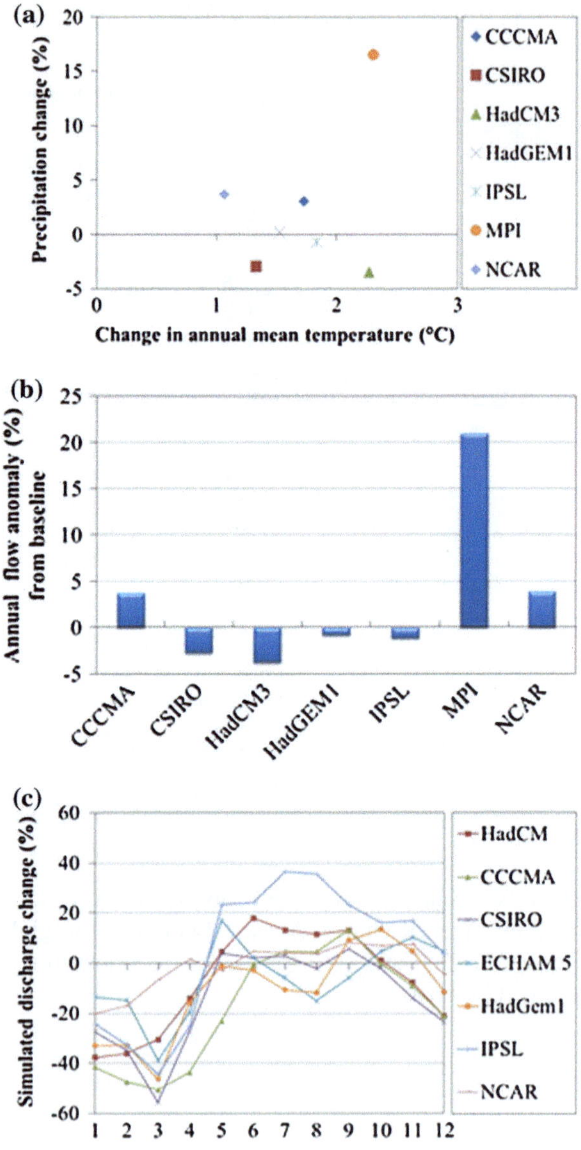

Fig. 4 Projected changes (%) in **a** annual climate, **b** annual discharge, and **c** monthly discharge under SRES A1B across 7 GCMs

in annual precipitation are small (within ± 4 %) for most GCMs except MPI GCM which shows a high increase (16.6 %).

Projected changes in annual river discharge show a substantial variation between GCMs as shown in Figs. 4b. Six GCMs (CCCMA, CSIRO, HadCM3, HadGEM1, IPSL, and NCAR) suggest that the river flow will change by small amounts (−3.7 to 3.9 % change compared with the baseline) while the MPI GCM shows a large

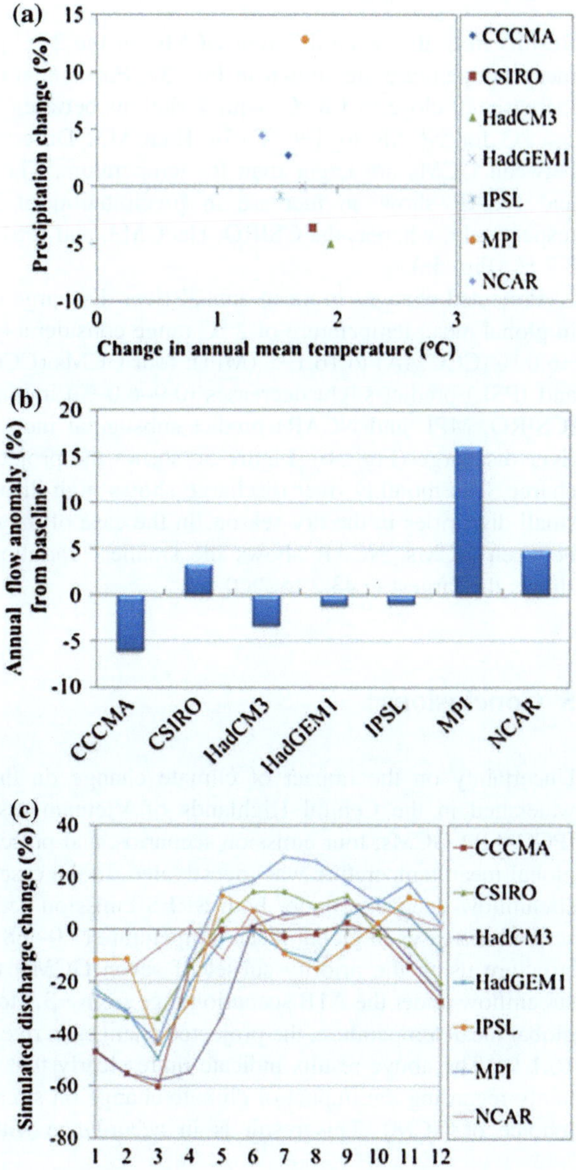

Fig. 5 Projected changes (%) in **a** annual climate, **b** annual discharge, and **c** monthly discharge under 2 °C warming scenario across 7 GCMs

increase of 21.0 % in river flow. Figure 4c shows that the projected increase or decrease in river discharge is evenly distributed over the year, high disparities in the wet season and small disparities in the dry season.

4.4 Uncertainty in the GCM Structure for a 2 °C Rise in Global Mean Air Temperature

Results from the seven different GCMs for the 2 °C prescribed warming in global mean temperature are shown in Fig. 5a. For annual temperature, all GCMs show increases of close to 1.8 °C, with variations between GCMs. The rise ranges from 1.4 °C for NCAR to 1.9 °C for HadCM3. Differences in annual precipitation between GCMs are larger than for temperature. The CCCMA, HadGEM1, MPI, and NCAR show an increase in precipitation of 3.0, 0.2, 16.6, and 3.7 %, respectively, whereas the CSIRO, HadCM3, and IPSL show decreases from 0.7 to 3.7 % (Fig. 5a).

Projected changes in mean annual river discharge under the prescribed increase in global mean temperature of 2 °C range considerably over the seven GCMs from −6.0 % (CCCMA) to 16.1 % (MPI); four GCMs (CCCMA, HadCM3, HadGEM1, and IPSL) predict slight decreases (0.9–6.0 %) in annual river flow, three GCMs (CSIRO, MPI, and NCAR) predict substantial increases (3.3–16.1 %) in annual river discharge (Fig. 5b). Figure 5c shows the projected changes in monthly discharge. The monthly river discharge shows high disparities in the wet season and small disparities in the dry season. In the case of the change in monthly discharge between GCMs, NCAR shows the smallest variation (−17.2–10.2 %) and MPI shows the largest (−43.2 to 28.0 %).

5 Conclusions

Uncertainty on the impact of climate change on the streamflow in the Srepok watershed in the Central Highlands of Vietnam associated with seven CMIP3/IPCC-AR4 GCMs, four emission scenarios, and prescribed increase of 0.5–6 °C in global mean temperature was investigated. In the case of a single GCM, HadCM3, streamflow decreases under both SRES emission scenarios (3.3–3.8 %) and prescribed increases in global mean temperature (2.0–9.8 %). In considering the GCM structure using the priority subset of seven GCMs, the projected changes in the streamflow under the A1B scenario range from −3.7 to 21.0 %. Under a 2 °C rise in global mean temperature, the projected changes in river discharge vary from −6.0 to 16.1 %. The above results indicate quite clearly that the greatest source of uncertainty regarding the impact of climate change on streamflow is the GCM structure (choice of GCM). This result is in accordance with the findings of Khoi and

Suetsugi (2012a, b) who conducted a similar study for the Be River Catchment in the south of Vietnam. The considerable disparity in projected streamflow produced by different GCMs emphasises the importance of using multi-model evaluations of climate change impacts on streamflow. This will help with future such studies in this area.

In the future, it will be necessary to perform other tests in various study areas of Vietnam to support the results and conclusions drawn from this research. Furthermore, the uncertainties of using different hydrological models and downscaling methods should be investigated to provide a useful guideline for evaluating the uncertainties in studies of climate change impacts on hydrology.

Acknowledgments This research is funded by the Vietnam National Foundation for Science and Technology Development (NAFOSTED) under grant number "105.06-2013.09" and by the Vietnam National University Ho Chi Minh City (VNU-HCM) under grant number "C2014-18-18."

References

Bae DH, Jung IW, Lettenmaier DP (2011) Hydrologic uncertainties in climate change from IPCC AR4 GCM simulations of the Chungju Basin, Korea. J Hydrol 401:90–105

IPCC (2007) The physical science basis: contribution of working group i to the fourth assessment report of the intergovernmental panel on climate change. Cambridge University Press, Cambridge

Kawasaki A, Takamatsu M, He J, Rogers P, Herath S (2010) An integrated approach to evaluate potential impact of precipitation and land-use change on streamflow in Srepok River Basin. Theor Appl GIS 18(2):9–20

Khoi DN, Suetsugi T (2012a) Hydrologic response to climate change: a case study for the Be River Catchment Vietnam. J Water Clim Change 3(3):207–224

Khoi DN, Suetsugi T (2012b) Uncertainty in climate change impacts on streamflow in Be River Catchment Vietnam. Water Environ J 26:530–539

Liu J, Williams JR, Wang X, Yang H (2009) Using MODAWEC to generate daily weather data for the EPIC model. Environ Model Softw 24:655–664

Kingston DG, Thompson JR, Kite G (2011) Uncertainty in climate change projections of discharge for Mekong River Basin. Hydroly Earth Syst Sci 15:1459–1471

Mitchell TD, Jones PD (2005) An improved method of constructing a database of monthly climate observations and associated high-resolution grids. Int J Climatol 25:693–712

MONRE–Vietnam Ministry of Natural Resources and Environment (2009) Climate Change, Sea Rise Scenarios for Vietnam, Hanoi

Moriasi DN, Arnold JG, Van Liew MW, Bingner RL, Harmel RD, Veith TL (2007) Model evaluation guidelines for systematic quantification of accuracy in watershed simulations. Trans ASABE 50:885–900

Neitsch AL, Arnold JG, Kiniry JR, Williams JR (2011) Soil and water assessment tool, theoretical documentation version 2009. Texas water resources institute technical report No. 406. Texas A&M University, Texas

Osborn TJ (2009) A user guide for ClimGem: a flexible tool for generating monthly climate datasets and scenarios, Climate Research Unit, University of East Anglia, Norwich

Thai TH, Thuc T (2011) Impacts of climate change on the flow in Hong—Thai Binh and Dong Nai river basins Vietnam National University. J of Sci Earth Sci 27:98–106

Thompson IR, Green AI, Kingston DG, Gosling SN (2013) Assessment of uncertainty in river flow projections for the Mekong River using multiple GCMs and hydrological models. J Hydrol 486:1–30

Todd MC, Taylor RG, Osborn TJ, Kingston DG, Arnell NW, Gosling SN (2011) Uncertainty in climate change impacts on basin-scale freshwater resources—preface to the special issue: the QUEST-GSI methodology and synthesis of results. Hydrol Earth Syst Sci 15:1035–1046

Assessment of the Impact of Climate Change on Water Availability in the Citarum River Basin, Indonesia: The Use of Statistical Downscaling and Water Planning Tools

I. Putu Santikayasa, Mukand S. Babel and Sangam Shrestha

Abstract Climate change would significantly affect the water resources system globally as well as at basin level. Moreover, future changes in climate would affect water availability, run-off, and the flow of the river. This study evaluates the impact of possible future climate change scenarios on the hydrology of the catchment area of the Citarum River Basin, Indonesia. The water evaluation and planning (WEAP) tool was used for hydrological modelling in the study area. The statistical downscaling model (SDSM) was used to downscale the daily temperatures and precipitation in the four sub-catchments of the study area. The global climate variables for A2 and B2 scenarios obtained from Hadley Centre Coupled Model version 3 (HadCM3) was used. After model calibration and testing of the downscaling procedure, the streamflow and evapotranspiration (ET) were projected for three future periods: short term (2010–2039), midterm (2040–2069), and long term (2070–2099). The impacts of climate change on the basin hydrology are assessed by comparing the present and future streamflow and the estimated ET. The temperature is projected to increase in the future. The results of the water balance study suggest increasing precipitation and run-off as well as potential ET losses over the sub-catchment in the study area.

I.P. Santikayasa (✉) · M.S. Babel · S. Shrestha
Water Engineering and Management, School of Engineering and Technology, Asian Institute of Technology (AIT), P.O. Box 4, Klong Luang, Pathum Thani 12120, Thailand
e-mail: psantika@gmail.com

M.S. Babel
e-mail: msbabel@ait.ac.th

S. Shrestha
e-mail: sangam@ait.ac.th

I.P. Santikayasa
Department of Geophysics and Meteorology, Bogor Agricultural University, Bogor, Indonesia

S. Shrestha et al. (eds.), *Managing Water Resources under Climate Uncertainty*, Springer Water, DOI 10.1007/978-3-319-10467-6_3

Keywords WEAP · Statistical downscaling · Climate change · Water planning

1 Introduction

The impact of climate change on water resources has been widely assessed in the past. The increase in greenhouse gasses and the impact of the future climate in different regions of the world have been investigated by research on the projection of future responses to global climate change. Various researches related to the general circulation model (GCM) experiments and studies indicate that increasing global temperature in the future will be connected with increased carbon monoxide concentration. As a result, the climatic processes are likely to be affected, including hydrological-related hazards such as flood and drought. Therefore, an assessment on the future impact of climate change in the hydrological processes in a basin is essential.

To assess the impact of climate change on hydrological processes in the basin, hydrological modelling is widely used. The hydrological model governs the hydrologic cycle and incorporates the physical laws of water movement and the parameters associated with the characteristics of a catchment area. Therefore, the hydrological model has been found to be very useful for such an impact assessment study. In the past, hydrology modelling has been used to assess the impact of climate change on hydrology in the basin (Bae et al. 2008; Christensen et al. 2004; Evans and Schreider 2002; Fujihara et al. 2008). The studies used the GCM outputs corresponding to a specified climate change scenario as the input to hydrology model to quantify the change in streamflow in the future.

In exploring ways to improve the assessment of climate change in hydrology, it is always important to consider future planning of the basin development. From the aspect of decision making, the incorporation of water planning in the assessment processes is very beneficial in terms of cost and time. Generally, the incorporation of impact assessment into climate change on hydrology in the basin and water planning is required to achieve the goal of integrated water resources management. The water evaluation and planning (WEAP) model is the hydrological model which incorporates the hydrological process and water planning. WEAP is widely used to assess the impact of climate change on the future possible climate (Alemayehu et al. 2010; Mehta et al. 2013; Purkey et al. 2007; Rosenzweig et al. 2004).

However, the GCM output has a too coarse spatial resolution as the most hydrological model needs the basin scale input of climate parameters. To resolve the resolution gap between the GCM output and hydrological model, spatial downscaling has been proposed. The downscaling approach is used to convert GCM outputs into the basin level as the input for hydrological modelling. The statistical downscaling method involves bridging the resolution gap by establishing the empirical relationship between the GCM output scale and local climate variables at a location (station). Statistical downscaling has been used widely to assess the impact of climate change on the hydrology in the river basin region (Christensen et al. 2004; Fowler et al. 2007; Fowler and Kilsby 2007; Hay and Clark 2003; Hay et al. 2002; Wilby et al. 2000; Wood et al. 2004). Statistical downscaling has a number of advantages over the use of the raw GCM output because of the stochasticity of the downscaling model, its ability to reproduce the unique meteorological characteristics of the individual stations, and finally being less data intensive than dynamical methods such as nested or regional climate modelling (Wilby et al. 1999).

This chapter aims to present an innovative approach to assess the impact of climate change on water availability using a combination of the downscaling technique and water planning tools. The future climate parameters were projected from the global circulation model (GCM) output using the statistical downscaling approach and hydrology parameters using the water planning tool. The approach is applied to the Citarum River Basin, Indonesia.

2 Study Area and Methodology

2.1 Study Area

The Citarum River Basin, located in West Java, Indonesia, has been selected as the study area. Streamflow in the basin comes mainly from rainfall during the Australian monsoon wet season of November to April with an average annual precipitation range of 1,500–4,000 mm. Furthermore, the study area is characterised as a tropical area with the average temperature of 24.7–27.3 °C and humidity of about 80–95 %.

The soils were mostly developed from volcanic and deposited materials with a slope of 0.5–1.5 %. In the upper stream, forest predominantly covered the area. In the middle and downstream, the area is covered by the forest, forested area, estate plantation, paddy rice fields, and annual crops. Vegetable crops were planted in the volcanic slope area. Paddy was planted in the terraced paddy field in the slopping area when sufficient water is available in the basin. The irrigated paddy field is found in the flat area of the upper, middle, and downstream part of the catchment.

As a study case, four sub-catchments, namely Citarum, Cikarang, Cibeet, and Bekasi, were investigated (Fig. 1).

2.2 Statistical Downscaling Using SDSM 4.2

2.2.1 Model Development

The statistical downscaling model (SDSM), a hybrid stochastic regression weather generator, was developed by Wilby (2002) to assess the impact of climate change on a local scale. It facilitates the rapid development of multiple, low-cost, single–

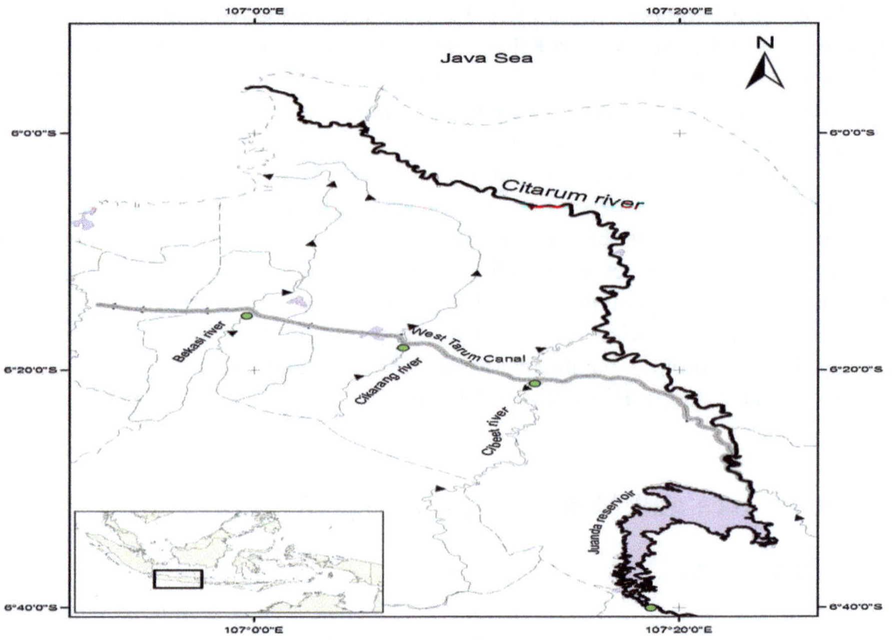

Fig. 1 The sub-catchment of the study area and the gauge location

site scenarios of daily surface weather variables under current and future climate forcing.

This study employed the output of the Hadley Centre Coupled Model version 3 (HadCM3) GCM developed by the Hadley Centre, UK, as the predictor. Two climate scenarios (A2 and B2) from HadCM3 were selected for the following reasons: (1) the availability of output for more than one SRES emission scenario; (2) the availability of primary downscaling variables that are produced daily; (3) the availability of unbroken daily series for the entire period from 1961 to 2010; and (4) the realism of the GCM behaviour in respect to NCEP reference data.

The SDSM establishes the empirical relationship function (*F*) between the GCM output and local climate variables. The GCM outputs are the predictors, and the local climate variables are the predictands. The function is a deterministic/stochastic one which is conditioned by predictors and predictands as shown in the following equation:

$$a = F(b) \tag{1}$$

where *a* is the predictand and *b* is the predictor

SDSM mainly contains four processes to generate a future series of climate variables: (1) identification of predictands and predictors; (2) model calibration; (3) weather generators; and (4) generation of a future series of climate variables.

Identification of predictands and predictors uses the quality control module in the SDSM. The module checks the performance of the predictand (e.g., temperature and precipitation) data to identify errors in the data records, the outliners, and missing data. In this study, the four-root transform was applied to precipitation data since the distribution of precipitation is skewed. For the analysis, the monthly model and the unconditional process were applied to temperature. The conditional process was applied to precipitation.

2.2.2 Model Calibration and Validation

The calibration model in the SDSM computes the parameters of multiple regression equations by using the optimisation algorithm (least squares) for a set of probable predictors and each predictand. A monthly model type is selected in which different model parameters are derived for each month. From the 30 years of data representing the current climate (1961–1990), the 20 years of data (1961–1980) are used for calibrating the regression model, whereas 10 years of data (1981–1990) are used to validate the model. The coefficient of determination (r^2) is used to evaluate the performance of SDSM. Here, r^2 is a comparison of the explained variance of modelled data with the total variance of observed data.

As the predictors, the atmospheric predictor variables from the National Center for Environmental Prediction (NCEP) reanalysis were used (Kalnay et al. 1996). NCEP and HadCM3 contain 26 daily predictors (which describe atmospheric circulation, thickness, and moisture content at the surface, 850 and 500-hPa levels) for the regions covering the study area during the period 1961–2099. The predictors over the area which overlays the study area were employed for the A2 and B2 scenarios of the IPCC Special Report on Emission Scenarios (SRES).

For the validation process, the weather generator module was used to simulate the climate for the years 1981–1990 in the study area. The comparison of observed data and climate simulation is facilitated by summary statistics and frequency of analysis in the SDSM. During the calibration and validation of downscaling models, for the evaluation of the performance of the regression models for temperature, the mean values of observed and NCEP simulated data are compared. For precipitation, the mean daily precipitation and daily precipitation variability for each month are used to evaluate the performance of the model.

The scenario generator module produces an ensemble of downscaled synthetic daily weather series based on the atmospheric predictor variables supplied by the HadCM3. In this study, about 10 ensembles of temperature and precipitation are downscaled for the A2 and B2 scenarios using the corresponding set of predictor variables. As HadCM3 has year lengths of 360 days, the downscaled temperature and precipitation will each have a year of 360 days. Therefore, the output from the DSM has been analysed to convert the number of days from 360 into 365 days in a year.

2.2.3 Hydrological Modelling Using WEAP

The WEAP package tool is used to model the hydrological process in the study area. WEAP is a computer modelling package design, which is used for simulation of the water resources system and trade-off analysis. It operates on the basic principle of a water balance model that defines the processes on a watershed level. The supply is defined as the amount of precipitation that falls on the watershed, which is depleted through the watershed processes, water uses, and accretion to the downstream. Moreover, it enables integrated assessment of watershed climate, hydrology, and land use, infrastructure, and water management priorities. WEAP has been used to model the impact of climate change, land use, and adaptation scenarios on water resources (Joyce et al. 2011; Purkey et al. 2007; Wilby et al. 2002).

The model is spatially continuous (lumped model) represented by a set of catchments that covers the entire river basin in the study area, considering them to be a complete network of rivers, reservoirs, channels, ground-surface water interaction, and demand points. Furthermore, the model includes four methods to simulate the catchment processes (evapotranspiration (ET), run-off, infiltration, and irrigation demands): (1) rainfall run-off; (2) Irrigation Demands Only method of the FAO crop requirements approach; (3) soil moisture method; and (4) MABIA method.

2.2.4 Model Setup

The WEAP model for the Citarum watershed was built to assess the impact of future climate change on water availability. The model was run on a monthly time cycle. For spatial data analysis, the geographic information system (GIS) application is used to analyse the land use, elevation, and watershed information to generate data input variables of the WEAP model. Land use and streamflow data were provided by the Coordinating Agency for Surveys and Mapping (Bakosurtanal), Indonesia. Elevation data were extracted from the digital elevation model (DEM) from the U.S. Geological Survey. Historical climate data were provided by the Meteorological, Climatological, and Geophysical Agency, Indonesia.

For this particular study, the soil moisture method of WEAP is used to simulate the hydrological processes in the catchments. The soil moisture method is one-dimensional, two-layer, soil moisture accounting, which uses empirical functions to describe ET, surface run-off, sub-surface run-off, and deep percolation for a watershed unit. The deep percolation within the watershed unit can be transmitted to a surface water body as base flow or directly to groundwater storage if the appropriate link is made between the watershed unit node and a groundwater node.

The model considers the movement of water through two vertical soil layers. The first layer represents water retained near the surface, which is available to plant roots; the second layer is deeper, and water from this layer can be transmitted as base flow or groundwater recharge. The main parameters of this model include the water-holding capacity for both layers as well as water movement between them. For each sub-catchment, the model computes the water balance due to inflows, outflows, and storage changes in each layer. A watershed is firstly divided into sub-catchments and then further into the N fractional area, j of N. The climate is assumed to be uniform over each fractional area where the continuous mass balance equation is as follows:

$$\mathrm{Sw}_j \frac{\mathrm{d}z_{1,j}}{\mathrm{d}t} = P_\mathrm{e}(t) - \mathrm{PET}(t) k_{c,j}(t) \left(\frac{5z_{1,j} - 2z_{1,j}^2}{3} \right) - P_\mathrm{e}(t) z_{1,j}^{\frac{\mathrm{LAI}_j}{2}} - f_j f k_{s,j} z_{1j}^2 - (1 - f_j) k_{s,j} z_{1j}^2 \quad (2)$$

where

$z_{1,j} \in [0.1]$ is the relative soil water storage, a fraction of the total effective water storage in the root zone layer in area j (dimensionless)

Sw_j is the soil water-holding capacity of the area j (mm);

$P_\mathrm{e}(t)$ is the effective precipitation (mm);

$\mathrm{PET}(t)$ is the reference potential ET (mm/day);

$k_{c,j}(t)$ is the crop coefficient for area j;

LAI_j is the leaf area index for area j which depends on land cover;

$P_\mathrm{e}(t) z_{1,j}^{\mathrm{LAI}_j}$ is the surface run-off;

$f_j f k_{s,j} z_{1j}^2$ is the interflow from the first soil layer for area j;

f_j is the partition coefficient related to the land cover type, soil, and topography for area j, which divides the flow into horizontal f_j and vertical $1 - f_j$ flows; and

$k_{s,j}$ is the saturated hydraulic conductivity of the root zone layer for area j (mm/time).

2.2.5 Catchment Model Calibration

In order to assess the impact of future climatic change on hydrology parameters, a rainfall run-off model is usually required. The model was calibrated based on the assumption that the change in processes represented by the parameters will be small in comparison with the changes affecting the climatic conditions. Therefore, on this study, the hydrology model was calibrated and validated using the observed climate and time series of natural stream flow.

The statistical indicators, namely the Nash-Sutcliffe efficiency (NSE) criterion, the coefficient of determination (R2), and the percent bias (PBIAS), are used to assess model accuracy in simulating the stream flow. The R^2 values are a measure of the relationship between observed and simulated values. NSE indicates model performance, and PBIAS is the deviation of data being evaluated expressed as a percentage (Moriasi 2007). PBIAS is calculated as follows:

$$\text{PBIAS} = \frac{1}{N}\sum_{i=1}^{N}(Qs_i - Qm_i) \times 100\,\% \tag{3}$$

where QM and Qs are the measured and simulated streamflows in the months *i*th, respectively, and *N* is the number of months in the analysis period.

2.2.6 Assessment of the Impact of Climate Change to the Hydrological Component Using the Hydrological Model

To analyse the future impact of climate change on ET and streamflow, the WEAP model is used for the hydrological processes using the A2 and B2 climate projection from the HadCM3 and SDSM.

The comparison of ET and streamflow in baseline and future scenarios is performed in order to quantify the change in hydrology of the watershed as an impact of climate change. To assess the baseline conditions, the validated WEAP is used to generate the streamflow and ET in the four sub-basins of the study area during the period (1961–1990). Similar steps were applied to generate the streamflow and ET during future periods in the 2020s (2011–2040), 2050s (2041–2070), and 2080s (2071–2990) using downscaled temperature and precipitation corresponding to A2 and B2 climate projections. In this study, it was assumed soil cover and crop pattern to be similar for all future periods to ensure that the projections for the future are entirely dependent on the climate change scenario.

3 Results

3.1 Statistical Downscaling

3.1.1 SDSM Model Performance

The SDSM performance was evaluated by downscaling the temperature and precipitation of the study area. The model performance results of temperature and precipitation are shown in Figs. 2 and 3, where a scatter plot is used to identify the performance of the model output compared to the observed data. During the calibration period, overall, the temperature is accurately simulated. During the validation periods, the high temperature was projected as below the observed data. Similarly, precipitation is simulated accurately during the calibration period, while

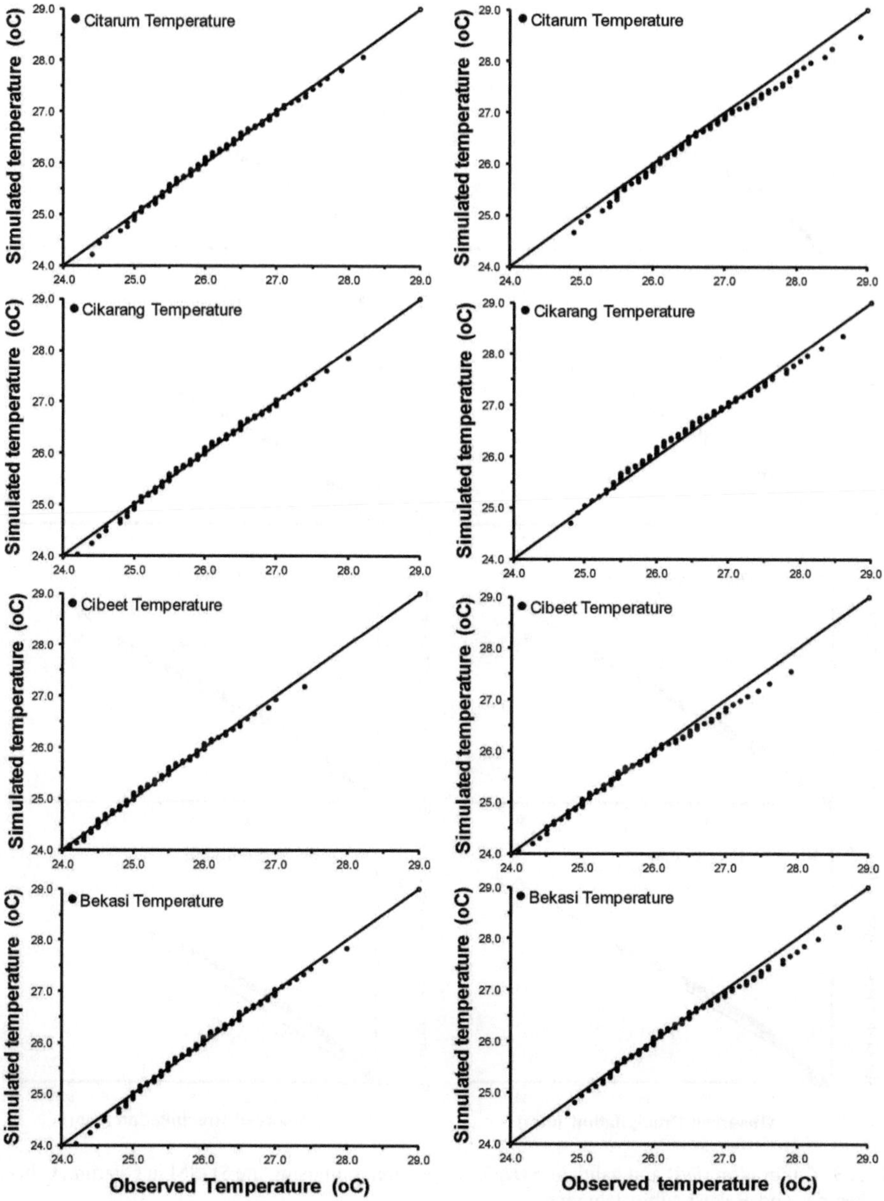

Fig. 2 Calibration (*left*) and validation (*right*) of the temperature in Citarum, Cibeet, Cikarang, and Bekasi sub-catchments

it was simulated below the observed data during validation periods. However, as the model results show they closely followed the observation temperature and precipitation, we utilised those results.

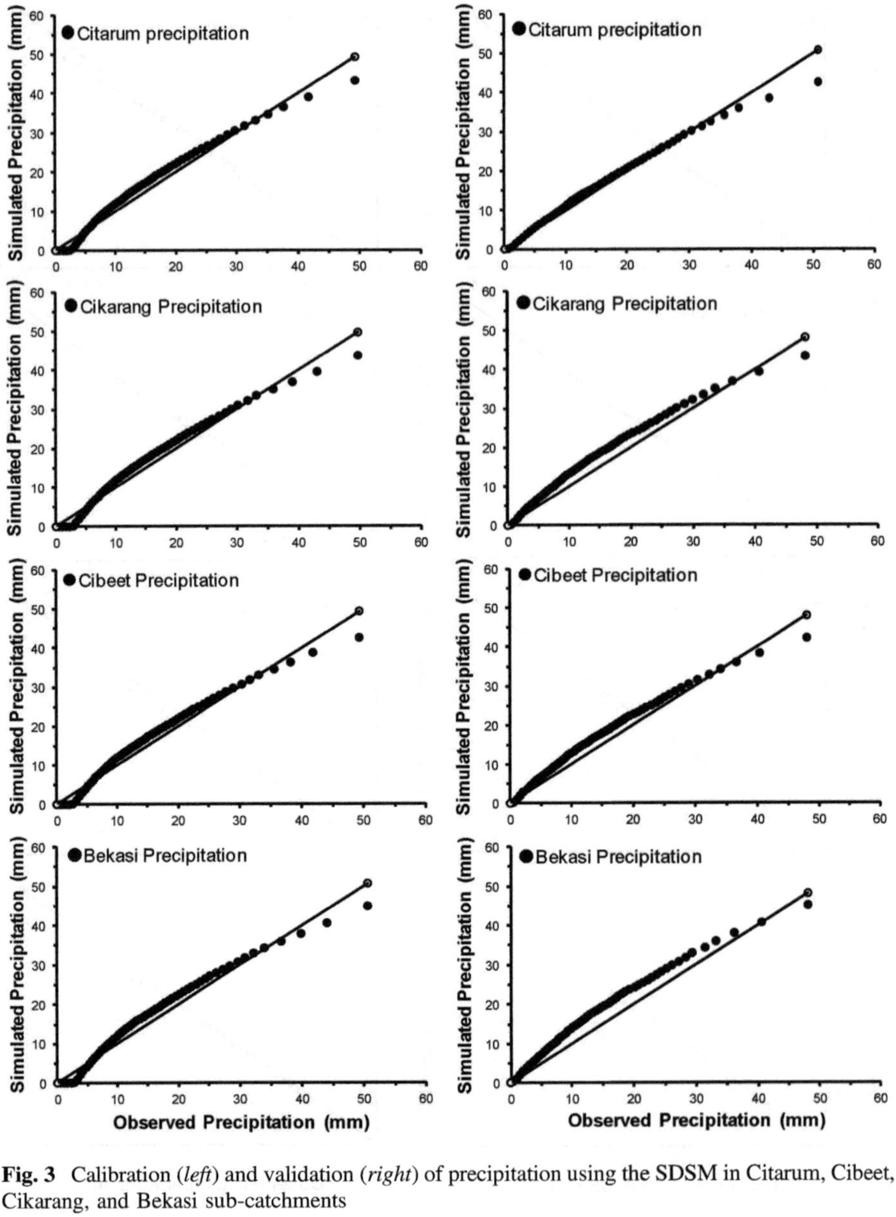

Fig. 3 Calibration (*left*) and validation (*right*) of precipitation using the SDSM in Citarum, Cibeet, Cikarang, and Bekasi sub-catchments

As shown in Figs. 2 and 3, the temperature was simulated using r^2 of 0.89, 0.93, 0.85, and 0.82 at Citarum, Cibeet, Cikarang, and Bekasi sub-catchments, respectively. On the other hand, the precipitation was simulated using r^2 of 0.85, 0.83, 0.75, and 0.78 at Citarum, Cibeet, Cikarang, and Bekasi sub-catchments, respectively.

3.1.2 Projection of Temperature and Precipitation

The temperature and precipitation are projected to increase in the two scenarios (A2 and B2) in all sub-catchments (Figs. 4 and 5). The temperature at the Cikarang sub-catchment in the A2 scenario increased by 0.33, 0.91, and 1.63 °C, respectively, in the 2020s, 2050s, and 2080s. Under the B2 scenario, the temperature increased by 0.33, 0.67, and 1.13, respectively, in the 2020s, 2050s, and 2080s. However, the average increased temperature in all basins is about 0.14, 0.73, and 1.46 °C under the A2 scenario and is about 0.15, 0.50, and 0.97 °C under the B2 scenario in the 2020s, 2050s, and 2080s, respectively.

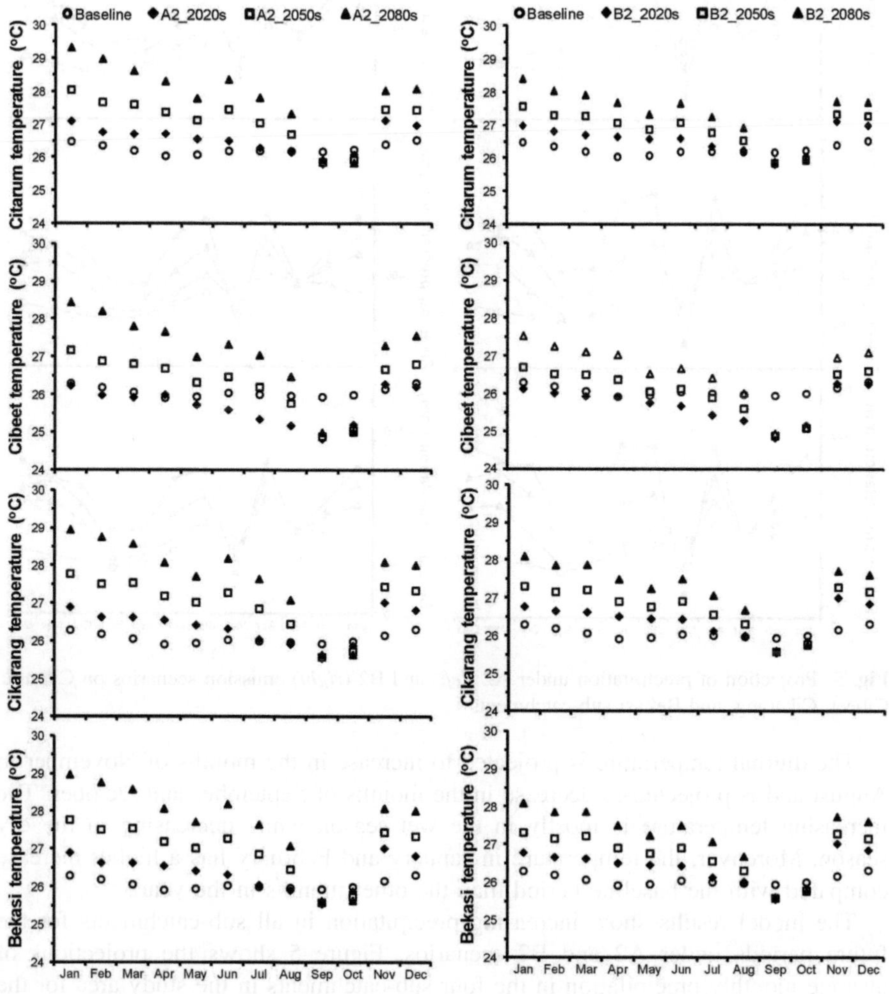

Fig. 4 Projection of temperature under A2 (*left*) and B2 (*right*) emission scenarios on Citarum, Cibeet, Cikarang, and Bekasi sub-catchments

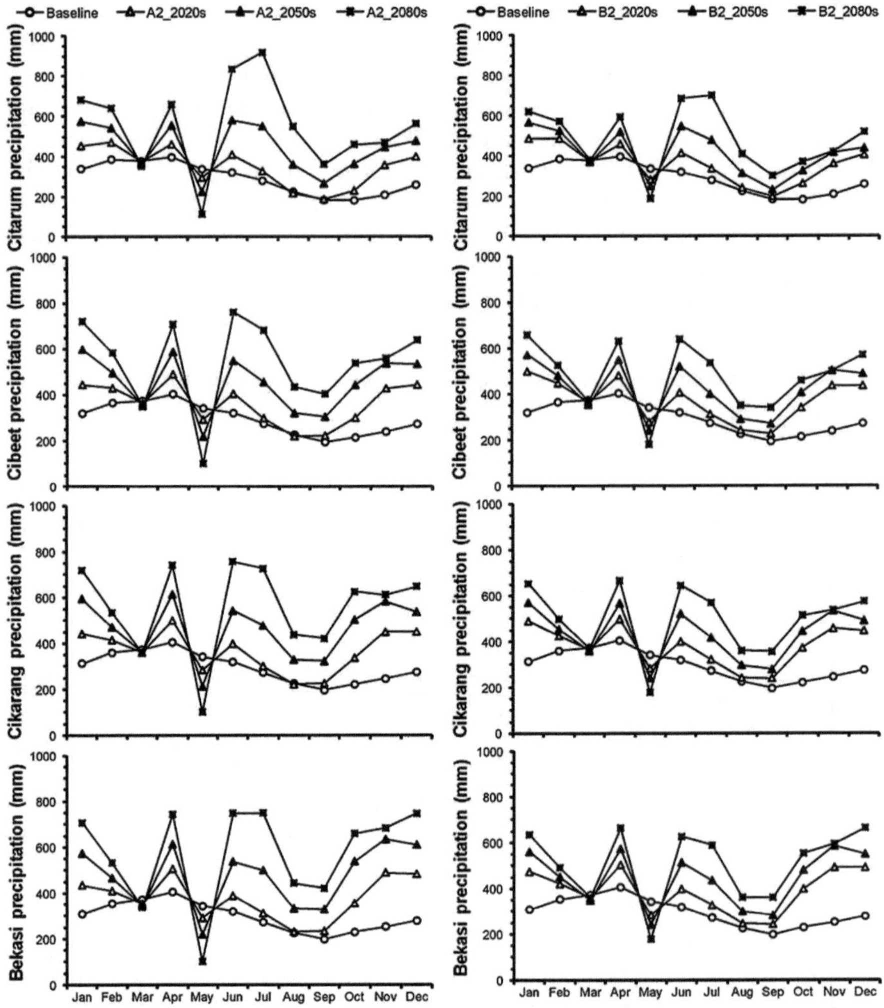

Fig. 5 Projection of precipitation under A2 (*left*) and B2 (*right*) emission scenarios on Citarum, Cibeet, Cikarang, and Bekasi sub-catchments

The diurnal temperature is projected to increase in the months of November to August and is projected to decrease in the months of September and October. The increasing temperature is mostly in the wet season while decreasing in the dry season. Moreover, the temperature in January and February has a higher increase compared with the baseline period than the other months in the year.

The model results show increasing precipitation in all sub-catchments for the future periods under A2 and B2 scenarios. Figure 5 shows the projections of average monthly precipitation in the four sub-catchments in the study area for the baseline A2 and B2 scenarios. The precipitation is projected to increase about 23–88 % under the A2 scenario and is about 27–66 % under the B2 scenario. The increasing precipitation is mostly about 26, 60, and 93 % of the Bekasi sub-catchment for the 2020s, 2050s, and 2080s under the A2 scenario. Under the B2 scenario in the same sub-catchment, precipitation is projected to increase about 30, 49, and 71 % for the 2020s, 2050s, and 2080s, respectively. In the 2020s, the increasing precipitation under the A2 scenario is lower than in the B2 scenario. However, in the 2080s, the increasing projected precipitation under the A2 scenario is higher than the B2 scenario.

3.2 Hydrology Model Performance

The WEAP model is calibrated at the four streamflow gauging stations located in the catchment. The calibration of the WEAP model for the study area is carried out by comparing the simulated monthly streamflow with the observed data in the outlet (Fig. 1).

A monthly streamflow has been used for calibration and future scenario analyses in this study considering the water residence time during which all flows are assumed to occur (Purkey et al. 2007). Observed monthly discharge for the period 1994–2009 was used to calibrate the Citarum sub-catchment. The observed streamflow for the period 1987–2005 was used to calibrate the Cibeet, Cikarang, and Bekasi sub-catchments. Monthly ET values for the period are calculated using the Penman–Monteith method incorporated in the WEAP model (Yates et al. 2005). The calibration was done by adjusting the parameters to achieve good agreement between observed and simulated streamflow. The crop coefficient (Kc) parameter is calibrated using the ranges provided by the Food and Agriculture Organization (FAO) (Allen et al. 1998). Monthly effective precipitation is calibrated based on land-use-specific run-off coefficients and basin-wide average values provided by PJT-II (2003).

Results indicate the reasonable ability of the WEAP model to simulate long-term monthly time series of streamflow for the 30-year period. Good fit statistics, NSE criterion, the correlation coefficient (R), and the percentages bias (PBIAS) are used to assess model accuracy in simulating observed streamflow at each of the four stations. During the calibration period, the N_{SE} (0.52–0.66) and R values (0.73–0.81) indicate the model's ability to adequately represent hydrological conditions in the basin. PBIAS results for the gauging stations ranged from −0.34–1.5 % and indicate water balance errors, but fall within the ±15 % range which suggests a good model performance (Moriasi 2007). Overall, since the statistical results are satisfactory, the model was suitable for use in exploring potential changes in streamflow due to climate change.

In the Citarum sub-catchment, monthly streamflows from years 1994 to 2009 were simulated with a PBIAS of −0.24 %, Nash-Sutcliffe of 0.58, and $R = 0.78$. At the Cibeet sub-catchment, the monthly flows from years 1987 to 2005 were simulated with a PBIAS of −0.34 %, Nash-Sutcliffe of 0.64, and $R = 0.81$. At the Cikarang sub-catchment, the monthly flows of years 1987–2005 were simulated

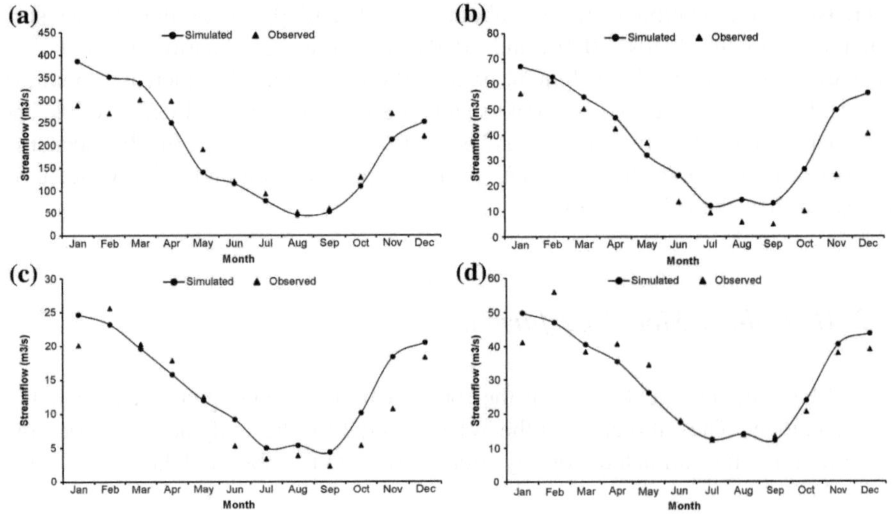

Fig. 6 The comparison of simulated and observed streamflow at (A) Citarum, (B) Cbeet, (C) Cikarang, (D) Bekasi sub-catchments

with a PBIAS of 1.5 %, Nash-Sutcliffe of 0.52, and $R = 0.73$. At the Bekasi sub-catchment, the monthly flow of years 1987–2005 was simulated with a PBIAS of 0.08 %, Nash-Sutcliffe of 0.66, and $R = 0.79$ (Fig. 6).

3.3 Impact of Climate Change on the Hydrology in the Study Area

3.3.1 Change in Streamflow

The A2 and B2 scenarios produce a wide range of changes in the hydrology of the sub-catchment. As shown in Table 1, for three periods, streamflow is projected to increase in the future for both scenarios. Under A2 scenario, the annual mean flow

Table 1 The change of future streamflow compared with the baseline period

Sub-catchment	Baseline (mm)	Change of future streamflow (mm)					
		2020s		2050s		2080s	
		A2	B2	A2	B2	A2	B2
Citarum	2,707	514.9	652.0	1,783.2	1,317.0	3,142.8	2,078.0
Cibeet	1,933	464.4	579.5	1,570.0	1,245.1	3,193.9	2,078.7
Cikarang	2,356	490.2	607.3	1,722.3	1,252.1	3,158.9	2,111.6
Bekasi	1,746	602.1	824.0	1,780.4	1,433.0	3,418.2	2,114.9

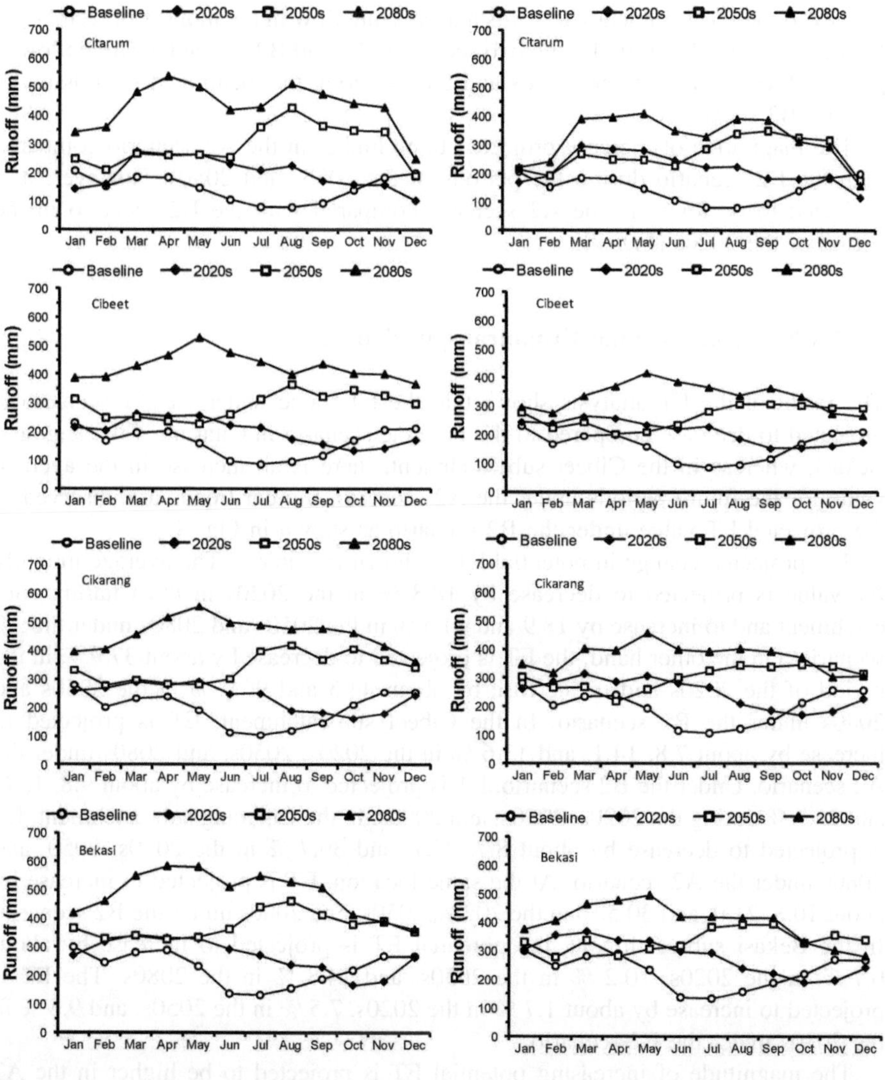

Fig. 7 The projection of streamflow under A2 (*left*) and B2 (*right*) scenarios in the Citarum, Cibeet, Cikarang, and Bekasi sub-catchments

is projected to increase by about 24, 81, and 151 % in the period of 2020s, 2050s, and 2080s, respectively, compared with the baseline period. Under the B2 scenario, the annual mean flow of the baseline periods was 2,185 mm which is projected to increase by about 31, 61, and 98 % during the periods of the period of 2020s, 2050, and 2080s, respectively.

Figure 7 shows the projected changes of the average monthly streamflow in the future periods under A2 and B2 scenarios compared with the baseline periods. It shows that the projected streamflow shows an increase in all months of the year.

However, there is a tendency towards a decreasing flow in the months of October to December in the 2020s under the two scenarios A2 and B2, while the streamflow is projected to increase for the 2050s and 2080s during the mentioned periods under A2 and B2 scenarios.

The magnitude of increase projected to be higher in the A2 scenario compared with the B2 scenario during the periods of the 2050s and 2080s. However, it is projected to be lower in the A2 scenario compared with the B2 scenario in the period of the 2020s (Fig. 7).

3.3.2 Change in Potential Evapotranspiration

The result of the ET analysis shows that the ET value under the A2 scenario is projected to decrease compared to the baseline scenario in Citarum, Cikarang, and Bekasi, whereas in the Cibeet sub-catchment, there is an increase in the average value for the future periods under the A2 scenario. Similar trends are observed in the projected ET value under the B2 scenario as shown in Fig. 8.

The projected change in potential ET is shown in Table 2. The average monthly ET value is projected to decrease by 18.8 % in the 2020s in the Citarum sub-catchment and to increase by 18.9 and 44.5 % in the 2050s and 2080s under the A2 scenario. On the other hand, the ET is projected to decrease by about 37.9 % in the period of the 2020s and to increase by about 44.5 and 96.9 % in the 2050s and 2080s under the B2 scenario. In the Cibeet sub-catchment, ET is projected to increase by about 7.8, 14.1, and 11.6 % in the 2020s, 2050s, and 2080s under the A2 scenario. Under the B2 scenario, ET is projected to increase by about 9.8, 18.6, and 25.7 % during the 2020s, 2050s, and 2080s. In the Cikarang sub-catchment, ET is projected to decrease by about 8.7, 27.1, and 39.7 % in the 2020s, 2050, and 2080s under the A2 scenario. At the same location, ET is projected to increase by about 10.8, 21.0, and 30.5 % in the 2020s, 2050s, and 2080s under the B2 scenario. In the Bekasi sub-catchment, the potential ET is projected to increase by about 0.7 % in the 2020s, 10.2 % in the 2050s, and 14.8 % in the 2080s. The ET is projected to increase by about 1.7 % in the 2020s, 7.5 % in the 2050s, and 9.8 % in the 2080s under the B2 scenario.

The magnitude of increasing potential ET is projected to be higher in the A2 scenario compared with the B2 scenario during the periods of the 2050s and 2080s. However, it is projected to be lower in the A2 scenario compared with the B2 scenario for the period of the 2020s (Fig. 8). The average annual ET is projected to increase by about 21.5, 146.8, and 206.1 mm in the periods of the 2020s, 2050s, and 2080s, respectively, under the A2 scenario and increase by about 45.7, 130.6, and 190.5 mm in the periods of the 2020s, 2050s, and 2080s under the B2 scenario.

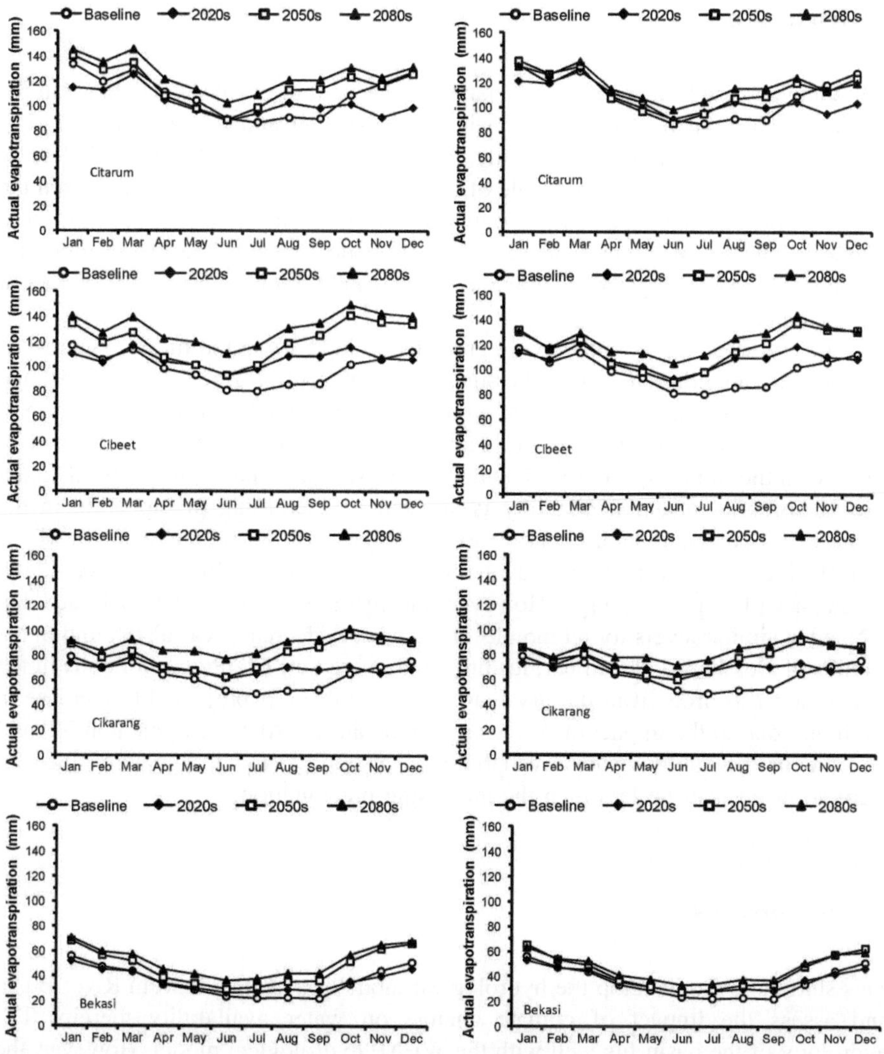

Fig. 8 The projection of potential evapotranspiration under A2 (*left*) and B2 (*right*) scenarios in the Citarum, Cibeet, Cikarang, and Bekasi sub-catchments

Table 2 The change in future potential evapotranspiration compared with the baseline period

Sub-catchment	Baseline (mm)	Change in future potential evapotranspiration (mm)					
		2020s		2050s		2080s	
		A2	B2	A2	B2	A2	B2
Citarum	427.8	−80.9	-37.9	80.6	44.5	190.5	96.9
Cibeet	1,180.0	91.7	115.9	166.6	219.0	136.3	303.6
Cikarang	764.5	66.4	82.2	207.0	160.3	303.8	233.3
Bekasi	1,311.3	8.7	22.6	133.1	98.7	193.7	128.2

4 Discussion

The study assessed the impact of climate change on future water availability in the Citarum River Basin using the statistical downscaling and the WEAP tool. The future climate was downscaled using the SDSM to downscale the global scale output of the GCM to the local scale of the basin. The hydrological model used the WEAP application. This study not only demonstrates the use of the application to assess the impact of climate change, but also to demonstrate the use of a combination of a downscaling application and water planning tool to assess the impact of climate change.

Temperature and precipitation are projected to increase in the future. The results of this study are consistent with the study by IPCC (2007), Boer et al. (2005), and Naylor et al. (2007), suggesting that an increasing temperature and precipitation in tropical areas as well as in Indonesia. The water availability was projected to increase in the future by 21–157 % under the A2 scenario and about 30–94 % under the B2 scenario. Water availability is affected by increasing precipitation in the future. The water availability projection for this study is in line with the IPCC (2007) which reported that water availability will increase in Southeast Asia due to the impact of climate change. However, the increase in precipitation is not distributed at similar levels for all months of the year. A decrease was also found in the months of October to December, leading to an increased likelihood of floods in the wet season and drought in the dry season. Potential ET is projected to increase in the future due to the impact of increasing temperatures. Although potential ET will increase in the future, water availability is also projected to increase since an increase in ET will be less than the increasing precipitation.

5 Conclusions

This study aims to develop the hydrological modelling of the Citarum River Basin and assess the impact of climate change on water availability therein. The hydrology of the basin fits well with the WEAP hydrological model. However, the model is underestimated when simulating the streamflow in the basin. This is due to the lack of capability in the hydrological model for extreme events. However, the model simulated high agreement with the streamflow.

To project future climates, the variables obtained from the HadCM3 GCM output are downscaled using the SDSM application. The downscaled output is then used as input for hydrological assessment. The GCM output was downscaled because the scale of GCM output is too coarse to use directly on the hydrology model. The statistical downscaling of temperature and precipitation performed as well as the high explained variance and R^2. Future projection of the temperature indicates an increase compared with the baseline period. There is a larger increase of temperature under A2 than for the B2 scenario. The average temperature is projected to increase by about 0.14, 0.73, and 1.46 °C under the A2 scenario and by about 0.15, 0.50, and 0.97 °C under the B2 scenario in the 2020s, 2050s, and 2080s, respectively. The precipitation values for the wet season are projected to increase in all of the sub-catchments in the study area under both scenarios. The increasing precipitation is about 23, 55, and 88 % in the 2020s, 2050s, and 2080s under A2 scenario. Under the B2 scenario, the precipitation is projected to increase by about 27, 36, and 46 % for the 2020s, 2050s, and 2080s, respectively. In the 2020s, the increasing precipitation under the A2 scenario is lower than under the B2 scenario. However, in the 2080s, the increasing precipitation under the A2 scenario is higher than under the B2 scenario.

The performance of the hydrological model in the climate change impact assessment was measured by comparing the baseline flow with the observed flow. The validated model, when provided by the future climate variables, generated future streamflow of the study area. The impacts of climate change on the hydrology are then investigated by comparing the streamflow and ET during the baseline period (1961–1990) and the future periods (2020s, 2050s, and 2080s).

The increasing water availability is projected to be larger under the A2 scenario compared with the B2 scenario during the periods of the 2050s and 2080s. However, it is projected lower in the A2 scenario compared with the B2 scenario in the period of the 2020s. The projected water availability will increase by about 517, 1,714, and 3,228 mm in the 2020s, 2050s, and 2080s, respectively, under the A2 scenario and increase by about 665, 1,311, and 2,095 mm in the 2020s, 2050s, and 2080s under the B2 scenario. The average monthly ET is projected to increase by about 21.5, 146.8, and 206.1 mm in the 2020s, 2050s, and 2080s, respectively, under the A2 scenario and increase by about 45.7, 130.6, and 190.5 mm in the 2020s, 2050s, and 2080s under the B2 scenario.

The result of the study provides a useful input for future water planning in the study area. The study area is commonly covered by an irrigation area when about 80 % of the water was used as irrigation. Effective planning of water resources is one of the focuses of water resources development in the study area. Therefore, this study provides a useful tool to assess the impact of climate change on future water availability as input for future water planning. Furthermore, this study focuses on development of the impact assessment of climate change on future water availability using the freely available tools. The results have provided input for the use of decision-makers on how to manage water resources under future climate change.

References

Alemayehu T, McCartney M, Kebede S (2010) The water resource implications of planned development in the Lake Tana catchment, Ethiopia. Ecohydrology Hydrobiol 10:211–221

Allen RG, Pereira LS, Raes D, Smith M (1998) Crop evapotranspiration. Guidelines for computing crop water requirements. FAO Irrigation and drainage paper 56. FAO, Rome, 300:6541

Bae D, Jung IW, Chang H (2008) Potential changes in Korean water resources estimated by high-resolution climate simulation. Clim Res 35:213–226

Boer R, Dasanto BD, Perdinan, Martinus D (2005) Hydrology balance of Citarum watersheds under current and future climate. Technical Report of AIACC Project integrated assessment of climate change impacts, adaptation and vulnerability in watershed areas and communities in Southeast Asia. Indonesia component (AIACC AS21). Bogor Agricultural University, Indonesia, p 17

Christensen N, Wood A, Voisin N, Lettenmaier D, Palmer R (2004) The effects of climate change on the hydrology and water resources of the Colorado river Basin. Clim Change 62:337–363

Evans J, Schreider S (2002) Hydrological impacts of climate change on inflows to Perth, Australia. Clim Change 55:361–393

Fowler H, Kilsby C, Stunell J (2007) Modelling the impacts of projected future climate change on water resources in north-west England. Hydrol Earth Syst Sci 11:1115–1126

Fowler HJ, Kilsby CG (2007) Using regional climate model data to simulate historical and future river flows in northwest England. Clim Change 80:337–367

Fujihara Y, Tanaka K, Watanabe T, Nagano T, Kojiri T (2008) Assessing the impacts of climate change on the water resources of the Seyhan river Basin in Turkey: use of dynamically downscaled data for hydrologic simulations. J Hydrol 353:33–48

Hay LE, Clark MP (2003) Use of statistically and dynamically downscaled atmospheric model output for hydrologic simulations in three mountainous basins in the western United States. J Hydrol 282:56–75

Hay LE et al (2002) Use of regional climate model output for hydrologic simulations. J Hydrometeorology 3:571–590

IPCC (2007) Climate change 2007: summary for policy maker of synthesis report. Core Writing Team, Cambridge University Press, Cambridge, MA

Joyce B, Mehta V, Purkey D, Dale L, Hanemann M (2011) Modifying agricultural water management to adapt to climate change in California's central valley. Clim Change 109:299–316

Kalnay E, Kanamitsu M, Kistler R, Collins W, Deaven D, Gandin L, Joseph D (1996) The NCEP/NCAR 40-year reanalysis project. Bull Am Meteorol Soc 77(3):437–471

Mehta VK, Haden VR, Joyce BA, Purkey DR, Jackson LE (2013) Irrigation demand and supply, given projections of climate and land-use change, in Yolo County, California. Agric Water Manage 117:70–82

Moriasi DN, Arnold JG, Van Liew MW, Bingner RL, Harmel RD, Veith TL (2007) Model evaluation guidelines for systematic quantification of accuracy in watershed simulations. Trans ASABE 50(3):885–900

Naylor RL, Battisti DS, Vimont DJ, Falcon WP, Burke MB (2007) Assessing risks of climate variability and climate change for Indonesian rice agriculture. Proc Nat Acad Sci 104:7752–7757

Purkey D, Huber-Lee A, Yates D, Hanemann M, Herrod-Julius S (2007) Integrating a climate change assessment tool into stakeholder-driven water management decision-making processes in California. In: Craswell E, Bonnell M, Bossio D, Demuth S, Giesen N (eds) Integrated assessment of water resources and global change. Springer, Netherlands, pp 315–329

Rosenzweig C, Strzepek KM, Major DC, Iglesias A, Yates DN, McCluskey A, Hillel D (2004) Water resources for agriculture in a changing climate: international case studies. Glob Environ Change 14:345–360

Wilby RL, Hay LE, Leavesley GH (1999) A comparison of downscaled and raw GCM output: implications for climate change scenarios in the San Juan River basin, Colorado. J Hydrol 225:67–91

Wilby RL, Dawson CW, Barrow EM (2002) SDSM—a decision support tool for the assessment of regional climate change impacts. Environ Model Softw 17:145–157

Wilby RL et al (2000) Hydrological responses to dynamically and statistically downscaled climate model output. Geophys Res Lett 27:1199–1202

Wood AW, Leung LR, Sridhar V, Lettenmaier DP (2004) Hydrologic implications of dynamical and statistical approaches to downscaling climate model outputs. Clim Change 62:189–216

Yates D, Sieber J, Purkey D, Huber-Lee A (2005) WEAP21–a demand, priority, and preference-driven water planning model—Part 1: model characteristics. Water Int 30:487–500

Impact of the Uncertainty of Future Climates on Discharge in the Nam Ou River Basin, Lao PDR

Manisha Maharjan and Mukand S. Babel

Abstract Climate change is likely to increase the global mean temperature in future decades, increasing evapotranspiration and affecting precipitation and discharge. However, there is always uncertainty in future climate projection which might be due to the use of different general circulation models (GCMs), greenhouse gas emission scenarios, downscaling techniques, or other factors. This uncertainty in future climates may affect quantification of discharge. This chapter aims to describe the impact of climate change and uncertainty on future climate projection, specifically on discharge in the Nam Ou River Basin, Lao PDR. The Long Ashton Research Station Weather Generator (LARS-WG) was used to downscale daily precipitation, and maximum and minimum temperatures using 8 GCMs for future periods 2011–2030, 2046–2065, and 2080–2099 under scenarios A1B and A2, respectively. Probability density functions (PDFs) were constructed to estimate the uncertainty of future climates. The maximum and minimum temperatures are expected to increase and change for precipitation in future and are observed to be multidirectional. The temperature and precipitation projection due to the GCMs varied by a higher range under both scenarios for the 2090s compared to the 2020s and 2055s. The inter-model variability and variance of the future projections increased in the latter part of the century compared to the early and mid-centuries. In order to assess the impact of these uncertainties in future climates, the soil and water assessment tool (SWAT) model was used. An increase in annual discharge of 5.7 % is observed during the 2055s and 3.13 % during the 2090s under A1B scenario and of 9.23 % during the 2090s under A2 scenario. At the same time, the discharge is estimated to decrease by 0.33 % during the 2020s under A1B and by 1.15 % during the 2055s under A2 scenario. Thus, an increase and decrease in discharge is expected for future periods when all GCMs are considered under A1B and A2 scenarios.

M. Maharjan (✉) · M.S. Babel
Water Engineering and Management, School of Engineering and Technology,
Asian Institute of Technology, P.O. Box 4, Klong Luang, Pathum Thani 12120, Thailand
e-mail: maneesha064@gmail.com

M.S. Babel
e-mail: msbabel@ait.ac.th

S. Shrestha et al. (eds.), *Managing Water Resources under Climate Uncertainty*,
Springer Water, DOI 10.1007/978-3-319-10467-6_4

Keywords Climate change · LARS-WG · GCMs · Uncertainty · Probability density function · Soil and water assessment tool · Discharge

1 Introduction

Climate change is a major issue nowadays. An increase in global temperature of 1.8–4 °C by the end of twenty-first century has been predicted by IPCC (2007). An increase in temperature will lead to an increase in evapotranspiration, altering the precipitation pattern, and finally changing run-off of the river (Lu 2005). Climate warming has many impacts on various hydrological and ecological processes (Dunn et al. 2012; Wang et al. 2013). The change in precipitation pattern and behaviour of hydrological cycles under spatiotemporal scales affect run-off and flow regimes (Beyene et al. 2010). Thus, it is well evidenced that climate change will have a significant impact on the global hydrological cycle (Kundzewich et al. 2007). The effect of climate change on the hydrology of river basins is evaluated by using hydrological models with climate projections derived from different general circulation models (GCMs) forced under greenhouse gas emission scenarios (GHGES). This approach has been used in many studies for various types of river basins (Gosling et al. 2010; Luo et al. 2013; Menzel and Burger 2002; Minville et al. 2008; Nijssen et al. 2001; Nohara et al. 2006). The average annual river discharge is predicted to be higher in high northern latitudes and large parts of the tropical monsoon region (Vliet et al. 2013). A more significant increase in discharge is seen under A1B than under A2, which is due to the change in projected rainfall patterns and meteorological forcing.

Climate predictions are highly uncertain due to the fact that the climate system is very sensitive to changing GHG concentration and difficult to quantify accurately (Stott and Kettleborough 2002). Various uncertainties are associated with climate change and its projections (Thompson et al. 2013). These uncertainties are introduced during the assessment of climate change impacts on hydrological processes (Gosling et al. 2011; Nawaz and Adeloye 2006). Uncertainty is related to the use of definitions in GHGES for GCMs and their development. The structural uncertainty within these models causes different climate models to project alternative climate projections under various GHGES. In addition, the downscaling of GCMs from coarser to finer regional scales also produces uncertainty during hydrological model application. However, the hydrological models also play a significant role in increasing uncertainty while transferring different climatological processes to hydrological processes. The uncertainty produced due to GCMs and GHGES can be reduced to some extent by using a dynamical approach for climate projections. Another approach could be the use of statistical methods in establishing an empirical relationship between GCM output, climate variables, and local climate (Fowler et al. 2007; Maraun et al. 2010).

Statistical downscaling is the method mostly used to produce climate change scenarios for impact assessment studies and hence is widely accepted (Chen et al. 2011, Chiew et al. 2010). This method can be categorised into three methods: weather typing, stochastic, and regression. A stochastic weather generator, Long Ashton Research Station–Weather Generator (LARS-WG) (Semenov and Barrow 1997) has been analysed in different climates and has demonstrated good performance in diverse climatic conditions (Semenov and Stratonovitch 2010). For instance, LARS-WG has been used to assess climate change impacts in the Clutha River in New Zealand by Hashmi et al. (2009); small wetlands in Southern and Central Saskatchewan, Canada, by Zhang et al. (2011); the Manicouagan River Basin of Central Quebec, Canada by Chen et al. (2011); and many others. The uncertainty of the downscaling method has also been studied in many basins (Chen et al. 2011 and 2013; Khan et al. 2006). Details of the model LARS-WG and its properties are well described in Semenov and Stratonovitch (2010). This stochastic weather generator incorporates 15 GCMs used in IPCC-AR4 for future climate projections.

The main objective of this study is to quantify the impact of uncertainty in future climates on the discharge of the Nam Ou River Basin: the sub-basin of Lower Mekong River Basin. This investigates how the uncertainty introduced in future climate projections under various GHGES and GCMs affects the run-off in the river basin.

2 Study Area and Data

2.1 Study Area

This study has been conducted in the Nam Ou River Basin of Northern Lao PDR. It is one of the tributaries of the Mekong River Basin. The Nam Ou River Basin lies within the latitudes 21°17′17″ and 22°30′40″ N and longitudes 101°45′47″ and 103°11′57″ E (Fig. 1). The altitude varies from 263 to 2035 m above mean sea level. The drainage area of the basin is approximately 26,000 km^2 with an annual rainfall of 1,700 mm. It has two different seasons: a wet season from May to October and a dry season from November to April. It constitutes 60 % of woodland and shrubland as land use.

2.2 Observed Data

The observed precipitation data was obtained from the Mekong River Commission (MRC) Secretariat Phnom Penh, Cambodia, for 11 stations (Dien Bien, Lai Chau, Luang Prabang, Muong Namtha, Muong Ngoy, Muong Te, Oudomxay, Phong Saly, Quynh Nhai, Tuan Giao, and Xieng Ngeun) for period 1980–2003 (Fig. 1).

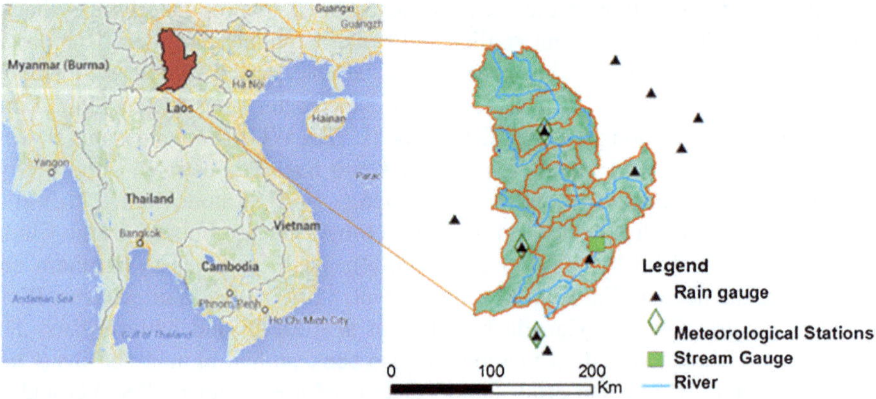

Fig. 1 Study area and locations of rain gauge and meteorological stations

The observed precipitation at the stations was interpolated and aggregated to 19 sub-basins using the MQUAD program in the Decision Support Framework of the MRC by Shrestha et al. (2013).

Similarly, the meteorological data such as maximum and minimum temperature, wind speed, humidity, and solar radiation were obtained from three stations (Luang Prabang, Oudomxay, and Phongsaly) for the period 1992–2003. The daily maximum and minimum temperature for 1980–1991 was derived by using the statistical relationship between observed data and 0.5° gridded global daily temperature data from the Santa Clara University for 1992–1999.[1]

A digital elevation model (DEM) of 250m resolution, land use map of year 2000 and soil map were obtained from the MRC Secretariat, Phnom Penh, Cambodia. The daily discharge data for the period 1992–2003 from the Muong Ngoy gauging station were also collected from MRC.

3 Methodology

3.1 Future Climate Projections

Long Ashton Research Station–Weather Generator (LARS-WG), a stochastic weather generator model developed by Mikhail Semenov in the UK, is used to project the future climates, i.e. precipitation and temperature. Twenty years' (1981–2000) daily data are used in this study for projecting future climates. Three steps are performed to develop synthetic weather data: site analysis, Q-test, and climate change scenario generation as described in Semenov and Stratonovitch

[1] SCU (http://www.engr.scu.edu/~emaurer/global_data/.

(2010). Two special range of emission scenarios (SRES) A1B and A2 are used for projecting precipitation, and maximum and minimum temperature for three periods 2011–2030 (2020s), 2046–2065 (2055s), and 2080–2099 (2090s). The SRES A1B and A2 represent medium and high emissions of greenhouse gases (GHGs) with respect to the prescribed concentrations in the SRES.

Though there are 15 GCMs incorporated in LARS-WG, all GCMs are not available for the projection of both A1B and A2 scenarios in future periods. Thus, 8 GCMs out of 15, using both scenarios, are incorporated in this study (MIHR being an exception). The selected GCMs (Table 1) are developed by different research centres, and their resolution varies from 1.4° × 1.4° to 4° × 5°. The change in climate projection is represented in relation to the baseline period 1981–2000 (1990s). The observed precipitation, and maximum and minimum temperature are used to determine the statistical parameters of the present baseline period. The synthetic weather data is generated using those statistical parameters such that it represents the baseline period. From the statistical analysis of observed and synthetic generated data, the performance of LARS-WG is analysed. Uncertainty analysis of future climate projections under various GCMs is carried out by constructing probability density functions (PDFs).

The performance of LARS-WG for precipitation, and maximum temperature (T_{max}) and minimum temperature (T_{min}) was evaluated using parameters such as the difference in mean value, standard deviation, and Kolmogorov–Smirnov test (K–S test) along with p value. The comparison of monthly mean and standard deviation of observed and generated synthetic data for the baseline period 1981–2000 was carried out. The observed monthly mean of precipitation is compared with generated mean by using t-statistics with p value. The standard deviation is compared by using f-statistics with p value. It is seen that the p value for both monthly mean precipitation and standard deviation is above the significance level of confidence 0.05 and therefore is acceptable, indicating that the generated data truly represent the observed data.

3.2 Simulation of the Rainfall-Run-off Model

In this study, the soil and water assessment tool (SWAT) is used to simulate discharge of the Nam Ou River Basin. SWAT is a process-based continuous hydrological model which can evaluate small or large basins by division into different sub-basins (Neitsch et al. 2009). SWAT requires weather, topography, land use, soil characteristics, etc. SWAT divides sub-basins into hydrological response units (HRUs) with homogeneous land use, soil type, and slope. The hydrology at each HRU is determined by using water balance equation which includes precipitation, run-off, evapotranspiration, percolation, and return flow components. The precipitation occurring can be either intercepted or infiltrated into the soil as a run-off. Some precipitation can be lost by evapotranspiration.

Table 1 General circulation models (GCMs) used in this study

No.	Research centre	Country	GCM	Model acronym	Grid resolution	Emission scenarios	Time periods
1.	Centre National de Recherches Meteorologiques	France	CNRM-CM3	CNCM3	1.9° × 1 .9	SRA1B, SRA2	B, T1, T2, T3
2.	Geophysical Fluid Dynamics Lab	USA	GFDL-CM2.1	GFCM21	2° × 2.5°	SRA1B, SRA2,	B, T1, T2, T3
3.	Institute for Numerical Mathematics	Russia	INM-CM3.0	INCM3	4° × 5°	SRA1B, SRA2	B, T1, T2, T3
4.	Institute Pierre Simon Laplace	France	IPSL-CM4	IPCM4	2.5° × 3.75°	SRA1B, SRA2	B, T1, T2, T3
5.	Max-Planck Institute for Meteorology	Germany	ECHAM5-OM	MPEH5	1.9° × 1.9°	SRA1B, SRA2	B, T1, T2, T3
6.	National Center for Atmospheric Research	USA	CCSM3	NCCCS	1.4° × 1.4°	SRA1B, SRA2	B, T1, T2, T3
7.	National Institute for Environmental Studies	Japan	MRI-CGCM2.3.2	MIHR	2.8° × 2.8°	SRA1B	B, T1, T2, T3
8.	UK Meteorological Office	UK	HadCM3	HADCM3	2.5° × 3.75°	SRA1B, SRA2	B, T1, T2, T3

Climate is the main factor responsible for water balance and energy components. Surface run-off can be estimated by using two different methods: the SCS-curve number method (SCS 1972) and Green and Ampt Infiltration Method (1911). In this study, surface run-off is estimated for each HRU using the SCS-curve number method (SCS 1972) given by

$$Q_{\text{surface}} = \frac{(R_{\text{day}} - I_a)^2}{(R_{\text{day}} - I_a + S)} \qquad (1)$$

where

Q_{surface}	Accumulated run-off (mm)
R_{day}	Rainfall depth for the day (mm)
I_{a}	Initial abstraction (mm) = 0.2S
S	Retention parameter (mm)

The retention number is given by

$$S = 25.4\left(\frac{1000}{\text{CN}} - 10\right) \qquad (2)$$

where, CN = curve number (function of permeability of soil).

The peak run-off rate (maximum rate of run-off that occurs with a certain rainfall event) is calculated by using a modified rational method. SWAT allows calculation of the evapotranspiration by three methods, namely Penman–Monteith, Hargreaves, and Priestley–Taylor. The Penman–Monteith method (Monteith 1965) is used to estimate the evapotranspiration for this study. The Manning equation is used to calculate the flow rate and velocity.

Flow routing in the channel reach is done by using the variable storage routing method. The input data for the model are DEM, land use map, soil map, slope, and climate data. The SWAT model is calibrated for discharge during the period 1992–1999 and validated for 2000–2003. A two-year warm-up period is used for considering the possible errors in initial state variables.

Three good fit measurements are used to evaluate the simulation performance of the SWAT model. They are Nash–Sutcliffe efficiency, (Nash and Sutcliffe 1970), Pearson's correlation coefficient (R^2), and percentage bias (PBIAS). The Nash–Sutcliffe efficiency (NS) is a normalised statistical method which determines the relative magnitude in the variance of the residual compared to the measured variance. It is calculated by the given equation:

$$\text{NS} = 1 - \left[\frac{\sum_{i=1}^{n}(Q_i^{\text{obs}} - Q_i^{\text{sim}})^2}{\sum_{i=1}^{n}(Q_i^{\text{obs}} - Q_i^{\text{mean}})^2}\right] \qquad (3)$$

where Q_i^{obs} is the observed discharge, Q_i^{sim} is the simulated discharge, and Q_i^{mean} is the mean discharge. NS value can be within ∞ to 1, but a NS equal to 1 fits perfectly. Another measure is Pearson's correlation coefficient R^2, the value of which ranges from 0 to 1, with 1 being the perfect fit.

The percentage bias (PBIAS) is used to measure where the average tendency of simulated discharge is larger or smaller than the observed data. It is calculated as in the equation:

$$\mathrm{PBIAS} = \left[\frac{\sum_{i=1}^{n}(Q_i^{\mathrm{obs}} - Q_i^{\mathrm{sim}}) \times 100}{\sum_{i=1}^{n}(Q_i^{\mathrm{obs}})}\right] \tag{4}$$

The optimal value of PBIAS is 0. The positive value shows underestimation, and the negative value shows the overestimation bias of the model.

Performance indicators NS and $R^2 > 0.6$ (Benaman et al. 2005; Santhi et al. 2001; and Setegn et al. 2010) and PBIAS < 15 % (Santhi et al. 2001 and Van Liew et al. 2007) were kept as decision criteria for model simulation.

4 Results and Discussion

4.1 Performance Analysis of LARS-WG

The downscaled precipitation for the baseline period (1990s) is compared with the observed mean precipitation using parameters such as difference in mean, difference in standard deviation, and with t test along with p value (Table 2). The statistics are compared and used for the future projection of precipitation.

The maximum and minimum temperature for the baseline period (1990s) is downscaled by comparing different parameters as shown in Tables 3 and 4, respectively. The monthly mean and standard deviation of $T_{\max}$ and $T_{\min}$ are compared along with t-statistics and p value. The p value obtained from the t test of both observed and generated precipitation, using maximum as well as minimum temperatures, signifies a good performance of LARS-WG in downscaling.

4.2 Future Climate Projections

The future projection of precipitation shows that there is multidirectional change under different GCMs. Changes in both positive (increase from baseline) and negative (decrease from baseline) directions are seen under A1B and A2 scenarios (Table 5). The maximum variation in precipitation as projected by different GCMs is observed under A2 for the 2090s. During the 2055s, the precipitation projection shows that it reduces by 84.46 mm under A2. Similarly, the changes in maximum and minimum temperature for the future are displayed in Table 5, considering all changes in the GCMs are averaged. The temperature in the 2090s shows the

Table 2 Performance analysis of LARS-WG for precipitation downscaling

Month	Jan	Feb	Mar	Apr	May	Jun	Jul	Aug	Sep	Oct	Nov	Dec
Observed mean (mm)	21.78	27.17	59.22	136.87	267.46	365.9	491.18	404.31	229.73	122.43	72.42	22.52
Generated mean (mm)	23.67	23.16	62.92	143.32	263.76	352.11	489.24	399.28	223.65	130.54	72.68	20.89
Difference	−1.89	4.01	−3.7	−6.45	3.7	13.79	1.94	5.03	6.08	−8.11	−0.26	1.63
Standard deviation (observed)	19.36	22.71	42.24	73.22	116.83	145.40	124.87	127.14	64.61	67.49	62.15	21.38
Standard deviation (generated)	23.03	22.67	40.91	56.44	79.85	78.23	106.99	73.06	79.83	47.64	48.21	32.05
Difference	−3.67	0.03	1.34	16.78	36.98	67.17	17.89	54.08	−15.23	19.85	13.94	−10.67
t-statistics	−0.28	0.56	−0.28	−0.31	0.12	0.37	0.05	0.15	0.27	−0.44	−0.02	0.19
p value	0.78	0.58	0.78	0.76	0.91	0.71	0.96	0.88	0.79	0.66	0.99	0.85

Table 3 Performance analysis of LARS-WG for downscaling of maximum temperature

Month	Jan	Feb	Mar	Apr	May	Jun	Jul	Aug	Sep	Oct	Nov	Dec
Observed mean (mm)	27.7	30.9	33.2	34.6	33.9	32.7	31.7	31.7	31.7	30.7	28.6	26.0
Generated mean (mm)	28.1	30.5	33.9	34.8	34.0	32.8	31.6	31.5	31.6	30.5	28.7	26.1
Difference	−0.4	0.4	−0.7	−0.2	−0.1	−0.1	0.1	0.2	0.1	0.2	−0.1	−0.1
Standard deviation (observed)	1.53	1.37	1.85	1.59	1.17	0.72	0.64	0.92	1.08	1.03	1.19	1.79
Standard deviation (generated)	0.57	0.64	0.76	0.77	0.76	0.33	0.23	0.26	0.29	0.41	0.49	0.61
Difference	0.96	0.73	1.09	0.82	0.41	0.39	0.41	0.66	0.79	0.62	0.70	1.18
t-statistics	−1.03	1.25	−1.52	−0.47	−0.30	−0.48	0.03	0.90	0.32	0.96	−0.28	−0.06
p value	0.31	0.22	0.14	0.64	0.76	0.63	0.97	0.37	0.75	0.34	0.78	0.95

Table 4 Performance analysis of LARS-WG for downscaling of minimum temperature

Month	Jan	Feb	Mar	Apr	May	Jun	Jul	Aug	Sep	Oct	Nov	Dec
Observed mean (mm)	14.52	15.08	18.31	21.35	22.64	24.06	23.73	23.48	22.9	20.85	18.29	14.39
Generated mean (mm)	14.56	15.09	18.58	21.19	22.67	23.97	23.79	23.44	22.71	20.75	18.38	14.47
Difference	−0.04	−0.01	−0.27	0.16	−0.03	0.09	−0.06	0.04	0.19	0.1	−0.09	−0.08
Standard deviation (observed)	0.91	1.19	1.93	1.72	1.42	0.30	0.43	0.47	0.30	0.95	1.19	1.45
Standard deviation (generated)	0.48	0.45	0.43	0.43	0.35	0.10	0.14	0.11	0.23	0.40	0.37	0.40
Difference	0.43	0.74	1.50	1.29	1.07	0.2	0.29	0.36	0.07	0.55	0.82	1.05
t-statistics	−0.15	−0.05	−0.61	0.43	−0.10	1.25	−0.60	0.35	2.2	0.43	−0.31	−0.23
p value	0.88	0.95	0.54	0.67	0.92	0.22	0.55	0.73	0.03	0.66	0.76	0.82

Table 5 Change in annual precipitation, and maximum and minimum temperature under A1B and A2 scenarios during three future periods: 2020s, 2055s, and 2090s

Period	Annual precipitation			T_{max} (°C)			T_{min} (°C)		
Baseline	1626.55 mm			27.88 °C			17.63 °C		
	Change in precipitation (mm)			Change in T_{max} (°C)			Change in T_{min} (°C)		
	B1	A1B	A2	B1	A1B	A2	B1	A1B	A2
2020s	6.77	−24.48	−19.98	0.60	0.62	0.57	0.65	0.67	0.58
2055s	35.18	85.69	−84.46	1.46	1.84	1.76	1.50	1.89	1.75
2090s	78.65	96.71	131.26	1.90	3.03	3.49	2.12	3.04	3.50

maximum deviation from the baseline period under both A1B and A2 scenarios. The temperature during the 2020s is less altered than in the 2055s and 2090s periods. The result shows that the maximum and minimum temperature is expected to increase by 3.5 °C by the end of the century, considering the average change from all the GCMs used in the study.

4.3 Change in Annual Future Climates

The change in future climates is uncertain when projections are carried out using different climate models such as GCMs and different GHGES. The difference in projection of future climates becomes wider under different GCMs and GHGES. Figures 2, 3, and 4 demonstrate the range of change occurring in the projection of precipitation, and maximum and minimum temperature, respectively, for future periods under various GCMs averaged and GHGES used.

In case of precipitation, the average change is minimal in the 2020s under A1B and A2 scenarios. The change in mean annual precipitation increases in the 2055s and 2090s except under A2 scenario. The range of change as projected by different GCMs becomes wider with time. Projection of precipitation is more complex than that of temperature, since different GCMs can result in an increase as well as decrease in precipitation at a particular site (Girvetz et al. 2009). Figure 2 shows that change in precipitation can vary from 68 to −99.03 mm during 2020s, from 237.74 to −625.62 mm during 2055s and from 420.33 to −179.57 mm during respectively (positive value refers to increment and negative value refers to decrement).

Results from the future projection of temperature exhibit an increase in temperature during future periods (Figs. 3 and 4). The projected maximum and

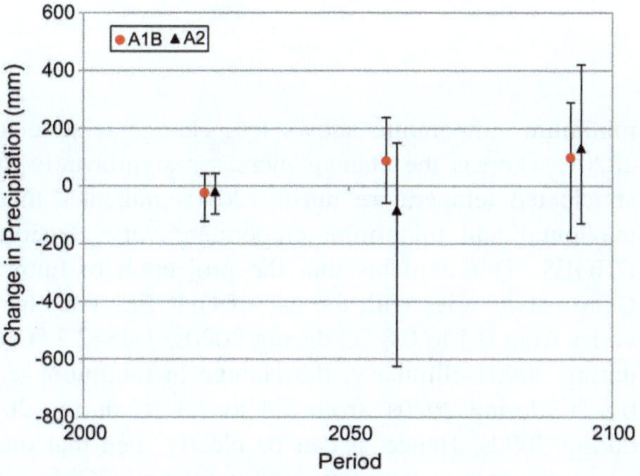

Fig. 2 Change in annual mean precipitation as projected under A1B and A2 scenarios during the 2020s, 2055s, and 2090s in the Nam Ou River Basin

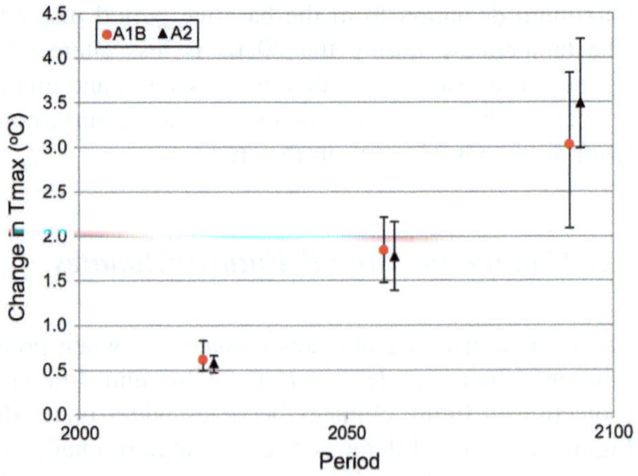

Fig. 3 Change in annual mean maximum temperature as projected under A1B and A2 scenarios during the 2020s, 2055s, and 2090s in the Nam Ou River Basin

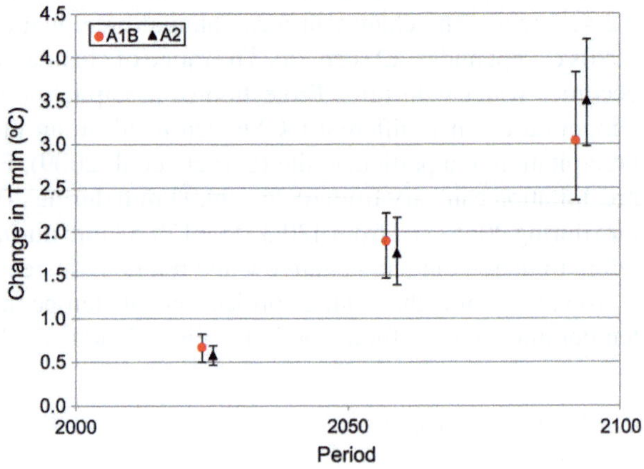

Fig. 4 Change in annual mean minimum temperature as projected under A1B and A2 scenarios during the 2020s, 2055s, and 2090s in the Nam Ou River Basin

minimum temperature shows less change relative to the baseline period during 2020s, whereas the change increases significantly during 2055s and 2090s. The anticipated temperature during 2090s indicates that the mean changes in both maximum and minimum temperature vary considerably between two different GHGES. This explains that the projection of future temperature under different GCMs also varies with the use of GHGES. The change in maximum temperature varies from 0.5 to 0.8 °C during 2020s, 1.45–2.3 °C during 2055s, and 2.1–4.3 °C during 2090s. Similarly, the change in minimum temperature varies from 0.5 to 0.8 °C during 2020s, from 1.4 to 2.3 °C during 2055s, and from 2.4 to 4.2 °C during 2090s. Hence, it can be clearly seen that the range of projection of temperature increases with time under different GCMs and GHGES. This indicates that the choice of emission scenarios highly influences climate prediction during the latter part of the century.

4.4 Uncertainty of Annual Precipitation, and Maximum and Minimum Temperature

Figures 5 and 6 represent uncertainty in annual mean precipitation under 8 GCMs, A1B and A2 scenarios, respectively. As shown in Fig. 5, the annual mean precipitation of the baseline period lies in between 1,552 and 1,719 mm with a median value of 1,626 mm. The precipitation is projected to vary between 1,500 and 1,790 mm during 2020s; between 1,565 and 1,870 mm during 2055s; and between 1,455 and 1,912 mm during 2090s under A1B scenario. Similarly, under A2 scenario, the median value of annual mean precipitation is anticipated to be between 1,533 and

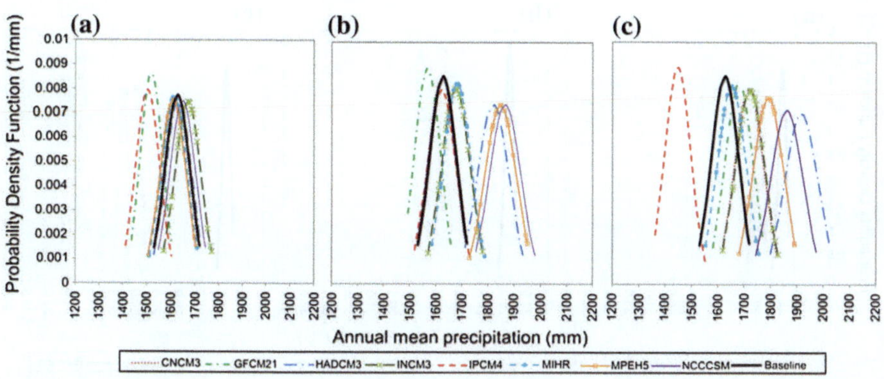

Fig. 5 Probability density functions of the annual mean precipitation for the baseline period (1981–2000) and future periods (2020s, 2055s, and 2090s) under the A1B scenario in the Nam Ou River Basin. **a** 2020s. **b** 2055s. **c** 2090s

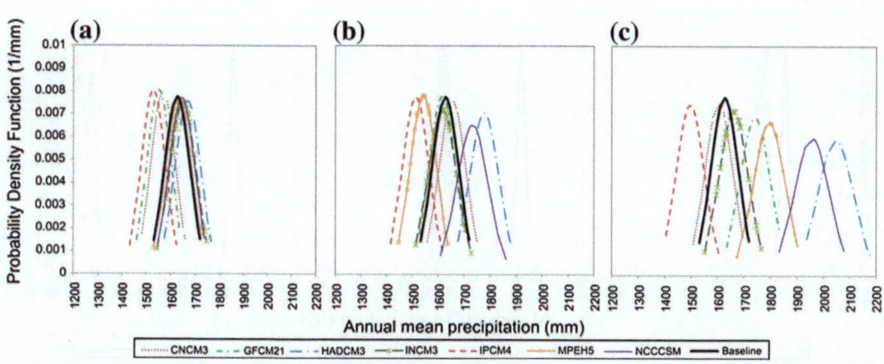

Fig. 6 Probability density functions of the annual mean precipitation for the baseline period (1981–2000) and future periods (2020s, 2055s, and 2090s) under the A2 scenario in the Nam Ou River basin. **a** 2020s. **b** 2055s. **c** 2090s

1,655 mm during 2020s; between 1513 and 1,784 mm during 2055s; and between 1,507 and 2,054 mm during 2090s. It can be seen that the latter part of the century is expected to be wetter than the early and mid-century under both the A1B and A2 scenarios. However, there are some GCMs which also predicted lower precipitation in future than for the baseline period. This demonstrates that the projection of precipitation varies by a wider range in the mid- and latter part of the century for different GCMs under A1B and A2 scenarios.

Figures 7 and 8 display uncertainty in the projection of maximum temperature under A1B and A2 scenarios, respectively, for three future periods, while Figs. 9 and 10 display the same for minimum temperature. Each GCM suggests an increase in annual mean T_{max} and T_{min} for all three future periods under A1B and A2

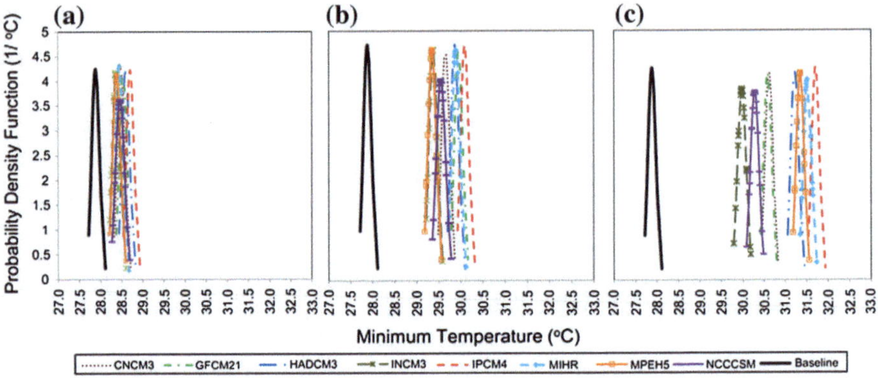

Fig. 7 Probability density functions of the annual mean maximum temperature for the baseline period (1981–2000) and future periods (2020s, 2055s, and 2090s) under the A1B scenario in the Nam Ou River Basin. **a** 2020s. **b** 2055s. **c** 2090s

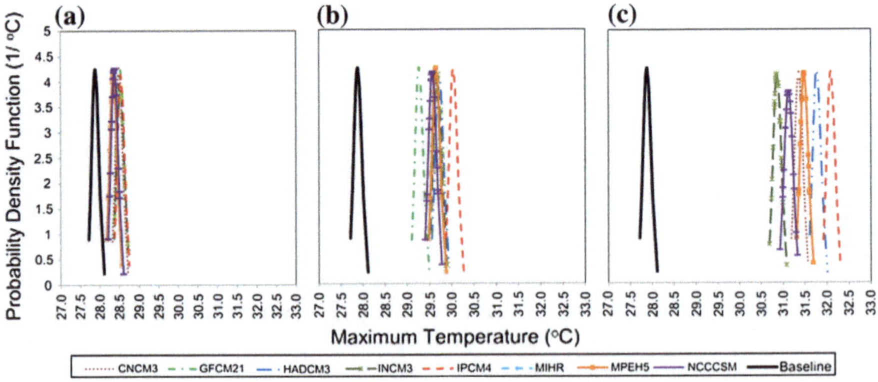

Fig. 8 Probability density functions of the annual mean maximum temperature for the baseline period (1981–2000) and future periods (2020s, 2055s, and 2090s) under the A2 scenario in the Nam Ou River Basin. **a** 2020s. **b** 2055s. **c** 2090s

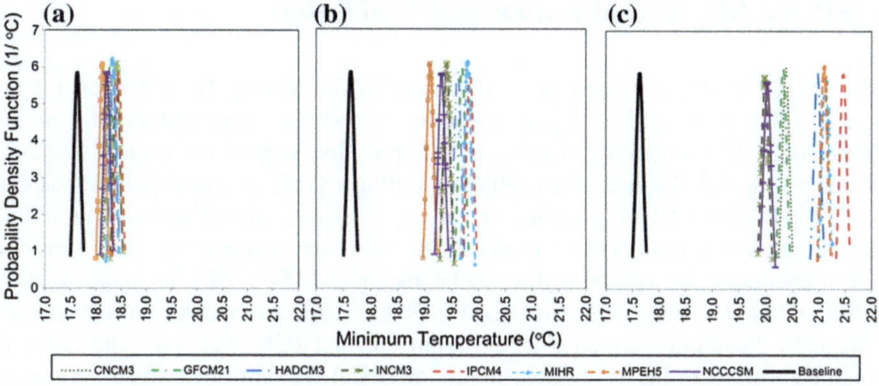

Fig. 9 Probability density functions of the annual mean minimum temperature for the baseline period (1981–2000) and future periods (2020s, 2055s, and 2090s) under the A1B scenario in Nam Ou River Basin. **a** 2020s. **b** 2055s. **c** 2090s

scenarios. The magnitude of increase varies from one GCM to another for future periods. Under the A1B scenario, the median value of T_{max} varies between 28.4 and 28.7 °C for the 2020s, between 29.37 and 30.11 °C for the 2055s, and between 29.9 and 31.7 °C for the 2090s. Meanwhile, the median T_{max} ranges between 28.4 and 28.5 °C for the 2020s, between 29.3 and 30.05 °C for the 2055s, and between 30.8 and 32.09 °C for the 2090s under the A2 scenario. Figures 9 and 10 show the uncertainty of annual mean T_{min} for three future periods under A1B and A2, respectively. The IPCM4 GCM predicted a higher increase in temperature under both GHGES. The study found that the temperature is predicted to increase to a higher extent during the 2090s, followed by the 2055s.

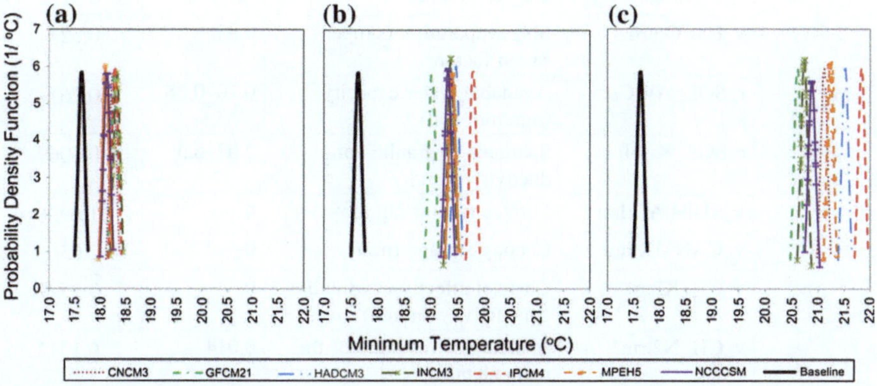

Fig. 10 Probability density functions of the annual mean minimum temperature for the baseline period (1981–2000) and future periods (2020s, 2055s, and 2090s) under the A2 scenario in the Nam Ou River Basin. **a** 2020s. **b** 2055s. **c** 2090s

5 SWAT Model Calibration and Validation

The SWAT model is calibrated for discharge at the Muong Ngoy Station for the period 1992–1999 with a 2-year warm-up period and validated for the period 2000–2003. The sensitivity analysis of different flow-related parameters was conducted using the Sequential Uncertainty Fitting (SUFI-2) algorithm (Abbaspour et al. 2004, 2007). SUFI-2 enables sensitivity analysis, calibration, validation, and uncertainty analysis of SWAT models. On the basis of these, the most sensitive flow parameters are the base flow alpha factor (ALPHA_BF), recharge to deep aquifer (RCHRG_DP), curve number (CN2), hydraulic conductivity of the channel (CH_K2), Manning's n value of the main channel (CH_N2), etc. The different parameters used in the calibration of the model along with their fitted values are shown in Table 6.

Figures 11 and 12 compare observed discharges with the simulated values for calibration and validation, respectively. The calibration result shows that the simulated values fit the observed data for the calibration period with $R^2 = 0.64$, NS = 0.64, and PBIAS = 5.12 %. The model validation also shows an acceptable

Table 6 Fitted parameters during calibration of the SWAT model

Variable	Parameter name	Description and units	Initial value or range	Fitted parameter value
Discharge	v_ALPHA_BF.gw	Base flow alpha factor	0.048	0.8125
	v_RCHRG_DP.gw	Deep aquifer percolation fraction	0.05	0.0850
	v_GW_DELAY.gw	Groundwater delay time (days)	31	51.7870
	r_CN2.mgt	Curve number	48.91–80	0.0189
	v_ESCO.hru	Soil evaporation compensation factor	0.95	0.8271
	r_SOL_AWC.sol	Available water capacity (mm/mm soil)	0.10–0.28	0.261
	r_SOL_K.sol	Saturated hydraulic conductivity (mm/h)	2.07–6.5	0.6069
	v_SURLAG.hru	Surface run-off lag (days)	4	11.6938
	v_CANMX.hru	Canopy storage (mm)	0	2.0367
	v_CH_K2.rte	Channel effective hydraulic conductivity (mm/h)	0	3.1527
	v_CH_N2.rte	Manning's "n" value of the main channel	0.014	0.1912

Note The qualifier (r_) refers to relative change in the parameter where the value from the SWAT database is multiplied by 1 plus a factor in the given range; the qualifier (v_) refers to the substitution of a parameter by a value from the given range

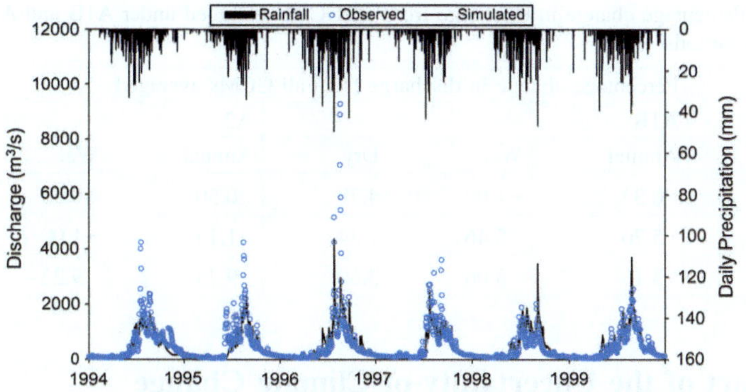

Fig. 11 Comparison of observed and simulated daily discharge for the calibration period 1994–1999 with daily precipitation

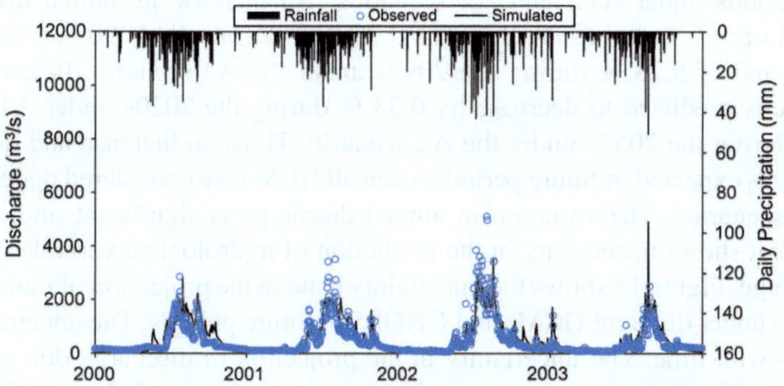

Fig. 12 Comparison of observed and simulated daily discharge for the validation period 2000–2003 with daily precipitation

result of $R^2 = 0.74$, NS = 0.72, and PBIAS = −14.25 %. The model is capable of predicting the discharge for the Nam Ou River Basin quite well, and the total runoff volumes of the observed data are also matched well by the simulated result. The peak flows are not matched well except for 1998 and 1999 during the calibration period and for 2000 and 2003 during the validation period. This discrepancy in peak flows might be attributed to the precipitation data and possible errors in the observed discharge data, especially during high flows. Rossi et al. (2009) also reported that from the case study in the Mekong River Basin errors in gauging stations can contribute to unreliable matching of hydrographs, mainly at locations along the Mekong's tributaries. The difficulty in characterising some floodplain processes might result in flow acceleration.

Table 7 Percentage change in discharge from all GCMs averaged under A1B and A2 scenarios and future periods

Periods	Percentage change in discharge from all GCMs averaged					
	A1B			A2		
	Annual	Wet	Dry	Annual	Wet	Dry
2020s	−0.33	−1.09	4.79	0.50	0.58	−0.07
2055s	5.70	5.46	7.34	−1.15	−1.00	−2.21
2090s	3.13	3.06	3.60	9.23	9.23	9.20

6 Impact of the Uncertainty of Climate Change on Discharge

Table 7 shows the percentage change in discharge from the average of all GCMs in future periods under A1B and A2 scenarios. An increase in annual discharge is observed of 5.7 % during the 2055s and 3.13 % during the 2090s under the A1B scenario and of 9.23 % during the 2090s under the A2 scenario. In addition, the discharge is predicted to decrease by 0.33 % during the 2020s under A1B and by 1.15 % during the 2055s under the A2 scenario. Thus, an increase and decrease in discharge is expected in future periods when all GCMs are considered under the A1B and A2 scenarios. The variation in annual discharge is significant under different GCMs as it shows uncertainty in the prediction of hydrological variables under climate change. Figure 13 shows the uncertainty band in the projection of monthly mean discharge under different GCMs and GHGES in future periods. The uncertainty band increases with time. The uncertainty in the projection of discharge due to different GCMs and GHGES is less during the 2020s compared to the 2055s and 2090s. The uncertainty range is quite high during the months of July to September. This reveals that the wet season in the future is expected to be wetter compared to dry seasons. The highest discharge during the 2020s reached 1,750 m^3/s from the projections of all

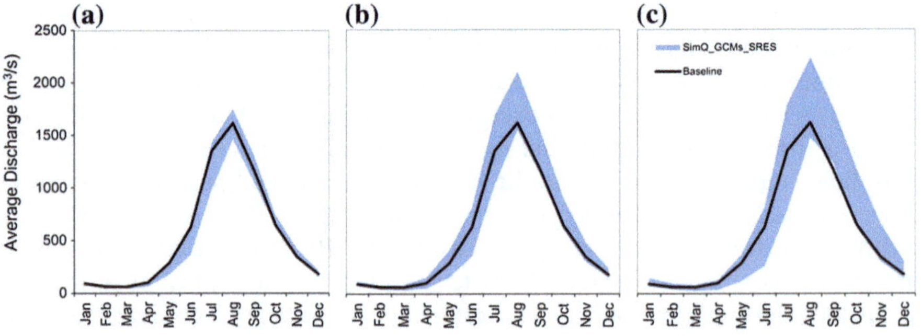

Fig. 13 Envelopes of simulated discharge under different GCMs and GHGES for three future periods using the plot of simulated discharge for the baseline period 1981–2000 for comparison. (SimQ_GCMS_SRES represents the result of simulated discharge under all the GCMs and GHGES considered)

GCMs and GHGES. The discharge reached the maximum limit of 2,100 m^3/s during the 2055s and 2,230 m^3/s during the 2090s. It depicts that future discharge in the Nam Ou River Basin is expected to be too high by the end of the century.

7 Conclusions

This study assesses the impact of the uncertainty of climate change on discharge in the Nam Ou River Basin. In this study, a stochastic weather generator LARS-WG is used to generate future precipitation and temperature. Eight GCMs incorporated in LARS-WG are used for the projection of future climates under two GHGES, A1B and A2 for the three periods: 2020s, 2055s, and 2090s, respectively. The hydrological model SWAT is used to simulate present and future changes in discharge in the river basin. Calibration and validation of the discharge suggests that the SWAT model can be used for the simulation of the run-off under present and future climates. Results show that high uncertainties exist in all projected future precipitation and temperatures when different GCMs and GHGES are used. Due to this, it is difficult to predict the discharge accurately. It can be seen from this study that the projections of precipitation have a significant effect on the estimation of the run-off in the river basin. The change in precipitation in the future is observed to increase as well as decrease depending upon the GCMs used. The uncertainty in the prediction of the climate has a direct impact on the run-off in the basin. Due to such uncertainty, it becomes difficult to quantify the impact of climate change in the hydrological variables. In addition, the uncertainty of climate variables increases with time. The inter-annual variability of the future projection of precipitation and temperature is observed in this study. The results indicate quite clearly that the choice of GCMs and GHGES is critical for any climate change impact study on hydrology. Though the selection of GCMs and emission scenarios is quite difficult, the appropriate GCMs can be selected based on their ability to reproduce the observed historical mean annual rainfall. The identification of skilled GCMs mainly focuses on rainfall estimation since the GCMs show very similar patterns of change in temperature and may help to reduce the uncertainty. This also emphasises the need for better understanding of uncertainties in future projections and the development of efficient techniques to reduce them. However, the results of this research can be useful for policy and decision makers in their planning and management strategies for climate change and its adaptation in the Nam Ou River Basin.

References

Abbaspour KC, Johnson CA, Van Genuchten MT (2004) Estimating uncertain flow and transport parameters using a sequential uncertainty fitting procedure. Vadose Zone J 3:1340–1352

Abbaspour KC, Yang J, Maximov I, Siber R, Bogner K, Mieleitner J, Zobrist J, Srinivasan R (2007) Modelling hydrology and water quality in the Pre-Alpine/Alpine Thur watershed using SWAT. J Hydrol 333:413–430

Benaman J, Christine AS, Douglas AH (2005) Calibration and validation of soil and water assessment tool on an agricultural watershed in upstate New York. J Hydrol Eng ASCE 10 (5):363–374

Beyene T, Lettenmaier DP, Kabat P (2010) Hydrologic impacts of climate change on the Nile River Basin: implications of the 2007 IPCC scenarios. Clim Change 100:433–461

Chen J, Brissette FP, Leconte R (2011) Uncertainty of downscaling method in quantifying the impact of climate change on hydrology. J Hydrol 401:190–202

Chen X, Yang T, Yang XY, Xu C-Y, Yu ZB (2013) Uncertainty intercomparison of different hydrological models in simulating extreme flows. Water Resour Manage 27:1393–1409

Chiew FHS, Kiromo DGC, Kent DM, Frost AJ, Charles SP, Timbal B, Nguyen KC, Fu G (2010) Comparison of runoff modeled using rainfall from different downscaling methods for historical and future climates. J Hydrol 387:10–23

Dunn SM, Brown I, Sample J, Post H (2012) Relationships between climate, water resources, land use and diffuse pollution and the significance of uncertainty in climate change. J Hydrol 434–435:19–35

Fowler HJ, Blenkinsop S, Tebaldi C (2007) Linking climate change modeling to impact studies: recent advances in downscaling techniques of hydrological modeling. Int J Climatol 27: 1547–1578

Girvetz EH, Zganjar C, Raber GT, Maurer EP, Kareiva P et al (2009) Applied climate-change analysis: the climate wizard tool. PLoS ONE 4(12):e8320. doi:10.1371/journal.pone.0008320

Gosling SN, Bretherton D, Haines K, Arnell NW (2010) Global hydrology modelling and uncertainty: running multiple ensembles with a campus grid. Philos Trans R Soc A 368:1–17

Gosling SN, Taylor RG, Arnell NW, Todd MC (2011) A comparative analysis of projected impacts of climate change on river runoff from global and catchment-scale hydrological models. Hydrol Earth Syst Sci 15:279–294

Green WH, Ampt GA (1911) Studies on soil physics, 1. The flow of air and water through soils. J Agric Sci 4:11–24

Hashmi MZ, Shamseldin AY, Melville BW (2009) Downscaling of future rainfall extreme events: a weather generator based approach. 18th World IMACS/MODSIM Congress. Cairns, Australia

IPCC (2007) Climate Change 2007: The physical science basis. contribution of working group I to the fourth assessment report of the intergovernmental panel on climate change. In: Solomon S, Qin D, Manning M, Marquis M, Averyt K, Tignor MM et al., (eds.) Cambridge University Press, Cambridge, United Kingdom and New York, NY, USA

Khan MS, Coulibaly P, Dibike Y (2006) Uncertainty analysis of statistical downscaling methods. J Hydrol 319:357–382

Kundzewicz ZW, Mata LJ, Arnell NW, Döll P, Kabat P, Jiménez B, Miller KA, Oki T, Sen Z, Shiklomanov IA (2007) Freshwater resources and their management. Climate change 2007: Impacts, adaptation and vulnerability. contribution of working group II to the fourth assessment report of the intergovernmental panel on climate change, In: Parry ML, Canziani OF, Palutikof JP, van der Linden PJ, Hanson CE (eds.), Cambridge University Press, Cambridge, UK, pp 173–210

Lu XX (2005) Spatial variability and temporal change of water discharge and sediment flux in the lower Jinsha tributary: impact of environmental changes. River Res Appl 21:229–243

Luo Y, Ficklin DL, Liu X, Zhang M (2013) Assessment of climate change impacts on hydrology and water quality with a watershed modeling approach. Sci Total Environ 450–451:72–82

Maraun D, Wetterhall F, Ireson AM, Chandler RE, Kendon EJ, Widmann M, Brienen S, Rust HW, Sauter T, Themeßl M, Venema VKC, Chun KP, Goodess CM, Jones RG, Onof C, Vrac M, Thiele-Eich I (2010) Precipitation downscaling under climate change. Recent developments to bridge the gap between dynamical models and the end user. Rev Geophys 48 (3)

Menzel L, Burger G (2002) Climate change scenarios and runoff response in the Mulde catchment (Southern Elbe, Germany). J Hydrol 267(1–2):53–64

Minville M, Brissette F, Leconte R (2008) Uncertainty of the impact of climate change on the hydrology of a nordic watershed. J Hydrol 358:70–83

Monteith JL (1965) Evaporation and the environment: in the state and movement of water in living organisms. In: XIXth Symposium Social for Experimental Biology, Cam-bridge University Press, Swansea, pp 205–234

Nash JE, Sutcliffe JV (1970) River flow forecasting through conceptual models. Part I—a discussion of principles. J Hydrol 10:282–290

Nawaz NR, Adeloye AJ (2006) Monte Carlo assessment of sampling uncertainty of climate change impacts on water resources yield in Yorkshire, England. Clim Change 78:257–292

Neitsch SL, Arnold JG, Kiniry JR, Williams JR (2009) soil and water assessment tool theoretical documentation version 2009

Nijssen B, O'Donnell GM, Hamlet AF, Lettenmaier DP (2001) Hydrologic sensitivity of global rivers to climate change. Clim. Change, 50: 143–175

Nohara D, Kitoh A, Hosaka M, Oki T (2006) Impact of climate change on river discharge projected by multimodel ensemble. J Hydrometeorol 7:1076–1089

Rossi CG, Srinivasan R, Jirayoot K, Le Duc T, Souvannabouth P, Binh N, Gassman PW (2009) Hydrologic evaluation of the Lower Mekong River Basin with the soil and water assessment tool model. Int Agric Eng J 18:1–13

Santhi C, Arnold JG, Williams JR, Dugas WA, Srinivasan R, Hauck LM (2001) Validation of the SWAT model on a large river basin with point and nonpoint sources. J Am Water Resour Assoc 37(5):1169–1188

Semenov MA, Barrow EM (1997) Use of stochastic weather generator in the development of climate change scenarios. Clim Change 35:397–414

Semenov MA, Stratonovitch P (2010) The use of multi-model ensembles from global climate models for impact assessments of climate change. Clim Res 41:1–14

Setegn SG, Dargahi B, Srinivasan R, Melesse AM (2010) Modeling of sediment yield from Anjeni-Gauged watershed, Ethiopia using SWAT model. J Am Water Resour Assoc 46(3): 514–526

Shrestha B, Babel, MS, Maskey S, Van Griensven A, Uhlenbrook S, Green A, Akkharath I (2013) Impact of climate change on sediment yield in the Mekong River basin: a case study of the Nam Ou basin, Lao PDR. Hydrol Earth Syst Sci, 17: 1–20

Soil Conservation Service (1972) Section 4: Hydrology In National engineering handbook. SCS

Stott PA, Kettleborough JA (2002) Origins and estimates of uncertainty in predictions of twenty first century temperature rise. Nature 416:723–726

Thompson JR, Green AJ, Kingston DG, Gosling SN (2013) Assessment of uncertainty in river flow projections for the Mekong River using multiple GCMs and hydrological models. J Hydrol 486:1–30

Van Liew MW, Veith TL, Bosch DD, Arnold JG (2007) Suitability of SWAT for the conservation effects assessment project: A comparison on the USDA-ARS experimental watersheds. J Hydrol Eng 12(2): 173–189

van Vliet MTH, Franssen WHP, Yearsley JR, Ludwig F, Haddeland I, Lettenmaier DP, Kabat P (2013) Global river discharge and water temperature under climate change. Glob Environ Change 23:450–464

Wang SJ, Jiao ST, Xin HJ (2013) Spatio-temporal characteristics of temperature and precipitation in Sichuan Province, Southwestern China, 1960 to 2009. Quatern Int 286:103–115

Zhang H, Huang GH, Wang D, Zhang X (2011) Uncertainty assessment of climate change impacts on the hydrology of small prairie wetlands. J Hydrol 396:94–103

Integrated Modelling of Climate Change and Urban Drainage

Ashish Shrestha, Mukand Singh Babel and Sutat Weesakul

Abstract Patterns of climate variables are changing under climate change resulting in increased frequency and intensity of extreme events. The implications are thus observed in existing natural and man-made systems. Man-made systems, mainly storm water management systems, are prone to it's functional failure due to recurrent events of extreme rainfall. Most urban drainage systems designed under stationary climate consideration are operating over capacity well ahead of their design period. There are several approaches and tools available to model climate change and urban drainage. This chapter discusses the integrated approach of climate change and urban drainage modelling to assess the direct implications of climate change on hydraulic performance of urban drainage.

Keywords Climate change · Mike urban · Urban drainage · IDFs · LARS-WG

1 Introduction

The increasing concentration of carbon dioxide in the atmosphere has been the principal reason for global warming and climate change. Projections from climate models showing increasing frequency, intensity, and volume of extreme precipitation have been shown by many studies (Trenberth 1999; Emori and Brown 2005; Boo et al. 2006). The effects of climate change can be seen in large and small

A. Shrestha (✉) · M.S. Babel · S. Weesakul
Asian Institute of Technology, Klong Luang, Pathum Thani, Thailand
e-mail: ashish.shrs@gmail.com

M.S. Babel
e-mail: msbabel@ait.ac.th

S. Weesakul
e-mail: sutat@ait.ac.th

S. Shrestha et al. (eds.), *Managing Water Resources under Climate Uncertainty*,
Springer Water, DOI 10.1007/978-3-319-10467-6_5

municipalities both urban and rural in nature with distinct types of municipal infrastructure. Hence, water and sewage networks need to accommodate more intense precipitation (Mehdi et al. 2006). The possible impacts of climate change on urban drainage are increased combined sewer overflows (CSO) causing environmental pollution; pollution of drinking water sources; infiltration of sewers into groundwater; increased pollutant levels in the waste water treatment plants (WWTPs); flooding in pumping stations; surcharging of manholes causing surface flooding; incremental sediment volume in storm water collection systems; and decreased infiltration capacity in the infiltration basin (Berggren et al. 2007). Apart from climate change, there are other factors, such as an increase in population, new developments, and the condition of infrastructural assets that determine the functional effectiveness of urban drainage. However, considering the intense impacts of climate change, there is need for detail studies for proper analysis of the drainage system. There are many approaches to analyze the impact of climate change on the hydraulic functions of urban drainage. Shrestha et al. (2014) approached the analysis of impact of climate change on surface flooding in the peri-urban area of Pathum Thani, Thailand, with the application of a one-dimensional PCSWMM (James et al. 2002) urban drainage model. The study included the application of rainfall and temperature data generated from the ECHAM4 general circulation model (Roeckner et al. 1992, 1996) with data sets from the PRECIS regional climate model for long-term simulation as well as the theoretical increments of rainfall percentage to discern the performance scenarios of urban drainage. The indicators used were flooded nodes, maximum hours of critical conduit flow, and total flood volume. Urban drainage models can also be integrated with a short-term weather forecasting model to predict real time flooding scenarios.

This chapter presents the course of the integrated climate change and urban drainage models with examples of case study results and discussions in Bangkok, Thailand. The worst floods experienced in central Thailand during October 2011 impacted on some 2.3 million people, causing US $25 billion in damage. The following discussions are part of study done in the Sukhumvit area of Bangkok. The integrated modelling approach includes the application of series of models which is summarised in Fig. 1. The first element is climate modelling, in order to project the daily time series of rainfall as an output. The second element of the process is the generation of intensity–duration–frequency curves (IDFs) to project

Fig. 1 Integrated modelling for urban drainage and climate change

rainfall events. Thirdly, there is the application of the urban drainage model. Hence, hydrological and hydrodynamic models are integrated to study climate change impacts on urban drainage performance.

2 Climate Modelling for the Projection of Climate Variables

Climate modelling is essential to project the climate variables in order to study possible impacts of climate change on urban drainage and surface flooding. Global circulation models (GCMs) characterise the fundamental background of physics in the mathematical expressions which simulate the behaviour of the climate systems, which are used to project the future climate with the variation in atmospheric variables under different scenarios defined by the Intergovernmental Panel on Climate Change (IPCC). The GCM outputs are in a coarse resolution of 150–300 km, and the regional climate model (RCM) has a resolution of 12–50 km (Sunyer et al. 2012). Climate modelling and downscaling of climate variables using different techniques are the general practice of projecting climate variables under different emission scenarios. There are essentially two widely practised downscaling techniques: dynamic and statistical downscaling. The statistical downscaling approach is comparatively inexpensive compared to dynamic downscaling (Wilby et al. 2004). Nguyen (2005) studied downscaling methods using the statistical downscaling model, SDSM (Wilby et al. 2004), and stochastic Long Ashton Research Station Weather Generator (LARS-WG) to evaluate climate change and variability on a hydrological regime at basin scale. Results showed that the LARS-WG was better suited to precipitation while the regression-based SDSM was best for temperature projection. Both the LARS-WG and SDSM can project temperature accurately and LARS-WG is far superior in producing precipitation statistics such as mean and standard deviations (Irwin et al. 2012). The LARS-WG incorporates predictions from 15 GCMs used in the IPCC Assessment Report 4 (IPCC 2007). Climate predictions are also available for the Special Report on Emission Scenarios, SRES (IPCC 2000; Semenov and Stratonovitch 2010) and emission scenarios SRB1, SRA1B, and SRA2 for most of the GCMs (Semenov and Stratonovitch 2010). The weather generation process follows two steps: firstly, analysis of the observed weather parameters such as temperature, precipitation and solar radiation, followed by the generation of synthetic daily weather data utilising weather parameters.

In the study of Sukhumvit (part of Bangkok), the LARS-WG was used with observed daily data from 1981 to 2010, for climate projection of the single site station at Bangkok Metropolis Station belonging to the Thai Meteorological Department. Some of the analysis presented below shows the accuracy of the climate projections. However, only rainfall has been considered as a climate variable (Fig. 2).

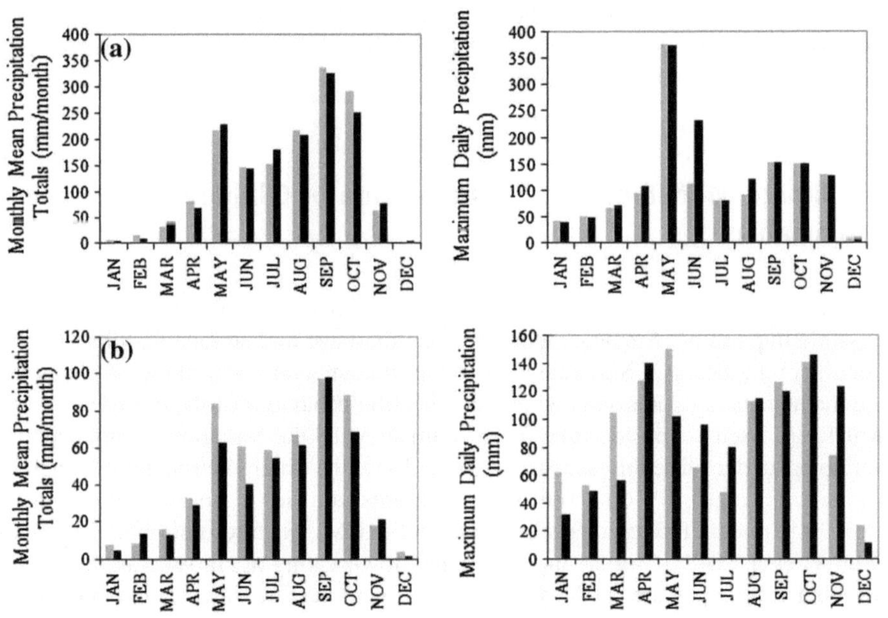

Fig. 2 **a** Weather projection 1981–2000. Comparing monthly the total mean and daily maximum for 12 months (*Black bar* synthetic, *Grey Bar* observed). **b** Weather projection 2001–2010. Comparing monthly the total mean and daily maximum for 12 months (*Black bar* synthetic, *Grey bar* observed)

The root-mean-square error (RMSE) for the monthly total mean in the first and second projection periods was observed as 16.45 mm/month and 10.76 mm/month, respectively, with a corresponding Efficiency Index (EI) of 0.97 and 0.9. While for maximum daily precipitation, RMSE was 35.34 mm/month and 29.05 mm/month and the efficiency index was 0.84 and 0.62 for the first and second projections, respectively. RMSE and EI analysis provide evaluation of the model for future projections.

2.1 Rainfall Disaggregation for Urban Hydrological Study

Rainfall disaggregation is essential to obtain high temporal resolution from the coarser time series. In most of the developing and developed countries, collecting continuous rainfall data of short durations require huge resources. While there are many methods and models developed to disaggregate rainfall, it still require considerably research efforts to develop the robust methodology for rainfall disaggregation. Many hydrological applications need short duration rainfall. The flood studies require short duration rainfall for continuous simulation tools for the design and management of hydro-systems (Koutsoyiannis 2003).

A disaggregation model developed by Koutsoyiannis and Onof (2001) and Koutsoyiannis (2003), based on the Barlett Lewis Rectangular Pulse process theory, later named Hyetos, was applied for the case of Sukhumvit, Bangkok. The Hyetos method requires the estimation of Bartlett Lewis Rectangular Pulse, BLRP parameters from the historical rainfall data, and to disaggregate the coarser rainfall series in the same station. The methodology can be applied to disaggregate the series from daily to sub-daily (hourly or minutes). The estimation of parameters is done with the application of certain equations to resemble the statistics of the observed and simulated series such as mean, standard deviation, lag 1-autocovariance, and proportionately dry. The equations and details are explained in Koutsoyiannis (2003).

3 Intensity–Duration–Frequency (IDF) and Design Storms

Intensity–Duration–Frequency (IDF) is a mathematical relationship between the rainfall intensity *i*, duration *d*, and return period *T* (Koutsoyiannis et al. 1998). The IDF curve is the most convenient form of rainfall information. Duration and intensity are two inversely proportional elements in the IDF curve which varies for different return periods (Butler and Schutze 2005). Rainfall intensity estimates are necessary for hydrological analysis, design, and planning problems. The IDF curve is an important tool for the design of urban drainage (Solaiman and Simonovic 2011).

The IDF can be expressed mathematically as $i = f(T,d)$. The generalised IDF relationship can be expressed as

$$i = \frac{a(T)}{b(d)} \tag{1}$$

here *a* and *b* are constants, *d* is the arbitrary time duration (typically from a few minutes to several hours or a few days), and *T* is the return period. This expression has the advantage of a separable functional dependence of *i* on *T* and *d*.

The function *b*(*d*) is:

$$b(d) = (d + \theta)^{\eta} \tag{2}$$

where θ and η are parameters to be estimated ($\theta > 0$, $0 < \eta < 1$)

$$a(T) = \lambda T^{k} = c + \lambda \ln T \tag{3}$$

This equation is old, empirical, and most commonly used in computation due to its simplicity; otherwise, if the maximum rainfall intensity has the Gumbel distribution, then *k* and λ depends on the return period *T* (Koutsoyiannis et al. 1998). Other distribution functions are gamma distribution, log Pearson III distribution, lognormal distribution, exponential distribution, and Pareto distribution. It is well known that IDFs are being used for hydrological application. However, the study by Mailhot and Duchesne (2010), for developing IDF under non-stationary conditions, explains that urban drainage infrastructures which had been designed under stationary climate situations for its particular design period are being affected as the rainfall pattern, and intensity and frequency of extreme rainfall modifies under climate change impacts. The design criteria will be revised as part of global adaptation strategies. However, there are uncertainties which deter the development of unambiguous guidelines for the design criteria. Solaiman and Simonovic (2011) developed a probability-based IDF curve using the Gumbel probability distribution for the City of London with methodology consisting of 11 Atmosphere-Ocean General Circulation Models (AOGCM) and disaggregation by algorithm to consider short duration rainfall shapes from historical data to be projected in similar shapes for the future. The study showed that rainfall patterns will change in future for London.

In Sukhumvit, IDF curves were developed for 3-hourly observed data and hourly future projected and disaggregated data from 1980 to 2010, using graphical corrections by fitting the ratio between observed and model rainfall intensities at different return periods and times. The corrections were necessary to correct the underestimation of intensities from the LARS-WG and Hyetos outputs. Correction for intensities were developed by Shrestha (2013), using equation $y = 1.58x^{-0.14}$, providing the factor for correction where x is time (h). Some studies by the Bangkok Metropolitan Administration (BMA) in Bangkok, Thailand, for past rainfall and flood events suggest that when the 3-h rainfall depth exceeds 100 mm, surface flooding occurs. Hence, two GCMs, GFCM21 and HADCM3, are selected, showing a maximum 3-h rainfall depth (Fig. 3).

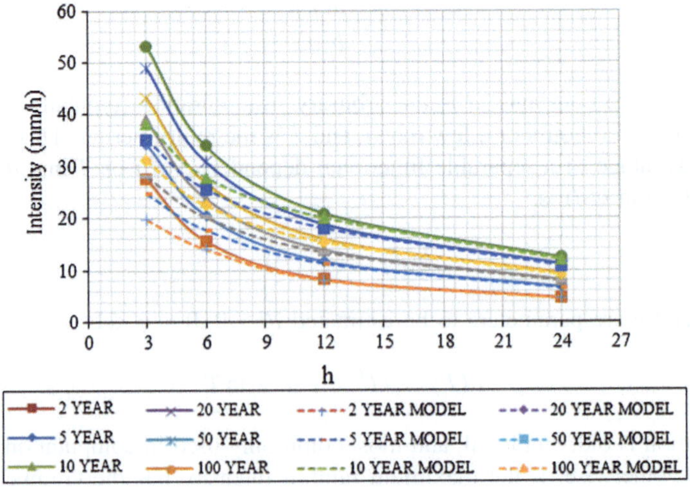

Fig. 3 IDFs for 1980–2010, developed through two data sets. (*Solid line* Observed (O), *Dash line*: Model (M))

Table 1 Ratio and difference between observed and simulated rainfall intensities

Duration		Return period						Correction factor
		2 Yr	5 Yr	10 Yr	20 Yr	50 Yr	100 Yr	
3 h	Ratio = O/M	1.39	1.39	1.39	1.389	1.388	1.387	1.355
6 h	Ratio = O/M	1.111	1.156	1.177	1.192	1.208	1.218	1.229
12 h	Ratio = O/M	1.025	1.033	1.036	1.039	1.041	1.042	1.116
24 h	Ratio = O/M	1.022	1.027	1.029	1.030	1.031	1.032	1.013

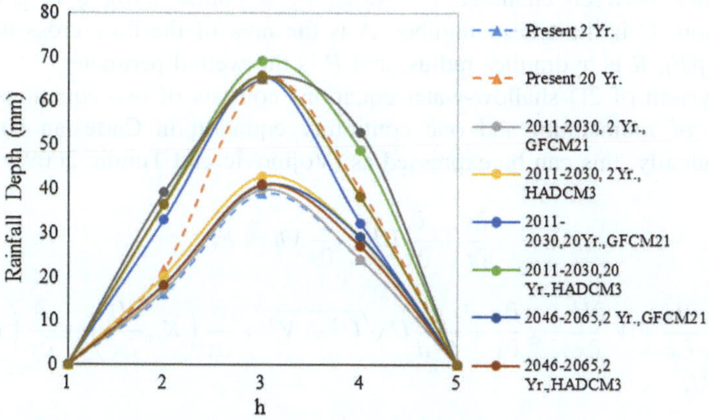

Fig. 4 Generated design storms for urban drainage modelling

Design storms are patterns of precipitation used for hydrological design purposes and derived from the IDF relationship. Most common methods for developing design storms are the triangular hyetograph, Chicago design storm, and alternating block (Chow et al. 1988) Table 1.

Some examples of design storm generated using the alternative block method (Chow et al. 1988) are shown in the figure below (Fig. 4).

4 Urban Drainage Modelling

In the urban catchment, modelling of urban drainage can be done using a one-dimensional (1D) model as the general scientific practice, but in the case of urban flooding, there is an interaction of drainage and street channel systems which require a two-dimensional (2D) model.

The 1D modelling of flow in the conduits or open channels follows the Saint Venant equation. The equation can be represented mathematically as (Vojinovic and Tutulic 2009):

$$\frac{\partial A}{\partial t}+\frac{\partial Q}{\partial x}=F_s \tag{4}$$

$$\frac{\partial Q}{\partial t}+\frac{\partial}{\partial x}\left(\frac{\beta Q^2}{A}\right)+gA\frac{\partial h}{\partial x}+g\frac{Q|Q|}{C^2\mathrm{AR}}=0 \tag{5}$$

here h is water depth, Q is discharge, β is the velocity distribution coefficient, x is the distance between chainage, t is time, F_s is course term, g is gravitational acceleration, C is the Chezy number, A is the area of the flow cross the section which is $f(h)$, R is hydraulics radius, and P is the wetted perimeter.

The system of 2D shallow-water equations consists of two equations for conservation of momentum and one continuity equation in Cartesian coordinates. Mathematically, this can be expressed as (Vojinovic and Tutulic 2009):

$$\frac{\partial s}{\partial t}+\frac{\partial}{\partial x}Uh+\frac{\partial}{\partial x}Vh=F_s \tag{6}$$

$$\frac{\partial U}{\partial t}+U\frac{\partial U}{\partial x}+V\frac{\partial U}{\partial x}+g\frac{\partial s}{\partial x}+\frac{g}{C^2d}U\sqrt{U^2+V^2}+\frac{\partial}{\partial x}\left(K_{xx}\frac{\partial U}{\partial x}\right)+\frac{\partial}{\partial y}\left(K_{yy}\frac{\partial U}{\partial y}\right)=F_sU_s \tag{7}$$

$$\frac{\partial V}{\partial t}+U\frac{\partial V}{\partial x}+V\frac{\partial V}{\partial y}+g\frac{\partial s}{\partial y}+\frac{g}{C^2d}V\sqrt{U^2+V^2}+\frac{\partial}{\partial x}\left(K_{xx}\frac{\partial V}{\partial x}\right)+\frac{\partial}{\partial y}\left(K_{yy}\frac{\partial V}{\partial y}\right)=F_sV_s \tag{8}$$

here s is the water surface elevation, U and V are depth-averaged velocities, K_{xx} and K_{yy} are eddy viscosities, and U_s and V_s are the velocities at the source.

4.1 Integrating Models

Hydrodynamic modelling tools require a rainfall hyetograph or time series for the simulation of rainfall–runoff processes. Physical processes like evaporation, infiltration, and runoff are numerically computed during the simulation. The integrated models are interlinked with input and output to each other. In the previous section, a climate change model is applied together with rainfall disaggregation to obtain a time series of future rainfall on an hourly scale. IDF curves were subsequently generated and design storms for future time and return periods were projected. However, the condition of drainage networks and sediment deposition could also play a significant role.

When a storm event occurs in a city, runoff produced by roofs and terraces is generally directly conveyed to the underground sewer network, while runoff produced by roadways, parks, squares, etc. circulates over the urban surfaces to reach the inlet structures of the drainage system. Therefore, it is necessary to estimate the hydraulics of the surface drainage structures and to model them through a coupled 1D/2D approach (Russo et al. 2011). The results obtained by Mark et al. (2004) showed that the greatest inaccuracy of a 1D model is to assume flow as 1D. When the flow path can be identified, and as long as overland flow stays on the street, 1D/1D modelling is sufficient but in the event of extreme flooding, a 1D/2D model is much preferred. The two important objectives of 1D/2D modelling are firstly, to assess performance of the combined sewer system and economic evaluation of flood damage to assist decision makers, and secondly the ability to predict the extent of flooding using an equation such as Navier–Stokes by using the depth-average method (Seyoum et al. 2012). 2D modelling can provide abundant information on the dynamics of flooding, which will help in flood risk management (Néelz and Pender 2010). Other studies conducted in 2D modelling focus particularly on methods such as adjusted conveyance and storage characteristics, parallel computing techniques, and grid coarsening to reduce the computation time while maintaining the level of accuracy and detail (Vojinovic et al. 2012; Chen et al. 2012a, b).

Surface flooding can be simulated using two approaches. A 1D/1D approach by defining the pipe network as a 1D model interlinked with an overland 1D model defined by a cross section of open channels. The 1D/2D approach represents overland flow conditions and interaction with sewer networks more accurately. However, a 1D/2D model is computationally more time-consuming than the former.

The input file required for a 1D/2D model is a working pipe flow model, Digital Elevation Model (DEM) in a raster data set and a number of defined couplings between the 1D and 2D model. The selection of a 2D model can be based on different types of solvers, namley single grid using a rectangular cell solver, single grid using a rectangular multi-cell solver, and a flexible mesh solver. The DEM consists of grid cells of a defined size which have different elevations. The elevation threshold for the DEM, flooding and drying depth, bed resistance, and eddy viscosity should be specified prior to running the simulation. The flow parameter in the coupled nodes and basins is governed by orifice, weir, or exponential function.

There are a number of hydrodynamic model packages applicable for urban drainage and urban flood studies, the most popular being: the Storm Water

Management Model (SWMM) developed by the United States Environmental Protection Agency (EPA), Mike Packages (MOUSE, Mike 11, Mike 21, Mike Urban, Mike Flood) developed by the Danish Hydraulic Institute (DHI).

Shrestha et al. (2014) applied a 1D PCSWMM model to simulate possible climate change impacts on surface flooding in the peri-urban area of Pathum Thani, Thailand.

5 Performance Assessment

Performance assessment of the urban water infrastructure is essential for the planning, design, and operation of water utilities. The boundary conditions for assessment of urban drainage performance can be simple or complex depending on population projection, pipe conditions, etc. The important boundary conditions when considering climate change implications are changes in climate variables. While precipitation is significant in studying the performance of urban drainage and surface flooding, temperature and evaporation have much less effect during surface flooding. In the study by Shrestha et al. (2014), applying the one-dimensional PCSWMM model to study possible climate change impacts on surface flooding in the peri-urban area of Pathum Thani, Thailand. It was found that when considering the evaporation rate, the reduction in flood volume was only up to 1.08 %. The simulation results showed that climate change had a significant impact on the existing drainage capacity.

When the hydraulic gradient of the pipe system exceeds that of the ground level, the surcharging of manholes occurs. Sewer length is considered to be surcharged when it is running at full capacity and under pressure, with manholes filling and spilling over. Surcharge can occur when there is hydraulic overload of the system or a blockage in the downstream flow. Surface flooding can possibly occur as a consequence of surcharging, with manhole covers being forced off under pressure from inside pipes. Surcharging can also be understood as the flow in a pipe when it exceeds the designed flow level. The design of the collection system usually involves gravity flow under a certain pipe gradient. If the elevation for pipe grades towards the outfall is not sufficient then intermediate pumps are arranged. Increasing the amount of storm water results in more pumping, which is one implication that could be studied. When surface flooding occurs from surcharging, the duration of the flooding is critical, particularly in urban settlements. The area and duration of the flood could result in economic loss.

Simulation for urban drainage in Sukhumvit was achieved using modelling tools of the coupled 1D/2D approach of Mike Urban and Mike Flood. A digital elevation

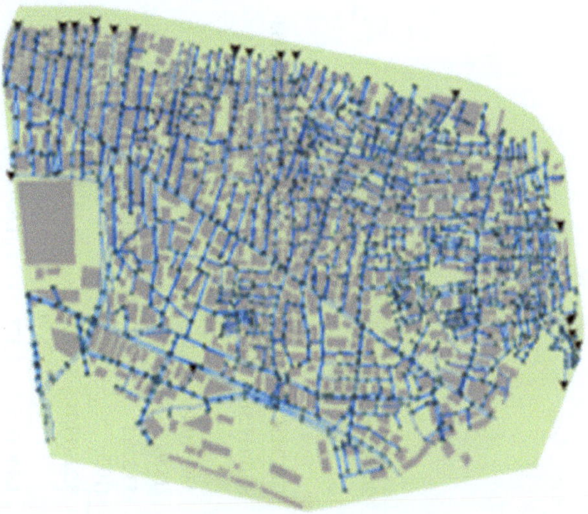

Fig. 5 Pipe flow model over the overland land flow model

model (DEM) of 10-by-10-m resolution was prepared by taking the spot elevation consisting of streets and buildings (Boonya-aroonnet et al. 2001; Chingnawan 2003; Nguyen 2009). The model consisted of 3,487 manholes, 3,858 pipes, 22 outlets, and 26 pumps, representing the actual situation of storm water management in the catchment. Manholes are connected with 2,019 numbers of catchments with total area of 2,049 Ha.

Calibration and validation of an urban drainage model is essential for accurate simulation. Figure 5 shows the validation of the model in Sukhumvit, conducted using rainfall data from a rain event and the measured water depth at two stations. The RMSE/EI in the first and second stations was obtained as 0.24 m/0.60 and 0.04 m/0.82.

The simulation results of Sukhumvit, Bangkok, are summarised in Table 2 below. Results are organised using the sequence of base case and future timescale of 2011–2030 and 2046–2065 under the SRA2 scenario, due to the fact that more carbon dioxide emissions are presumed for this storyline making it the worst case scenario (IPCC 2000). Simulations are made for 2- and 20-year return periods to signify frequent and extreme case scenarios. The results showed that total outflow volume, flood volume, flood area, surcharging nodes, and links flowing over capacity increase with the increasing precipitation. The flood maps are shown in Fig. 6 for maximum flooding (Fig. 7).

Table 2 Simulation results showing performance of urban drainage

Criterion	Baseline		2011–2030				2046–2065			
Return periods (Yr.)	2	20	2		20		2		20	
GCMs	–	–	GFCM21	HADCM3	GFCM21	HADCM3	GFCM21	HADCM3	GFCM21	HADCM3
Total 3-h rainfall depth (mm)	79.1	126.7	80.9 (2.3)	92.3 (16.7)	131.1 (3.5)	154.7 (22.1)	88.2 (11.5)	86.1 (8.5)	157.2 (24.1)	140.4 (10.8)
Total outflows (m^3)	275,617.3	375936.2	331,704.1 (3.3)	339,945.6 (17.2)	387,589.6 (3.1)	429,193.7 (24.1)	339,510.4 (11.9)	338,459.4 (10)	430,374.2 (24.1)	404,846.1 (9.8)
Flooded volume (m3)	42,900	83,127	50,879.0 (6.7)	54,499.7 (27.7)	77,914.6 (3.3)	95,008.6 (18.6)	53,326.1 (10.3)	52,975.5 (12)	96,070.3 (16.3)	84,443.4 (11.2)
Area flooded (km^2)	2.69	4.58	3.01 (11.8)	3.55 (31.9)	5.11 (11.5)	5.87 (21.8)	3.22 (19.7)	3.47 (28.9)	5.67 (23.7)	5.21 (13.7)
Links over capacity (Qmax/Qfull) > 1 (Nos.)	1,344	1,620	1,356 (0.9)	1,436 (6.8)	1,621 (0.06)	1,679 (3.6)	1,406 (4.6)	1,375 (2.3)	1,629 (0.5)	1,628 (0.5)
Surcharging nodes (Nos.)	280	455	283 (1.1)	330 (17.8)	486 (6.8)	557 (22.4)	322 (15)	315 (12.5)	552 (21.3)	518 (14)

*() Refers to percentage increased from baseline condition

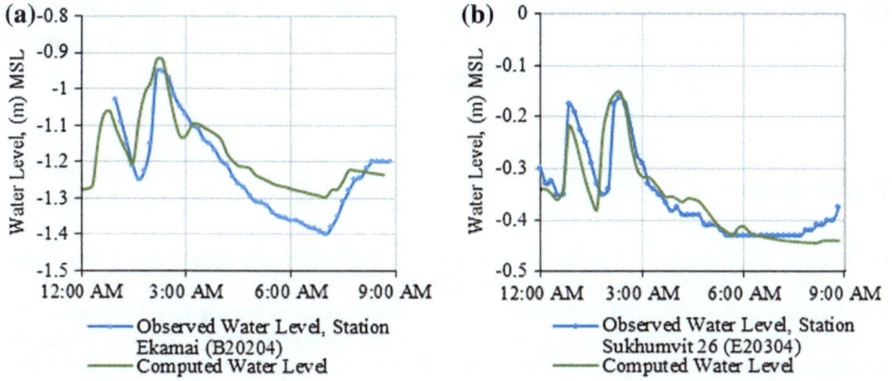

Fig. 6 Simulations showing observed and computed water levels at two stations (Ekamai and Sukhumvit 26)

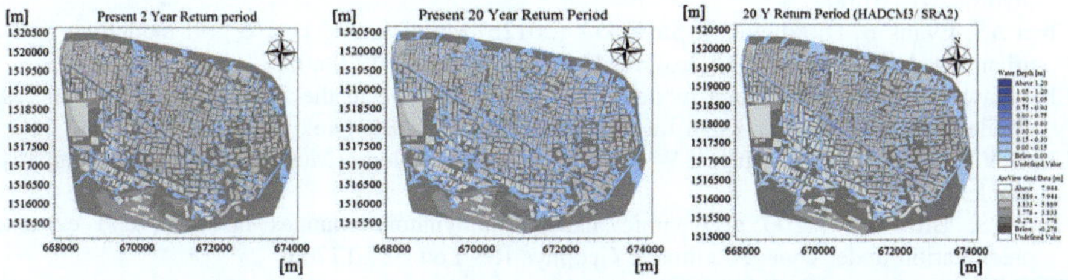

Fig. 7 Flood scenarios showing flood depth and area during present 2-year and present 20-year return periods, and during 2011–2030 future 20-year return period. Note: figures are not to scale and figures are in a UTM 47 coordinates system of degree decimal

6 Conclusion

The integrated model is the most appropriate approach for quantifying the climate change impacts on urban drainage performances. This approach could be applied to study the implication of climate change under uncertainties. The existing system of storm water management could be studied, and this type of study will also be crucial for new designs of urban water infrastructure. Many storm water modelling tools are available today. These models can be integrated with short-term weather projection models to predict the surface flooding scenarios in real time. This might provide water managers and engineers with crucial information for planning urban water infrastructures, including measures for flood protection. There is still a need for research efforts in the generation of high temporal resolution precipitation data through rainfall disaggregation. Climate change mitigation measures could further be studied using a similar approach. The condition of the existing drainage network could be assessed and further damage can be quantified. The evidence provided in this chapter further underscores the need to update the urban drainage design criteria. It also highlights the need for adaptation measures and reassessment of storm water management design and planning processes.

References

Berggren K, Olofsson M, Viklander M, Svensson G (2007) Tools for measuring climate change impacts on urban drainage systems. NOVATECH, pp 239–246

Boo KO, Kwon WT, Baek HJ (2006) Change of extreme events of temperature and precipitation over Korea using regional projection of future climate change. Geophys Res Lett 33:L01701

Boonya-Aroonnet S, Weesakul S, Mark O (2001) Modelling of urban flooding in Bangkok. Global Solutions for Urban Drainage, American Society of Civil Engineers

Butler D, Schutze M (2005) Integrating simulation models with a view to optimal control of urban wastewater systems. Environ Model Softw 20(4):415–426. doi:10.1016/j.envsoft.2004.02.003

Chen AS, Evans B, Djordjević S, Savić DA (2012a) A coarse-grid approach to representing building blockage effects in 2D urban flood modelling. J Hydrol 426–427:1–16. doi:10.1016/j.jhydrol.2012.01.007

Chen AS, Evans B, Djordjević S, Savić DA (2012b) Multi-layered coarse grid modelling in 2D urban flood simulations. J Hydrol doi:10.1016/j.jhydrol.2012.06.022

Chingnawan S (2003) Real time modelling of urban drainage in the Sukhumvit area, Bangkok, Thailand. Master Thesis, Asian Institute of Technology, Bangkok

Chow VT, Maidment DR, Mays LW (1988) Applied hydrology. McGraw-Hill Book Company, Maidenherd

Emori S, Brown SJ (2005) Dynamic and thermodynamic changes in mean and extreme precipitation under changed climate. Geophys Res Lett 32:L17706

IPCC Working Group III (2000) IPCC special report emissions scenarios. intergovernmental panel on climate change. Geneva, IPCC (Intergovernmental Panel on Climate Change)

IPCC (2007) ipcc fourth assessment report: climate change 2007 (AR4)—working group 1 report, the physical science basis. Geneva: IPCC (Intergovernmental Panel on Climate Change). http://www.ipcc.ch/index.htm#.UvBRgf3d2N8

Irwin SE, Sarwar R, King LM, Simonovic SP (2012) Assessment of climatic vulnerability in the Upper Thames River Basin: downscaling with LARS WG. Water resources research report: The University of Western Ontario, Department of Civil and Environmental Engineering

James W, Robert C, Huang J (2002) PCSWMM: Integration of the stormwater management model with GIS, the web and a group decision support system. Presented at fifth International Eco-city conference: Shenzhen, China. August 19–23

Koutsoyiannis D (2003) Rainfall disaggregation methods: theory and applications. In: Workshop on statistical and mathematical methods for hydrological analysis. Rome, Italy

Koutsoyiannis D, Onof C (2001) Rainfall disaggregation using adjusting procedures on a Poisson cluster model. J Hydrol 246(1–4):109–122. doi:10.1016/S0022-1694(01)00363-8

Koutsoyiannis D, Kozonis D, Manetas A (1998) A mathematical framework for studying rainfall intensity-duration-frequency relationships. J Hydrol 206:118–135

Mailhot A, Duchesne S (2010) Design criteria of urban drainage infrastructures under climate change. J Water Resour Plann Manage Am Soc Civil Eng, pp 201–208

Mark O, Weesakul S, Apirumanekul C, Aroonnet SB, Djordjević S (2004) Potential and limitations of 1D modelling of urban flooding. J Hydrol 299(3–4):284–299. doi:10.1016/j.jhydrol.2004.08.014

Mehdi B, Mrena C, Douglas A (2006) Adapting to climate change: an introduction for Canadian Municipalities. Canadian climate impacts and adaptation research network (CCIARN). Natural Resources: Canada, Ottawa

Néelz S, Pender G (2010) Benchmarking of 2D hydraulic modelling packages. Heriot Watt University, Edinburgh

Nguyen VT (2005) Downscaling methods for evaluating the impacts of climate change and variability on hydrological regime at basin scale. Role of Water Sciences in Transboundary River Basin Management

Nguyen HQ (2009) Rainfall forecasting and a real time hydrologic information system for Bangkok, Thailand. Doctoral dissertation: Asian Institute of Technology, Bangkok

Roeckner E, Arpe K, Bengtsson L, Brinkop S, Dümenil L, Esch M, Kirk E, Lunkeit F, Ponater M, Rockel B, Sausen R, Schlese U, Schubert S, Windelband M (1992) Simulation of the present day climate with the echam model: impact of model physics and resolution, mpi report (1987–2004) 93. Max-Planck-Institut für Meteorologie, Hamburg

Roeckner E, Arpe K, Bengtsson L, Christoph M, Claussen M, Dümenil L, Esch M, Giorgetta M, Schlese U, Schulzweida U (1996) The atmospheric general circulation model echam-4: model description and simulation of present day climate, MPI Report (1987–2004) 218. Max-Planck-Institut für Meteorologie, Hamburg

Russo B, Pouget L, Sunyer D, Malgrat P, Theias H (2011) Flood risk assessment through a 2D-1D coupled modeling approach including impact of climate change in Barcelona. In the Proceedings of 12th international conference on urban drainage. Porto Alegre/Brazil September

Semenov MA, Stratonovitch P (2010) The use of multi-model ensembles from global climate models for impact assessments of climate change. Climate Research 41:1–14

Seyoum SD, Vojinovic Z, Price RK, Weesakul S (2012) Coupled 1D and non inertia 2d flood inundation model for simulation of urban flooding. J Hydraul Eng Am Soc Civil Eng. doi:10.1061/(ASCE)HY.1943-7900

Shrestha A (2013) Impact of climate change on urban flooding in Sukhumvit area of Bangkok. Master Thesis, Asian Institute of Technology, Bangkok

Shrestha A, Chaosakul T, Priyankara DPMP, Chuyen LH, Myat SS, Syne NK, Irvine KN, Koottatep T, Babel MS (2014) Application of PCSWMM to Explore possible climate change impacts on surface flooding in a Peri-urban area of Pathum Thani, Thailand. Journal of Water Management Modeling, ISSN 2292-6062: CHI 2014. doi:10.14796/JWMM.C377

Solaiman TA, Simonovic SP (2011) Development of probability based intensity duration-frequency curves under climate change. Water Resources Research Report: The University of Western Ontario, Canada

Sunyer MA, Madsen H, Ang PH (2012) A comparison of different regional climate models and statistical downscaling methods for extreme rainfall estimation under climate change. Atmos Res 103:119–128. doi:10.1016/j.atmosres.2011.06.011

Trenberth KE (1999) Conceptual framework for changes of extremes of the hydrological cycle with climate change. Clim Change 42:327–339

Vojinovic Z, Tutulic D (2009) On the use of 1D and coupled 1D-2D modelling approaches for assessment of flood damage in urban areas. Urban Water J 6(3):183–199. doi:10.1080/15730620802566877

Vojinovic Z, Seyoum S, Salum MH, Price RK, Fikri AK, Abebe Y (2012) Modelling floods in urban areas and representation of buildings with a method based on adjusted conveyance and storage characteristics. J Hydroinformatics 1:19. doi:10.2166/hydro.2012.181

Wilby R, Dawson C, Barrow E (2004) SDSM—a decision support tool for the assessment of regional climate change impacts. Environ Model Softw 17(2):145–157. doi:10.1016/S1364-8152(01)00060-3

Estimating the Impacts and Uncertainty of Climate Change on the Hydrology and Water Resources of the Koshi River Basin

Anshul Agarwal, Mukand S. Babel and Shreedhar Maskey

Abstract In this study, projections of temperature and precipitation in future periods and their impacts on hydrology and water resources of the Koshi River Basin in Nepal were investigated. The statistical downscaling model Long Ashton Research Station Weather Generator (LARS-WG) was used to downscale low-resolution data from ten general circulation models (GCMs) and three IPCC SRES scenarios (B1, A1B, and A2). The physically based hydrological model Soil and Water Assessment Tool (SWAT) was used to analyse the impacts of climate change on hydrology. LARS-WG simulated the baseline period (1981–2000) climate quite satisfactorily. Changes in climate and hydrological variables are presented at monthly and annual scales for three future periods: 2011–2030, 2046–2065, and 2080–2099. The results indicate that the Koshi basin tends to become warmer in the future as projected by all GCMs under three SRES scenarios. Changes in precipitation and streamflow are not univocal and vary depending on the GCM, GHGES. The difference in the projection of flow varies by as much as −35 to 51 % under the A1B scenario during the 2055s. The maximum increase in flow is projected during spring season with increase of 23 and 25 % during the 2055s and 2090s, respectively, under the A1B scenario. Similarly, the range of projections for all water balance components is very large. The water balance components: surface flow, baseflow, and water yield may decrease or increase in future periods, as GCMs do not agree on the direction of change. The potential ET and actual ET are projected to increase as projections from all GCMs and scenarios indicate, although a great deal of uncertainty exists in the magnitude of change.

A. Agarwal (✉)
Regional Integrated Multi-Hazard Early Warning System, PO Box 4, Klong Luang, Pathum Thani 12120, Thailand
e-mail: anshul.agarwal@ait.asia

A. Agarwal · M.S. Babel
Asian Institute of Technology, PO Box 4, Klong Luang, Pathum Thani 12120, Thailand
e-mail: msbabel@ait.asia

S. Maskey
UNESCO-IHE Institute for Water Education, PO Box 3015, 2601, DA Delft, The Netherlands
e-mail: s.maskey@unesco-ihe.org

S. Shrestha et al. (eds.), *Managing Water Resources under Climate Uncertainty*, Springer Water, DOI 10.1007/978-3-319-10467-6_6

The potential ET is projected to increase in the range of 6–24 % in the 2090s. There is high variability among the models and scenarios for projections, and the variability increases with future time periods.

Keywords Climate change · GCM · GHGES · LARS-WG · SWAT · Uncertainty

1 Introduction

The global climate projections in the Fourth Assessment Report of IPCC 2007 concluded that global mean atmospheric temperature is likely to increase between 1.8 and 4.0 °C by the end of this century. This projected change in temperature is likely to intensify the hydrological cycle and average mean water vapour, and precipitation is likely to increase. As a result, hydrological systems are anticipated to experience, not only changes in the average availability of water, but also changes in extremes (Zhang et al. 2011). Mountains being repositories of biodiversity, water and other ecosystem services are among the most fragile environments. The various global changes are creating enormous pressures on the mountains (Sharma et al. 2007). The assessment of future climate is based on different greenhouse gas (GHG) emission scenarios which are the product of very complex dynamic systems, determined by driving forces such as demographic growth, socio-economic development, and technological change (Anandhi et al. 2008). Global climate models (GCMs) are used to estimate the consequences of these developments on the climate in future periods. The outputs of GCMs are usually available at resolutions of 100's of km, and thus, they do not resolve sub-grid scale features and topographic effects that are of significance to many impact studies (Moriondo and Bindi 2006). Future climate projections on a much finer scale are required for impact studies at regional scales (Tisseuil et al. 2010).

To bridge the gap between scales of climate information provided by GCMs and those required for impact studies at regional or local scales, downscaling is used. Basically, two fundamental approaches exist for downscaling: dynamical downscaling (DD) and statistical downscaling (SD). In the dynamical approach, a higher resolution climate model (RCM) is embedded within a GCM. In the statistical approach, various statistical methods are used to establish empirical relationships between GCM output, climate variables, and local climate (Maraun et al. 2010). SD methods are easier and more readily applied to develop higher resolution climate scenarios for impact studies and thus more widely adopted (Chiew et al. 2010). The SD techniques can be grouped into three categories: weather typing method, stochastic weather generators, and regression methods.

The Long Ashton Research Station Weather Generator (LARS-WG) is a stochastic weather generator developed by Semenov and Barrow (1997) for statistical downscaling. Several studies (such as Hashmi et al. 2011) have compared the performance of LARS-WG with other statistical downscaling techniques and have concluded that LARS-WG can be adopted with confidence for climate change studies. LARS-WG has been applied in climate change impact studies in many research studies, such as in the Saguenay watershed in northern Quebec, Canada (Dibike and Coulibaly 2005); in Montreal, Canada (Nguyen 2005) and in different locations in Europe (Semenov and Stratonovitch 2010). The description of the latest version of LARS-WG, called LARS-WG 5 and its capabilities, is given in Semenov and Stratonovitch (2010). LARS-WG 5 incorporates climate projections from 15 GCMs used in the IPCC-AR4.

A large number of uncertainties exist at the various stages of future climate projections and hydrological analysis, which impose a challenge for impact analysis studies. Uncertainty in climate projections comes mainly from GCMs, SRES scenarios, downscaling methods, and the internal variability of climates (Hawkins and Sutton 2010). The models used for impact analysis (e.g. hydrological models) also bring uncertainty to the assessment of the impacts of climate change (Bastola et al. 2011; Minville et al. 2008). An estimate of the uncertainty in climate projections is potentially valuable for policy makers and planners (Stott and Kettleborough 2002).

This study was conducted in the Koshi River Basin. The Koshi flows through China, Nepal, and India and is one of the largest tributaries of the Ganges. The Koshi River Basin consists of seven major sub-basins (Sun Koshi, Indrawati, Dudh Koshi, Tama Koshi, Bhote Koshi, Arun, and Tamor) all originating from the Himalayas. The Sun Koshi joins the Indrawati and then moves on south-eastwards to collect the following rivers: Tama Koshi, Bhote Koshi, Dudh Koshi and join with Arun and Tamor at Tribeni. About 82 % of the Arun catchment lies in the Tibet Autonomous Region of China. The Sun Koshi and Tama Koshi also have remarkably large catchment areas in the Tibet Autonomous Region of China (WWF 2009). The Koshi basin, characterised as highly varied in climate and geographical features, spans latitudes between 26° 51′N and 29° 79′N and longitudes between 85° 24′E and 88° 57′E. The elevation of the basin ranges from about 65 m in the Terai to over 8,000 m in the high Himalayas (Fig. 1). A large part of the Koshi basin (almost 65 %) is above 4,000 m in elevation.

Many studies confirm that major parts of the Koshi and other river basins in Nepal are undergoing warming trends as well as changes in the precipitation pattern (Agrawala et al. 2003; Bartlett et al. 2010; Chhetri 2010; Shrestha and Devkota 2010). Despite progress in understanding the impact of climate change on water resources, there has been a lack of research in investigating the element of uncertainty. To address the issue, this study analysed future climates using projections from multi-models and multi-scenarios. Projections from different GCMs for various scenarios were then used to investigate the climate change impacts on water resources. This analysis will help in managing water more efficiently and making the necessary plans for adaptation to changing climatic conditions in the Koshi basin.

In this study, the climate projections for the Koshi River Basin were investigated using data from multiple GCMs for three emission scenarios. The impact of climate change on the hydrology of the Koshi basin was analysed using the Soil and Water Assessment Tool (SWAT) model. The uncertainty of GCMs and SRES scenarios in

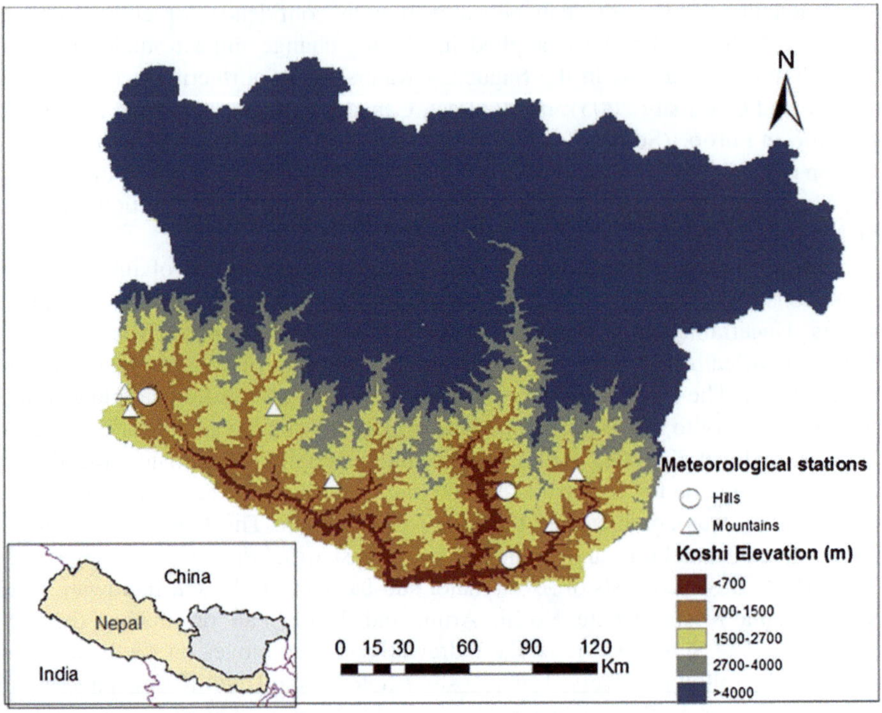

Fig. 1 Elevation range and meteorological stations in the Koshi basin

future climate projections and the hydrology of the Koshi basin were then estimated. The uncertainty resulting from downscaling methods and hydrological model parameters were not analysed in this study.

2 Materials and Methods

The observed daily climate data for the period 1971–2008 were obtained from the Department of Hydrology and Meteorology (DHM), Nepal. The future climate data were downloaded from the IPCC-data distribution centre website.[1] Data from ten GCMs included in IPCC AR4 were considered in this study. The data for these GCMs for selected SRES scenarios are available through LARS-WG5. The GCMs used in the study are listed in Table 1.

Three emission scenarios (B1, A1B, and A2), which represent low, medium, and high emissions of GHG with respect to the prescribed concentrations relative to SRES, were considered in this study. A stochastic weather generator LARS-WG

[1] http://www.ipcc-data.org/.

Table 1 Global climate models used in the study

No.	Model	Research centre	Resolution		Scenarios	Vintage
			Atmospheric	Ocean		
1	ECHAM 5	MPI, Germany	$1.87^0 \times 1.87^0$	$1.5^0 \times 1.5^0$	B1, A1B, A2	2005
2	MRI-CGCM2.3	MRI, Japan	$2.8^0 \times 2.8^0$	$2.5^0 \times 2.0^0$	B1, A1B	2003
3	HadCM3	Hadley Centre UK	$2.5^0 \times 3.75^0$	$1.25^0 \times 1.25^0$	B1, A1B, A2	2000
4	CGCM 3.1	CCCMA Canada	$2.8^0 \times 2.8^0$	$1.4^0 \times 1.0^0$	A1B	2005
5	MK3	CSIRO, Australia	$1.9^0 \times 1.9^0$	$1.9^0 \times 1.9^0$	B1, A1B	2001
6	CNCM3	CNRM, France	$1.9^0 \times 1.9^0$	$2.0^0 \times 2.0^0$	A1B, A2	2004
7	IPCM4	IPSL, France	$2.5^0 \times 3.75^0$	$2.0^0 \times 2.0^0$	B1, A1B, A2	2005
8	GFCM21	GFDL, USA	$2.0^0 \times 2.5^0$	$1.0^0 \times 1.0^0$	B1, A1B, A2	2005
9	CCSM3	NCAR, USA	$1.4^0 \times 1.4^0$	$1.0^0 \times 1.0^0$	B1, A1B, A2	2005
10	INCM3	INM, Russia	$4.0^0 \times 5.0^0$	$2.5^0 \times 2.0^0$	B1, A1B, A2	2004

(Semenov and Barrow 1997) was used in this study. A description of the latest version of model LARS-WG 5 and its capabilities is given in, Semenov and Stratonovitch (2010). The model was downloaded from Rothamsted Research.[2] LARS-WG can generate synthetic daily datasets of precipitation, minimum and maximum temperature, and solar radiation based on observed weather, and generally 20 or 30 years of daily climate data are used in order to capture real climate variability and seasonality. Based on the relative monthly changes in mean daily precipitation, wet and dry series duration, temperature and temperature variability between current and future periods predicted by the GCM, and local station climate variables are adjusted proportionately to represent climate change.

The data required to develop the SWAT model for the Koshi basin were compiled using global data sources. A Digital Elevation Model (DEM) of 90 m resolution was obtained from the Shuttle Radar Topography Mission (SRTM). The soil map used was from the Soil Terrain Database (SOTER), which shows that the Koshi basin has 11 soil types, with Gleyic Leptosols (soil texture clay loam) and Gleyic Phaeozems (soil texture sand) as the dominant soil types. Land cover data were obtained from MODIS land cover type products MCD12Q1 available at the spatial resolution of 500 m for the period 2001–2010. The weather data used for developing the SWAT model is: daily precipitation, minimum and maximum temperature, relative humidity, solar radiation, and wind speed. A large part of the Koshi basin lies in an elevation above 3,000 m where observation stations are not available. To overcome this deficit, a temperature lapse rate of −5.7 °C/km was

[2] http://www.rothamsted.ac.uk/mas-models/larswg/download.php.

incorporated and precipitation data was used from Asian Precipitation Highly Resolved Observational Data Integration Towards Evaluation of Water Resources (APHRODITE) which contains gridded data and is available at a spatial resolution of 25 km.

The SWAT model was developed for the Koshi basin using the data mentioned above. The model performance to simulate the flow was evaluated for the Koshi River and three major tributaries, Sun koshi, Arun, and Tamur. The hydrological data from the DHM, Nepal was available for the years from 1990 onwards. Data for the period 1990–2000 were considered for model calibration and 2001–2008 were considered for model validation. Three statistical parameters: coefficient of determination (R^2), percentage volume error (VE), and Nash–Sutcliffe efficiency (NSE) were used in this study to analyse model performance during calibration and validation.

The calibrated and validated SWAT model was used to analyse the impacts of future climate change on hydrology and water resources in the Koshi basin. The minimum and maximum temperature and precipitation was downscaled using LARS-WG for 10 temperature and 60 precipitation stations located in the Koshi basin were used to evaluate the hydrological parameters and water balance components in the future periods. The solar radiation and relative humidity for the future periods were simulated in SWAT using the WXGEN weather generator. WXGEN uses rainfall and temperature data from each scenario based on the assumption that the occurrence of rain on a given day has a major impact on the relative humidity and solar radiation on that day (Ficklin et al. 2009). The analysis in this study focuses on three periods over the twenty-first century; an early-century period 2011–2030 (2020s), a mid-century period 2046–2065 (2055s), and a late-century period 2080–2099 (2090s). The period 1981–2000 is considered as the baseline period.

3 Results and Discussion

3.1 Model Calibration and Validation

3.1.1 LARS-WG

The performance of the LARS-WG model was tested against the historical data (daily T_{min} and T_{max}) for the baseline period (1971–2000). The mean monthly bias values for T_{min}, T_{max}, and precipitation for ten meteorological stations located in the Koshi basin (Fig. 1) are presented in Fig. 2. The LARS-WG performance was evaluated for all 60 precipitation stations used in this study although results are presented here for only ten stations. The bias values for T_{min} and T_{max} were mostly close to zero, with a narrow range of ±0.5 °C. The mean monthly bias for precipitation was also found to be satisfactory for all the stations. The higher bias values were obtained during the months of the monsoon season (JJAS) compared to the remaining months. The mean annual bias was close to zero, which may be because of the positive bias in certain months being balanced out by the negative

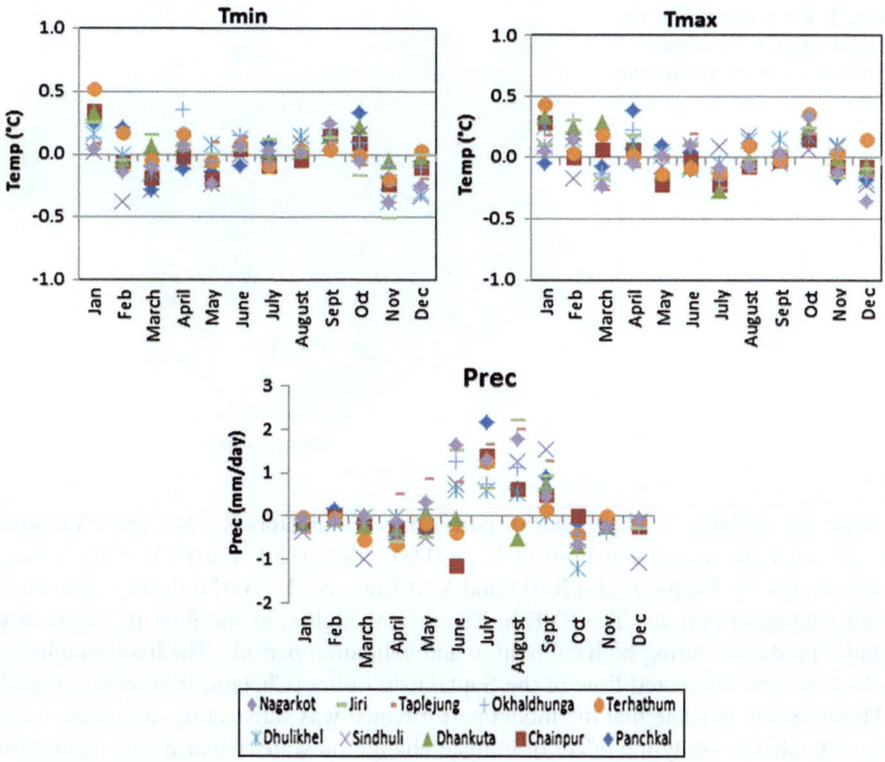

Fig. 2 Bias in mean monthly T_{min} (°C), T_{max} (°C), and precipitation between the observed and LARG-WG simulated data for the baseline period (1971–2000) for ten stations in the Koshi basin. The bias is calculated as observed minus simulated

bias. The results indicate that LARS-WG simulated temperature and precipitation agreed well with the observed values, and thus, the model can be used for downscaling future climate data in this study.

3.1.2 SWAT

The SWAT model was calibrated using streamflow measured at the Saptakoshi outlet (Chatara) and its major tributaries (Arun, Tamor, and Sun Koshi) (Fig. 3). The available observed flow data were split for calibration (1990–2000) and validation (2001–2008) purposes. The most sensitive parameters were chosen in the calibration procedure based on literature review and a preliminary sensitivity analysis of the parameters. The sensitive parameters for the Koshi basin are presented in Table 2. The value of the sensitive parameters was adjusted within the appropriate ranges as defined in SWAT documentation to obtain the best calibration for the model. The calibration and validation results for the SWAT model are

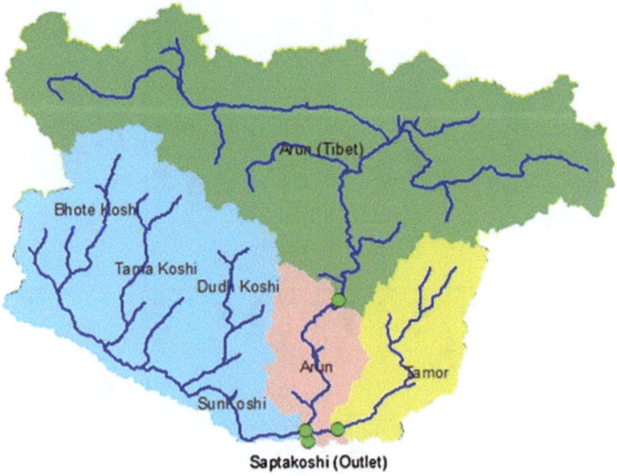

Fig. 3 River network of the Koshi basin and calibration points considered in this study

presented in Table 3. The values of performance indicators R^2, NS, and PVE were well under the acceptable limit of $R^2 > 0.60$, NS > 0.50, and PVE < 15 % (recommended by Santhi et al. (2001) and Van Liew et al. (2007)) during calibration and validation periods. The PVE in Tamur was higher, as the flow in Tamur was under predicted during both calibration and validation periods. The hydrographs for observed and simulated flow of the Saptakoshi outlet (Chatara) is shown in Fig. 4. These results indicate that the model performance was satisfactory, and thus, it can be extended to study the effect of climate change on water balance and streamflow of the Koshi basin. The model parameters calibrated for the past periods are assumed to remain valid for the future period simulations. The land use in the Koshi basin is also assumed to remain unchanged in the future periods for this study.

Table 2 Parameters considered for the SWAT model calibration

Parameters	Description	Default value	Range
CN2	Curve number	38–91	35–98
ESCO	Soil evaporation compensation factor	0	0–1
SOL_AWC	Available water capacity of the soil layer (mm H_2O/mm soil)	0.21	0–1
RCHRG_DP	Deep aquifer percolation fraction	0.05	0–1
GW_REVAP	Groundwater "revap" coefficient	0.02	0.02–0.20
GW_Delay	Groundwater delay from soil to channel (days)	31	0–500
GWQMN	Threshold depth of water in the shallow aquifer required for return flow to occur (mm H_2O)	0	0–500
Alpha_bf	Baseflow alpha factor (days)	0.048	0–1
CH_K2	Channel effective hydraulic conductivity (mm/h)	0	0–500
SOL_K	Saturated hydraulic conductivity (mm/h)	51.6	0–2,000

Table 3 Calibration and validation statistics for the Koshi River Basin and its tributaries

	Calibration (1990–2000)			Validation (2001–2008)		
	R2	NS	PVE (%)	R2	NS	PVE (%)
Sun Koshi	0.84	0.82	3.3	0.85	0.83	−2
Arun	0.60	0.58	−12	0.61	0.51	0
Tamur	0.84	0.57	−39	0.88	0.66	−36.4
Saptakoshi	0.87	0.85	−12.3	0.84	0.83	−10.7

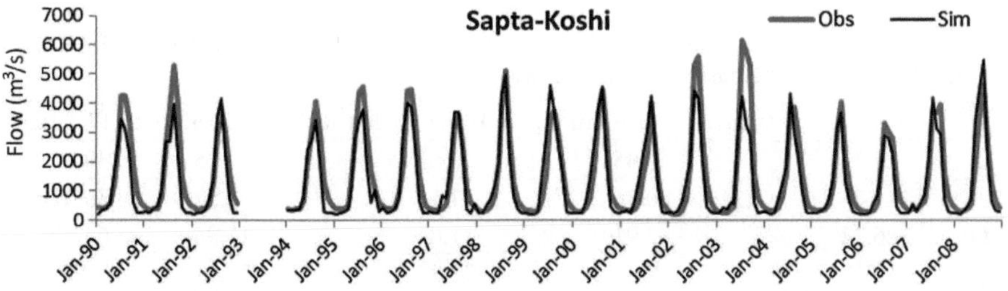

Fig. 4 Observed and simulated monthly streamflow at the Saptakoshi outlet during calibration and validation periods

3.2 Baseline Period Simulations

The performance of the SWAT model was analysed using the baseline period observed and simulated data. This analysis helps to check the consistency of weather data simulated using LARS-WG for its application to simulate the hydrological parameters of the Koshi basin. The annual water balance components of the Koshi basin using observed and simulated weather data are presented in Table 4. The annual surface runoff and baseflow were simulated with a difference of 0.2 and 1.5 %, respectively. The total water yield was simulated with a difference of less than 1 %. The difference between observed and simulated values of actual ET and potential ET is also less than 1 %. This indicates that the baseline period weather data simulated from LARS-WG performs well for the Koshi basin. This increases the confidence of using LARS-WG simulated climate data to analyse impacts of climate change on the water resources of the Koshi basin. The annual water balance of the Koshi basin suggests that the total water yield accounts for nearly 67 % of the total precipitation. The evapotranspiration and deep percolation represent 23 and 10 % of the annual precipitation. The monthly water balance of the Koshi basin presented in Table 5 indicates that the majority of precipitation (75 %), surface runoff (65 %), and water yield (76 %) are concentrated during the four months of the monsoon season (JJAS). Minimum flow is observed during the winter season (DJF), which is less than 5 % of the total annual flow.

Table 4 Annual average water balance components of the Koshi basin simulated using observed and LARS-WG simulate climate for the baseline period (1981–2000)

	Rain	Surf Q	Baseflow	WY	ET	PET
Simulated using observed climate	1,033.9	244.7	431.6	676.3	274.8	747.1
Simulated using LARS-WG simulated climate	1,041.9	243.1	425.5	668.6	276.7	754.2
Difference (%)	−0.8	0.7	1.4	1.1	−0.7	−1.0

Table 5 Monthly water balance components of the Koshi basin simulated using LARS-WG simulated climate for the baseline period (1981–2000)

	Precipitation (mm)	Surf Q (mm)	Water Yield (mm)	ET (mm)	PET (mm)
Jan	12.2	0.45	9.62	4.25	30.7
Feb	15.74	0.36	7.11	5.57	38.01
Mar	26.72	2.91	11.64	11.95	64.33
Apr	46.29	7.47	22.82	21.3	80.7
May	79.33	15.23	44.9	32.79	91.08
Jun	153.28	33.33	91.38	39.64	85.18
Jul	248.33	66.05	156.26	46.66	83.36
Aug	236.67	66.91	154.62	44.46	79.42
Sept	148.48	38.26	100.15	33.72	67.74
Oct	51.73	10.14	39.87	21.06	58.86
Nov	13.12	1.43	17.4	10.19	42.31
Dec	9.98	0.55	12.79	5.14	32.54

3.3 Climate Projections for Future Periods

The climate projections for three future periods (2020s, 2055s, and 2090s) relative to the baseline period (1981–2000) are presented in this section. The downscaled temperature and precipitation projections of 25 ensembles were used to analyse the range of projections for three future periods. The box-whisker plots are used to represent the uncertainty arising in projections from GCMs and the scenarios. The upper and lower boundaries of the boxes represent the 25th and 75th percentiles respectively, while the line in the box shows the median values. The ticks outside the boxes show the maximum and minimum value of the projected changes. The results are presented and discussed for changes in mean monthly values of climate variables (T_{min}, T_{max}, and precipitation) projected by ten GCMs under the A1B scenario, and in annual values projected by all GCMs under the three scenarios.

3.3.1 Temperature

The change in mean monthly temperature projected by ten GCMs under the A1B scenario, relative to the baseline period, is shown in Fig. 5. The change in annual temperature, as projected by the GCMs under the three scenarios: B1, A1B, and A2 for three future periods (relative to the baseline period), is shown in Fig. 6. Projections of all the GCMs under the A1B scenario (Fig. 5) indicate an increase in both $T_{\min}$ and $T_{\max}$ in each month for all three future periods. While the GCMs do not agree on the magnitude of change, they do project that the range of temperature change within each month will increase with the time horizon. The projections of annual $T_{\min}$ and $T_{\max}$ during the 2020s, as shown in Fig. 6, are within the narrow range of 0.5–1.5 °C with almost similar median values. This indicates that $T_{\min}$ and $T_{\max}$ projections during the 2020s are expected to be similar, irrespective of the scenario that may follow. The differences between projected values of $T_{\min}$ and $T_{\max}$ become greater in accordance with the choice of the scenario during the mid-century and the late-century periods. This heightened difference is because of the significant increase in differences among the different emission scenarios themselves. The differences among the median values of projections justify this finding: during the 2090s for $T_{\max}$, the median value as projected under the A2 scenario is 1.6 °C higher than as projected under the B1 scenario. This may influence runoff in the basin and monthly water availability especially during the dry season.

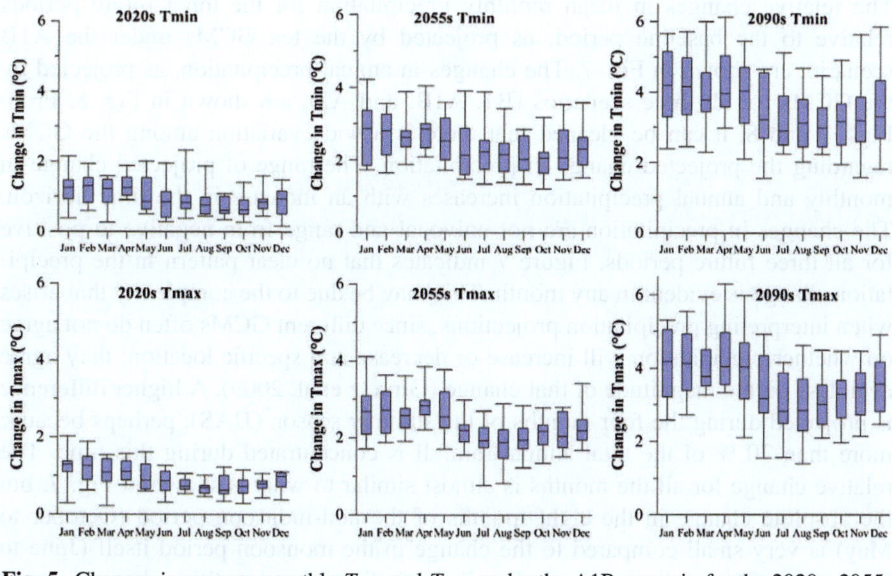

Fig. 5 Changes in average monthly $T_{\min}$ and $T_{\max}$ under the A1B scenario for the 2020s, 2055s, and 2090s relative to the baseline period

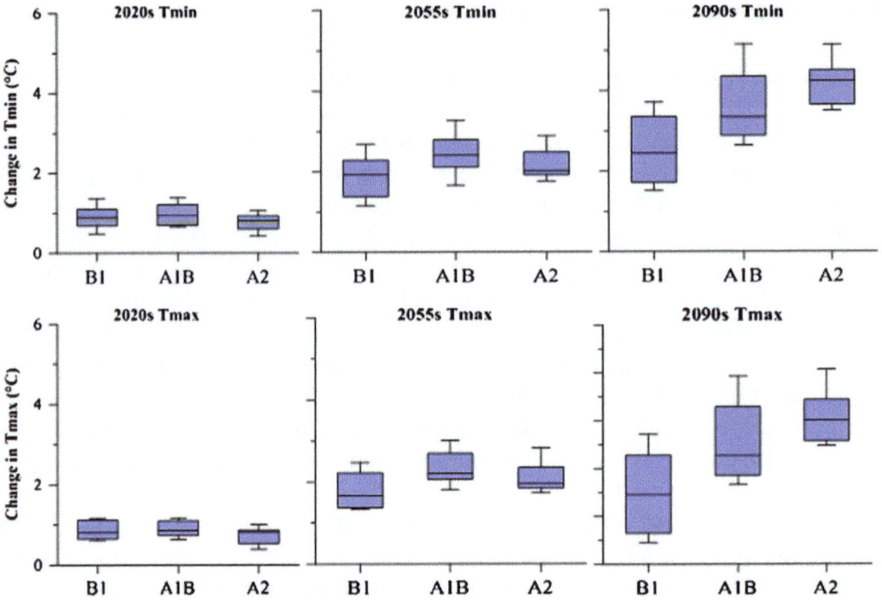

Fig. 6 Changes in annual average $T_{\min}$ and $T_{\max}$ for the 2020s, 2055s, and 2090s relative to the baseline period

3.3.2 Precipitation

The relative changes in mean monthly precipitation for the three future periods relative to the baseline period, as projected by the ten GCMs under the A1B scenario, are shown in Fig. 7. The changes in annual precipitation, as projected by the GCMs for all three scenarios (B1, A1B, and A2), are shown in Fig. 8. From Figs. 7 and 8, it can be gleaned that there is a wide variation among the GCMs regarding the projected change in precipitation. The range of projected change in monthly and annual precipitation increases with an increase in the time horizon. The changes in precipitation are not univocal and range from negative to positive for all three future periods. Figure 7 indicates that no clear pattern in the precipitation change is evident in any month. This may be due to the complexity that arises when interpreting precipitation projections, since different GCMs often do not agree on whether precipitation will increase or decrease at a specific location; they agree even less on the magnitude of that change (Girvetz et al. 2009). A higher difference is projected during the four months of the summer season (JJAS), perhaps because more than 70 % of the total annual rainfall is concentrated during this time. The relative change for all the months is almost similar to what is shown in Fig. 7, but the absolute change in the eight months of the non-monsoon period (October to May) is very small compared to the change in the monsoon period itself (June to September). The median values in Fig. 7 also indicate that a positive change is more

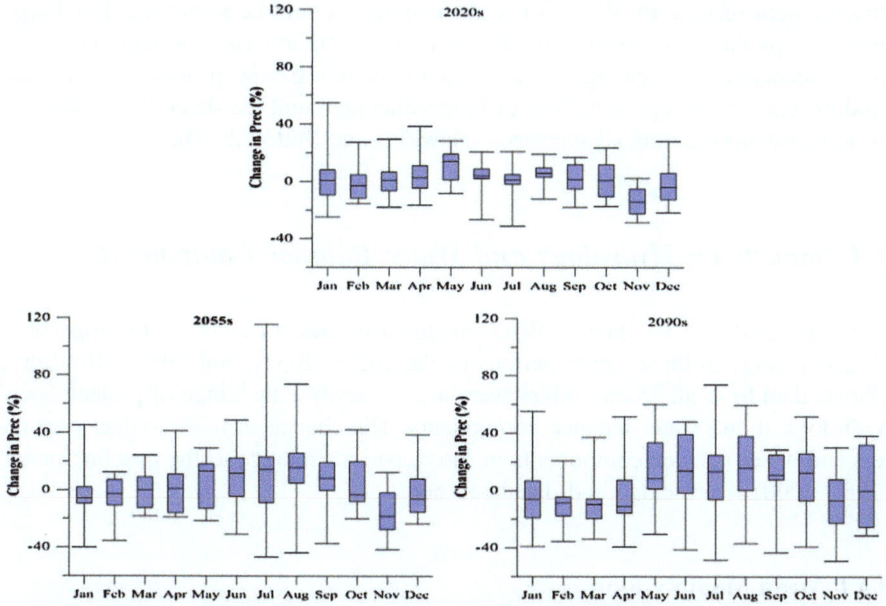

Fig. 7 Changes in average monthly precipitation under the A1B scenario for the 2020s, 2055s, and 2090s relative to the baseline period

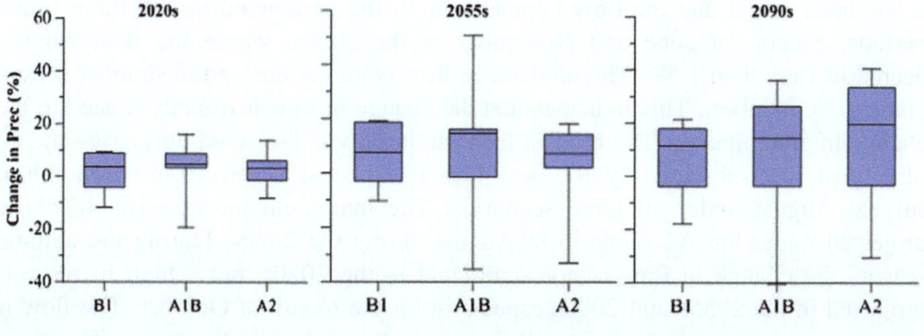

Fig. 8 Changes in annual precipitation for the 2020s, 2055s, and 2090s relative to the baseline period

likely in the summer season while a negative change is more likely in the winter season (DJF). The changes in annual precipitation, as projected under the three scenarios (Fig. 8), also indicate that the difference in projections increase with the time horizon. The median value indicates a positive change in annual precipitation under all three scenarios. Also, the median value for the three scenarios is closer to each other during the early-century period but shows significant differences during the mid- and late-century periods. It is important to note here that the scenario uncertainty is estimated based on three IPCC SRES scenarios, and with the development of new RCP scenarios, the estimates could be improved. For longer lead-time predictions, assessments of the relative importance of model and uncertainty scenarios may change as a result of including new processes in climate models (e.g. better representation of biogeochemical and ice-sheet feedbacks) and improved understanding of scenarios (Hawkins and Sutton 2010).

3.4 Impacts on Hydrology and Water Balance Components

The calibrated and validated SWAT model was used to analyse the impacts of climate change in three future periods of the 2020s, 2055s, and 2090s. The future climate data from all 25 ensembles were used to analyse the range of projections of hydrological and water balance components. The change in hydrological parameters and water balance components in future periods relative to the baseline period (1981–2000) is presented and discussed here.

3.4.1 Mean Monthly Flow

The flow during the baseline period and the change (in percentage) in the flow during the three future periods are presented in Table 6. The values in Table 6 represent the mean value of change from all GCMs under the respective scenarios. It has been found that the flow increases in all the months during all three future periods, except for June and November in the 2020s, where the flow slightly decreases (less than 1 %). The increase in flow is higher during the summer season (June to September). This indicates that the change in flow is directly related to the change in precipitation. The highest increase in flow is projected during the month of August, and subsequently, the peak flow is expected to shift from the month of July to August under all three scenarios. The maximum increase (of 48 %) is projected under the A2 scenario in August during the 2090s. During the autumn season, the change in flow is not significant in the 2020s, but a high increase is projected in the 2055s and 2090s especially for the month of October. The flow is also projected to increase during all three months of the winter season. The maximum increase of flow (9 %) is projected during the winter season months under the A2 scenario during the 2090s. The spring season flow shows a significant increase during all three future periods. This increase might be due to the increase in temperature, which in turn may cause early snow melt.

3.4.2 Changes in Seasonal and Annual Flow

The changes in seasonal and annual flows, as projected by ten GCMs under the A1B scenario, for all three future periods (with respect to the baseline period's seasonal and annual flow), are shown in Fig. 9. The solid blue bar in this figure

Table 6 Mean monthly flow at Chatara during the baseline period and changes in the flow (in percentage) during the three future periods

	Baseline (m^3/s)	2020s			2055s			2090s		
		B1	A1B	A2	B1	A1B	A2	B1	A1B	A2
		% change with respect to baseline								
Jan	380.6	3.3	3.3	4.9	6.2	6.0	4.3	5.1	7.5	8.9
Feb	340.5	3.3	3.6	5.1	4.7	5.6	3.8	6.9	6.7	8.7
Mar	364.5	1.8	4.1	3.5	4.6	8.0	4.4	5.5	8.3	12.2
Apr	524.1	5.9	12.3	7.7	8.5	14.2	11.1	12.6	14.1	16.8
May	928.4	8.3	14.4	3.1	9.3	12.4	13.0	9.9	20.1	12.1
Jun	2,002.2	2.9	6.9	−0.9	6.8	17.4	6.7	7.4	20.9	14.4
Jul	3,286.8	4.2	5.1	8.4	12.3	26.3	6.6	16.5	25.3	34.9
Aug	3,249.3	9.8	10.5	13.6	25.6	27.7	12.0	25.6	29.7	48.4
Sep	2,307.6	0.7	4.8	3.0	12.0	17.1	3.9	11.8	20.9	33.0
Oct	1,001.8	1.7	4.0	1.5	10.3	12.2	4.4	11.2	19.6	27.4
Nov	546.2	0.3	−0.9	−1.0	4.2	3.2	0.9	2.9	6.6	11.5
Dec	441.4	3.2	1.5	3.3	6.2	5.3	5.0	4.1	6.7	9.0

represents the mean of the ten GCMs under the A1B scenario. The results indicate that there are likely chances of an increase in seasonal as well as annual runoff in all three future periods as the majority of GCMs and their mean indicates. Among the ten GCMs considered here, only two CSMK and IPCM indicate a decrease in flow. The maximum relative increase (of 12 %) is projected for spring during the 2020s, while increases of 23 and 25 % are projected, respectively, for the summers of the 2055s and 2090s. The range of change in seasonal and annual flow (not shown here) varies from negative to positive under all three scenarios. The uncertainty in projection increases with the increase in time periods with much higher differences during the 2090s compared to the 2020s.

3.4.3 Water Yield

The changes in monthly and annual water yield for the three future periods relative to the baseline period are shown in Figs. 10 and 11. Both monthly and annual water yield during the early-century period (2020s) are expected to change by only a small amount as median values indicate. The middle 50 % range (box in Figs. 10 and 11) indicates that the differences in the GCMs' projections are within the narrow range, with a maximum difference projected for April during the 2020s. The difference in projections of water yield increases with the time horizons. The median value indicates a positive change during the 2055s and 2090s with the exception of April and May in the 2090s. A smaller difference between the median values and narrow range of projections in Fig. 11 also indicates that the relative annual change is dominantly affected by GCMs rather than by SRES scenarios

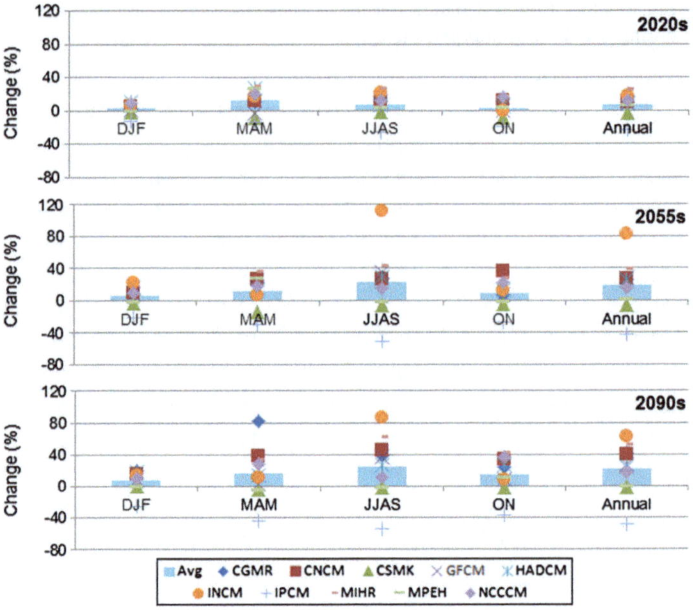

Fig. 9 Projected changes in seasonal and annual runoff at the Saptakoshi outlet according to ten GCMs under the A1B scenario for three future periods relative to the baseline period. The blue solid bar represents the mean of ten GCMs

during the 2020s. During the 2055s, both scenarios and GCMs are affected by the change in water yield. The median value indicates an increase during the 2055s under all three scenarios with the highest increase under the A1B scenario. The projections for precipitation as shown in Figs. 7 and 8 also indicate a higher chance of increase during the 2055s. During the 2090s, both GCMs and scenarios show a dominant effect on the water yield in the basin, with a more likely chance of increase.

3.4.4 Potential ET

The relative changes in potential ET for three future periods with respect to the baseline period are shown in Fig. 12. The PET value is projected to increase during all three future periods according to all the GCMs, with the exception of INCM during the monsoon season. INCM indicates a decrease in PET which might be because of its temperature projections which show the least increase during the monsoon season among all the GCMs under the A1B scenario. Temperature increase, as projected by all the GCMs, is the main factor causing an increase in PET. The average monthly PET varies significantly with the change in temperature. Results indicate that the increase in PET is highest in the months of the winter season (DJF) which are also expected to show a maximum increase in temperature,

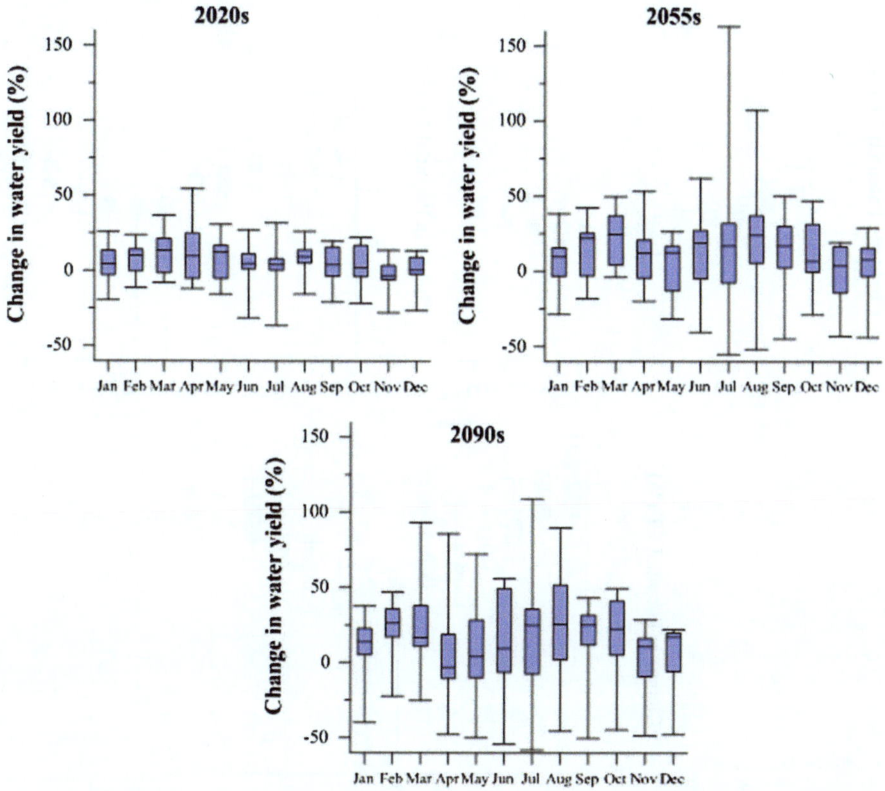

Fig. 10 Changes in average monthly water yield under the A1B scenario for the 2020s, 2055s, and 2090s relative to the baseline period

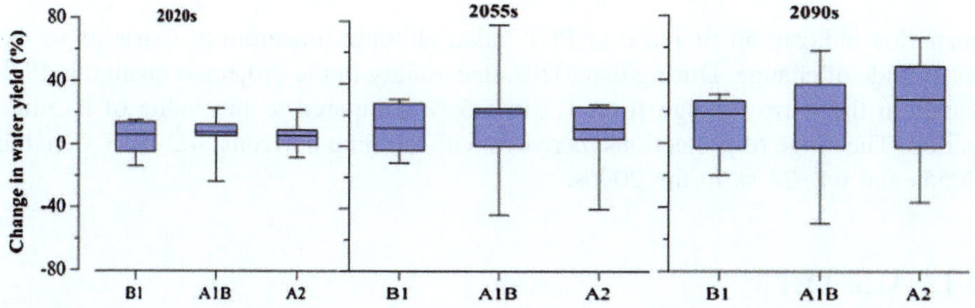

Fig. 11 Changes in annual average water yield for the 2020s, 2055s, and 2090s relative to the baseline period

while the minimum increase is projected for the summer season (JJAS) during all three future periods. The relative change in average annual PET (Fig. 13) also indicates an increase during all three future periods. All GCMs under three

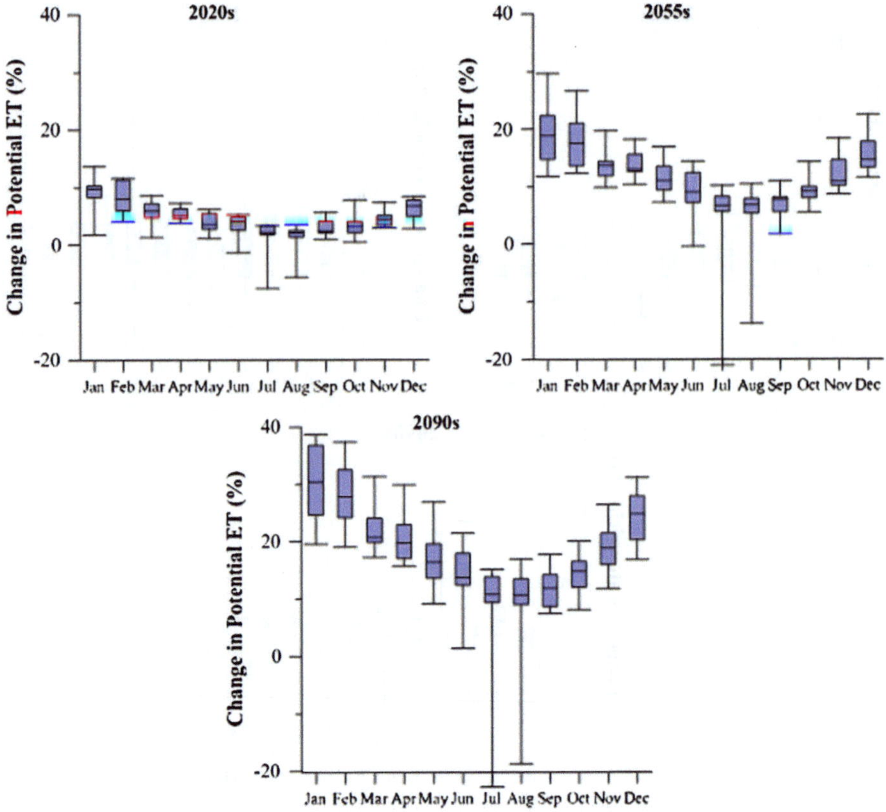

Fig. 12 Changes in average monthly potential ET under the A1B scenario for the 2020s, 2055s, and 2090s relative to the baseline period

scenarios indicate an increase in PET value although uncertainty exists as to the magnitude of change. During the 2020s, uncertainty in the projected change in PET varies in the narrow range from 1.7 to 5.5 % compared to the value of baseline period. The range of projections increases with the time horizons: 4.2–13.7 % in the 2055s and 6.8–24 % in the 2090s.

3.4.5 Actual ET

The relative change in mean monthly and annual value of actual ET is more likely to increase during all the three future periods, as middle 50 % values indicate in Figs. 14 and 15. Two GCMs (INCM and IPCM) indicate a decrease in ET values in certain months, especially during the summer season. An analysis of the change in temperature and precipitation shows INCM's projected minimum increase in temperature and IPCM's projected maximum decrease in precipitation among the ten GCMs

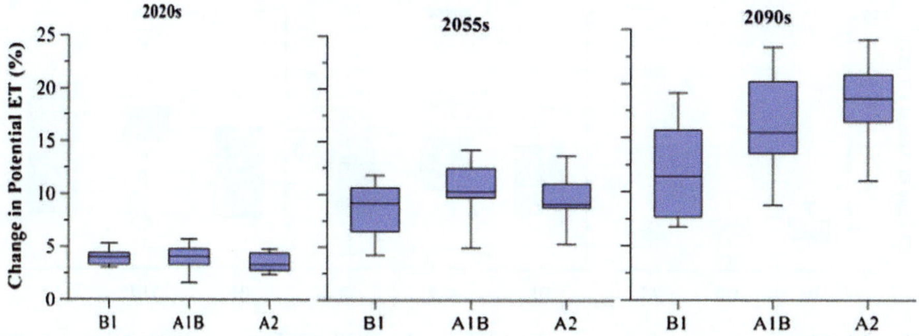

Fig. 13 Changes in annual average potential ET for the 2020s, 2055s, and 2090s relative to the baseline period

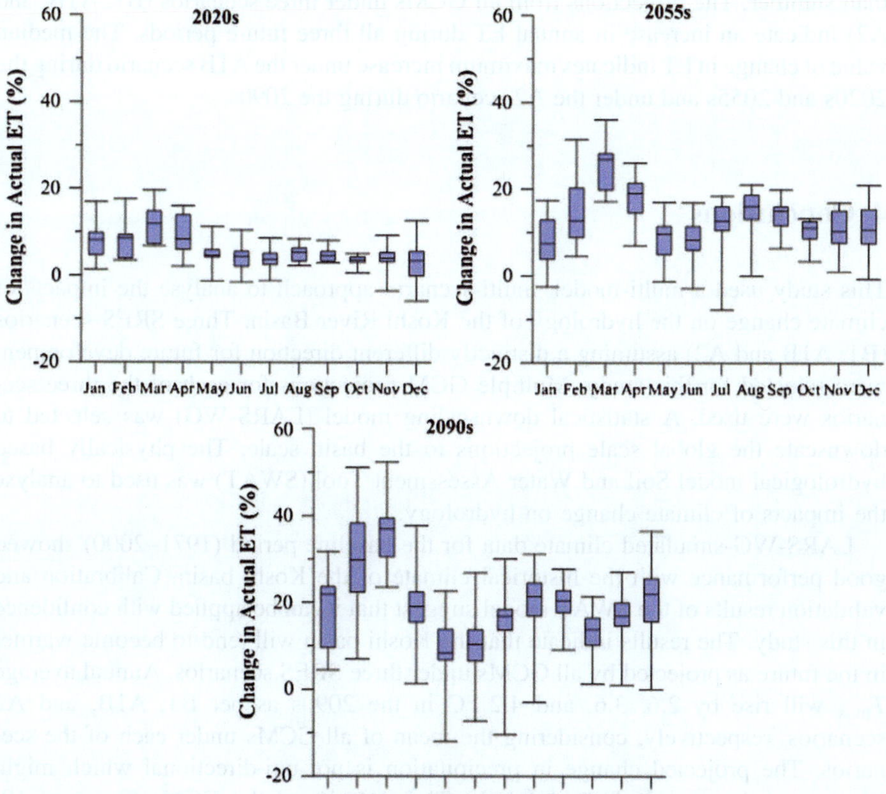

Fig. 14 Changes in average monthly actual ET under the A1B scenario for the 2020s, 2055s, and 2090s relative to the baseline period

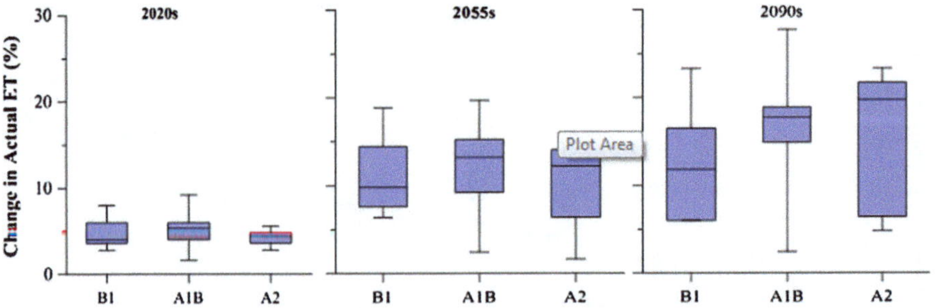

Fig. 15 Changes in annual average actual ET for the 2020s, 2055s, and 2090s relative to the baseline period

under the A1B scenario. The highest relative change in ET is projected during the months of February, March, and April. The increase in ET is also higher during winter than summer. The projections from all GCMs under three scenarios (B1, A1B, and A2) indicate an increase in annual ET during all three future periods. The median value of change in ET indicates maximum increase under the A1B scenario during the 2020s and 2055s and under the A2 scenario during the 2090s.

4 Conclusions

This study used a multi-model, multi-scenario approach to analyse the impacts of climate change on the hydrology of the Koshi River Basin. Three SRES scenarios (B1, A1B and A2) assuming a distinctly different direction for future development were selected for this study. Multiple GCM projections for each of the three scenarios were used. A statistical downscaling model (LARS-WG) was selected to downscale the global scale projections to the basin scale. The physically based hydrological model Soil and Water Assessment Tool (SWAT) was used to analyse the impacts of climate change on hydrology.

LARS-WG-simulated climate data for the baseline period (1971–2000) showed good performance with the historical climate of the Koshi basin. Calibration and validation results of the SWAT model suggest that it can be applied with confidence in this study. The results indicate that the Koshi basin will tend to become warmer in the future as projected by all GCMs under three SRES scenarios. Annual average T_{max} will rise by 2.6, 3.6, and 4.2 °C in the 2090s as per B1, A1B, and A2 scenarios, respectively, considering the mean of all GCMs under each of the scenarios. The projected change in precipitation is not uni-directional which might increase or decrease in future periods. The majority of the GCMs (8 out of 10) considered here and the mean value of projections from all GCMs under each of the three scenarios indicated an increase in future precipitation. The maximum difference in precipitation projections is simulated under the A1B scenario with a range varying from 38 to 50 % in the 2055s. A large uncertainty range exists in the projections of flow with results indicating both decrease and increase for all three future periods. The mean monthly and annual flow is expected to increase as projections from the majority of GCMs and the mean value of all GCMs indicates. The maximum increase in flow is projected during the spring season with increases of 23 and 25 % during the 2055s and 2090s, respectively, under the A1B scenario. This high increase in spring flow indicates the need for more flood control measures in the Koshi basin. The water balance components: surface flow, baseflow, and water yield may decrease or increase in future periods, as GCMs do not agree on the direction of change. The potential ET and actual ET are projected to increase as projections from all GCMs and scenarios indicate, although major uncertainty exists as to the magnitude of change. The potential ET is projected to increase in the range of 6–24 % in the 2090s.

There is high variability among the models and scenarios for projections, and the variability increases with future time periods. Although inter-model variability exists in each of the scenarios, for the early-century period, the differences among the scenarios are much less. During the mid- and late-century periods, inter-model variability in all climatic and hydrological variables increases, resulting in higher projection uncertainty. The multi-model and multi-scenario approach presented in this paper helps in understanding the uncertainty linked to future climate projections and the impact on water resources. Although different downscaling techniques and hydrological model parameters also bring uncertainty in projections, it is expected that the large variability induced by different GCMs and GHG emission scenarios, as shown in this paper, will dwarf those induced by other sources. The impacts of climate change on the hydrology and water resources of the Koshi basin will further affect all water use sectors which may be analysed considering the findings from this study.

Acknowledgment This study was funded in part through the UNESCO-IHE Partnership Research Fund (UPaRF) under the AGloCAP (Adaptation to Global Changes in Agricultural Practices) project and in part through the Asian Institute of Technology (AIT) research fellowship. The authors would like to thank the Department of Hydrology and Meteorology (DHM), Nepal for the meteorological data used in this study, which was made available for the AGloCAP project.

References

Agrawala S, Raksakulthai V, Aalst MV, Larsen P, Smith J, Reynolds J (2003) Development and climate change in nepal: focus on water resources and hydropower. Organisation for Economic Co-operation and Development (OECD). http://www.oecd.org/dataoecd/6/51/19742202.pdf

Anandhi A, Srinivas VV, Nanjundiah RS, Kumar DN (2008) Downscaling precipitation to river basin in India for IPCC SRES scenarios using support vector machine. Int J Climatol 28:401–420

Bartlett R, Bharati L, Pant D, Hosterman H, McCornick P (2010) Climate change Impacts and adaptation in Nepal. Colombo, Sri Lanka: International Water Management Institute. 35p. (IWMI Working Paper 139). doi:10.5337/2010.227

Bastola S, Murphy C, Sweeney J (2011) Role of hydrological model uncertainty in climate change impact studies. Adv Water Resour 34(5):562–576

Chhetri MBP (2010) Effects of climate change: the global concern. In: Proceedings of the 4th AMCDRR 25-28 October 2010 Republic of Korea. http://www.iawe.org/WRDRR/2010/Meena.pdf

Chiew FHS, Kirono DGC, Kent DM, Frost AJ, Charles SP, Timbal B, Nguyen KC, Fu G (2010) Comparision of runoff modeled using rainfall from different downscaling methods for historical and future climates. J Hydrol 387:10–23

Dibike YB, Coulibaly P (2005) Hydrologic impact of climate change in the Saguenay watershed: comparison of downscaling methods and hydrologic models. J Hydrol 307:145–163

Ficklin DL, Luo Y, Luedeling E, Zhang M (2009) Climate change sensitivity assessment of a highly agricultural watershed using SWAT. J. Hydrology 374:16–29

Girvetz EH, Zganjar C, Raber GT, Maurer EP, Kareiva P et al (2009) Applied climate-change analysis: the climate wizard tool. PLoS ONE 4(12):e8320. doi:10.1371/journal.pone.0008320

Hawkins E, Sutton R (2010) The potential to narrow uncertainty in projections of regional precipitation change. Clim Dyn 37:407–418

IPCC (2007) Climate Change 2007: the physical science basis. In: Solomon S, Qin D, Manning M, Chen Z, Marquis M, Averyt KB, Tignor M, Miller HL (eds) Contribution of WG I to AR4 of IPCC. Cambridge University Press, ISBN 978-0-521-88009-1 (pb: 978-0-521-70596-7)

Maraun D, et al (2010) Precipitation downscaling under climate change. Recent developments to bridge the gap between dynamical models and the end user. Rev Geophys doi:10.1029/2009RG000314

Minville M, Brissette F, Leconte R (2008) Uncertainty of the impact of climate change on the hydrology of a nordic watershed. J Hydrol 358:70-83

Moriondo M, Bindi M (2006) Comparison of temperatures simulated by GCMs, RCMs And statistical downscaling: potential application in studies of future crop development. Clim Res 30:149–160

Nguyen VTV (2005) Downscaling methods for evaluating the impacts of climate change and variability on hydrological regime at basin scale. Role of water sciences in transboundary river basin management, Thailand

Santhi C, Arnold JG, Willams JR, Dugas WA, Srinivasan R, Hauck LM (2001) Validation of the SWAT model on a large river basin with point and nonpoint sources. J Am Water Resour Assoc (JAWRA) 37(5):1169–1188

Semenov MA, Barrow EM (1997) Use of stochastic weather generator in the development of climate change scenarios. Clim Change 35:397–414

Semenov MA, Stratonovitch P (2010) The use of multi-model ensembles from global climate models for impact assessments of climate change. Clim Res 41:1–14

Sharma E, Bhuchari S, Xing M, Kothyari BP (2007) Land use change and its impact on hydro-ecological linkages in Himalayan watersheds. Tropical Ecol 48(2):151–161

Shrestha AB, Devkota LP (2010) Climate change in the Eastern Himalayas: Observed trends and model projections. Climate change impact and vulnerability in the Eastern Himalayas-Technical report 1. ICIMOD, Kathmandu

Stott PA, Kettleborough JA (2002) Origins and estimates of uncertainty in predictions of twenty first century temperature rise. Nature 416:723–726

Tisseuil C, Vrac M, Lek S, Wade AJ (2010) Statistical downscaling of river flows. J Hydrol 385:279–291

Van Liew MW, Veith TL, Bosch DD, Arnold JG (2007) Suitability of SWAT for the conservation effects assessment project: a comparison on USDA-ARS experimental watersheds. J. Hydrol Eng 12(2):173–189

WWF (2009) Climate change impact on discharge at Koshi River Basin using regional climate and hydrological models. Final report on a collaborative project of WWF-Nepal/DHM

Zhang H, Huang GH, Wang D, Zhang X (2011) Uncertainty assessment of climate change impacts on the hydrology of small prairie wetlands. J Hydrol 396:94–103

Uncertainty Analysis of Statistically and Dynamically Downscaled Precipitation Data: A Study of the Chaliyar River Basin, Kerala, India

Santosh G. Thampi and N.R. Chithra

Abstract Although assessment of the anticipated impacts of projected climate change is very much required for many sectors, non-availability of climate data on the local scale is a major limiting factor. In this background, statistical downscaling has a lot of scope. However, the downscaled data have to be thoroughly analysed in order to assess the uncertainty associated with it. In this study, a detailed uncertainty analysis was performed on statistically and dynamically downscaled monthly precipitation data in the Chaliyar River Basin, in Kerala, India. The mean and variance of the downscaled and observed data for each month were compared. The Wilcoxon signed-rank test, Levene's test, Brown–Forsythe test, and the nonparametric Levene's test were performed on the downscaled precipitation data at 5 % significance level. Results showed that the error is not significant in the case of the statistically downscaled data using predictors generated from the reanalysis data. In the case of statistically downscaled data from the predictions of the general circulation model (GCM), error in the mean is significant in some months, probably due to uncertainty in the GCM predictors, whereas the error in the variance is insignificant. For dynamically downscaled data, the error in the mean as well as the variance is not significant. Uncertainty analysis is required to be performed on the downscaled data before its use in impact assessment.

1 Introduction

The Earth's atmosphere is currently undergoing changes unprecedented in human history and, although changes as large as those being witnessed now have occurred in the geological past, relatively few have happened as fast as today's changes

---

S.G. Thampi (✉) · N.R. Chithra
Department of Civil Engineering, National Institute of Technology, Calicut 673601, Kerala, India
e-mail: santoshthampi@yahoo.com

N.R. Chithra
e-mail: chithranr@nitc.ac.in

S. Shrestha et al. (eds.), *Managing Water Resources under Climate Uncertainty*, Springer Water, DOI 10.1007/978-3-319-10467-6_7

(McGuffie and Henderson-Sellers 2005). General circulation models (GCMs) are comprehensive models of the climate system constructed by discretising equations representing the basic laws that govern the behaviour of the atmosphere. Any study on projected climate change and assessment of likely impacts requires data from GCMs, which are the only reliable source for future climate scenarios. But GCMs perform well only at a coarse resolution—2.5°–3°, whereas a hydrologist may require climate data at a much finer scale: river basin scale or station scale. Also, the coupling of GCMs with hydrological models is challenging for various reasons. Due to limitations in the understanding of local climate processes, data, and computing power, GCMs are typically run on large-sized grid cells. Also, in areas where coasts and mountains significantly influence weather, scenarios based on global climate models are unable to capture local level details needed for assessing impacts at regional scales (Still et al. 1999; Bradley et al. 2006). Furthermore, at coarse resolutions, extreme events like cyclones or heavy rainfall episodes are either not captured, or the predicted intensities are unrealistically low (Viviroli et al. 2011). To overcome these tribulations, downscaling methods are developed which yield local-scale surface weather from regional-scale atmospheric variables provided by GCMs.

Downscaling techniques can be broadly classified into (i) dynamical downscaling and (ii) statistical downscaling. In dynamical downscaling, regional climate models simulate climate features dynamically when supplied with time-varying atmospheric conditions modelled by a GCM bounding a specified domain (Wilby and Fowler 2011). Detailed information at spatial scales ranging from 10 to 20 km may be obtained from these models at temporal scales of hours or less. However, these models are computationally demanding and only a few simulations can be afforded, thereby limiting their use for long simulations and extensive hypothesis testing. The alternate approach is statistical downscaling, in which quantitative relationships are derived between the large-scale atmospheric variables (predictors), and the local surface variables (the predictands). It is a computationally efficient technique compared to dynamical downscaling and is based on the following assumptions: (i) The predictors selected are reproduced realistically by the particular GCM, (ii) these predictors account for most of the observed variations in the predictands, and (iii) the derived relationships will be valid under the changed climate scenarios (Wilby et al. 2004). Statistical downscaling is a two-step process consisting of: (i) development of a statistical relationship between the local climate variables such as surface air temperature and precipitation, and large-scale predictors such as pressure fields, wind speed, and humidity and (ii) the application of these relationships to the output of GCM experiments to predict local climate characteristics in future periods. It is a practical approach for addressing the current needs of the climate change research community, especially in countries with limited RCM data availability. Application of the transfer functions derived from observations to downscale GCM climate change experiments allows development of regional climate change scenarios at sub-GCM grid-scale resolutions with greater reliability. Statistical downscaling is used in the assessment of climate change impacts at regional and local scales, when sufficient observed data are available to derive statistical relationships. Empirical statistical downscaling is cheap to run and hence can be applied to results from a number of different coupled GCMs to obtain an idea of the uncertainties associated with the GCMs, but its accuracy has to be evaluated.

Uncertainty analysis is required to be performed to quantitatively evaluate whether the downscaled data reproduces the current observed characteristics of the climatic variable. This will enable identification of the most robust downscaling method under present conditions. Several studies have been conducted to assess the model performance by comparing the results obtained using different models and observations (Lambert and Boer 2011). Also, several regional-scale uncertainty analyses have been performed to assess model performance in different regions (Stainforth et al. 2007; Smith and Chandler 2010; Fu et al. 2013). Though the statistical distribution of climate data is not known, scientific-based uncertainty quantification has always involved a degree of expert belief (Cooke 2013). In this study, uncertainty analysis was performed on dynamically and statistically downscaled monthly precipitation data for the Chaliyar River Basin, Kerala, India. The mean and variance of the downscaled data were compared with that of the observed data for each month. This was done by performing nonparametric hypothesis tests such as the Wilcoxon signed-rank test, Levene's test, Brown–Forsythe test, and the nonparametric Levene's test on the downscaled precipitation data.

2 Study Area and Data

The study area is the Chaliyar River Basin in Kerala, India, located in a humid tropical setting and situated between 11° 10′N and 11° 30′N latitudes and 75° 50′E and 76° 30′E longitudes. A major part of precipitation in the tropics appears to be convective, and GCMs show deficiencies in simulating convective precipitation (Maraun et al. 2010). Hence, GCM data downscaled for such regions should be thoroughly analysed for the uncertainty associated with it before it is used in impact assessment studies. The data used in this study are the observed values of monthly precipitation at the Kottaparamba Observatory of the Centre for Water Resources Development and Management (CWRDM), Kozhikode (1980–2007) and the State Seed Farm, Puduppadi, Kozhikode (1965–2007), NCEP/NCAR reanalysis data, predictions from GCM simulations, and dynamically downscaled monthly precipitation data using PRECIS. NCEP/NCAR reanalysis data (Kalnay et al. 1996) were extracted for the period from January 1980 to December 2007 for latitudes ranging from 10°N to 12.5°N and longitude ranging from 75°E to 77.5°E at a spatial resolution of 2.5°. The T63 resolution of CGCM3.1 for the future scenario A2 during the period January 2001 to December 2007 on a monthly time scale, for four grid points (latitudes ranging from 9.77°N to 12.56°N and longitudes ranging from 75.94°E to 78.75°E) were downloaded from the web site http://www.cccma.ec.gc.ca/data/cgcm3/cgcm3.shtml. Figure 1 shows the Chaliyar River Basin in Kerala with two rain gauge stations and its location in the PRECIS domain.

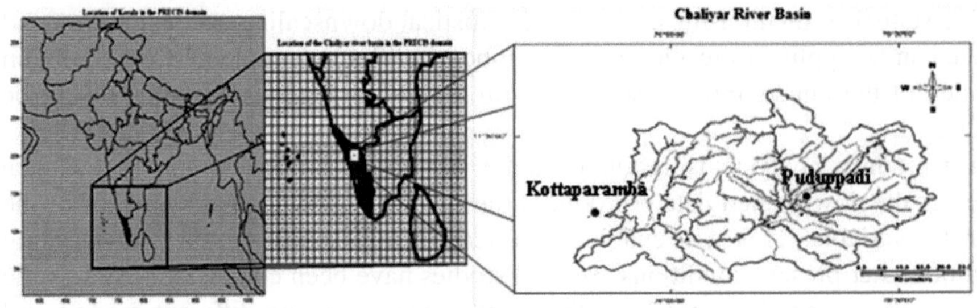

Fig. 1 Chaliyar River Basin and its location in the PRECIS domain

3 Methodology

Monthly precipitation data for the study area were downscaled statistically using ANN-based models, and the dynamically downscaled data were collected from IITM Pune. Exploratory data analysis (EDA) was performed on the monthly precipitation data to decide upon the methods of data analysis, parametric, or nonparametric. Hypothesis tests for the equality of two population means and the equality of two population variances were carried out to compare the mean and variance of the downscaled data with that of the observed data.

3.1 Artificial Neural Network (ANN)-Based Statistical Downscaling

Downscaling methods based on artificial neural networks (ANN) are very widely used owing to their superior capability in capturing nonlinear relationships between the predictors and the predictands (Crane and Hewitson 1998; Hewitson and Crane 1996; Fistikoglu and Okkan 2011). An ANN is a massively parallel, distributed, information processing system having certain performance characteristics resembling the biological neural networks of the human brain (Haykin 1994). ANN-based transfer functions have been widely used for downscaling climatic variables (Gardner and Dorling 1998; Murphy 1999; Haylock et al. 2006; Dibike and Coulibaly 2006; Fistikoglu and Okkan 2011). In general, these are complex nonlinear regression models structured between the predictors and the predictands. The input layer of an ANN consists of potential predictors; the second is a hidden layer that transforms the input nonlinearly to output, and the output layer consists of the predictand, in this case, precipitation. The Levenberg-Marquardt feed forward network with back propagation algorithm was used in this study (Hagan and Menhaj 1994).

The predictors that are physically related to the predictand precipitation were identified from the available NCEP reanalysis data. The potential predictors were selected based on the correlation between the NCEP predictors, viz. geopotential height, air temperature, eastward wind, northward wind, specific humidity, at various pressure levels and the predictand precipitation. Also, in order to ensure that the predictors are computed well by the GCM, the correlation between the concurrent NCEP and GCM predictors was also estimated. This correlation was estimated using the product moment correlation coefficient (Pearson 1896). Changes in the predictor–predictand relationship during various seasons were recognised and included in the study; three seasonal models were developed using three separate sets of potential predictors.

Precipitation in the study area is monsoon driven, and the three predominant seasons are pre-monsoon (April and May), south-west monsoon (June–September), and north-east monsoon (October–December). Observed data at the two stations showed that less than 1 % of the annual precipitation occurs during the months January to March, and hence, these months were considered as dry periods. ANN-based transfer functions were derived between the potential NCEP predictors and the observed precipitation for the training period (1980–2000). The normalised mean square error (NMSE) was used as an index for assessing the performance of the model. The trained networks were further validated using observed data for the period 2001–2007. The validated networks were used to predict monthly precipitation using the GCM predictors.

3.2 Dynamical Downscaling Using PRECIS

The Indian Institute of Tropical Meteorology (IITM), Pune, in collaboration with the Hadley Centre for Climate Prediction and Research has developed climate scenarios for India (Rupa Kumar et al. 2006). Monthly precipitation data predicted by the GCM for the period 2001–2007 have been downscaled using the regional climate model (RCM), Providing Regional Climate for Impacts Studies (PRECIS); these data were obtained from IITM and used in this study. PRECIS is a hydrostatic, primitive equation grid point model with 19 levels described by a hybrid vertical coordinate system (Simmons and Burridge 1981), and the version with a horizontal resolution of 0.44° × 0.44° was used in this study (Simon et al. 2004). The lateral boundary conditions were derived from HadAM3H, a global atmosphere only model developed at the Hadley Centre, with a horizontal resolution of 3.75° × 2.5°. Hadley Centre models have been applied successfully in the tropics (Hassell and Jones 1999).

3.3 Uncertainty Analysis

Uncertainties in downscaled climate data result from the concept on which the downscaling models are based and the GCM data used (Khan et al. 2006). In this study, the first and second moments of the observed and downscaled data are used to assess the performance of dynamical and statistical downscaling models. Statistical downscaling was station based, and the downscaled data at the stations (Kottaparamba and Puduppadi) were compared with the observed data. Since dynamical downscaling was river basin based, the downscaled data were compared with the average value of the corresponding observed data.

3.3.1 Exploratory Data Analysis

A parametric or nonparametric approach can be used for statistical testing, depending on the results of EDA.

Classical methods of statistical inference depend heavily on the assumption that the data are outlier-free and nearly normal (Khan et al. 2006). Results obtained would be misleading if the data do not satisfy these assumptions. To check whether the data contain outliers or it really follow a normal distribution, EDA was carried out. Density function plot, box plot, histogram, and normal quantile–quantile plots (*qq*-plots) were used for this purpose. Histogram and box plots are useful to get an idea of the shape of a distribution. *qq*-plots are the most sensitive tools used for this purpose. A *qq*-plot is a plot of the percentiles of a standard normal distribution against the corresponding percentiles of the observed data. If the observations follow a normal distribution, the resulting plot will be roughly a straight line with a positive slope. Box plots provide a graphical representation of the median (the middle horizontal line), the quantiles (the top and bottom lines), and the extremes (whiskers extending from the box) (Levin and Rubin 1998). Based on the results of the EDA, distribution-free or nonparametric tests were performed.

3.3.2 Hypothesis Tests

Wilcoxon Signed-Rank Test

It is a nonparametric test used to compare two related samples. It can be used to test the null hypothesis of no difference in the median in paired samples (Wilcoxon 1945). In this test, the difference in each pair of data (D_i) is calculated, its absolute value is sorted from the smallest to the largest, and ranks (R_i) are assigned to each of these values. The test statistics are defined as:

$$W = \left| \sum_{i=1}^{N_r} [\operatorname{sgn}(D_i) \cdot R_i] \right| \tag{1}$$

where $\operatorname{sgn}(D_i)$ is sign of the difference D_i and N_r is the reduced sample size by excluding pairs with zero difference. As N_r increases, the sampling distribution converges to a normal distribution, and hence, for $N_r \geq 10$, the z-score can be calculated as:

$$z = \frac{W - 0.5}{\sigma_W}, \quad \sigma_W = \sqrt{\frac{N_r(N_r + 1)(2N_r + 1)}{6}} \tag{2}$$

The hypothesis is tested by comparing the z-score computed with the critical values of the distribution for $N_r \geq 10$. When $N_r < 10$, if $W \geq W_{\text{critical},N_r}$, the null hypothesis is rejected.

Levene's Test

Levene's test is used to assess the equality of variances for a variable calculated for two or more groups (Levene 1960). Although in this test the population under consideration is assumed to be approximately normally distributed, it is less dependent on this assumption, and hence, this method can be used to test data even when it is not normally distributed. In this test, the absolute difference of each data pair (D_i) is calculated and a one-way analysis of variance (ANOVA) is performed on these differences. The null hypothesis can be stated as:

$$H_0 : \sigma_1 = \sigma_2$$

where σ_1 and σ_2 are the variance of observed data and downscaled data, respectively, and the test statistics is defined as:

$$W = \frac{\frac{\sum_{i=1}^{k} N_i(\overline{D}_i - \overline{D})^2}{(k-1)}}{\frac{\sum_{i=1}^{k}\sum_{j=1}^{N_i}(D_{ij} - \overline{D}_i)^2}{(N-k)}} \tag{3}$$

where k is the number of groups, N_i is number of observations in group i, $\overline{D}_i$ is the average of N_i absolute deviations, $\overline{D}$ is average of all N absolute deviations, $D_{ij} = |X_{ij} - \overline{X}_i|$, the absolute deviation of jth observation from ith group mean, and N is the total number of observations from all groups. The null hypothesis is rejected at a significance level α if $W \geq F_{\alpha,df_1,df_2}$, where $df_1 = k - 1$ and df_2 is $N - k$. p-values can also be used for stating the rejection criteria. It is the largest significance level at which the null hypothesis is accepted. This means that at any significance level greater than the p-value, the null hypothesis is rejected, and at any significance level less than the p-value, the null hypothesis is accepted.

Brown–Forsythe Test

The Brown–Forsythe test (Brown and Forsythe 1974) is a statistical test for the equality of group variances based on performing an ANOVA on the absolute deviations from the median. Hence, the test procedure is similar to that of Levene's test with the median being used in place of the mean. It is very robust against many types of non-normal data, and the statistical power of the method is good. The power of a statistical test is a measure of its ability to reject the null hypothesis when it is actually false or it is the probability of not committing a Type II error, i.e. non-rejection of a null hypothesis when the alternate hypothesis is true (Johnson 2011).

Nonparametric Levene's Test

This is a newly developed test for the equality of variances, which utilises ranks (Nordstokke and Zumbo 2010). The test involves pooling data from all the groups, ranking the scores, placing the rank values back into their original groups, and conducting the Levene's test on the ranks.

$$\mathrm{ANOVA}\left(\left|R_{ij} - \overline{X_J}\right|\right) \tag{4}$$

where R_{ij} is calculated by pooling the values from each of the j groups and ranking the scores. Analysis of variance is conducted on the absolute value of the mean of the ranks for each group $\left(\overline{X}_J\right)$ subtracted from each individual's rank $\left(R_{ij}\right)$. Nordstokke et al. (2011) compared the Brown–Forsythe test with the nonparametric Levene's test across a large number of conditions and found that it varied in terms of its degree of distributional symmetry and the nonparametric Levene's test outperformed the Brown–Forsythe test consistently when the population distributions being sampled were asymmetric to varying degrees.

4 Results

4.1 Statistical Downscaling Using NCEP Predictors

ANN models for various seasons were derived by connecting potential NCEP predictors and the predictand precipitation using data for the training period (1980–2000), and these models were validated subsequently using data for the validation period (2001–2007). The result of validation at the Kottaparamba station is presented in Fig. 2. The observed and downscaled data for various months are presented using box plots, enabling comparison of the median, quantiles, and whiskers. Figure 3 presents a comparison of the downscaled data with the observed data at the Puduppadi station during the south-west monsoon and the north-east monsoon periods. In this study area, about 75 % of the annual rainfall occurs during the south-west monsoon period (June to September) and during these months the performance of the model appears to be better. Figure 4 shows the mean bias (*MB*) calculated on a monthly basis from the observed and NCEP downscaled data using the following equation:

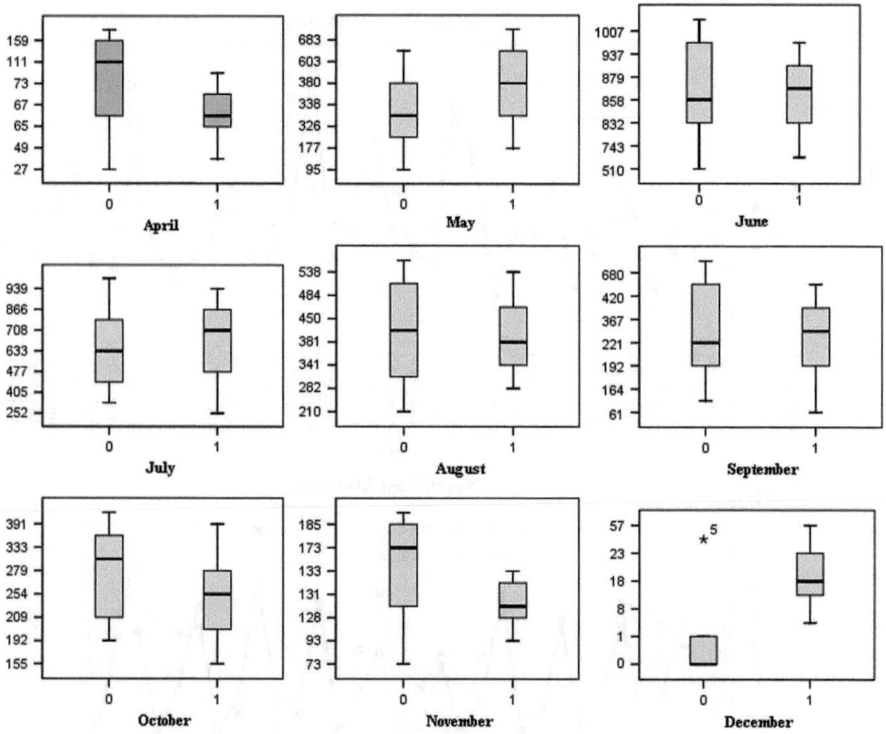

Fig. 2 Box plots of NCEP downscaled data for the validation period

$$\mathrm{MB} = \frac{1}{N}\sum_{i=1}^{N}\left|x_{e,i} - x_{o,i}\right| \tag{5}$$

where $x_{o,i}$ is the ith observed value (monthly), $x_{e,i}$ is the ith predicted value, and N is the amount of data.

4.2 Statistical Downscaling Using GCM Predictors

The validated ANN models were used to downscale monthly precipitation using the predictors simulated by the GCM, CGCM3 for the A2 scenario. Re-gridding was performed to the GCM predictors relative to the NCEP predictors since the grid spacing of the NCEP data and the GCM data were different, based on the guidelines outlined by Skelly and Sellers (1996).

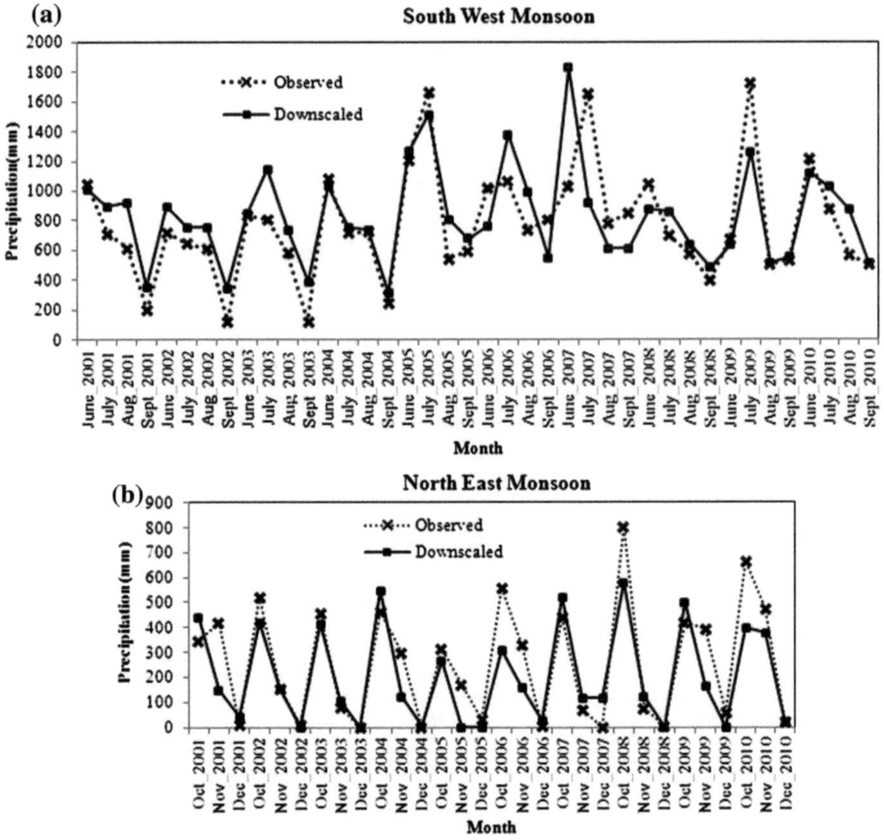

Fig. 3 Comparison of NCEP downscaled data with the observed data at the Puduppadi station during **a** south-west monsoon and **b** north-east monsoon

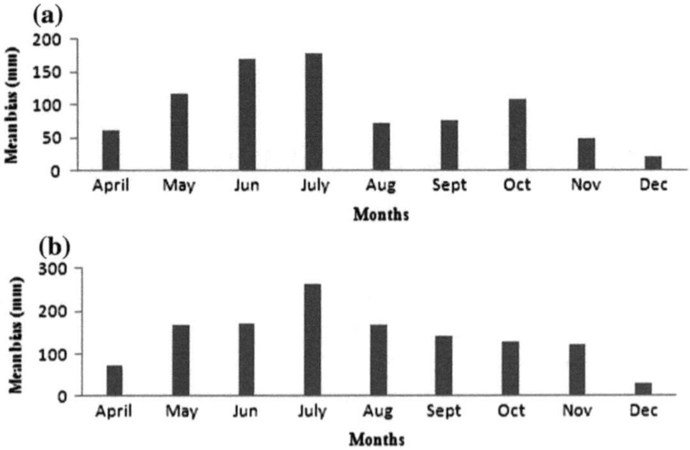

Fig. 4 Mean bias of the statistically downscaled precipitation data using NCEP predictors in the validation period at the **a** Kottaparamba station and **b** Puduppadi station

4.3 Exploratory Data Analysis

Graphical EDA was performed with the observed data for all months. EDA plots for April, August, and September are presented in Fig. 5. From the histogram, density plots, and *qq*-plots, it can be interpreted that the distribution is not normal and is skewed. The box plots indicate that the data contain outliers. It can be concluded that the monthly precipitation data possess outliers and the distribution is not normal; hence, a nonparametric approach should be adopted for the hypothesis tests.

4.4 Hypothesis Test for the Equality of Two Population Means

The Wilcoxon signed-rank test was used to test whether the difference in values of the median of the observed and the downscaled monthly precipitation is zero. Both statistically and dynamically downscaled data were tested at a 5 % significance

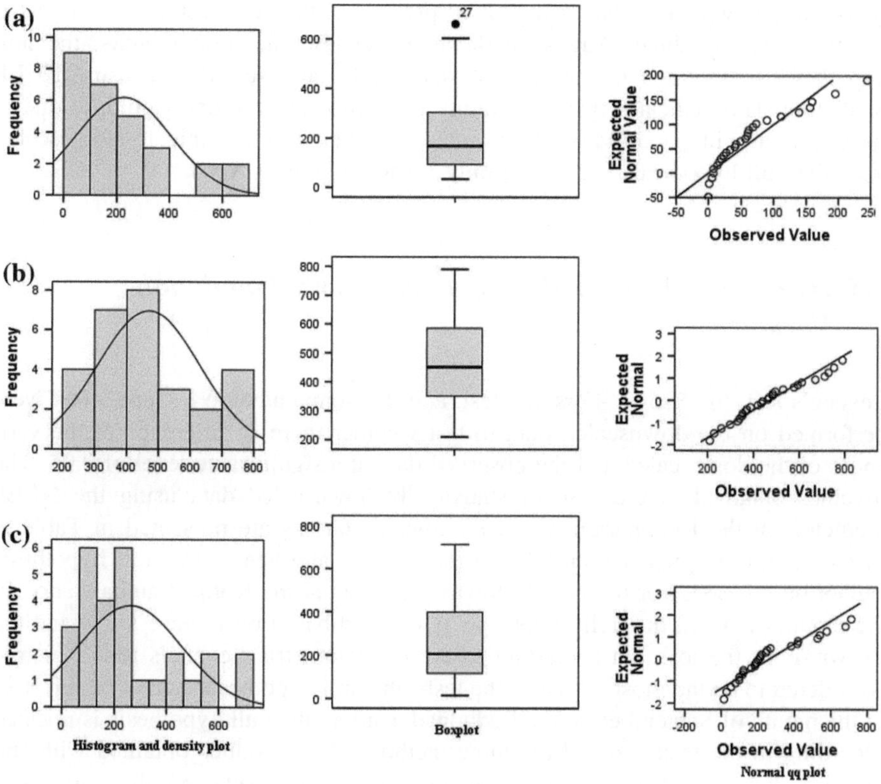

Fig. 5 EDA plots of observed monthly precipitation data for the months of **a** April, **b** August, and **c** September at the Kottaparamba station

Table 1 The results (p-values) of the Wilcoxon signed-rank test

Month	NCEP downscaled data (statistical) at Kottaparamba	NCEP downscaled data (statistical) at Puduppadi	GCM downscaled data (statistical) at Kottaparamba	Dynamically downscaled data—Chaliyar River Basin
April	0.128	0.169	0.499	0.028
May	0.091	0.386	0.398	0.176
June	0.866	0.114	0.018	0.176
July	0.075	0.721	0.176	0.128
August	0.499	0.575	0.018	0.128
September	0.176	0.386	0.499	0.063
October	0.398	0.074	0.237	0.310
November	0.176	0.047	0.018	0.499
December	0.018	0.859	0.686	0.917

level. The p-values for different months are presented in Table 1. It can be seen that the p-values are greater than 0.05, and the null hypothesis is accepted in most of the months; in other words, the error in the mean is not significant. In the case of statistically downscaled data using GCM predictors, the null hypothesis is rejected in the months of June, August, and November. But in these months, the null hypothesis is accepted in the case of statistically downscaled data using NCEP predictors. This indicates that uncertainty in the downscaled data could be due to the uncertainty in the GCM (CGCM3) data. In the case of dynamically downscaled data, the null hypothesis is rejected only for the month of April.

4.5 Hypothesis Test for the Equality of Two Population Variances

Levene's test, the Brown–Forsythe test, and the nonparametric Levene's test were performed on the downscaled data to test whether there is difference in the variances of the downscaled and the observed data at a significance level of 0.05. The p-values obtained in the case of statistically downscaled data using the NCEP predictors at the Kottaparamba and Puduppadi stations are presented in Table 2. The p-values are greater than 0.05 in most cases, and hence, the null hypothesis cannot be rejected. For the NCEP downscaled data at the Kottaparamba station in the month of April, the null hypothesis is rejected based on Levene's test and the Brown–Forsythe test, but according to the nonparametric Levene's test, which is considered to be the most robust of the tests, the null hypothesis cannot be rejected. In the month of September, at the Puduppadi station, the null hypothesis is rejected according to the results of all the three methods. The p-values obtained with statistically downscaled data using GCM predictors at the Kottaparamba station and dynamically downscaled data using PRECIS are presented in Table 3. It can be seen

Table 2 *p*-values of the statistical tests conducted for the equality of variances of the observed and downscaled data using NCEP predictors at the Kottaparamba and Puduppadi stations

Month	Kottaparamba			Puduppadi		
	Levene's	Brown–Forsythe	Nonparametric Levene's	Levene's	Brown–Forsythe	Nonparametric Levene's
April	0.006	0.008	0.109	0.283	0.269	0.196
May	0.713	0.803	1	0.225	0.592	0.657
June	0.150	0.161	0.289	0.277	0.291	0.876
July	0.499	0.587	1	0.088	0.367	0.552
August	0.214	0.252	0.332	0.250	0.251	0.296
September	0.296	0.577	0.413	0.01	0.011	0.005
October	0.345	0.417	0.704	0.351	0.526	0.951
November	0.03	0.057	0.031	0.006	0.007	0.095
December	0.814	0.79	0.465	0.446	0.679	0.444

Table 3 *p*-values of the statistical tests conducted for the equality of variances of the observed and downscaled data using dynamical downscaling and statistical downscaling with GCM predictors at the Kottaparamba station

Months	Statistical downscaling using GCM predictors			Dynamical downscaling using PRECIS		
	Levene's	Brown–Forsythe	Nonparametric Levene's	Levene's	Brown–Forsythe	Nonparametric Levene's
April	0.618	0.331	0.475	0.039	0.149	0.737
May	0.511	0.097	0.475	0.074	0.179	0.864
June	0.019	0.019	0.586	0.223	0.905	0.410
July	0.087	0.814	0.621	0.216	0.272	0.509
August	0.064	0.414	0.348	0.068	0.688	0.780
September	0.438	0.745	0.489	0.024	0.079	1
October	0.209	0.738	0.601	0.209	0.738	0.601
November	0	0.198	1	0.309	0.660	0.733
December	0.903	0.903	0.485	0.821	0.768	0.167

that the null hypothesis is accepted in all but a few of the cases. Based on these results, it can be concluded that the error in variance is not significant in the case of both dynamically and statistically downscaled data.

5 Conclusions

Uncertainty in both dynamically and statistically downscaled monthly precipitation data in the Chaliyar River Basin was assessed and compared in this study. As the data contain outliers and the distribution is not normal, nonparametric methods have been used. For evaluating the error in the mean of the data, the Wilcoxon signed-rank test was used, and for variance, Levene's test, the Brown–Forsythe test, and the nonparametric Levene's test were employed. Uncertainty in the downscaled data was assessed by comparing it with the observed data at a significance level of

5 %, and the *p*-values were determined. It showed that the error is not significant in most of the months. In the case of statistically downscaled data using the GCM predictors, in the months of June, August, and November the error in mean is significant at a 5 % significance level. This may be due to uncertainty in the GCM predictors, since in the same months, the error in the mean was found to be insignificant when the analysis was performed with statistically downscaled data using the NCEP predictors. Dynamically downscaled data performed better when considering error in the mean alone, whereas when error in the variance is also considered, both dynamically and statistically downscaled data performed equally well since error in the variance was found to be insignificant in both cases. These results may be refined further by statistically downscaling precipitation data using predictors simulated by other GCMs.

Any study for assessment of the impact of climate change requires downscaled GCM data at the river basin scale. RCM simulations may not be available for many locations, and statistical downscaling will be the only option available. However, the data should be thoroughly assessed for uncertainty before using it in these studies. In regions for which results of at least a few RCM simulations are available, these may be compared with statistically downscaled data from a number of GCM predictors obtained by performing a simple statistical analysis. This study highlights the necessity for performing an uncertainty analysis of the downscaled data.

References

Bradley RS, Vuille M, Diaz HF, Vergara W (2006) Threats to water supplies in the tropical Andes. Science 312(5781):1755–1756

Brown MB, Forsythe AB (1974) Robust tests for equality of variances. J Am Stat Assoc 69:364–367

Cooke RM (2013) Uncertainty analysis comes to integrated assessment models for climate change…and conversely. Clim Chang 117:467–479

Crane RG, Hewitson BC (1998) Doubled CO_2 precipitation changes for the Susquehanna basin: downscaling from the genesis general circulation model. Int J Climatol 18:65–76

Dibike YB, Coulibaly P (2006) Temporal neural networks for downscaling climate variability and extremes. Neural Network 19(2):135–144

Fistikoglu O, Okkan U (2011) Statistical downscaling of monthly precipitation using NCEP/NCAR reanalysis data for Tahtali River Basin in Turkey. J Hydrol Eng 16(2):157–164. doi:10.1061/(ASCE)HE.1943-5584.0000300

Fu G, Liu Z, Charles SP, Xu Z, Yao Z (2013) A score based method for assessing the performance of GCMs: A case study of southeastern Australia. J Geophys Res D 118:4154–4416

Gardner MW, Dorling SR (1998) Artificial neural networks (the multi-layer perceptron)—a review of applications in the atmospheric sciences. Atmos Environ 32:2627–2636

Hagan MT, Menhaj MB (1994) Training feed forward network with Marquardt algorithm. IEEE Trans Neural Network 5(6):989–993

Hassell D, Jones RG (1999) Simulating climatic change of the southern Asian monsoon using a nested regional climate model. Technical note no. 8, Hadley Centre, Bracknell, UK, 16 p

Haykin S (1994) Neural networks: a comprehensive foundation. MacMillan, New York

Haylock MR, Cawley GC, Harpham C, Wilby RL, Goodess CM (2006) Downscaling heavy precipitation over the United Kingdom: a comparison of dynamical and statistical methods and their future scenarios. Int J Climatol 26:1397–1415
Hewitson BC, Crane RG (1996) Climate downscaling: techniques and application. Clim Res 7:85–95
Johnson RA (2011) Miller and Freund's probability and statistics for engineers. Pearson Prentice Hall, New Jersey
Kalnay E, Kanamitsu M, Kistler R, Collins W, Deaven D, Gandin L, Iredell M, Saha S, White G, Woollen J, Zhu Y, Chelliah M, Ebisuzaki W, Higgins W, Janowiak J, Mo KC, Ropelewski C, Wang J, Leetmaa A, Reynolds R, Jenne R, Joseph D (1996) The NCEP/NCAR 40-year reanalysis project. Bull Am Meteorol Soc 77(3):437–471
Khan MS, Coulibaly P, Dibike Y (2006) Uncertainty analysis of statistical downscaling methods. J Hydrol 319:357–382
Lambert SJ, Boer GJ (2011) CMIP1 evaluation and intercomparison of coupled climate models. Clim Dynam 17:83–106
Levene H (1960) Contributions to probability and statistics. Stanford University Press, California
Levin RI, Rubin DS (1998) Statistics for management. Prentice Hall, New Jersey
Maraun D, Wetterhall F, Ireson AM, Chandler RE, Kendon EJ, Widmann M, Brienen S, Rust HW, Sauter T, Themeßl M, Venema VKC, Chun KP, Goodess CM, Jones RG, Onof C, Vrac M, Thiele-Eich I (2010) Precipitation downscaling under climate change: recent developments to bridge the gap between dynamical models and the end user. Rev Geophys 48:RG3003
McGuffie K, Henderson-Sellers A (2005) A climate modelling primer. Wiley, England
Murphy JM (1999) An evaluation of statistical and dynamical techniques for downscaling local climate. J Clim 12:2256–2284
Nordstokke DW, Zumbo BD (2010) A new nonparametric test for equal variances. Psicologica 31:401–430
Nordstokke DW, Zumbo BD, Cairns SL, Saklofske DH (2011) The operating characteristics of the nonparametric Levene test for equal variances with assessment and evaluation data. Pract Assess Res Eval 16(5)
Pearson K (1896) Mathematical contributions to the theory of evolution III regression heredity and panmixia. Philos Trans R Soc Lond 187:253–318
Rupa Kumar K, Sahai AK, Krishna Kumar K, Patwardhan SK (2006) High-resolution climate change scenarios for India for the 21st century. CurrSci 90:334–345
Simmons AJ, Burridge DM (1981) An energy and angular-momentum conserving finite-difference scheme and hybrid coordinates. Mon Weather Rev 109:758–766
Simon W, Hassell D, Hein D, Jones R, Taylor R (2004) Installing and using the Hadley Centre regional climate modelling system PRECIS. Met Office Hadley Centre, Exeter, UK
Skelly WC, Henderson-Sellers A (1996) Grid box or grid point: what type of data do GCMs deliver to climate impacts researchers? Int J Climatol 16:1079–1086
Smith I, Chandler E (2010) Refining rainfall projections for the Murray Darling Basin of south-east Australia—the effect of sampling model results based on performance. Clim Chang 102: 377–393
Stainforth DA, Allen MR, Tredger ER, Smith LA (2007) Confidence, uncertainty and decision support relevance in climate predictions. Phil Trans R Soc A 365:2145–2161
Still CJ, Foster PN, Schneider SH (1999) Simulating the effects of climate change on tropical mountain cloud forests. Nature 398(6728):608–610
Viviroli D, Archer DR, Buytaert W, Fowler HJ, Greenwood GB, Hamlet AF, Huang Y, Koboltschnig G, Litaor MI, Lopez-Moreno JI, Lorentz S, Schadler B, Schreier H, Schwaiger K, Vuille M, Woods R (2011) Climate change and mountain water resources: overview and recommendations for research, management and policy. Hydrol Earth Syst Sci 15:471–504. doi:10.5194/hess-15-471-2011
Wilby RL, Fowler HJ (2011) Regional climate downscaling. In: Fung F, Lopez A, New M (ed) Modelling the impact of climate change on water resources, 1st edn. Blackwell, West Sussex, pp 34–85

Wilby RL, Charles SP, Zorita E, Timbal B, Whetton P, Mearns LO (2004) The guidelines for use of climate scenarios developed from statistical downscaling methods. Supporting material of the Intergovernmental Panel on Climate Change (IPCC), prepared on behalf of Task Group on Data and Scenario Support for Impacts and Climate Analysis (TGICA). http://ipccddc.cru.uea.ac.uk/guidelines/StatDown Guide.pdf
Wilcoxon F (1945) Individual comparisons by ranking methods. Biometrics 1:80–83

Assessment of the Impact of Projected Climate Change on Streamflow and Groundwater Recharge in a River Basin

Santosh G. Thampi and K.Y. Raneesh

Abstract In this work, a general approach for evaluating the impacts of projected climate change on streamflow and groundwater recharge in a river basin located in the humid tropical zone of India is presented. The projections of a GCM for two important climate variables, viz. rainfall and temperature for two scenarios, A2 and B2, are downscaled by a RCM to predict future climate in the river basin and then input into a physically based hydrologic model, SWAT, to evaluate the impact of projected climate change on streamflow in the river basin. Also, a simple conceptual semi-distributed model to assess the impact of projected climate change on direct groundwater recharge is developed and used to predict groundwater recharge in the A2 and B2 scenarios. This model is based on the water balance concept, linking the atmospheric and hydrogeologic parameters to different hydrological processes and estimates daily water table fluctuations. Results show that in the A2 scenario, for the southwest monsoon period, there is an average increase in temperature of 2 °C and a decrease in rainfall of 11.50 %. Predictions by SWAT indicate an increase in potential evapotranspiration of 1.14 % and a decrease in streamflow of 7.53 % from present-day average values. For the same period, in the B2 scenario, there is an average increase in temperature of 1 °C and a decrease in rainfall of 8.79 %. SWAT predictions show an increase in potential evapotranspiration of 1.12 % and a decrease in streamflow of 4.62 %. Similar trends are predicted for the north-east monsoon period also. Groundwater recharge is predicted to decrease by 7 and 4 % in the A2 and B2 scenarios, respectively.

Keywords Global warming · Regional climate model · SWAT · Watershed · Groundwater recharge

S.G. Thampi (✉)
Department of Civil Engineering, National Institute of Technology,
Calicut 673601, Kerala, India
e-mail: santoshthampi@yahoo.com

K.Y. Raneesh
Department of Civil Engineering, Vedavyasa Institute of Technology,
Calicut, Kerala, India
e-mail: kyraneesh@yahoo.com

S. Shrestha et al. (eds.), *Managing Water Resources under Climate Uncertainty*,
Springer Water, DOI 10.1007/978-3-319-10467-6_8

1 Introduction

The post-industrial revolution era has witnessed a tremendous growth in industrial and economic activities, resulting in the emission of large amounts of greenhouse gases into the atmosphere. Increasing greenhouse gas concentrations in the atmosphere is believed to be causing large-scale changes to the earth's climate system including the hydrological cycle. How water resources and water availability in different parts of the world may be affected by the projected climate change is a question of grave concern to the scientific community. Climate variability and change are expected to alter regional hydrologic conditions resulting in a variety of impacts on water resource systems throughout the world, primarily by changes in precipitation, temperature, and evapotranspiration, which are the primary input variables for the terrestrial part of the hydrological cycle. Climate reports from the Intergovernmental Panel on Climate Change (IPCC) conclude that freshwater systems are highly vulnerable to climate change (Cooley et al. 2009).

Large-scale global climate models (GCMs) are powerful tools for evaluating the impact of increasing greenhouse gas concentrations in the atmosphere on climate, although it is widely recognised that there is considerable uncertainty in the results. However, the coarse resolution of the GCMs prohibits their direct use at local or regional scales. Most river basins are of the same order of magnitude or smaller than the grid dimensions employed in the GCMs. In order to translate the changes in atmospheric conditions to variables that are of direct relevance to water resources management, the output from GCMs has to be applied to hydrological models. However, coupling the GCMs with hydrological models is quite challenging for several reasons. A major problem is the difference in the spatial scale employed for obtaining global climate projections and that required to be used by local or regional hydrologic models for water resources management. Due to limitations in the understanding of local climate processes, data, and computing power, GCMs are typically run on large sized grid cells (150–300 km) and consequently, local gradients in precipitation and temperature are smoothened out. However, for many local hydrological processes, these gradients are very important, and therefore, the climate projections from GCMs have to be downscaled before applying it to hydrological models. The need for downscaling is particularly important for mountainous regions, which provide important environmental services such as water supply for drier lowlands, and are considered quite sensitive to environmental changes (Viviroli et al. 2010).

Although many methods for downscaling large-scale climate projections have emerged, very few of these methods have been implemented in complex regions such as tropical mountain areas (Maraun et al. 2010). The application of statistical downscaling methods is hindered by the lack of long-term precipitation records needed to fit statistical models of precipitation occurrence and quantity as well as the low density of existing rain gauge stations. On the other hand, regional climate models (RCMs) operate at a typical resolution of 50 km or lower and can capture the spatio-temporal variability of climate in much greater detail than GCMs. RCMs are the best tools for dynamic downscaling of climate features in order to make predictions for a particular region (Jones et al. 2004). By providing more realistic simulations of present and future climate change compared to the coarser resolution GCMs, RCMs can provide a better understanding of the impact these changes will have on water resources. RCMs are based on the same model physics as the driving GCM and therefore do not require a dense observational network as is needed for statistical downscaling. However, careful model validation with observed data is a necessary prerequisite to ensure that the RCMs accurately portray present-day spatio-temporal climate variability, particularly in mountainous regions, where the seasonal distribution of precipitation is strongly influenced by the orographic effect. Although RCMs have been used successfully for climate change studies all over the world (Marengo et al. 2009; Soares and Marengo 2009), their application is still in the early stages (Urrutia and Vuille 2009). A major drawback of RCMs is that they are computationally expensive and rather complex to implement.

Linsley (1960) observed that determination of the response of a catchment to climate change is one of the most serious problems that water resource managers face. Nemec and Schaake (1982) performed simulations to analyse the sensitivity of water resource systems to climate change in a deterministic framework. Most predictions on freshwater resources or the hydrological response of river basins under climate change scenarios are expressed in terms of annual streamflows (Kamga 2001; Legesse et al. 2003; Vorosmarty et al. 2005; Messager et al. 2006; Kundzewicz et al. 2008) and are derived by forcing hydrological models with the projections from RCMs (Jacob 2001; de Wit et al. 2007; Mileham et al. 2009). In many studies, this is done without assessing the quality of the RCM data (Middelkoop et al. 2001; Arnell et al. 2003) although it is quite evident that reliability of the output of a spatially distributed hydrological model is strongly dependent on the quality of climate forcing data used. Uncertainty in the predictions arise from the inability of RCMs to simulate present-day climatic conditions accurately (Christensen et al. 2008), and hence, it is extremely important that the RCM output is validated with historical observations before it is input into a hydrological model. In general, bias correction of the RCM output has to be performed in order to obtain a reasonable match with the observed data (Shabalova et al. 2003; Kleinn et al. 2005; Leander and Buishand 2007). Vegetative growth is affected directly by the changing concentrations of atmospheric gases, primarily CO_2, and indirectly by its impact on the regional climate and local weather patterns (Thomson et al. 2005). Increasing CO_2 concentration in the atmosphere may have a significant impact on regional crop production, necessitating cultivation of appropriate crop varieties and adoption of proper agricultural management practices (Reilly et al. 2001). The direct impact of increasing CO_2 concentration in the atmosphere on plant growth, crop water requirements, and runoff is not yet fully understood and is a subject of active research (Allen et al. 1998; Maroco et al. 1999; Wigley and Jones 1985; Allen et al. 1991; McCabe and Wolock 1992). The impact of climate change on water resources and vegetation are closely linked and must be considered together.

As political and technical measures for adapting to projected climate change need to be evolved at a regional level, studies on the impact of climate change at this level are very important (Varis et al. 2004). This requires downscaling of climate data to the level of hydrological models. Quantitative estimation of the hydrologic impacts of climate change will be helpful in understanding potential water resource problems and making better planning decisions. The rapid pace of economic development coupled with the population explosion will further increase the conflict between water use and water supply in the future. Understanding the likely impacts of projected climate change on water resources is important to enable proper management and utilisation of this resource.

Groundwater significantly contributes to meeting the rapidly increasing urban, industrial, and agricultural water requirements (de Vries and Simmers 2002). Proper groundwater recharge assessment is key to the efforts needed to manage this resource (Sanford 2002). Estimation of groundwater recharge is one of the most challenging problems in water resources research; it depends on a number of factors and is further complicated by environmental changes including climate change (Maxwell and Kollet 2008), urbanisation (Foster et al. 1994) and change in land use (Favreau et al. 2009). Accurate assessment of recharge, including its spatial and temporal distribution, is required for estimation of the groundwater resource and the streamflow, water quality protection, aquifer replenishment, etc. (Potter and Bowser 1995). Conventional water balance methods and numerical unsaturated–saturated zone simulations have been used to estimate groundwater recharge. However, the errors involved in the estimation of recharge are generally large (Gee and Hillel 1988). Also, these methods cannot give useful results during the short periods of measurement that are often used (Walker et al. 1991).

It is projected that global climate change will have a strong impact on water resources in many parts of the world (Bates et al. 2008). About one-third of the global population depends on groundwater for meeting their drinking water requirements (Falkenmark 2005). India is the largest user of groundwater in the world, with an estimated use of 230 cubic km of groundwater every year; this is more than a quarter of the global usage. Groundwater supports about 60 % of irrigated agriculture and more than 80 % of rural and urban water supplies. By 2025, an estimated 60 % of the groundwater blocks in the country will be in a critical condition (Shah 2009). Aquifers respond to climate fluctuations much more slowly than surface storage; as a result, compared to surface storage, aquifers act as a more resilient buffer during dry spells, especially when they have large storage capacity. Hence, the demand for groundwater is likely to increase further in view of the projected climate change (Shah 2009). Variations in temperature, quantity, and intensity of precipitation, may have a marked influence on aquifer recharge (Eckhardt and Ulbrich 2003; Scibek and Allen 2006). Determining how these variations affect natural groundwater recharge is a question of crucial importance, as direct recharge constitutes a basic element in the water balance, and knowledge and evaluation of this component is absolutely essential for the efficient management of water resources (Sophocleous 1992).

In this study, the impact of projected climate change on streamflow at a river gauge station, viz. Kuniyil, in the Chaliyar River in Kerala was predicted by performing simulations using AVSWAT2000 (ArcView interface of the Soil and Water Assessment Tool—SWAT), a physically based distributed hydrological model. The model was initially calibrated and subsequently its performance was validated using observed streamflow records for the period 1987–2006. Predictions of temperature and precipitation in the river basin for climate change scenarios A2 and B2 from the RCM PRECIS were corrected for bias and then input to SWAT to compute streamflow and evapotranspiration in the post-climate change period, 2071–2100. Although the use of RCM output in a hydrological model to predict the impact of projected climate change is not new, climate change impact studies on watersheds in humid tropical zones are relatively few in number. Further, a simple conceptual semi-distributed model for assessing the impact of climate change on direct groundwater recharge in a humid tropical river basin was developed. The model is based on the water balance concept and links the atmospheric and hydrogeological parameters to different hydrological processes to estimate the daily water table fluctuation. Data for six years from 2000–2005 was used for model calibration and that for four years from 2006–2009 was used for validation. Thereafter, the impact of projected climate change on groundwater recharge in the river basin during the period 2071–2100 was predicted using the model for the climate change scenarios A2 and B2.

2 Materials and Methods

2.1 Study Area

The Chaliyar is the fourth longest river in the state of Kerala, India, with a length of about 170 km. The total area of the Chaliyar River Basin is located in the Malappuram and Kozhikode districts of Kerala and the Nilgiris District of Tamil Nadu. The geographical area of the part of the river basin in Kerala (Fig. 1) is 2,530 km^2, and it lies between latitude 11° 06′N and 11° 36′N and longitude 75° 48′E and 76° 33′E. The basin exhibits undulating topography with steep slopes. Physiographically, the basin can be divided into three regions: the lowlands, mostly comprising the coastal belt; the midlands, comprising mostly of lateritic formations; and the highlands, covered by hard rocks. The main river starts from the Elambalari hills at an altitude of 2,067 m above mean sea level (MSL). The river basin has a humid tropical climate. The south-west monsoon (June to August) contributes about 60 % and the north-east monsoon (September to November) contributes about 25 % to the annual rainfall, respectively. The remaining 15 % is received as pre-monsoon showers during the months April to May. December to March is the dry period. The mean annual precipitation in the basin is 3,012 mm and the mean maximum and minimum temperatures are 34 and 24 °C, respectively. The annual average relative

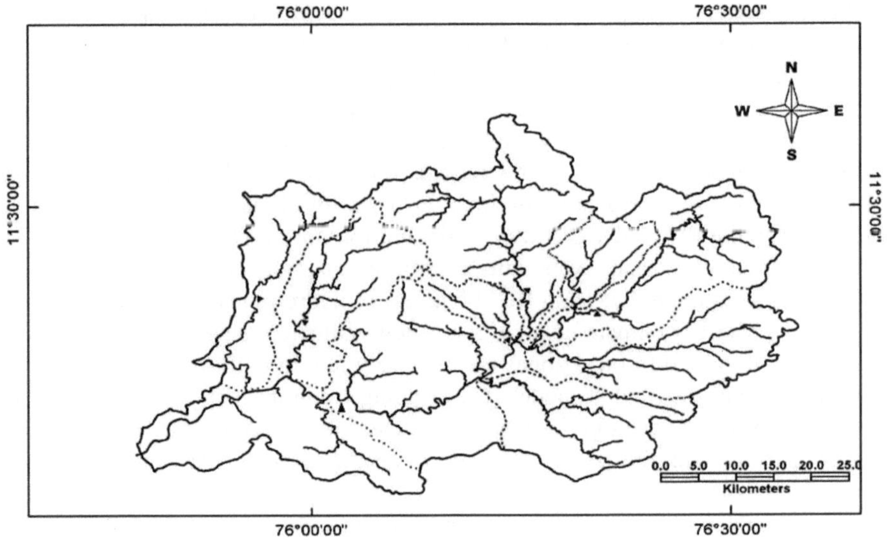

Fig. 1 Chaliyar River Basin

humidity ranges from 60 % (minimum) to 90 % (maximum) in summer, the corresponding values for winter being 65 and 85 %. The predominant land use is agriculture (60.04 %) and forests (38.74 %) with urban areas, pastures, waste lands, and rocky areas constituting less than 1 % of the total area. The elevation varies from about 20 m above the MSL in the lowlands to about 2,250 m in the highlands; the mean elevation being 338 m with a standard deviation of 458 m. The predominant soil type in the basin is loam (42.74 %), followed by clay (28.66 %), clay loam (24.18 %), and sandy loam (4.42 %). The soil/rock characteristics in the lowland, midland, and highland regions of the river basin are presented in Fig. 2. In the lowland coastal region, fine to medium grained alluvial deposits of 5–15 m

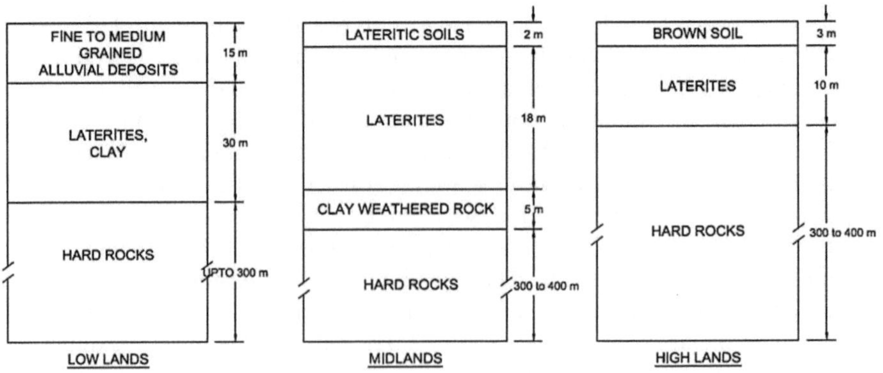

Fig. 2 Details of subsurface geology

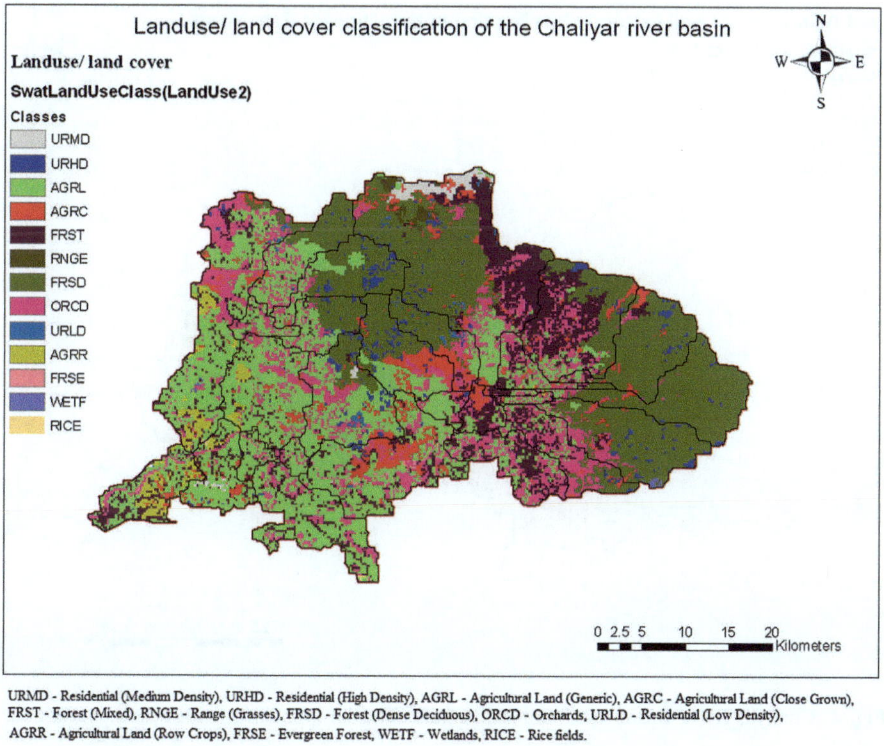

Fig. 3 Chaliyar River Basin—land use classification

thickness are present. Laterites, lithomargic clay, weathered, and hard rocks are located beneath the alluvial deposits. In the midland region, lateritic soil of 0.5–2.0 m thickness overlies laterite layers of 5–18 m thickness followed by lithomargic clay of less than 3 m thickness. This clay layer lies over a weathered rock layer of 1–5 m thickness underlain by hard rock with or without fractures. In the highland region, brown soil of less than 1–3 m thickness is found above laterites, and weathered and hard rock with or without fractures. The thickness of the laterites in the highland varies between 2 and 10 m. Land use and soil maps are also presented (Figs. 3 and 4, respectively).

2.2 Input Data

Streamflow data at the Kuniyil gauging station of the Central Water Commission (CWC) was collected for a period of 26 years from 1981 to 2006. Also, meteorological data including daily rainfall, temperature, relative humidity, and solar radiation at the Kottaparamba observatory of the Centre for Water Resources

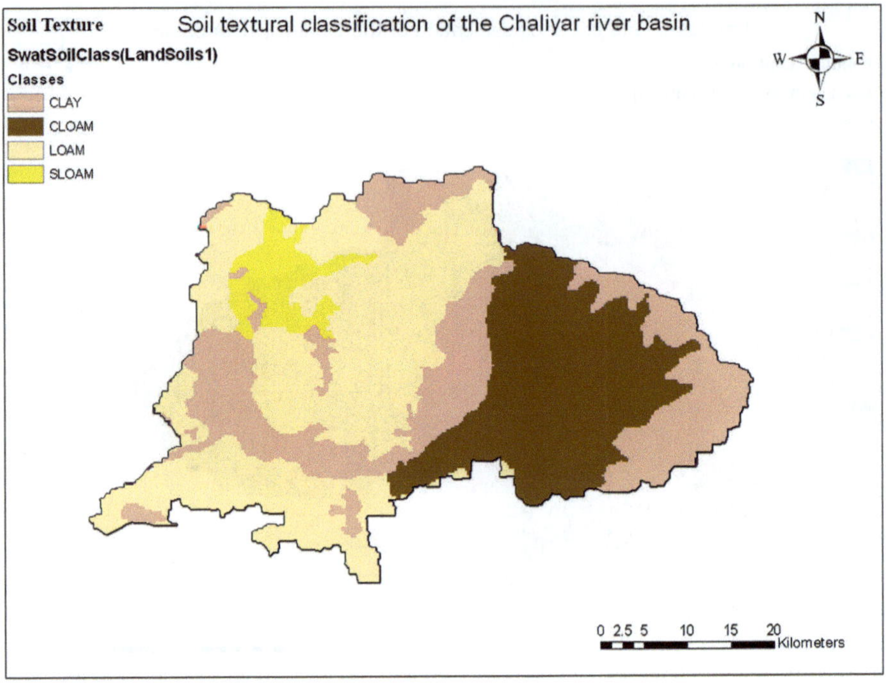

Fig. 4 Chaliyar River Basin—soil textural classification

Development and Management (CWRDM) were collected for the same period. Figure 5 shows the location of these stations. Meteorological data from a single station only was used in the simulation as this was the only station at which data was available. The total area of the river basin is covered in the following GTS maps (toposheets) published by the Survey of India (SoI): 49M14, 49M15, 49M16, 58A2, 58A3, 58A4, 58A6, 58A7, 58A8, 58A10, 58A11, and 58A12. These maps were scanned and digitised using the ArcGIS software, georeferenced and the georeferenced maps were digitised using the editor toolbar in ArcGIS. After preparing the contour layer, the slope map for the basin was generated. Figure 6 is the slope map of the Chaliyar River Basin generated using the ArcGIS software. Also, the digital elevation model (DEM) of the basin was derived in grid format using the ArcInfo software. The DEM (Fig. 7) was used for establishing sub-watershed boundaries and obtaining the point elevations.

For the estimation of groundwater recharge in the basin, data pertaining to observation wells in the river basin including their locations and position of the water table were collected from the Kerala State Groundwater Department. Six open wells each were identified for this study in the lowland, midland, and highland regions of the basin. The locations of these wells are shown in Fig. 8. A summary of the input data used in this study is presented in Table 1.

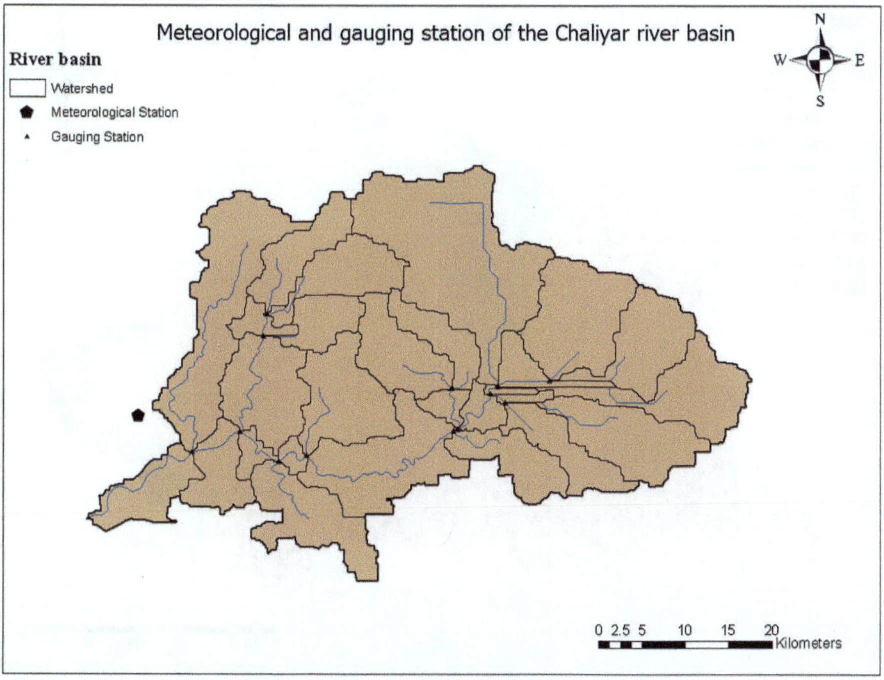

Fig. 5 Location of weather stations and river gauge station

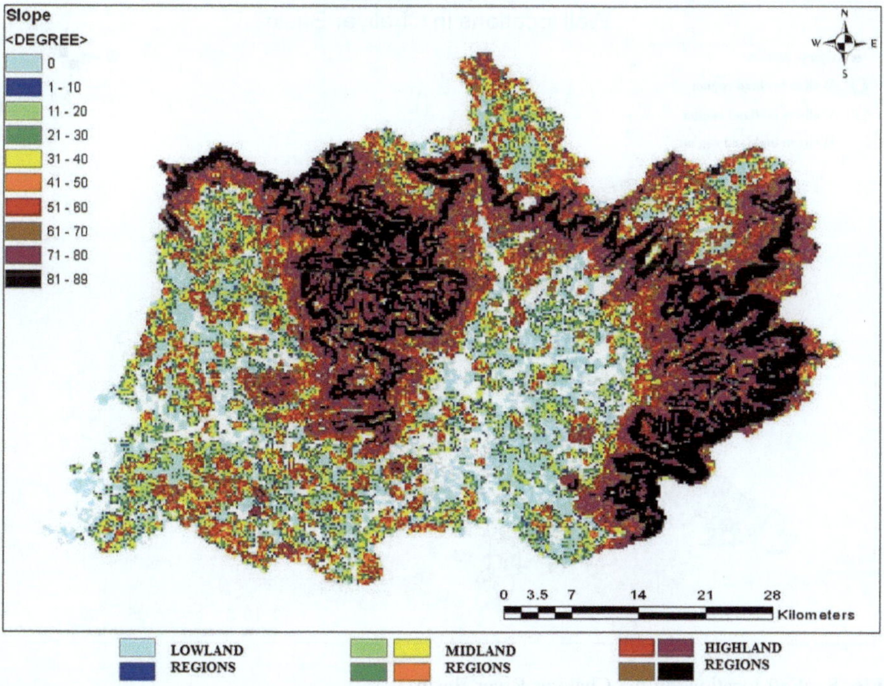

Fig. 6 Slope map of the Chaliyar River Basin

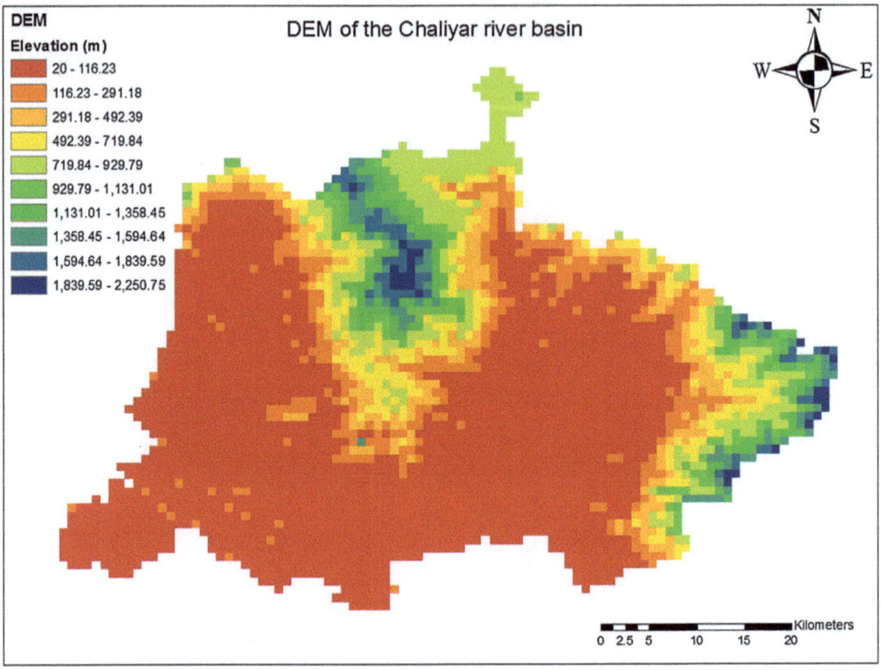

Fig. 7 DEM of the Chaliyar River Basin in grid format

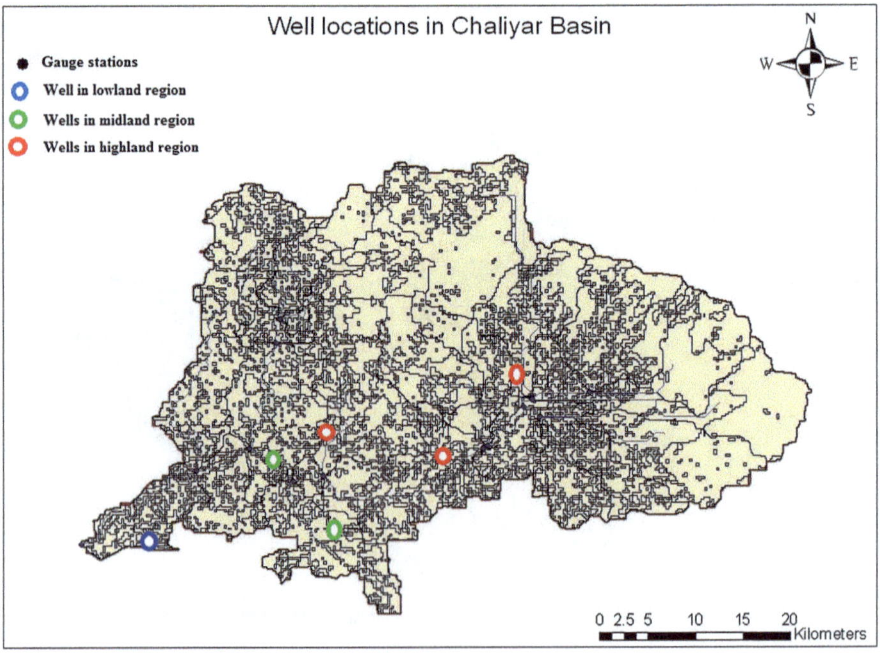

Fig. 8 Well locations in the Chaliyar River Basin

Table 1 Summary of input data

Data Type	Source	Scale	Data description/Properties
Topography	Survey of India	1:50,000	Elevation, spot height, drainage, boundary
Soil	Kerala State Land use Board	1:2,50,000	Soil classification and physical properties
Land use	Kerala State Land use Board	1:50,000	Land use classification—cropland, forest, pasture, etc.
Weather	Meteorological observatory of the CWRDM at Kottaparambu	–	Daily precipitation (1981–2006), pan evaporation (1981–2006), max. and min. relative humidity (1981–2006), max. and min. temperature (1981–2006)
Streamflow, water quality	Gauging station of the Central Water Commission at Kuniyil	–	Daily discharge (1981–2006), water level (1981–2006), water quality (1981–2006), sediment load (1981–2006)
Water table fluctuation	Kerala State Groundwater Dept.	–	Daily water level in observation wells for the above period

2.3 Soil and Water Assessment Tool (SWAT)

The hydrological model employed in this study is the Soil and Water Assessment Tool (SWAT), developed by the United States Department of Agriculture (USDA) to predict the impact of land management practices on water, sediment, and agricultural chemical yields in large complex watersheds with varying soils, land use, and management conditions over long periods of time. SWAT is in the public domain and is considered a standard tool in spatial decision support systems (Schuol and Abbaspour 2007; Keshta et al. 2009; Cibin et al. 2010). All major components of the hydrological cycle including rainfall, snow, interception storage, surface runoff, infiltration, evapotranspiration (ET), lateral flow, percolation, pond and reservoir water balances, shallow and deep aquifers, and channel routing are represented in the model. Generation of sediment based on a modified version of the Universal Soil Loss Equation (USLE) and routing of sediments in river channels are included. The crop growth model in SWAT is a simplification of the EPIC (Erosion Productivity Impact Calculator) with the concepts of phenological crop development based on daily accumulated energy units, harvest index for partitioning grain yield, Monteith's approach for potential biomass and water, and nutrient and temperature stress adjustments. It tracks the movement and transformation of several forms of nitrogen and phosphorus in the watershed and the movement and decay of pesticides. It has a modular setup and details on the modules can be found in the theoretical documentation (Neitsch et al. 2001). SWAT has been extensively used since 1993 mainly by hydrologists for analysing problems in watershed hydrology (Srinivasan et al. 1998; Santhi et al. 2001; Cao et al. 2006; Schuol and Abbaspour 2007; Keshta et al. 2009; Cibin et al. 2010; Alansi et al. 2009).

For modelling, a watershed may be partitioned into a number of sub-watersheds or sub-basins. The part of the Chaliyar River Basin in Kerala was delineated into a number of sub-basins using a 20-m DEM created in ArcInfo grid format and the stream network map. By partitioning, different areas of the watershed can be referenced to one another spatially. The input for each sub-basin is grouped or organised into the following categories: climate; HRUs; ponds/wetlands; groundwater; and the main channel or reach draining the sub-basin. HRUs are lumped land areas within the sub-basin that have unique land use, soil, and management combinations. Once the HRU distribution has been defined, the weather data to be used in the simulation is imported into SWAT. SWAT requires daily data of minimum and maximum temperatures, precipitation, solar radiation, wind speed, and relative humidity to perform simulations and uses long-term monthly averages of data. In addition, statistical parameters are generated from the long-term daily data to predict missing observations. Surface runoff from daily rainfall is estimated with the modified Soil Conservation Service Curve Number method (NRCS–CN). Channel routing is performed using either the variable-storage method or the Muskingum method. Evapotranspiration includes evaporation from rivers and lakes, bare soil, and vegetative surfaces, evaporation from within the leaves of plants (transpiration), and sublimation from ice and snow surfaces. The model computes evaporation from soils and plants separately as described by Ritchie (1972). Potential soil water evaporation is estimated as a function of potential evapotranspiration and leaf area index (area of plant leaves relative to the area of the HRU) and is assumed to be unaffected by microclimatic processes such as advection or heat-storage effects. Of the methods available for estimating potential evapotranspiration (Hargreaves et al. 1985; Priestley-Taylor 1972; Monteith 1965), the Penman–Monteith method was used in this study. Actual soil water evaporation was estimated by using exponential functions of soil depth and water content. Plant transpiration was estimated as a linear function of potential evapotranspiration and the leaf area index.

The categories specified in the land use/land cover map should be reclassified into SWAT land cover/crop types (Table 2) using any one of the three options available in the model. In this study, the SWAT default database land cover/crop-type or urban code for each category was selected when the land cover/land use map theme is loaded in the interface. The crop growth parameters for a land cover/crop type are stored in a database (crop.dat) file in SWAT. A code is used to represent the land cover/crop type in the database, and this is used by the GIS interface to link the land use/land cover maps to SWAT crop types. When adding a new crop species or land cover category, the four letter code for the new crop must be unique. The land cover/crop-type database contains information needed by SWAT to simulate the growth of a particular crop. The growth parameters in the crop growth database define plant growth under ideal conditions and quantify the impact of water stress on plant growth. Differences in growth between plant species are defined by parameters in the crop growth database. A number of parameters are specified for more than 70 crop/vegetation types in the crop database attached to SWAT: biomass–energy ratio, harvest index, base, and optimal temperature for plant growth, maximum leaf area index (LAI), fraction of growing season when

Table 2 SWAT land use/Land cover classification

Sl. No.	Area		Land use/Land cover	Land cover defined in SWAT database	
	Percentage	Hectare		Definition	Symbol
1	0.54	1,366.72	Residential area (rural)	Residential—medium/low density	URML
2	0.68	1,721.06	Residential area (urban)	Residential—medium/low density	URML
3	8.20	20,753.95	Double crop (Kharif and Rabi)	Rice	RICE
4	3.19	8,073.79	Agriculture plantation—arecanut	Agricultural land—generic	AGRL
5	14.71	37,230.57	Agriculture plantation—rubber	Agricultural land—close grown	AGRC
6	3.54	8,959.63	Agriculture plantation—banana	Agricultural land—row crops	AGRR
7	2.35	5,947.78	Agriculture plantation—cashew	Orchard	ORCD
8	2.54	6,428.66	Grass land	Range—grasses	RNGE
9	4.21	10,655.38	River/Water bodies	Water	WATR
10	21.23	53,732.49	Forest evergreen—dense	Forest—evergreen	FRSE
11	22.22	56,238.15	Forest deciduous—dense	Forest—deciduous	FRSD
12	16.59	41,988.79	Forest mixed	Forest—evergreen	FRST

LAI declines, maximum root depth, and potential heat units required for maturity of crop. The model differentiates between annual and perennial plants. The plant growth model is used to assess the removal of water and nutrients from the root zone, transpiration, and biomass/yield production. The model estimates stresses caused by water, nutrients, and temperature. It allows the user to define management practices adopted in every HRU, the start and the end of the growing season, the timing and amount of fertiliser, pesticide, and irrigation applications and the timing of tillage operations.

2.4 Groundwater Recharge Model

One of the most popular conceptual models employed in water balance studies is the 'tank model' or 'multiple box model', originally proposed by Sugawara and Funiyuki (1956) for the computation of runoff from the catchment area of a river, substituted by a combination of a number of storage type vessels. It is a kind of lumped, nonlinear, continuous model (Sugawara 1995). Alemaw and Chaoka (2003) proposed a continental distributed geographic information system (GIS)-based water

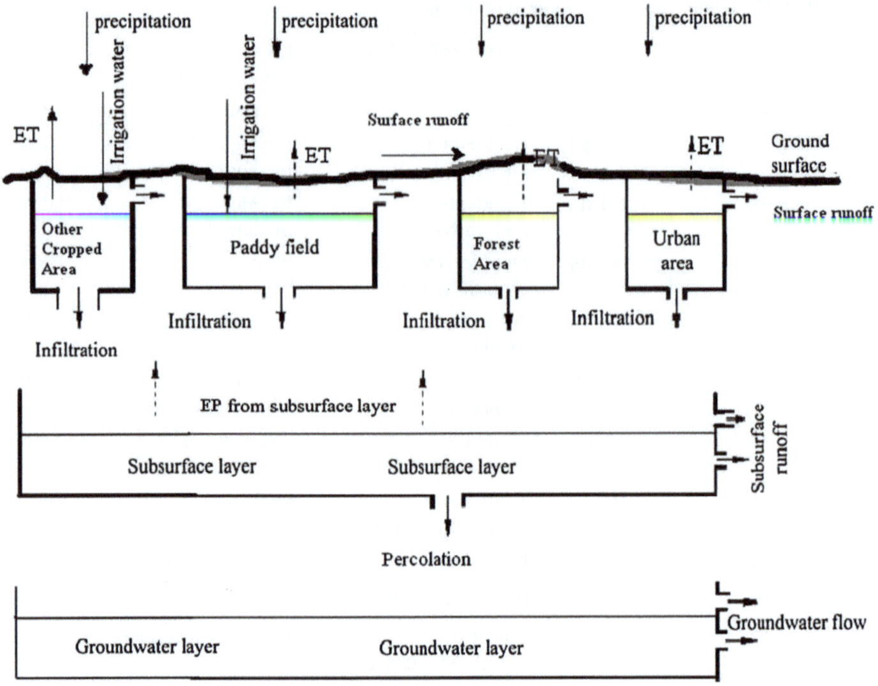

Fig. 9 Groundwater recharge model based on the water balance concept

balance model, taking into account surface and subsurface processes, to estimate the runoff from a matrix of specific georeferenced grids representing Southern Africa. Models developed for estimating groundwater recharge include that of Khazaei et al. (2003) for arid and semi-arid regions of South-East Iran and the semi-distributed model of He et al. (2008).

The model employed in this study for recharge assessment is a simple conceptual groundwater recharge model (Fig. 9) in which the aquifer is modelled as a series of three layers (Raneesh and Thampi 2013). The topmost surface layer is associated with land use such as paddy fields, farms, forest areas, urban areas. The middle layer is the subsurface layer, and the bottommost layer is the groundwater layer. The aquifer is assumed to function as vertical tank storage with many outlets characterised by outflow coefficients. Water balance is performed in the model at the level of the hydrologic response units (HRUs). The Chaliyar River Basin was divided into 79 HRUs. The model estimates direct groundwater recharge (i.e. from the infiltration of rainfall) based on changes in the soil moisture (Grindley 1967). It was assumed that recharge occurs when the effective precipitation (precipitation minus runoff) at the soil surface exceeds evapotranspiration and raises the soil moisture to field capacity. Daily time-stepping was employed because the averaging of weekly and monthly precipitation can mask daily recharge events (Howard and Lloyd 1979). The model was developed on a Microsoft Visual Basic 6.0 platform.

2.5 Regional Climate Models and Climate Change Scenarios

GCMs describe the global climate system, representing the complex dynamics of the atmosphere, oceans, and land by mathematical equations that balance mass and energy and predict the future in terms of temperature, precipitation, air pressure, and wind speed. These models are used to examine the impact of increased concentrations of greenhouse gases in the atmosphere on the long-term climate. However, being computationally intensive, they can only be run on supercomputers. Results vary with model attributes, including its components, resolution, flux-adjustment, and emission scenario forcings. Based on different projected emission scenarios, the likely range of change in global surface temperature is predicted to be between 1.1 and 6.4 °C (Meehl et al. 2007). For a future warmer climate, many of these models indicate that precipitation may decrease in the tropical regions (Meehl et al. 2007). Although wind speed, relative humidity, and sunshine hours influence evapotranspiration, predicted changes resulting from climate change on these variables are small and uncertain (IPCC 2007).

The Intergovernmental Panel on Climate Change (IPCC) published a set of emission scenarios in the Special Report on Emissions Scenarios (SRES) (Nakicenovic et al. 2000) to serve as the basis for assessment of the impacts of likely climate change in the future. To assess the hydrologic impacts of projected climate change, reliable climate models are required, in addition to projections of socio-economic developments and responses to climate change so as to enable the estimation of anthropogenic emission of greenhouse gases and aerosols. Anthropogenic emissions, coupled with any natural climate change trends, are referred to as emission scenarios. Among the scenarios outlined in the SRES, four marker scenarios, viz. A1, A2, B1, and B2 are most often used (Van Vuuren and O'Neill 2006). The A1 and B1 scenarios emphasise ongoing globalisation and project a homogeneous world, whereas the A2 and B2 scenarios lay emphasis on social, economic, and environmental development on a regional and local basis and project a heterogeneous world. In this study, scenarios A2 and B2 were adopted for investigating the hydrologic impact of climate change. The A2 scenario projects high population growth and slow economic and technological development, while the B2 scenario projects slower population growth, rapid economic development and lays more emphasis on environmental protection. The A2 scenario is characterised by lower trade flows, slower technological change, and uneven economic growth; the income gap between the industrialised and developing parts of the world does not narrow. The B2 world shows relatively better concern for environmental and social sustainability; government policies and business strategies at the national and local levels are increasingly influenced by environmentally aware citizens, with a trend towards local self-reliance and stronger communities. Human welfare, equality, and environmental protection all have high priority, and these are addressed through community-based social and technical solutions, though implementation rates vary across regions. Emission of greenhouse and other gases and other driving forces have been quantified in the Fourth Assessment Report of

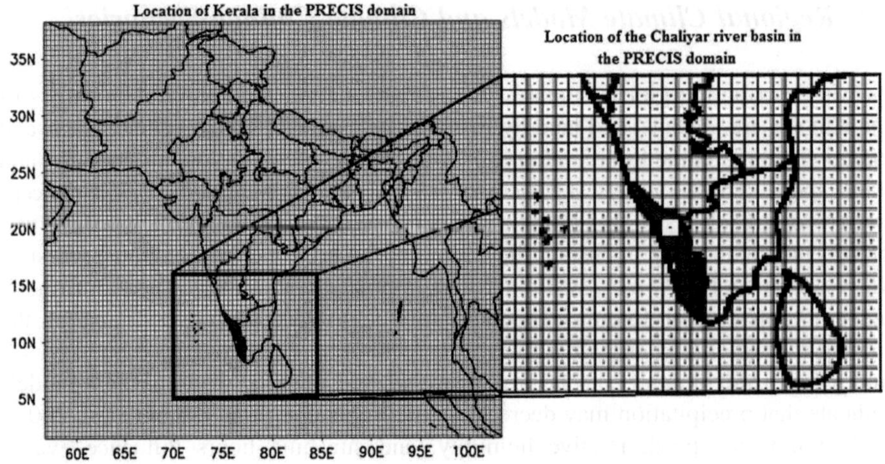

Fig. 10 Location of Kerala and the Chaliyar River Basin in the PRECIS domain

the IPCC (IPCC 2007) for use in climate simulations using GCMs. RCMs operate at a typical resolution of 50 km or lower and capture the spatio-temporal variability of climate in much greater detail than GCMs. By providing more realistic scenarios of anticipated climate change compared to coarser resolution GCMs, the RCMs provide a better understanding of the impact of these changes.

The Indian Institute of Tropical Meteorology (IITM), Pune, in collaboration with the United Kingdom Hadley Centre has developed future scenarios for India. Dynamical downscaling of the GCM output was performed using the RCM PRECIS (Rupa Kumar et al. 2006). The present version of PRECIS has a horizontal resolution of 50 km. Climate data from PRECIS for both the A2 and B2 scenarios, for the present (1981–2010) and future (2071–2100) periods, were collected from IITM and used in this study. PRECIS has been configured for a domain extending from about 1.5–38°N and 56–103°E (Fig. 10). The domain chosen is adequate to include almost all the regional-scale circulation mechanisms. The horizontal resolution is 1.24° latitude by 1.88° longitude in the driving GCM (HadAM3H) and 0.44° × 0.44° in the RCM (PRECIS). With a nominal resolution of 50 km × 150 km, the RCM provides a more realistic representation of orographic features over Southern Asia. The lateral boundary conditions for the RCM are specified using the output of the driving GCM. Although PRECIS captures local gradients better than the GCM, it is observed that a further increase in resolution is necessary to represent local gradients properly.

2.6 Bias Correction of the PRECIS Output

Several studies have identified the need to check and correct bias, if any, in the RCM output, before its use in impact studies (Shabalova et al. 2003; Kleinn et al. 2005; Leander and Buishand 2007). One problem with the use of output from

RCMs directly for hydrologic purposes is that the simulated precipitation differs systematically from the observed precipitation (Frei et al. 2003). Hay et al. (2002) proposed the use of gamma distribution to match the computed and the observed daily precipitation. It was observed that the corrected precipitation data did not have the day-to-day variability that exists in the observed data. Leander and Buishand (2007) observed that a relatively simple nonlinear correction, adjusting the bias both in the mean and variability, lead to better reproduction of observed daily precipitation than the commonly used linear scaling correction. But this method does not correct for the fraction of wet and dry days and lag inverse autocorrelation. They applied a power law transformation, which corrects the coefficient of variation and the mean of the precipitation values in a study on the Meuse basin. In this study, bias correction was performed using this method; the important statistics (coefficient of variation, mean, and standard deviation) of the PRECIS output and that of the observed data matched reasonably well. SWAT needs other meteorological forcing data as well (e.g. solar radiation, wind speed, and relative humidity). Unfortunately, predicted data were not available for these variables, and so simulations were performed using the SWAT simulated values. Bias correction was performed at a temporal resolution of one day, and on a grid size of 50 km × 50 km and the statistics (*CV*, mean and standard deviation) of the PRECIS, data were compared with those of the observed values. It was found that they matched reasonably well.

3 Results and Discussion

3.1 Bias Corrected Data

The future precipitation and temperature (2071–2100) after bias correction and the corresponding daily values for the present period (1981–2005) are presented in Figs. 11 and 12, respectively. Detailed analysis of the data reveals that precipitation in the future scenarios will be lower when compared to the present-day observed values during both southwest and northeast monsoon periods; however, in summer, precipitation in the future scenarios is likely to be higher in comparison with the observed present-day values. The projected annual average rainfall in the Chaliyar River Basin for the period 2071–2100 shows a reduction of about 20 and 10 % from the present-day annual average value in the A2 and B2 scenarios, respectively.

Both maximum and minimum temperatures will be higher in the A2 and B2 scenarios when compared with the present-day observed values, the increase being more in the A2 scenario. With other influencing factors, and land use and crop/vegetation patterns remaining constant, higher temperatures would lead to increased evapotranspiration. This coupled with reduced rainfall would imply lower streamflow.

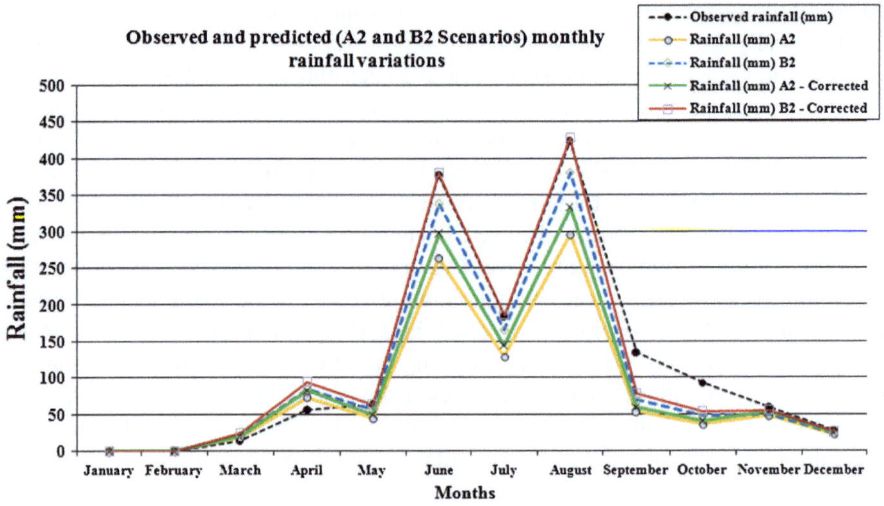

Fig. 11 Comparison of present and future daily precipitation

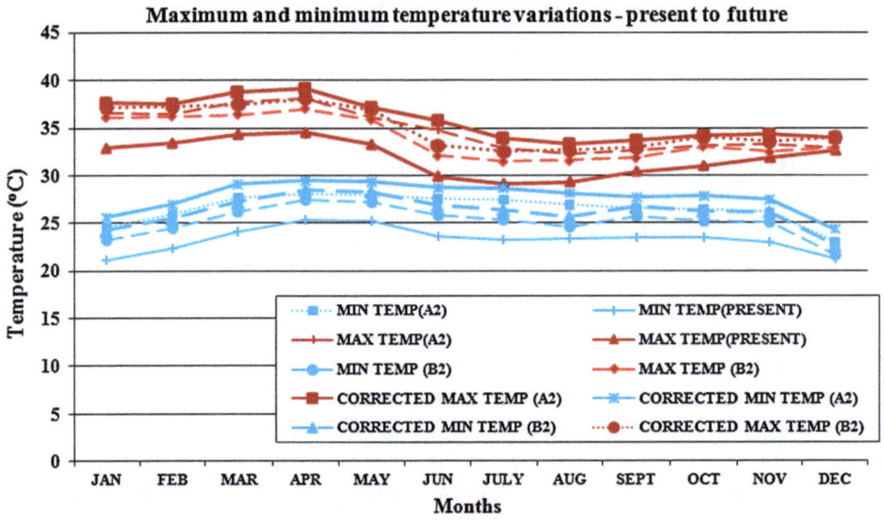

Fig. 12 Comparison of present and future daily temperature

3.2 Calibration and Validation of SWAT

Model calibration involves varying parameter values within reasonable ranges until the differences between the observed and computed values are minimised. Refsgaard and Storm (1996) distinguished three types of calibration methods, viz. the manual trial-and-error method; automatic or numerical parameter optimisation

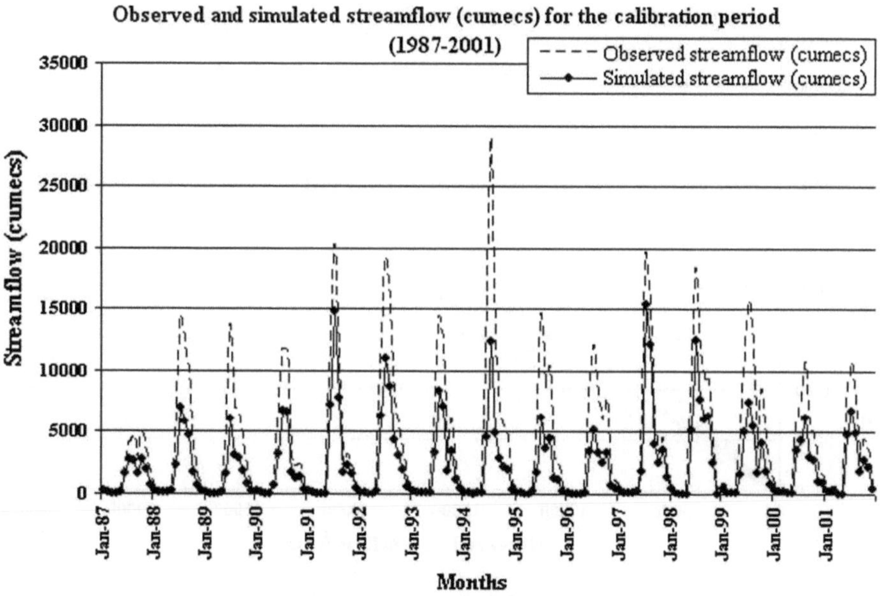

Fig. 13 Observed and computed annual streamflow—calibration

method; and a combination of these methods. Manual calibration is the most commonly used and is especially recommended in cases where a good graphical representation is strongly demanded for the application of simulation models (Haan and Skaggs 2003). However, it is very cumbersome, time-consuming and requires experience. In this study, the model was calibrated using observed streamflow records for a 15-year period from January 1987 to December 2001. The parameters were selected based on the study area and literature (Heuvelmans et al. 2004; Chu and Shirmohammadi 2004; Gosain et al. 2006). The model parameters were adjusted manually by trial and error based on certain statistical indicators and the characteristics of the study area. The statistical criteria used to evaluate the hydrologic goodness-of-fit were the coefficient of determination (R^2) and the Nash–Sutcliffe efficiency (E_{NS}) (Nash and Sutcliffe 1970); the latter being most widely used in SWAT calibrations (Gassman et al. 2007). To capture uncertainty in the values of parameters, two additional runs were performed: one with a combination of extreme parameter values that would result in maximum streamflow and one with a combination that would result in minimum flow. After calibration, model validation was performed using the data for the remaining five-year period (2002–2006). The results of calibration of SWAT (streamflow at the Kuniyil gauging station) and the plot of computed versus observed annual streamflow are presented in Figs. 13 and 14, respectively. It can be seen that the model slightly underestimates flow in all the years. The Nash–Sutcliffe efficiency for the calibration period was 0.667 and the R^2 value was 0.932. Figures 15 and 16 present the results of model validation. The Nash–Sutcliffe efficiency of the model in the

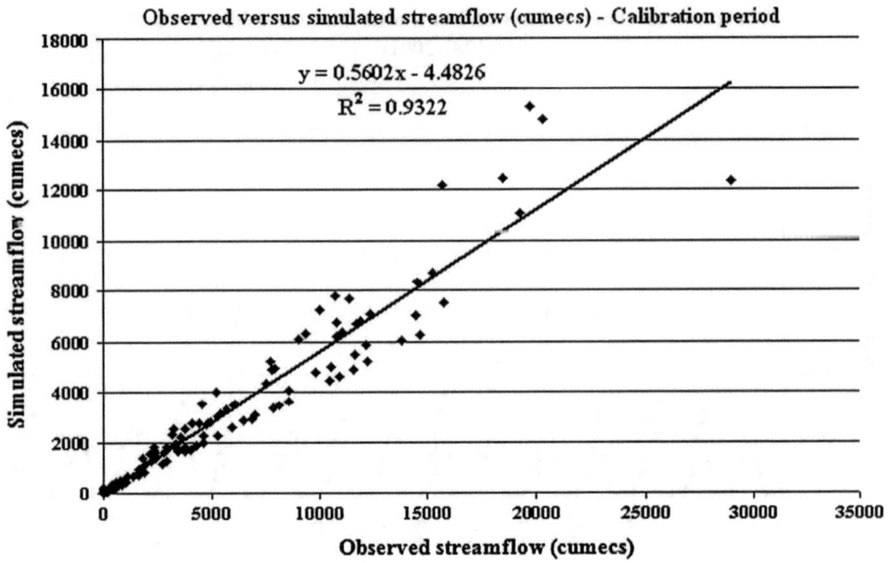

Fig. 14 Plot of observed versus computed streamflow—calibration

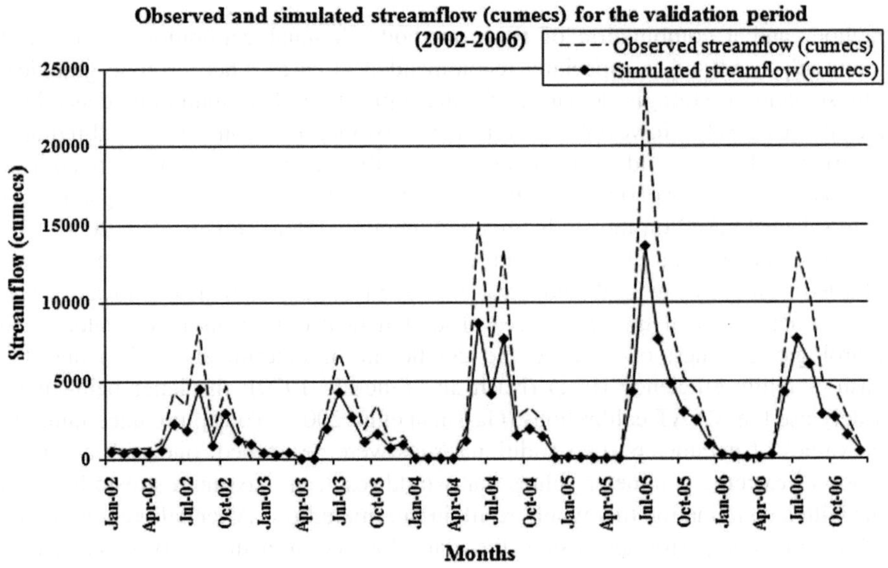

Fig. 15 Observed and computed streamflow—validation

validation period was 0.719 and the corresponding R^2 value for the validation phase was 0.996. Hence, it can be stated that the calibrated model performed reasonably well.

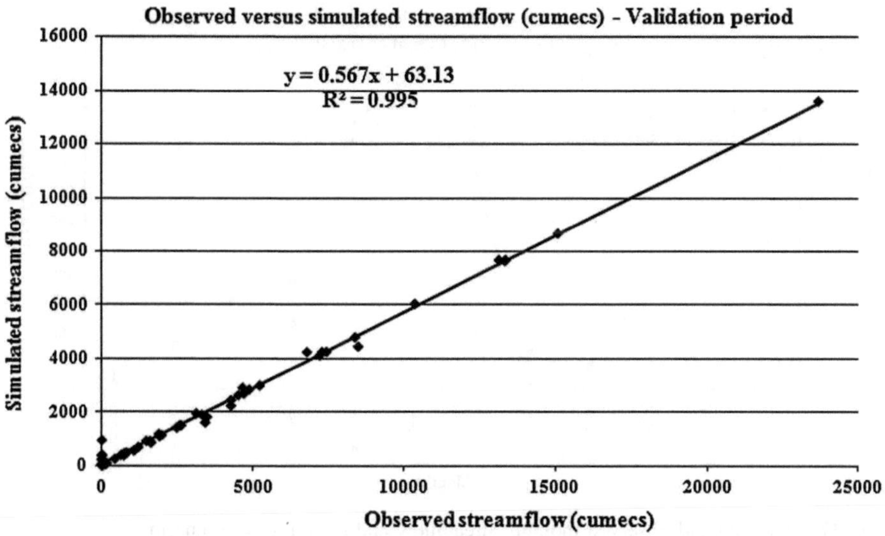

Fig. 16 Plot of observed versus computed streamflow—validation

Reliability and accuracy of results of model simulations depend on the availability of accurate meteorological and hydrologic data and good estimates of parameters. Arnold et al. (1998) cited spatial variability in precipitation data as one of the major limitations to large area hydrologic modelling. An important limitation of this study is the limited spatial resolution of hydrologic and meteorological data.

3.3 Streamflow in the Climate Change Scenarios

Streamflows in the future period (2071–2100), consequent to climate change, were predicted with the hydrological model SWAT using rainfall and temperature data for the A2 and B2 scenarios from PRECIS, after correction for bias. In the A2 scenario for the southwest monsoon period, an average increase in temperature of 2 °C and a decrease in rainfall of 11.50 % were observed. Predictions by SWAT indicate an increase in potential evapotranspiration of 1.14 % from the present-day average values. For the same period in the B2 scenario, an average increase in temperature of 1 °C and decrease in rainfall of 8.79 % was observed. SWAT predictions show an increase in potential evapotranspiration of 1.12 %. Streamflow showed a decreasing trend; 7.53 % decrease in the A2 and 4.62 % decrease in the B2 scenario. Similar trends were predicted in the northeast monsoon period also. In the A2 scenario, an average decrease in rainfall by 8.70 %, increase in temperature of 2 °C, increase in potential evapotranspiration by 1.09 %, and a decrease of 4.31 % in streamflow from the present-day average values were observed. In the B2 scenario, the average temperature increased by 1 °C, rainfall decreased by 4.73 %,

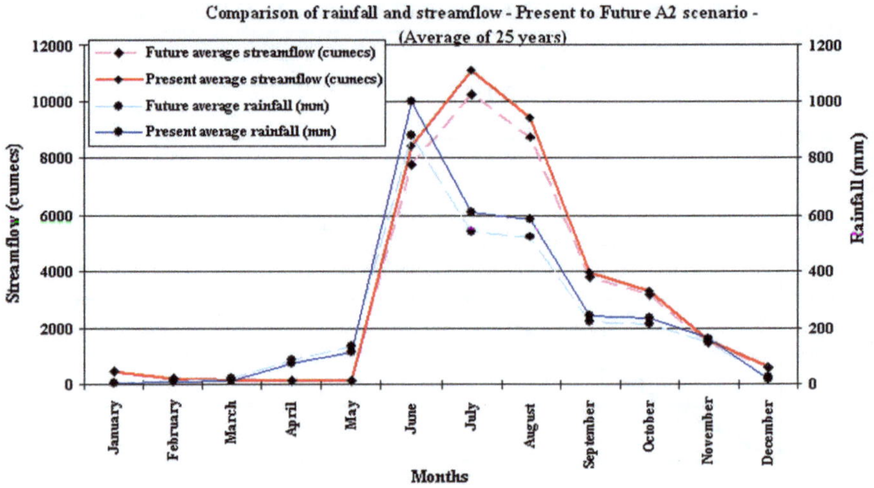

Fig. 17 Predicted and observed monthly streamflow and rainfall—A2 scenario

evapotranspiration increased by 1.01 %, and streamflow decreased by 3.01 %. The variation in monthly streamflow and rainfall from the present-day values in the A2 scenario is presented in Fig. 17, whereas the corresponding information for temperature and ET is presented in Fig. 18. The corresponding variations in the B2 scenario from the present-day values are presented in Figs. 19 and 20, respectively. A very small decrease is observed in streamflow in the Chaliyar River during the summer and pre-monsoon period due to predicted warming in the future even though there is a slight increase in the rainfall. In the A2 scenario, precipitation

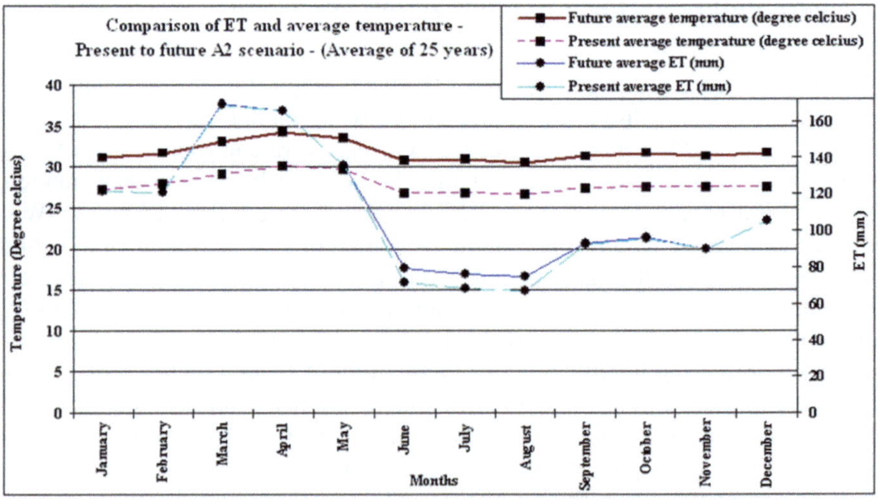

Fig. 18 Predicted and observed monthly average temperature and ET—A2 scenario

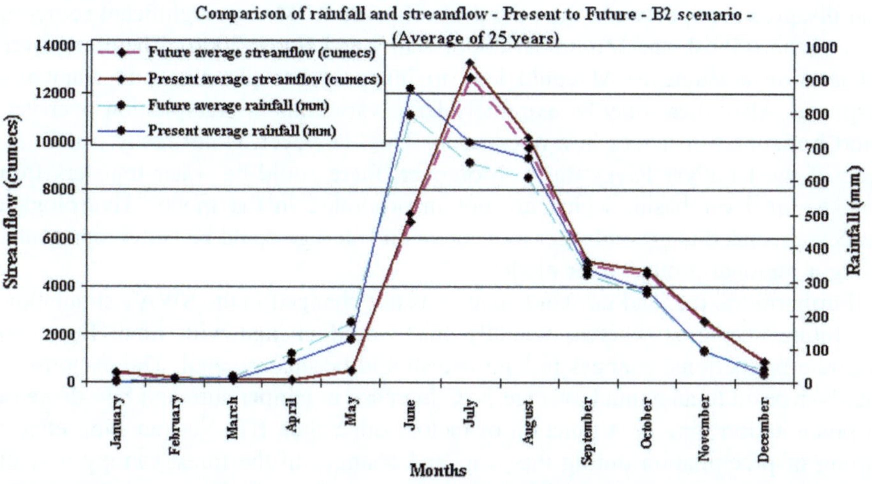

Fig. 19 Predicted and observed monthly streamflow and rainfall—B2 scenario

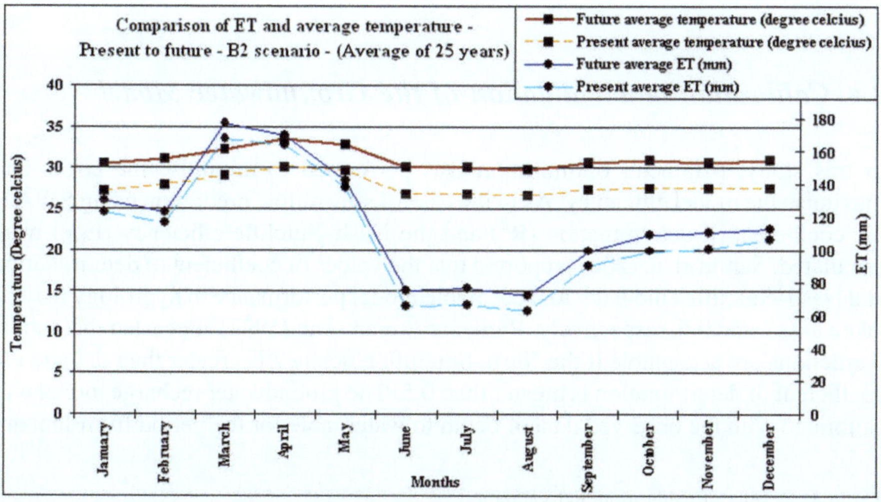

Fig. 20 Predicted and observed monthly average temperature and ET—B2 scenario

showed a 1.60 % change from the present-day average value; the corresponding value in the B2 scenario was 1.40 %. The average temperature increased by 3 and 2 °C from the present-day average values in the A2 and B2 scenarios, respectively. Evapotranspiration values showed an increase of 1.05 and 1.03 % in the A2 and B2 scenarios, respectively. Results indicate that although there is no drastic change in streamflow during the monsoon period, the occurrence of high flows will reduce significantly. It may be noted that the projections presented here are based on the predictions of a single RCM for the two emission scenarios; it is widely recognised that disagreements between the output of different RCMs is a significant source of uncertainty (Ghosh and Mujumdar 2008; Wilby and Harris 2006). Therefore, over-reliance on a single RCM could lead to inappropriate planning and adaptation response. Also, there may be extremely large variations in precipitation over very short horizontal distances in a river basin. This, however, is not really true in the case of the Chaliyar River Basin. Moreover, there could be water transfers from outside the river basin, which are not incorporated in the model. Hydrological processes related to groundwater movement and storage could be better represented using a regional groundwater model.

Furthermore, the land use/land cover was not changed in the SWAT simulations for future scenarios whereas actually this would change with time. For more accurate predictions, changes in land use should be incorporated. The response of the river basin to an annual average 3 °C increase in temperature and 8 % decrease in precipitation may be a function of factors other than ET. Net radiation effects, timing of precipitation during the year, and changes in the forest canopy may all play significant roles in controlling average streamflow (Graham et al. 1990). The analysis simply illustrates the likely trend and magnitude of streamflow changes in the river basin related to possible changes in temperature and precipitation.

3.4 Calibration and Validation of the Groundwater Model

In this study, parameter optimisation was performed to minimise the errors and maximise the model efficiency. As in the case of streamflow prediction using SWAT, the coefficient of determination (R^2) and the Nash–Sutcliffe efficiency (E_{NS}) were calculated. Santhi et al. (2001) reported that the values of coefficient of determination and Nash–Sutcliffe efficiency for acceptable model performance in hydrology may be taken as 0.7 and 0.6, respectively. Ramanarayanan et al. (1997) suggested that model predictions are acceptable if the Nash–Sutcliffe efficiency is greater than 0.4 and the coefficient of determination is greater than 0.5. The groundwater recharge model was calibrated with the observed data of depth to water table for the period from January

Table 3 Results of calibration and validation

Sl. No.	Designation of well	Region	Calibration period (2000–2005)		Validation period (2006–2009)	
			E_{NS}	R^2	E_{NS}	R^2
1	KKDOW 156	Lowland	0.64	0.84	0.74	0.85
2	KKDOW 012	Midland	0.65	0.77	0.78	0.79
3	KKDOW 157	Midland	0.77	0.86	0.77	0.89
4	KKDOW 013	Highland	0.62	0.72	0.63	0.77
5	MPMOW 016	Highland	0.76	0.81	0.78	0.83
6	MPMOW 012	Highland	0.69	0.85	0.71	0.86

2000 to December 2005. Initial values of the parameters were fixed based on data from literature (He et al. 2008). These were subsequently refined during calibration. After calibrating the model, it was validated using observed depth to water table values for the period from January 2006 to December 2009. Data on observations from six open wells each located in the lowland, midland, and highland regions of the Chaliyar River Basin were used for calibration and validation. The results are presented in Table 3. Results of the computations performed are presented in Figs. 21 and 22. It can be

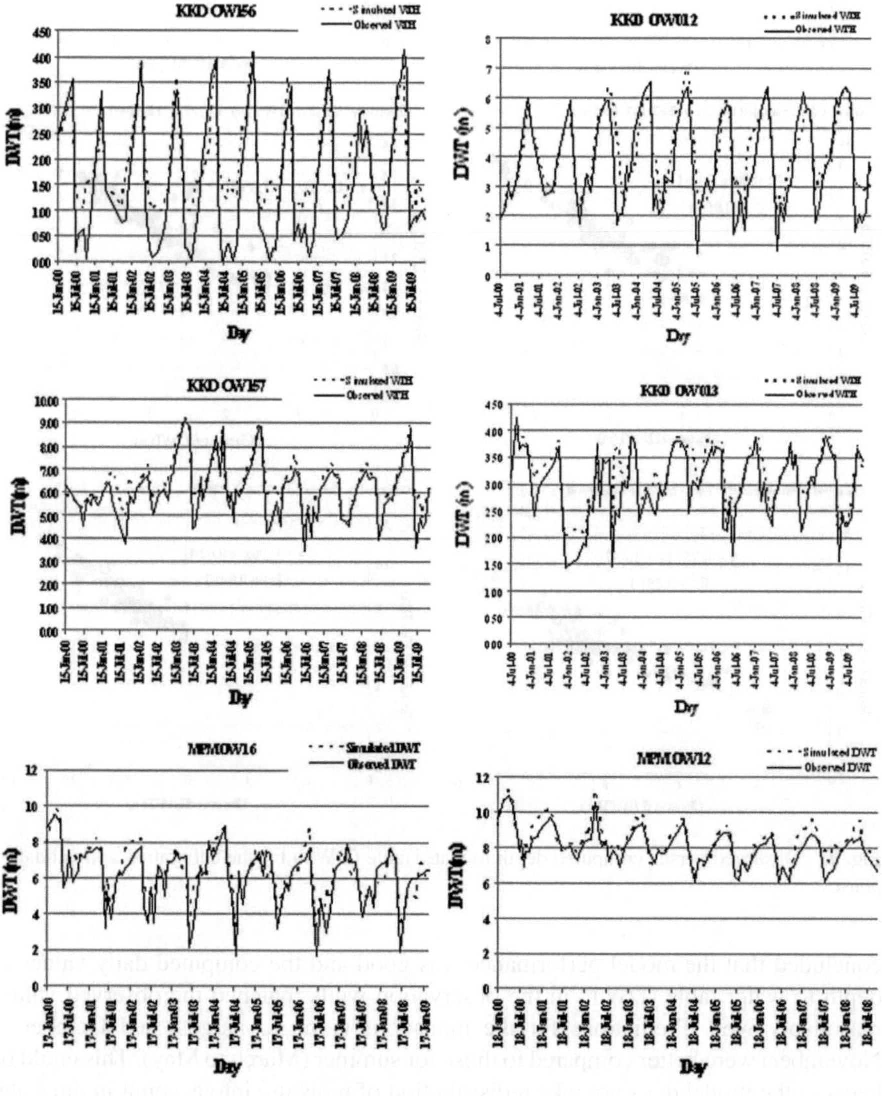

Fig. 21 Observed and computed depth to water table (DWT) values for the calibration and validation phases

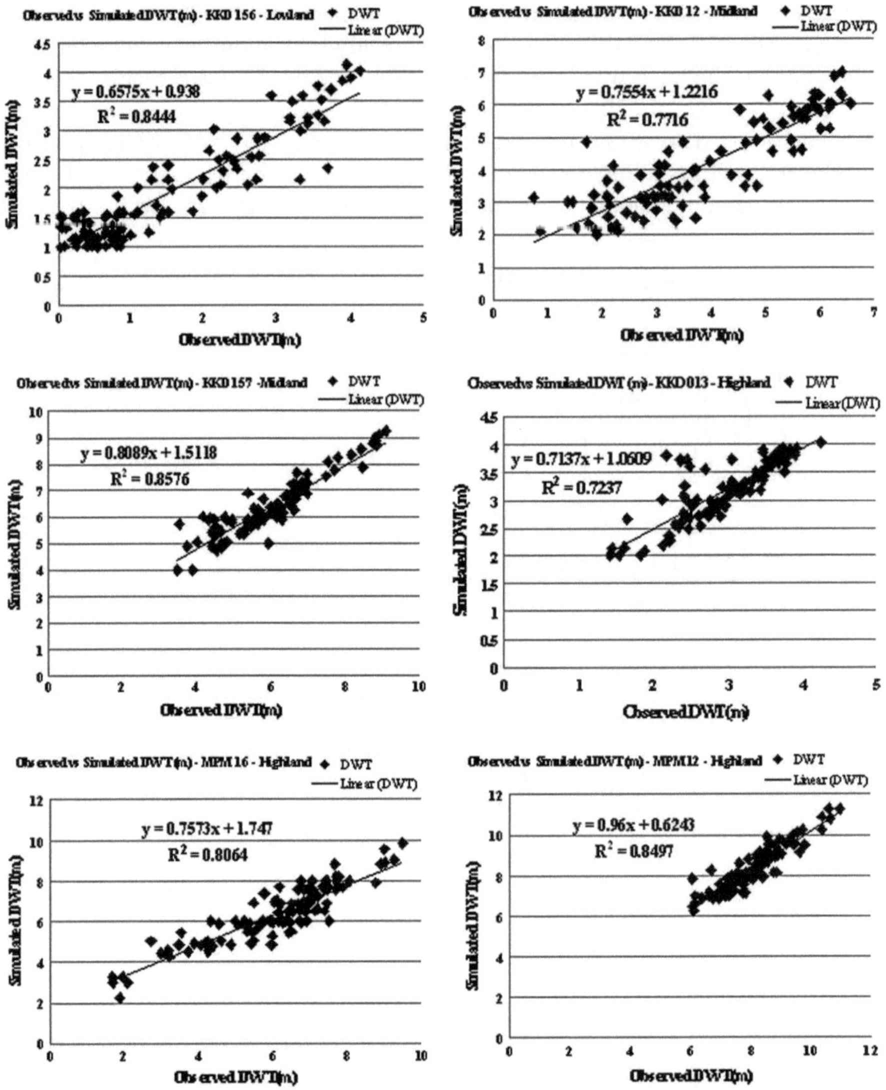

Fig. 22 Observed versus computed depth to water table (DWT) for the calibration and validation phases

concluded that the model performance was good and the computed daily values of depth to water table (DWT) in the observation wells matched the observed values reasonably well. Predictions for the monsoons (June to August and October to November) were better compared to those for summer (March to May). This could be because the model does not take redistribution of moisture into account in the water balance calculations.

3.5 Groundwater Recharge in Climate Change Scenarios

After validation, the water table levels in the future period (2071–2100) in the scenarios A2 and B2, were predicted using the groundwater recharge model. The rainfall and temperature data for this period generated by PRECIS was corrected for bias before it was input into the model. After correction for bias, the projected annual rainfall in the Chaliyar River Basin showed a decrease of 5 % in the A2 scenario and 3 % in the B2 scenario from the present-day annual average value. The maximum temperature increase was 3 and 1.5 °C from the present annual average values in the A2 and the B2 scenarios, respectively. The same pattern was observed in the case of the minimum temperature also (an increase of 3.5 °C in the A2 and 2 °C in the B2 scenarios). Potential evapotranspiration (*PET*) for this period was calculated using McCloud's formula (McCloud, 1970). Figure 23 presents a comparison between the depths to water table in selected wells in the basin. In the A2 scenario, there is an annual average increase in temperature by 3 °C, decrease in rainfall by 5 %, increase in potential evapotranspiration by 3 % and a reduction in groundwater recharge of 7 % whereas in the B2 scenario, annual average rainfall decreases by 3 %, temperature increases by 2 °C, potential evapotranspiration increases by 2 % and groundwater recharge reduces by 4 %. Highlights of predictions from this study are presented in Table 4. Reduction in recharge is more in the A2 scenario compared to that in the B2 scenario, since reduction in rainfall is more in the A2 scenario. It is observed that the net effect of climate change (decrease in rainfall and increase in temperature) is a lowering of the groundwater table. The maximum difference between the present-day and future depth to water table values in the two climate change scenarios occurs in the summer months when rainfall is minimal and the evaporative demand is at its peak. Soil water storage in the root zone in the future is lower than that under the present-day conditions, resulting in higher water stress for plants. Although the mean annual depth to water table values increased by 2 and 1.5 % only in the A2 and B2 scenarios, respectively, from the present-day values, cultivation of irrigated crops with the present water and land management practices may become difficult. This calls for appropriate and timely measures for land and water management/conservation.

The results point to the fact that managing groundwater storage will acquire greater significance in the future as it is expected to reduce groundwater recharge and increase the demand for groundwater. Since over and underestimation of various components of the hydrological cycle under different climate change scenarios impose restrictions on the informative value of the predicted changes in groundwater recharge rates, the resultant uncertainties should be kept in mind when using it for deriving possible adaption strategies for regional groundwater resources management. The results show that at least for some regions and scales of aggregation, regional climate models such as PRECIS may be instrumental in improving the simulations of global climate models. For local climates, RCMs may produce inappropriate or poor results when compared to GCMs, especially when precipitation

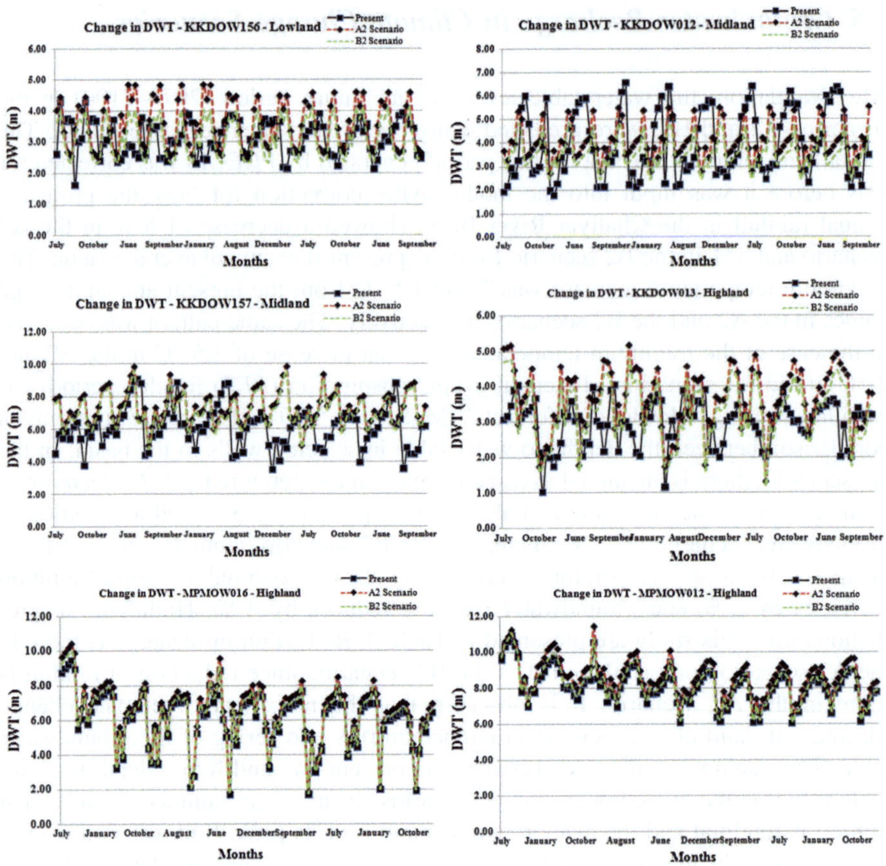

Fig. 23 Water table depths in the present period and in the A2 and the B2 scenarios

Table 4 Predictions from the groundwater recharge model

Sl. No.	Scenario	Change in temperature (°C)	Change in precipitation (%)	Change in ET (%)	Change in groundwater recharge (%)
1	A2	+3	−5	+3	−7
2	B2	+2	−3	+2	−4

is considered (Urrutia and Vuille 2009). For most other variables that are elevation-dependent (e.g. temperature), better resolution of RCMs yields significant improvement over GCMs (Urrutia and Vuille 2009).

4 Conclusion

In this study, the impact of projected climate change on streamflow at the Kuniyil gauging station and groundwater recharge in the Chaliyar River Basin was analysed for the climate change scenarios A2 and B2 outlined by the IPCC. Precipitation and temperature for the future period were generated using an RCM, PRECIS and corrected for bias. These were translated to streamflow using the hydrological model AVSWAT2000, and to depth to water table using a simple conceptual groundwater recharge model. A further increase in the resolution of predictions by PRECIS will improve representation of local gradients in climatic variables, thereby improving the accuracy of predictions of impact. Also, to assess uncertainties in downscaling, an ensemble of RCMs should be implemented. Although streamflow showed a declining trend in the two scenarios considered, it is not so severe as to adversely affect agricultural production in the river basin. Overall, the predicted rise in temperature will lead to an increase in PET. With rainfall predicted to decrease, direct groundwater recharge will reduce and if this is not augmented, the basin is likely to face scarcity of water in the future. In this study, land use/land cover and management practices were not changed in the simulations for the future period. Also, uncertainties exist in the predicted streamflows and depths to water table due to uncertainties in future emission scenarios and uncertainties in RCM projections.

Arnell et al. (2001) suggested that future scenarios provide extremely valuable insights into the sensitivity of hydrological systems to climate change. Wolock and McCabe (1999) stated that sensitivity studies of temperature and precipitation variations can provide important insight into the responses and vulnerabilities of different hydrological systems to climate change, especially when there is a great deal of uncertainty in the available GCM projections. When viewed in this background, the predicted streamflows and groundwater recharge can provide only a rough indication of likely changes in the future. In spite of this limitation, this work demonstrates the application of a powerful tool for use in such studies. The results of this study will be critically examined in the background of assumptions made and its impact on the results. Necessary measures to mitigate the impacts/adapt to the impacts may be incorporated into water resources management plans for the future so that sustainable water use can be achieved in the river basin.

Acknowledgments The authors are grateful to the Department for Environment, Food and Rural Affairs (DEFRA), Government of United Kingdom, for sponsoring the joint Indo–UK programme on Climate Change and the Ministry of Environment and Forests (MoEF), Government of India, for coordinating its implementation. Thanks are due to the Hadley Centre for Climate Prediction and Research, UK Meteorological Office, for making available the regional model—PRECIS. Support of the PRECIS simulation datasets is provided by the Indian Institute of Tropical Meteorology, Pune.

References

Alansi AW, Amin MSM, Abdul Halim G, Shafri HZM, Aimrun W (2009) Validation of SWAT model for stream flow simulation and forecasting in Upper Bernam humid tropical river basin. Malaysia Hydrol Earth Syst Sci Discuss 6:7581–7609

Alemaw BF, Chaoka TR (2003) A continental scale water balance model: a GIS approach for southern Africa. Phys Chem Earth 28:957–966

Allen LH, Valle RR, Jones JW, Jones PH (1998) Soybean leaf water potential responses to carbon dioxide and drought. Agron J 90:375–383

Allen RG, Gichuki FN, Rosenzweig C (1991) CO_2 induced climatic changes and irrigation water requirements. J Water Resour Plan Manag 117(2):157–178

Arnell NW, Liu C, Compagnucci R, da Cunha L, Hanaki K, Howe C, Mailu G, Shiklomanov I, Stakhiv E (2001) Hydrology and water resources. In: McCarthy JJ, Canziani O, Leary NA, Dokken DJ, and White KS (eds) Climate change 2001: impacts, adaptation, and vulnerability. Cambridge University Press, Cambridge, Chap. 4, pp 191–233

Arnell NW, Livermore MJL, Kovats S, Nicholls R, Levy P et al (2003) Socio-economic scenarios for climate change impacts assessments: characterising the SRES storylines. Glob Environ Change 14:3–20

Arnold JG, Srinivasan R, Muttiah RS, Williams JR (1998) Large area hydrologic modeling and assessment part I: model development. J Am Water Resour Assoc 34(1):73–89

Bates BC, Kundzewicz ZW, Wu S, Palutikof J (2008) Climate change and water. Technical paper of the Intergovernmental Panel on Climate Change, IPCC Secretariat, Geneva

Cao W, Bowden BW, Davie T (2006) Multi-variable and multi-site calibration and validation of SWAT in a large mountainous catchment with high spatial variability. Hydrol Process 20:1057–1073

Christensen J, Boberg F, Christensen O, Lucas-Picher P (2008) On the need for bias correction of regional climate change projections of temperature and precipitation. Geophys Res Lett 35: L20709. doi:10.1029/2008GL035694

Chu TW, Shirmohammadi A (2004) Evaluation of the SWAT model's hydrology component in the piedmont physiographic region of Maryland. Trans Am Soc Agric Eng 47(4):1057–1073

Cibin R, Sudheer KP, Chaubey I (2010) Sensitivity and identifiability of stream flow generation parameters of the SWAT model. Hydrol Process (Published online in Wiley InterScience). doi:10.1002/hyp

Cooley H, Christian-Smith Juliet, Peter H, Gleick Lucy Allen, Cohen Michael (2009) Understanding and reducing the risk of climate change for transboundary waters. Pacific Institute, California (43 pp)

de Vries JJ, Simmers I (2002) Groundwater recharge: an overview of processes and challenges. Hydrogeol J 10:5–17

de Wit M, van den Hurk B, Warmerdam P, Torfs P, Roulin E, van Deursen W (2007) Impact of climate change on low-flows in the River Meuse. Clim Change 82:351–372. doi:10.1007/s10584-006-9195-2

Eckhardt K, Ulbrich U (2003) Potential impacts of climate change on groundwater recharge and streamflow in a central European low mountain range. J Hydrol 284:244–252

Falkenmark M (2005) Water usability degradation—economist wisdom or societal madness? Water Int 30(2):136–146

Favreau G, Cappelaere B, Massuel S, Leblanc M, Boucher M, Boulain N, Leduc C (2009) Land clearing, climate variability, and water resources increase in semiarid southwest Niger: a review. Water Resour Res 45: W00A16. doi:10.1029/2007WR006785

Foster SSD, Morris BL, Lawrence AR (1994) Effects of urbanization on groundwater recharge. In: International proceedings on groundwater problems in urban areas. Telford, London, pp 43–63

Frei C, Christensen JH, Deque M, Jacob D, Jones RG, Vidale PL (2003) Daily precipitation statistics in regional climate models: evaluation and intercomparison for the European Alps. J Geophys Res 108(D3):4124. doi:10.1029/2002JD002287

Gassman PW, Reyes M, Green CH, Arnold JG (2007) The soil and water assessment tool: historical development, applications, and future directions. CARD working paper 07-WP 443. Center for Agricultural and Rural Development, Iowa State University
Gee GW, Hillel D (1988) Groundwater recharge in arid regions: review and critique of estimation methods. Hydrol Process 2:255–266
Ghosh S, Mujumdar PP (2008) Statistical downscaling of GCM simulations to streamflow using relevance vector machine. Adv Water Resour 31:132–146
Gosain AK, Rao Sandhya, Basuray Debajit (2006) Climate change impact assessment on hydrology of Indian River Basins. Curr Sci 90(3):346–353
Graham RL, Turner MG, Dale VH (1990) How increasing CO, and climate change affect forests. Bioscience 40(8):575–587
Grindley J (1967) The estimation of soil moisture deficits. Meteorol Mag 96:97–108
Haan PK, Skaggs RW (2003) Effects of parameter uncertainty on DRAINMOD predictions: hydrology and yield. Trans Am Soc Agric Eng 46(4):1061–1067
Hargreaves GL, Hargreaves GH, Riley JP (1985) Agricultural benefits for Senegal River Basin. J Irrig Drain Eng 111(2):113–124
Hay LE, Clark MP, Wilby RL, Gutowski WJ Jr, Leavesley GH, Pan Z, Arrit RW, Takle ES (2002) Use of regional climate model output for hydrologic simulations. J Hydrometeorol 3:571–590
He B, Takase K, Wang Y (2008) A semi-distributed groundwater recharge model for estimating water-table and water-balance variables. Hydrogeol J 16:1215–1228
Heuvelmans G, Muys B, Feyen J (2004) Evaluation of hydrological model parameter transferability for simulating the impact of land use on catchment hydrology. Phys Chem Earth 29:739–747
Howard KWF, Lloyd JW (1979) The sensitivity of parameters in the Penman evaporation equations and direct recharge balance. J Hydrol 41:329–344
IPCC (Intergovernmental Panel on Climate Change) (2007) Climate change 2007: the physical science basis. Contribution of WG I to the fourth assessment report of the Intergovernmental Panel on Climate Change. Cambridge University Press, Cambridge. Available at http://ipcc-wg1.ucar.edu/wg1/wg1-report.html
Jacob D (2001) A note to the simulation of the annual and inter-annual variability of the water budget over the Baltic Sea drainage basin. Meteorol Atmos Phys 77:61–73
Jones RG, Noguer M, Hassell DC, Hudson D, Wilson SS, Jenkins GJ (2004) Generating high resolution climate change scenarios using PRECIS. Met Office Hadley Centre, Exeter
Kamga FM (2001) Impact of greenhouse gas induced climate change on the runoff of the upper Benue River (Cameroon). J Hydrol 252:145–156
Keshta N, Elshorbagy A, Carey S (2009) A generic system dynamics model for simulating and evaluating the hydrological performance of reconstructed watersheds. Hydrol Earth Syst Sci 13:865–881
Khazaei K, Spink AEF, James W (2003) A catchment water balance model for estimating groundwater recharge in arid and semiarid regions of south-east Iran. Hydrogeol J 11:333–342
Kleinn J, Frei C, Gurtz J, Luthi D, Vidale P, Schar C (2005) Hydrologic simulations in the Rhine basin driven by a regional climate model. J Geophys Res 110:D04102. doi:10.1029/2004JD005143
Kundzewicz ZW, Mata LJ, Arnell N, Doll P, Jimenez B, Miller K, Oki T, Şen Z, Shiklomanov I (2008) The implications of projected climate change for freshwater resources and their management. Hydrol Sci J 53(1):3–10
Leander R, Buishand T (2007) Resampling of regional climate model output for the simulation of extreme river flows. J Hydrol 332:487–496. doi:10.1016/j.jhydrol.2006.08.006
Legesse D, Vallet-Coulomb C, Gasse F (2003) Hydrological response of a catchment to climate and land use changes in tropical Africa: case study South Central Ethiopia. J Hydrol 275:67–85
Linsley RK (1960) Computation of a synthetic streamflow record on a digital computer. In: Surface water (proceedings of the Helsinki symposium, 1960) IAHS Publ no 51, pp 526–538

Maraun D, Wetterhall F, Ireson AM, Chandler RE, Kendon EJ, Widmann M, Brienen S, Rust, HW, Sauter T, Themeßl M, Venema VKC, Chun KP, Goodess CM, Jones RG, Onof C, Vrac M, Thiele-Eich I (2010) Precipitation downscaling under climate change: recent developments to bridge the gap between dynamical models and the end user. Rev Geophys 48:1–34. RG3003. doi:10.1029/2009RG000314

Marengo JA, Jones R, Alves LM, Valverde MC (2009) Future change of temperature and precipitation extremes in South America as derived from the PRECIS regional climate modeling system. Int J Climatol 29:2241–2255

Maroco JP, Edwards GE, Ku MSB (1999) Photosynthetic acclimation of maize to growth under elevated levels of carbon dioxide. Planta 210:115–125

Maxwell EW, Kollet SJ (2008) Interdependence of groundwater dynamics and land-energy feedbacks under climate change. Nat Geosci 1:665–669

McCabe GJ, Wolock DM (1992) Sensitivity of irrigation demand in a humid-temperate region to hypothetical climatic change. Water Resour Bull 28(3):535–543

McCloud DE (1970) Water requirements for turf. Proc Florida Turfgrass Manage Conf 18:88–90

Meehl GA, Stocker TF, Collins WD, Friedlingstein P, Gaye AT, Gregory JM, Kitoh A, Knutti R, Murphy JM, Noda A, Raper SCB, Watterson IG, Weaver AJ, Zhao ZC (2007) Global climate projections. In: Solomon S, Qin D, Manning M, Chen Z, Marquis M, Averyt KB, Tignor M, Miller HL (eds) Climate change 2007: the physical science basis. Contribution of working group I to the fourth assessment report of the Intergovernmental Panel on Climate Change. Cambridge University Press, Cambridge

Messager C, Gallée H, Brasseur O, Cappelaere B, Peugeot C, Séguis L, Vauclin M, Ramel R, Grasseau G, Léger L, Girou D (2006) Influence of observed and RCM-simulated precipitation on the water discharge over the Sirba basin, Burkina Faso. Niger Clim Dynam 27 (2–3):199–214

Middelkoop H, Daamen K, Gellens D, Grabs W, Kwadijk J, Lang H, Parmet B, Schadler B, Schulla J, Wilke K (2001) Impact of climate change on hydrological regimes and water resources management in the Rhine Basin. Clim Change 49:105–128

Mileham L, Richard GT, Todd M, Tindimugaya C, Thompson J (2009) The impact of climate change on groundwater recharge and runoff in a humid, equatorial catchment: sensitivity of projections to rainfall intensity. Hydrol Sci J 54(4):727–738

Monteith JL (1965) Evaporation and the environment. The state and movement of water in living organisms. In: 19th Symposia of the Society for Experimental Biology. Cambridge University Press, London, pp 205–234

Nakicenovic N, Alcamo J, Davis G, de Vries B, Gaffin S, Gregory K, Grubler A, Dadi Z (2000) IPCC special report on emissions scenario. Cambridge University Press, Cambridge

Nash JE, Sutcliffe JV (1970) River flow forecasting through conceptual models: part I. A discussion of principles. J Hydrol 10(3):282–290

Neitsch SL, Arnold JG, Kiniry JR, Williams JR (2001) Soil and water assessment tool theoretical documentation, Version 2000. Blackland Research Center. Texas Agricultural Experiment Station, Temple, Texas. http://www.brc.tamus.edu/swat/swat2000doc.html

Nemec J, Schaake JS (1982) Sensitivity of water systems to climate variation. J Hydrol Sci 27:327–343

Potter KW, Bowser CJ (1995) Estimating the spatial distribution of groundwater recharge rates using hydrologic, hydrogeologic, and geochemical methods. Technical report. University of Wisconsin, Madison Water Resource Center, Madison, WRC-GRR-95.07

Priestley CHB, Taylor RJ (1972) On the assessment of surface heat flux and evaporation using large-scale parameters. Mon Weather Rev 100:81–92

Ramanarayanan TS, Williams JR, Dugas WA., Hauck LM, McFarland AMS (1997) Using APEX to identify alternative practices for animal waste management. ASAE Paper No. 972209. St. Joseph, Michigan, ASAE

Raneesh KY, Thampi SG (2013) A simple semi-distributed hydrologic model to estimate groundwater recharge in a humid tropical basin. Water Resour Manage 27(5):1517–1532
Refsgaard JC, Storm B (1996) Construction, calibration and validation of hydrological models. In: Abbott MB, Refsgaard JC (eds) Distributed hydrological modelling. Kluwer Academic, Dordrecht, pp 41–54
Reilly J, Tubiello F, McCarl B, Melillo J (2001) Climate change and agriculture in the United States, in climate change impacts on the United States: the potential consequences of climate variability and change. The National Assessment Synthesis Team. Cambridge University Press, Cambridge 618 pp
Ritchie JT (1972) A model for predicting evaporation from a row crop with incomplete cover. Water Resour Res 8:1204–1213
Rupa Kumar K, Sahai AK, Krishna Kumar K, Patwardhan SK, Mishra PK, Revadekar JV, Kamala K, Pant GB (2006) High-resolution climate change scenarios for India for the 21st century. Curr Sci 90(3):334–345
Sanford W (2002) Recharge and groundwater models: an overview. Hydrogeol J 10:110–120
Santhi C, Arnold JG, Williams JR, Dugas WA, Hauck L (2001) Validation of the SWAT model on a large river basin with point and nonpoint sources. J Am Water Resour As 37(5):1169–1188
Schuol J, Abbaspour KC (2007) Using monthly weather statistics to generate daily data in a SWAT model application to West Africa. Ecol Model 201:301–311
Scibek J, Allen DM (2006) Comparing the responses of two high permeability, unconfined aquifers to predicted climate change. Global Planet Change 50:50–62
Shabalova M, van Deursen W, Buishand T (2003) Assessing future discharge of the River Rhine using regional climate model integrations and a hydrological model. Clim Res 23:233–246
Shah T (2009) Climate change and groundwater: India's opportunities for mitigation and adaptation. Environ Res Lett 4(35005):1–13. doi:10.1088/1748-9326/4/3/035005
Soares WR, Marengo JA (2009) Assessments of moisture fluxes east of the Andes in South America in a global warming scenario. Int J Climatol 29:1395–1414
Sophocleous M (1992) Groundwater recharge estimation and regionalization: the Great Bend Prairie of central Kansas and its recharge statistics. J Hydrol 137:113–140
Srinivasan R, Ramanarayanan TS, Arnold JG, Bednarz ST (1998) Large area hydrologic modeling and assessment part II: model application. J Am Water Resour As 34(1):91–101
Sugawara M, Funiyuki M (1956) A method of revision of the river discharge by means of a rainfall model. Collection of research papers about forecasting hydrologic variables. The Geosphere Research Institute of Saitama University, Saitama, Japan, pp 14–18
Sugawara M (1995) Tank model. In: Singh VP (ed) Computer models of watershed hydrology. Water Resources Publications, Highlands Ranch
Thomson AM, Rosenberg NJ, Izaurralde RC, Brown RA (2005) Climate change impacts for the conterminous USA: an integrated assessment part 2: models and validation. Clim Change 69:27–41
Urrutia R, Vuille M (2009) Climate change projections for the tropical Andes using a regional climate model: temperature and precipitation simulations for the end of the 21st century. J Geophys Res 114:D02108. doi:10.1029/2008JD011021
Van Vuuren PD, O'Neill BC (2006) The consistency of IPCC's SRES scenarios to recent literature and recent projections. Clim Change 75:9–46
Varis O, Kajander T, Lemmelä R (2004) Climate and water: from climate models to water resources management and vice versa. Clim Change 66(3):321–344
Viviroli D, Archer DR, Buytaert W, Fowler HJ, Greenwood GB, Hamlet AF, Huang Y, Koboltschnig G, Litaor MI, López-Moreno JI, Lorentz S, Schädler B, Schwaiger K, Vuille M, Woods R (2010) Climate change and mountain water resources: overview and recommendations for research, management and politics. Hydrol Earth Syst Sci Discuss 7:2829–2895. doi:10.5194/hessd-7-2829-2010
Vorosmarty CJ, Douglas EM, Green PA, Revenga C (2005) Geospatial indicators of emerging water stress: an application to Africa. Ambio 34:230–236

Walker GR, Jolly ID, Cook PG (1991) A new chloride leaching approach to the estimation of diffuse recharge following a change in land use. J Hydrol 128:49–67

Wigley TM, Jones PD (1985) Influences of precipitation changes and direct CO_2 effects on streamflow. Nature 314:149–152

Wilby RL, Harris I (2006) A framework for assessing uncertainties in climate change impacts: low-flow scenarios for the River Thames. UK Water Resour Res 42:W02419. doi:10.1029/2005WR004065

Wolock DM, McCabe GJ (1999) Estimates of runoff using water-balance and atmospheric general circulation models. J Am Water Resour Assoc 35(6):1341–1350

Assessing the Impacts of Climate Change on Water Resources: Experiences From the Mediterranean Region

Ivan Portoghese, Michele Vurro and Antonio Lopez

Abstract The most critical impacts of climatic change on the Mediterranean society and environment are likely to be associated with water availability. The whole area is expected to become vulnerable to the scarcity and irregular availability of water resources. In the framework of the FP6 EU CIRCE project (http://www.circeproject.eu/), a Regional Assessment of Climate Change in the Mediterranean (RACCM) was produced in 2013 to support the design of adaptation and mitigation policies on the whole region including Europe, North Africa, and the Middle-East. To this end, a set of coupled models has been developed to produce regional climate change projections. These projections allow assessment of the response of the Mediterranean Sea to climate change over the period 1950–2050 under the A1B hypothesis and to a large extent, the associated uncertainty. Some results concerning the use of CIRCE scenarios for the assessment of impacts on water resources are presented with an emphasis on the methodological approach to bridge the scale gaps between climate model structure and the resolution of basin-scale hydrology. Focusing on Southern Italy, the sustainability of surface and groundwater resources is explored. Additional results concerning a coastal catchment in Lebanon are presented. In this case study regarding a snow-dominated hydrological system, the combined effect of changes in temperature and precipitation is highlighted. The above methodological approach set-up for the Mediterranean area could be adopted for regions in Southeast Asia where in order to undertake similar impact studies, the local scale of investigation plays a determinant role for water resources assessment and flood protection.

I. Portoghese (✉) · M. Vurro · A. Lopez
Water Research Institute of the National Research Council of Italy (IRSA-CNR), Viale F de Blasio 5, 70132 Bari, Italy
e-mail: ivan.portoghese@cnr.it

M. Vurro
e-mail: michele.vurro@ba.irsa.cnr.it

A. Lopez
e-mail: antonio.lopez@ba.irsa.cnr.it

S. Shrestha et al. (eds.), *Managing Water Resources under Climate Uncertainty*, Springer Water, DOI 10.1007/978-3-319-10467-6_9

1 Introduction

The traditional method for assessing water resource systems is based on the use of statistical analysis climatology and hydrology records under the hypothesis of statistical stationarity. As the evidence of global climate change continues to accumulate, the use of historical records as a proxy for possible future events becomes less appropriate (Wiley and Palmer 2008). In other words, the stationarity hypothesis in hydrological variables can no longer be invoked (e.g. Milly et al. 2008) because substantial anthropogenic change in the Earth's climate has altered mean and extreme values of precipitation, evapotranspiration, and river discharge.

Most of the potential impacts of climate change (change in precipitation and extreme events, sea level rise and groundwater contamination, glacial melt, and freshwater availability) are not dealt with by climate simulation models, but may be evaluated by adopting model cascades in which the altered atmospheric variables are used in the forcing of specific models designed to describe the land-surface hydrological processes.

Successful modelling of hydrological processes requires detailed representations of the physical processes controlling water and energy fluxes. The predictability of water cycle components is typically quantified incorporating theoretical understanding of the relevant processes with fruitful representations of available observations. Such a modelling approach also requires higher spatial resolution by one to two orders of magnitude than that currently used in dynamical weather forecasting models and climate change simulation models. Thus, formulating an effective modelling strategy, which encompasses the diverse temporal and spatial scales, is a high priority in impact studies.

A crucial step in the data assimilation of observations to improve climate model output is that observations of atmospheric, hydrologic, and land-surface variables are available at more than one scale, for example, point measurements of precipitation from rain gauges, and areal averages of precipitation from radar and satellites. Incorporating the knowledge coming from observations into model predictions is therefore necessary to create a framework by which observations at different scales can be optimally merged to produce the best conditional estimates of the investigated process, also taking into account the uncertainty at the scale of interest (e.g. the hydrological model's resolution).

The Mediterranean region is a critical area not only for political, social, and economic reasons but also from the climate point of view. It has a unique position at the border of the tropical zones and the mid-latitude areas, resulting in a complex interplay of interactions between the mid-latitude atmospheric dynamics and tropical processes. The delicate energy and hydrological balance of the Mediterranean Sea influence the Atlantic circulation and, ultimately, the world ocean circulation.

The central role of water resources in the Mediterranean region was therefore the basic motivation for undertaking an unprecedented investigation of the likely scenarios of regional climate change and their socio-economic consequences, which led to a huge effort by the European Commission under the Sixth Framework Programme. The research activity ranges from the detection and attribution of past climate change; the development of new models of the Mediterranean climate system; the assessment of probable impacts on the environment; the communities involved about 70 leading research institution in the EU; and the Mediterranean region under the CIRCE Project from 2007 to 2011.

As presented here in the two case studies, analysing the potential impacts of climatic change is an utmost important step towards adaptation and mitigation in the Mediterranean region. Among the main focal regions adopted in the CIRCE Project is Southern Italy, being subject to overexploitation of water resources and therefore particularly vulnerable to climate change.

Building on the CIRCE experience, investigation of the possible impacts on water resources was extended to the Middle-East and focused on a case study in Lebanon. In this second case study, the likely alterations in air temperature and precipitation patterns were investigated with regard to the snow processes, which regulate water resources in coastal Lebanon. This topic was the main objective of a biennial research programme from 2012 to 2013 for "Modelling Water Balance Using Remotely Sensed Data," which was funded under the Scientific Cooperation between the National Research Council of Italy (CNR) and the National Council for Scientific Research of Lebanon (CNRS-L).

2 Assessment of Climate Change Impacts on Water Resources: Challenges and Methodological Framework

The CIRCE models have been integrated from 1951 to 2050, with initial conditions obtained from a long spin-up run of the coupled systems. The simulations have been performed using observed radiative forcing (solar constant, greenhouse gases concentration, and aerosol distribution) during the first half of the simulation period and the IPCC SRES A1B scenario during the second half (2001–2050). The projections indicate that remarkable changes in the Mediterranean region climate might occur in the next few decades. A substantial warming (about 1.5 °C in winter and almost 2 °C in summer) and a significant decrease of precipitation (about 5 %) might affect the region in the 2021–2050 period compared to the reference period (1961–1990), in an A1B emission scenario. However, locally, the changes might be even larger.

Most of the CIRCE research findings were reported into the Regional Assessment of Climate Change in the Mediterranean (RACCM) region (Navarra and Tubiana 2013). This work provides the first comprehensive assessment of climate change and its impacts in the Mediterranean region, covering different sectors, from physical climate drivers, for example, temperature and precipitation, agriculture, and forests, from water resources to social impacts, evaluating policies, and determining costs of actions and inaction.

2.1 Methodology for the Assimilation of Climate Scenarios into Hydrological Models

Global climate models (GCMs) are the primary tool for understanding how the global climate may change in the future. However, they do not currently provide reliable information on scales below about 200 km (Meehl et al. 2007). Hydrological processes typically occur at finer scales (Parry et al. 2007). Consequently, basin-scale assessments of climate change impacts usually produce a large bias in the simulated hydrological processes whenever the raw output variables from a GCM are adopted (Mearns et al. 2003; Dibike and Coulibaly 2005). Hence, to reliably assess hydrological impacts of climate change, higher resolution scenarios are required.

Various downscaling techniques have been developed to bridge the scale gap between GCMs and finer scales required to assess hydrological impacts of climate change. Such techniques may be grouped into two downscaling approaches: the deterministic dynamical downscaling (DD) and statistical downscaling (SD). Although SD has been traditionally seen as an alternative to DD, recent works on SD have attempted to combine the benefits of these two approaches. In this context, a methodology has been proposed (Guyennon et al. 2013) to evaluate the relative performance of the selected GCM, DD, and SD and their combinations not only in terms of bias but also in terms of time variability, considering both the trend analysis and the non-stationarity.

The valuable objective of the study was to assess whether a DD processing performed before the SD permits more suitable climate scenarios to be obtained for basin-scale hydrological applications starting from GCM simulations. The case study presented here (Fig. 1) focuses on the Apulia region (South East of Italy, with a surface area about 20,000 km^2), characterised by a typical Mediterranean climate. The monthly cumulated precipitation and monthly mean of daily minimum and maximum temperature distribution were examined for the period 1953–2000. The fifth-generation ECHAM model from the Max-Planck-Institute for Meteorology was adopted as GCM. The DD was carried out with the Protheus system by the Italian ENEA (Artale et al. 2010), while the SD was performed through a monthly quantile-quantile correction.

In order to evaluate the relative performances of the DD and SD downscaling methods, the following four methods of data processing were compared with land observations: (1) direct output from the GCM control scenario (GCM); (2) DD applied to the GCM scenario (GCM-DD); (3) SD applied directly to the GCM scenario (GCM-SD); (4) SD applied to the DD of the GCM scenario (GCM-DD-SD). A spatial homogenisation through a statistical interpolation (SI) was performed before each comparison, as described below. Thus, data processing (1)–(4) refers to the SI performed on each processing output. Analogously, (ref) refers to the SI performed on the observations dataset. Data fluxes are schematised in Fig. 2.

The SD resulted in efficiently reducing the mean bias in the spatial distribution at both annual and seasonal scales, but it was not able to correct the drawback of the

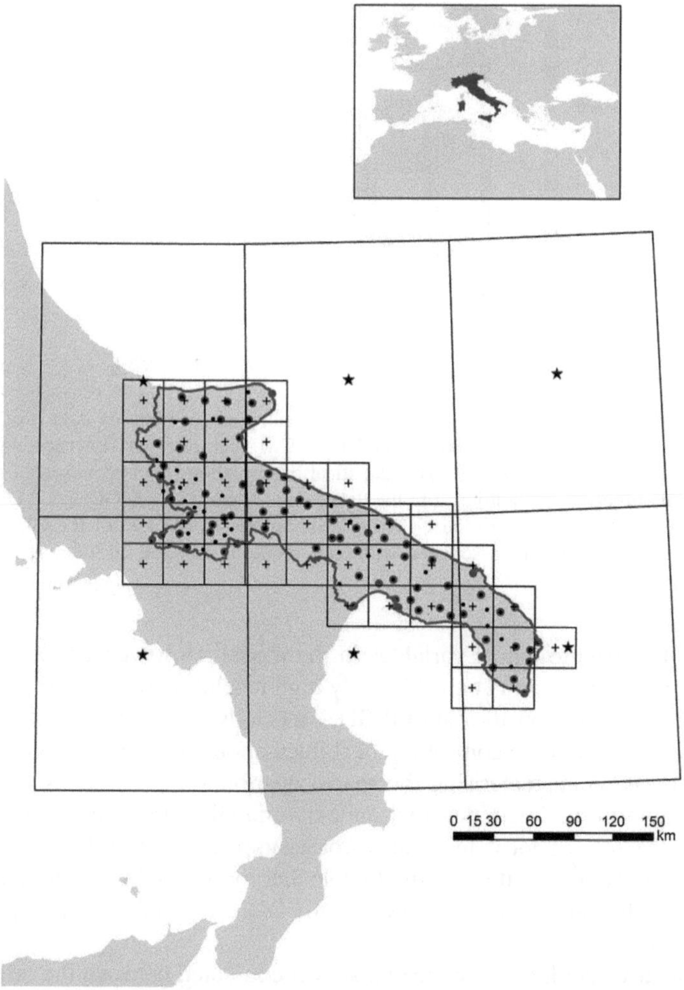

Fig. 1 Location of the Apulia region. The hydrological domain area is delimited by a *grey full line*. Locations of the temperature and precipitation sampling stations are shown with a *grey* and *black full circle*, respectively. GCM nodes are shown with *black stars*, and DD nodes are shown with *black crosses*. The *grid boxes* associated with GCM and DD nodes are delimited by *black full line*

modelled non-stationary components of the GCM dynamics. The DD provided a partial correction by enhancing the spatial heterogeneity of trends and the long-term time evolution predicted by the GCM. The best results were obtained through a combination of both DD and SD approaches.

Our analysis suggests that SD is a necessary step in the processing of climate simulation for obtaining reliable statistics at the local scale. For example, the quantile–quantile transform is confirmed as one of the best SD tools for removing

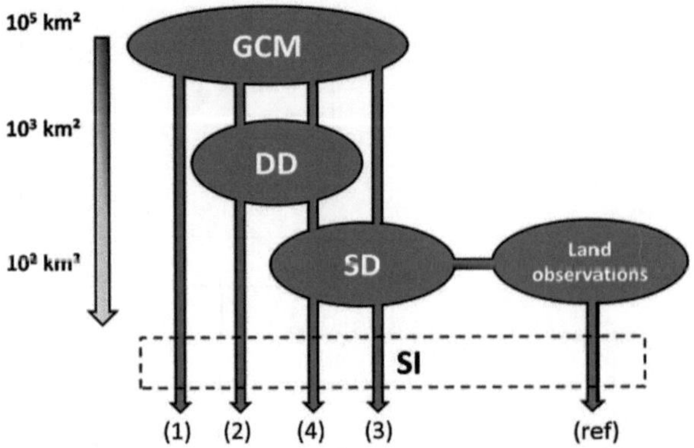

Fig. 2 Methodological framework representing the adopted methods of data processing. The *arrows* indicate the data fluxes, while models (GCM and relevant downscaling) and land observations are shown with *ellipses*. The statistical interpolation (SI) is represented by a *dashed rectangle*. Data processing resulting from the data flux is referred as: (*1*) GCM; (*2*) DD applied to GCM; (*3*) SD applied directly to the GCM; (*4*) SD applied to the DD of the GCM; (*ref*) land observations. The spatial scale associated with each model is reported on the *left*

bias from the meteorological variables in the model. However, an explicit modelling of the physical system at a sufficiently high resolution (hence, the DD) appears a necessary precondition to a skilful SD, especially during the seasons in which local processes have more control on local fluctuations of climate. In particular, DD plays a key role in characterising the spatial distribution of trends. Moreover, only the DD is able to modulate the inter-annual variability simulated by the GCM by enhancing the role of local feedbacks, for example, in the soil-vegetation–atmosphere system. However, it is worthy to note that for the GCM scenarios considered in this study, the correction introduced by the DD is not sufficient to reproduce the observed trends.

The spatial variability of the mean bias was computed between the reference and the data processing is reported in Fig. 3, in terms of percentiles (25th, 75th, 5th, and 95th), in Fig. 5. Figure 3 can be read as follows: the closer the mean bias to zero, the higher the ability of the data processing to reproduce the spatial mean condition for each variable; the narrower the distribution, the higher the ability of the data processing to reproduce the spatial heterogeneity of each variable.

The resulting complementarity of the two downscaling techniques, therefore, suggested that the combined DD-SD is a suitable choice for the generation of weather scenarios for impact modelling. In fact, the combined DD-SD presented the best results, both in terms of mean bias and spatial distribution of trends by retaining the improvements obtained by the DD in terms of climate non-stationarity as well as the intrinsic assimilation of observation datasets.

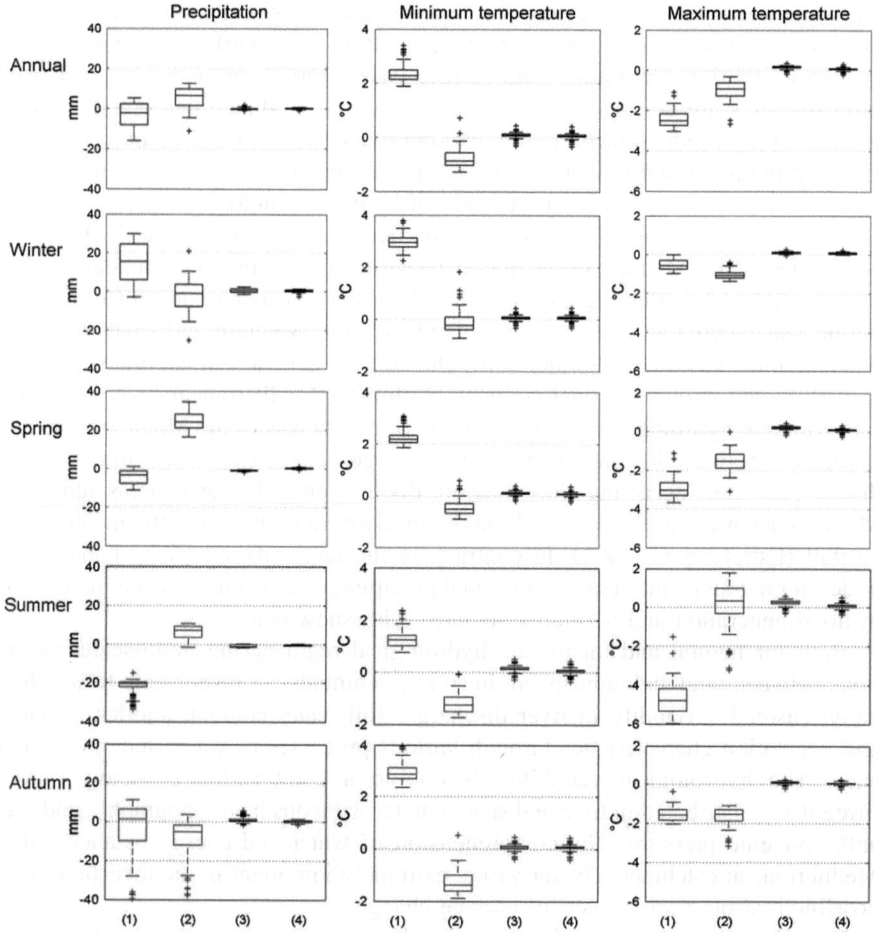

Fig. 3 Annual and seasonal spatial variability of the mean bias values for precipitation, minimum and maximum temperature computed between land observations (ref), GCM (*1*), GCM-DD (*2*), GCM-SD (*3*), and GCM-DD-SD (*4*)

3 Assessment of Climate Change Impacts in the Mediterranean Region: Some Relevant Examples at the Scale of Water Management

Understanding which processes are dominant or controlling at different scales in the different hydro-climatic regions of the Mediterranean is the key to undertaking meaningful impact investigation with the emphasis on reduction of model uncertainty. Nevertheless, any aprioristic definition of the dominant hydrological processes in a given catchment may be a mistake. Rather, it is advisable to characterise the basic interactions connecting the hydrological transformations at the scale of interest of the investigated problem. Precipitation, soil moisture, evaporation and evapotranspiration, streamflow and surface water storage, groundwater, and water infrastructures are deeply interlinked processes that modulate the terrestrial water balance in the Mediterranean region and consequently the water resources regime. The interpretation and modelling of such processes at suitable scales are the subjects of huge research efforts in the field of hydro-climatology.

Medium and small rivers deserve particular attention in the hydrological behaviour of the Mediterranean region because of their relative abundance, their strong relief in their hinterlands, as well as their highly variable run-off (from arid to humid watersheds) and response to episodic events. Medium and small rivers are those draining catchments <5,000 km^2. Due to their generally small drainage basins as well as the strong seasonal contrast of climate, Mediterranean rivers exhibit a rather unique hydrological character. In the Southern part of the Mediterranean, the differences between low and high water discharges can be extreme, with most water discharges often occurring during short flood events. In some areas along the Mediterranean coast, the recorded maximum daily rainfall is near the mean annual rainfall (Estrela et al. 2001). In contrast, in the larger river basins of the North, wide-ranging and continuous or seasonal precipitation is commonly the main factor in flood generation and is often associated with snowmelt.

Both for natural and man-made hydrological regimes, the multi-scale interactions between land and atmosphere in river catchments are responsible for the high daily/seasonal variability in river discharge, soil water content, aquifer recharge, and vegetation characteristics through various processes of water and energy balance. The hydrological variability is due to a combination of heavy rainfall (irregularly distributed in time and space), heterogeneous land topography, and high anthropogenic pressure. The comprehension of water and energy balance within Mediterranean catchments is, therefore, extremely important in the investigation of possible hydrological impacts of climate change.

It is, therefore, crucial to develop suitable model experiments to evaluate what the basin-scale response to the altered climate could be under a reasonable uncertainty framework. Such modelling experiments would be tailored to the specific hydrological behaviour of the river basin under investigation, as a response to the climatic, geomorphological, and other physical characteristics of catchments. An organisational principle to guide climate change impact studies would be useful, which by similarity with the biological sciences, could be assumed as a catchment classification system, considering variability in relevant characteristics, increasing human impacts on catchments, and unsteady climate conditions.

3.1 Southern Italy: Water Resources Projection at the Basin Scale

As a typical example of semi-arid Mediterranean area, the south-eastern coastal region of Italy was chosen to develop an impact assessment of climate change on water resources. The Apulia region with more than four million inhabitants has been exposed to a sequence of prolonged droughts in the past few decades, causing a general decrease in water supply and an increase in demand for irrigation. Moreover, in the past decade, the region has been recognised as being among those at the highest risk of desertification in Europe, due to the observed climatic trends and recently intensified agricultural practices. The climate is markedly Mediterranean, with mild wet winters and hot dry summers (the coldest month is January and the warmest is July). Climate variables and rainfall, in particular, exhibit a marked inter-annual variability, which makes water availability a permanent threat to the economic development and ecosystem conservation of the region. In addition, on average, rainfall has experienced a declining trend over the past four decades (Polemio and Casarano 2004). Due to the main carbonate nature of rocks (high substrate permeability and infiltration of rainwater), the region is generally poor in rivers and surface water. The most important river basins are located in the northern part of the region, where the morphological behaviour allows the presence of intermittent rivers. Instead, in the karst area, some basins related to a fossil hydrographical network present a superficial flow only during intense events. The basic features of water resources exploitation in the study region are summarised in Table 1.

The region is mainly dominated by agriculture; a vital economic resource for the region, with more than 70 % of the total area occupied by cropped land. The water resources derived from surface water bodies are limited, causing a major constraint to the social and economic development of the region. To overcome this problem, a great aqueduct was built at the beginning of the twentieth century, which supplies the region, collecting water from such carbonate Apennine springs as Cassano Irpino (mean annual discharge of 2.65 m^3/s) and Caposele. Moreover, many conveyance systems were built between 1960 and 1990 to transfer water from the

Table 1 Main features of water use and water supply for the Apulia region

Water use	Drinking	Agricultural	Industrial
Total 1.688 Mm^3	546 Mm^3 (32 %)	812/1.121 Mm^3 (~59 %)	142 Mm^3 (8 %)
Regional resources (%)	23	78	85
Extra-regional resource (%)	76	22	15
Surface water bodies (%)	54	24	15
Springs (%)	23	1	26
Groundwater bodies (%)	23	75	59

bordering regions in order to supply water for agricultural use, thus making the region very much reliant on external water resources. This infrastructure is the largest distribution network in Europe. Furthermore, a fast-growing trend in the last four decades towards irrigation farming has led to a massive exploitation of groundwater resources. As a result, the groundwater level has dramatically decreased in the river plain aquifers while sea water intrusion is observed in most of the coastal zones (Masciopinto 2005).

It is, therefore, crucial to investigate the possible impacts of climate projections in such a hydro-climatic context. Agriculture, water supply, and tourism are sectors vulnerable to climate change in the region. In this complex framework, the effect on regional the water balance was analysed.

The combined DD-SD downscaling method described in Sect. 2 was adopted in the case study region to evaluate the possible alteration in the hydrological regime of surface and groundwater bodies. The statistical transfer function, through the assimilation of reference observation records (extended over a few decades) as a paradigm for the variability mode, enables downscaling of the reference and scenario simulations from the computational nodes of the RCM to the scale of each available station.

Once a station-scale dataset for the reference (i.e. twentieth-century reference period) and scenario conditions: for the two obtained after the SD of the RCM simulations under reference and future emission scenarios, respectively, the atmospheric forcing was provided at a suitable scale to match the structural requirements of the adopted hydrological impact model (i.e. space–time resolution of the input data) through kriging spatial interpolation.

The adopted model named G-MAT (Portoghese et al. 2005) was proved suitable for the evaluation of hydrological water balance in semi-arid conditions. This model was originally developed for the sustainability assessment of water resources with particular emphasis on groundwater-dependent regions. It considers the major landscape features that determine the soil water balance such as vegetation activity through the season and the soil moisture storage and flux processes by adopting simple parameters. Moreover, the subsoil characteristics are considered as well in the model so that natural groundwater recharge can be evaluated, thus enabling further investigation of aquifer dynamics under different natural and anthropogenic forcing. The G-MAT yields natural groundwater recharge on a monthly basis, through the distributed application of the soil water balance equation, evaluated as the difference between the inflows (rainfall and irrigation) and the outflows (evapotranspiration and surface run-off). The spatial resolution of the water balance model is 1 km^2, thus assuring a feasible representation of the spatial heterogeneity of soil, subsoil, and vegetation features as well as a realistic description of catchment morphology. The monthly time step was chosen as a compromise between data availability over large domains and the uncertainty introduced by the various data manipulation necessary for the downscaling of climate scenarios through the observation-based corrections. The model scheme with an example of the output maps is reported in Fig. 4.

Fig. 4 Schematic structure of the G-MAT model (*left*) and example output maps representing the mean annual values of main water balance components

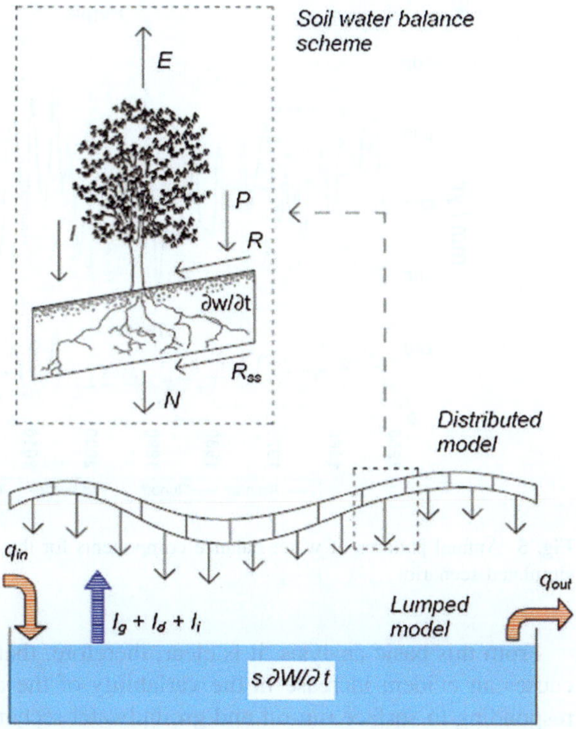

A comparison between the water balance reconstruction for the historical period and the twenty-first century is reported in Fig. 5 in which the inter-annual variability of precipitation shows an increased standard deviation in the annual values (Portoghese et al. 2013a, b). This increased variability of precipitation in the twenty-first century is intrinsically related to the dynamics of the global circulation model used as a lateral boundary condition to the RCM model, which is preserved by the downscaling algorithm. Consequently, the run-off and groundwater recharge controlled by the rainfall regime are similarly projected into the twenty-first century with an increased variability while preserving a decreasing trend. It is, therefore, arguable that the downscaled rainfall projections suffer from some degree of inconsistency inherited from the GCM and that both river run-off and groundwater recharge may be more rapidly decreasing in the twenty-first century, somewhat in agreement with the trend observed in the second half of the twentieth century.

Moreover, it is interesting to note that major water balance variables such as evapotranspiration (accounting for more than 60 % of the mean annual rainfall and irrigation) show a non-significant trend throughout the investigated periods, thus suggesting minor sensitivity of the simulated evapotranspiration processes to the projected perturbations in rainfall and temperature as a consequence of the soil moisture capacity, which acts as a limiting factor of plant transpirable water.

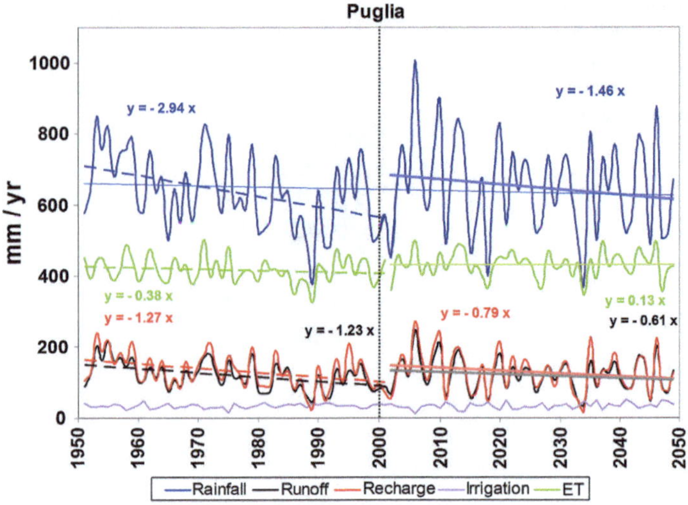

Fig. 5 Annual patterns of water balance components for the Apulia region in the historical and simulated scenario

From this basic analysis, it is clear, therefore, that the adopted climate scenario causes an evident increase in the variability of the available water resources corresponding to surface run-off and groundwater recharge while preserving the long-term average amount of each water balance component. Such a result is certainly a water management issue that has to be addressed in terms of adaptation to meet future water resources requirements.

The increased variability of the available water resources is even more severe from the aspect of drought occurrence (and conversely of extremely wet years). Drought events of a given return period (and conversely extremely wet event) are, therefore, expected to be more severe in terms of deviation from the mean values, which remain substantially unchanged between the historical and scenario periods.

Further details at the intra-annual scale of rainfall and consequent hydrological response are reported in Fig. 6. The main features revealed by precipitation comparison show slight increases in summer (+24 % and +32 % in July and August, respectively) and autumn (+5 % and +2 % in September and October, respectively) precipitation accompanied by an increase in rainfall variability, while some decrease in mean values is expected in the months corresponding to the late winter and April (−16 %).

Water resources availability and renewal rate are, respectively, related to monthly run-off and groundwater recharge estimations. A significant increase in the variability of summer and autumn values emerging from the comparison between the historical and the twenty-first century can be related to an increase in intense rainfall events, which are typical of late summer storms.

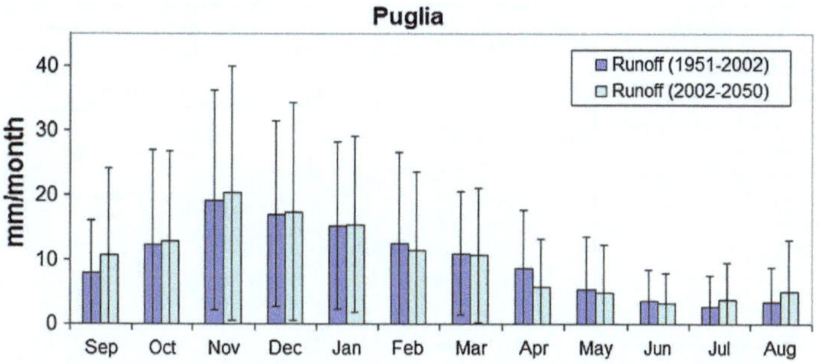

Fig. 6 Comparison between observations and twenty-first century simulations (after downscaling) of mean monthly precipitation, discharge, groundwater recharge, and irrigation (error bars according to respective standard deviations)

Concerning the irrigation water demand, the simulations performed for the historical and twenty-first century periods (Fig. 6) provided similar results in terms of mean monthly values, except for July and August, corresponding to a maximum water demand with corresponding reductions of −7 and −21 %, respectively. Nevertheless, as highlighted for the other water balance variables, the variability of irrigation requirements in the twenty-first century is expected to increase.

At the end of the data processing from climate simulation models to the hydrological impact model, various sources of uncertainty of each model component and data manipulation are combined, yielding a dataset of meteorological forcing (for reference and scenario conditions) that obviously implies a certain degree of uncertainty that should be quantified and taken into account when propagated through the impact model. Due to a relative amount of persistent uncertainty in the bias inherited from the GCM, the evaluation of impacts should preferably be undertaken through a comparative model simulation using reference and scenario conditions, which are both generated from the global model.

3.2 Case Study in Lebanon

The objective of this study is to develop basin-scale climate change scenarios (CCS) for a coastal watershed in Lebanon (Fig. 7). The Nahr Ibrahim watershed (NIW) north of Beirut is representative of the snowmelt-dominated watersheds located in Mount Lebanon (the so-called water tower of the Middle-East). To this end, the regional climate model providing regional climates for impact studies (PRECIS) developed by the Hadley Centre in the UK was adopted as a dynamic downscaling model of a GCM, thus providing spatially detailed projections and scenarios of future climate over the area of interest at a resolution of about 25 × 25 km. Daily simulations for precipitation (P), maximum and minimum temperatures (T_{max}, T_{min}) from PRECIS were adopted to evaluate the impact of climate change on the water balance of the NIW by taking into

Fig. 7 Map of Lebanon showing the location of the Nahr Ibrahim River in the Mount Lebanon region

account the time series for the recent past (1980–2000), present (2001–2011), near future (2012–2032), and the distant future (2080–2098) and comparing them with the available climate observations in order to assess possible future variations in precipitation and temperature. The available stations observations were also used to derive monthly regressions between climate and topographical elevations to be adopted as a simple interpolation tool to estimate the spatial distribution of T and P in

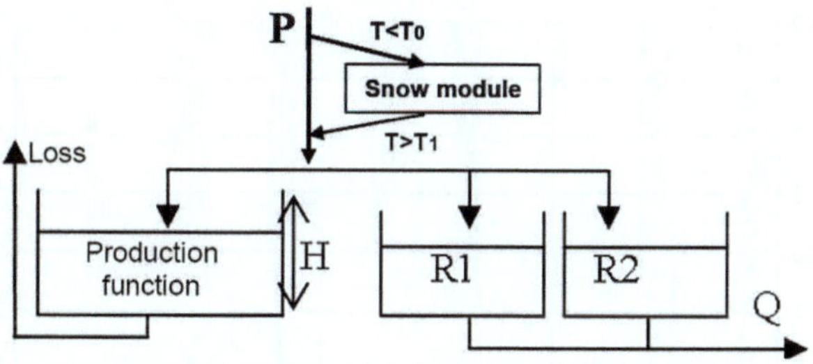

Fig. 8 Scheme of the adopted water balance model (modified from Hreiche et al. 2006)

NIW. Changes in climate variables were significant for the near future only for T while in the distant future, both P and T showed a remarkable decrease and increase, respectively. These alterations will correspond to a decrease in snow-covered areas and anticipated snowmelt, thus altering the water balance of NIW both in terms of mean values and variability.

To simulate the impacts on the streamflow regime, the basin-scale CCS was coupled with a conceptual water balance model named NIWaB, which was previously developed and calibrated (Portoghese et al. 2013a, b). The NIWaB is made up of three main parts: the snow-cover module, the water yield module, and the water transfer module (Fig. 8). Each of these modules is daily integrated providing a semi-distributed output in millimetre for the considered variables while the discharge for the considered sub-basins is yielded in m^3/s.

A consistent module of the model was developed to capture the space–time dynamics of the snow-cover and snowmelt contribution (Fig. 9a). The snow module was also validated by comparison with a MODIS-Terra snow product MOD10A2 over the watershed area. The combined validation using the discharge records and the MODIS snow retrieval product allowed assessment of model capability in representing the dominant hydrological processes and the consequent daily discharge despite the recognised limitations in the climate forcing adopted (Fig. 9b).

The overall capability of the developer model to adequately represent the main processes at catchment scale (Table 2) was a prerequisite for using the calibrated model for the assessment of climate change impacts on the hydrological regime of the NIW. The assumption that was made at this point is that the model structure and parameters are transferable from one period of data used for model calibration to another with a change in climate, even if this approach is being deeply questioned in the current scientific debate (Xu and Singh 2013).

Such an approach allowed assessment of the possible climate change impacts on the hydrological signature of NIW, which was strongly influenced by a shorter snowy period and a consequent enhanced seasonality of the river flows.

The climate change scenario for the case study was extracted from an RCM simulation specifically developed for Lebanon. To simulate the impacts on streamflow signatures, the climate scenario and the baseline simulation for the twentieth century were adopted as forcing of the NIWaB. Possible future alterations were investigated

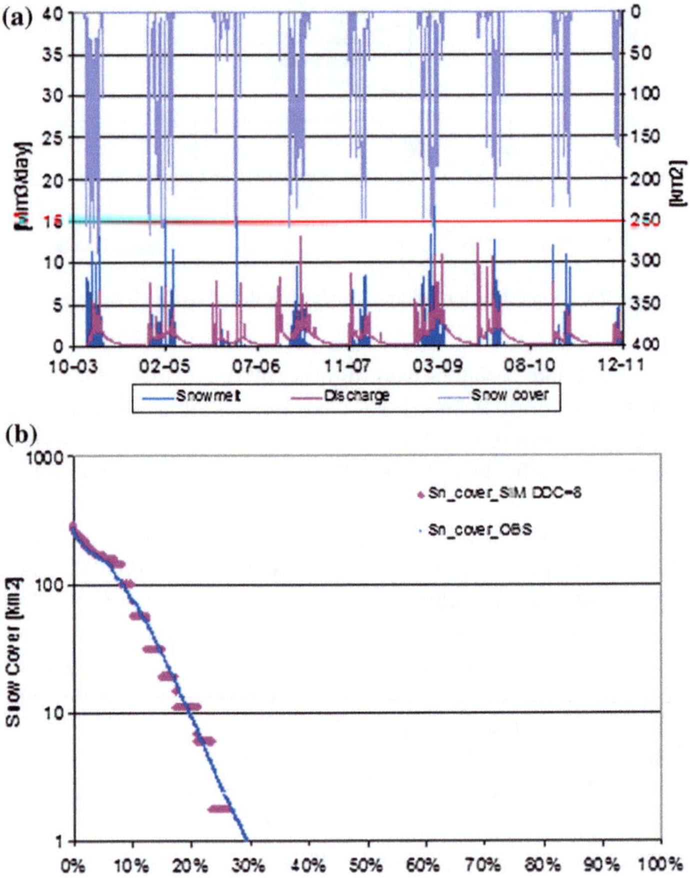

Fig. 9 Example of daily water balance simulation with the NIWaB model (**a**) and validation with MODIS snow-cover data (**b**)

Table 2 Predictive performance of the NIWaB model for the daily discharge and snow cover

	Discharge (Mm^3/day)	Discharge (Mm^3/day)	Snow cover (km^2)	Snow cover (km^2)
	Simulated 2003–2011	Observed 2003–2011	Simulated 2003–2011	Observed 2003–2011
Mean	0.99	0.99	22.4	19.3
Standard deviation	1.19	1.43	55.6	49.0
Coefficient of variation	1.20	1.44	2.48	2.54

by highlighting the expected impacts on the hydrological signature of the basin, which were strongly influenced by a shorter snowy period and a consequent enhanced seasonality of the river flows. Adopting a relative change approach with respect to the baseline water balance, a remarkable reduction in the mean annual water availability was found together with a sensible alteration in the discharge regime.

Concerning the first point, although it is worth remembering that the intrinsic uncertainty of the adopted modelling chain can be significant in the projection of future water resources, it is important to note that the mean annual discharge based on observation is 361 Mm^3 and the mean annual value given by the NIWaB model with climate forcing based on observations is 367 Mm^3. These values are both reasonably close to the 381 Mm^3 that was simulated with the NIWaB adopting the baseline climate simulation. It can be concluded, therefore, that the uncertainty in future water availability is, above all, related to the adopted climate scenario and only limited to the uncertainty of the water balance model. It is, therefore, prudent to develop a robust assessment on future water availability by using different climate scenarios produced with different climate models.

With regard to the discharge regime, it is considered that according to the discharge record, peak water availability in the spring season corresponds to peak water demand related to irrigation and tourism in the case study area, while the impact analysis highlighted that the delay between peak winter precipitation and peak river discharge in spring is expected to reduce. In other words, the hydrological regime is likely to change from a snowmelt-dominated regime to a rainfall dominated regime with clear consequences for water availability in the dry season.

These results may be helpful in assessing the opportunity for a new artificial reservoir as a strategy for climate change adaptation by enabling the storage of winter and spring flows as well as the regulation of flow peaks protecting lowland parts of the river basin from flooding.

4 Conclusion

With the aim of structuring reliable and feasible adaptation strategies for the Mediterranean region, the two case studies showed quite different results from the hydrological and socio-economic aspects, allowing representation of a wide range of situations. In the Italian case study, the high level of water infrastructures, including artificial reservoirs and deep pumping wells for groundwater exploitation, are representative of high water-demand societies in which a sound adaptation to climate change will be fundamentally based on optimising the available water resources through measures such as water-demand management, non-conventional water resources, and above all water resources protection.

The Lebanese case study, on the other hand, is representative of a socio-economic system characterised by fast-growing water demand for agricultural and domestic use in which the present and near future water policy is conditioned by cross-border issues and political instability. From the hydrological viewpoint, though rich in water resources, the Lebanon water supply is very limited due to poor water infrastructure, which cannot meet the increasing water demand in urban and rural areas of the country. In the Lebanese case study, the adaptation strategies will certainly start to boost the resilience of the water systems through appropriate design of new water storage and supply facilities.

Some general remarks can be summarised from the model experiments reported in the above case studies. First of all, it is important to develop suitable mathematical models to represent the non-linear processes such as the unsaturated soil processes, including the role of vegetation coverage, which dominate hydrological predictions in Mediterranean catchments. Thereafter, the downscaling and bias correction issues affecting climate models have to be thoroughly evaluated in order to compare the water balance signatures derived from observation-based climate forcing with those obtained using downscaled climate scenarios.

According to adopted climate model simulations, the evaluation of water resources availability in a water-scarce region has highlighted some peculiar responses in run-off and groundwater recharge to the predicted temperature and rainfall alterations. In particular, only slight decreasing trends were detected in the annual water balance components, but a marked increase in the variability of the hydrological system as a result of the increased rainfall variability is predicted for the twenty-first century. The increased variability of the hydro-systems in the Mediterranean is, therefore, confirmed as one of the main water management issues in the near future.

A positive point for the scientific and technical community is that climate change threats are pushing climate studies globally. Therefore, extraordinary efforts are being addressed to understand climate dynamics also in remote regions where data availability is often limited. Relevant advances have been achieved in recent years including new climate datasets that are becoming available from global data recovery programs and integration with climate simulation models.

Research funding is becoming more and more focused on applied sciences to investigate the possible impacts on altered climatic conditions. Development and cooperation initiatives are being launched between the EU and extra-EU.

Finally, it must be highlighted that the prediction of hydrological impacts on the catchment scale is still a highly uncertain practice due to the plurality of climate scenarios depending on different model chains and unpredictable evolution of socio-economic behaviour, which strongly influence water policy issues. Nevertheless, the urgent need to investigate climate related processes involving population security (e.g. flood and drought risks), food security, and climate-borne diseases is key to moving climate change research from science to the policy field.

Acknowledgments The authors wish to thank the organising committee of the SEA_EU-NET II project, funded by the European Commission under the FP7 Program, and the Asian institute of Technology for inviting the first author to take part in the expert workshop on "Managing Water Resources under Climate Uncertainty: Challenges and Opportunities" held in Bangkok, Thailand, on 17 and 18 October 2013.

References

Artale V, Calmanti S, Carillo A, Dell'Aquila A, Herrmann M, Pisacane G, Ruti PM, Sannino G, Struglia MV, Giorgi F, Bi X, Pal JS, Rauscher S (2010) An atmosphere-ocean regional climate model for the Mediterranean area: assessment of a present climate simulation. Clim Dynam 35:721–740. doi:10.1007/s00382-009-0691-8

Dibike YB, Coulibaly P (2005) Hydrologic impact of climate change in the Saguenay watershed: comparison of downscaling methods and hydrologic models. J Hydrol 307:145–163

Estrela T, Menéndez M, Dimas M, Marcuello C, Rees G, Cole G, Weber K, Grath J, Leonard J, Ovesen NB, Fehér J & Consult V (2001) Sustainable water use in Europe. Part 3: Extreme hydrological events: floods and droughts. Environmental issue report no. 21, EEA, Copenhagen, Denmark. available at http://www.eea.eu.int

Guyennon N, Romano E, Portoghese I, Salerno F, Calmanti S, Petrangeli AB, Tartari G, Copetti D (2013) Benefits from using combined dynamical-statistical downscaling approaches—lessons from a case study in the Mediterranean region. Hydrol Earth Syst Sci 17:705–720

Hreiche A, Bocquillon C, Najem W (2006) River flow simulation within ungauged catchments in Lebanon using a semi-distributed rainfall–runoff model. In: Voinov A, Jakeman A, Rizzoli A (eds) Summit on environmental modelling and software (Proceedings of iEMS 3rd Biennial meeting). International Environmental Modelling and Software Society, Burlington, USA

Masciopinto C (2005) Pumping-well data for conditioning the realization of the fracture aperture field in groundwater flow models. J Hydrol 309(1–4):210–228

Mearns LO, Giorgi F, Whetton PH, Pabon D, Hulme M, Lai M (2003) Guidelines for use of climate scenarios developed from regional climate model experiments, data distribution center of the intergovernmental panel on climate change. www.ipcc-data.org/guidelines/dgm_no1_v1_10-2003.pdf (last access: Feb 2013)

Meehl GA, Stocker TF, Collins W, Friedlingstein P, Gaye A, Gregory J, Kitoh A, Knutti R, Murphy JM, Noda A (2007) Global climate projections, climate change 2007: the physical science basis. In: Solomon S, Qin D, Manning M, Chen Z, Marquis M, Averyt KB, Tignor M, Miller HL (eds) Contribution of working group I to the fourth assessment report of the intergovernmental panel on climate change. Cambridge University Press, Cambridge, p 747–846

Milly PCD, Betancourt J, Falkenmark M, Hirsch RM, Kundzewicz ZW, Lettenmaier DP, Stouffer RJ (2008) Stationarity is dead: whither water management? Science 319:573–574. doi:10.1126/science.1151915

Navarra A, Tubiana L (eds) (2013) Regional assessment of climate change in the mediterranean, advances in global change research 50. doi:10.1007/978-94-007-5781-3_3, © Springer Science + Business Media Dordrecht

Parry ML, Canziani OF, Palutikof JP, van der Linden PJ, Hanson CE (2007) Climate change 2007: impacts, adaptation and vulnerability. In: Contribution of working group II to the fourth assessment report of the intergovernmental panel on climate change. Cambridge University Press, Cambridge, UK

Polemio M, Casarano D (2004) Rainfall and drought in Southern Italy (1821–2001). In: The basis of Civilization—Water Science? vol 286. IAHS Pub, pp 217–227

Portoghese I, Uricchio VF, Vurro M (2005) A GIS tool for hydrogeological water balance evaluation on a regional scale in semi-arid environments. Comput Geosci 31(1):15–27

Portoghese I, Bruno E, Dumas P, Guyennon N, Hallegatte S, Hourcade JH, Nassopoulos H, Pisacane G, Struglia MV, Vurro M (2013a) Impacts of climate change on fresh water bodies: quantitative aspects. In: Navarra A, Tubiana L (eds) Regional assessment of climate change in the mediterranean, Advances in global change research 50. doi:10.1007/978-94-007-5781-3_9, © Springer Science + Business Media Dordrecht

Portoghese I, Saqallah S, Vurro M, Darwish T, Shaban A, Khadra, R (2013b) modeling climate change impact on the water balance of a Coastal Watershed in Lebanon (in preparation)

Wiley MW, Palmer RN (2008) Estimating the impacts and uncertainty of climate change on a municipal water supply system. ASCE, J Water Res Plan Manag 134(3):239–246

Xu C, Singh VP (2013) Present achievements and key challenges in modeling hydrologic impacts of climate change. In: Proceedings of 2013 IAHR World Congress

References

Effect of Climate Change in Wastewater Treatment Plants: Reviewing the Problems and Solutions

Anastasios Zouboulis and Athanasia Tolkou

Abstract Climate change is considered to be one of the main challenges to urban wastewater systems in future decades. It is estimated that climate change has a dual effect on wastewater treatment (WWT) plants. The processes occurring in a wastewater treatment plant (WWTP) are subsequently affected by climate change; more extreme weather events and earlier snowmelt runoff will lead to more untreated sewer overflows, increased flooding, etc. Due to increased scarcity of water resources, wastewater reuse will become more necessary as climate change accelerates. On the other hand, during wastewater treatment, greenhouse gases (GHGs) including carbon dioxide (CO_2) from aerobic (oxidation processes), methane (CH_4) from anaerobic processes (3–19 % of global anthropogenic methane emissions), and nitrous oxide (N_2O) (3 % of N_2O emissions from all sources) associated with nitrification/denitrification (NDN) processes, as an intermediate product, can be emitted to the atmosphere. The various problems associated with climate change and WWT operation and the solutions that can be applied to deal with them are discussed in this chapter.

1 Introduction

1.1 Wastewater Treatment Plants (WWTPs)

Each particular wastewater treatment plant (WWTP) may be subjected to various operation conditions and restrictions, i.e. variable flow of incoming wastewater, quality of sewage, permitted levels of effluent, and other local guidelines.

A. Zouboulis (✉) · A. Tolkou
Department of Chemistry, Aristotle University of Thessaloniki, P.O. Box 116, 54124 Thessaloniki, Greece
e-mail: zoubouli@chem.auth.gr

A. Tolkou
e-mail: nancytolkou@gmail.com; tolkatha@chem.auth.gr

S. Shrestha et al. (eds.), *Managing Water Resources under Climate Uncertainty*, Springer Water, DOI 10.1007/978-3-319-10467-6_10

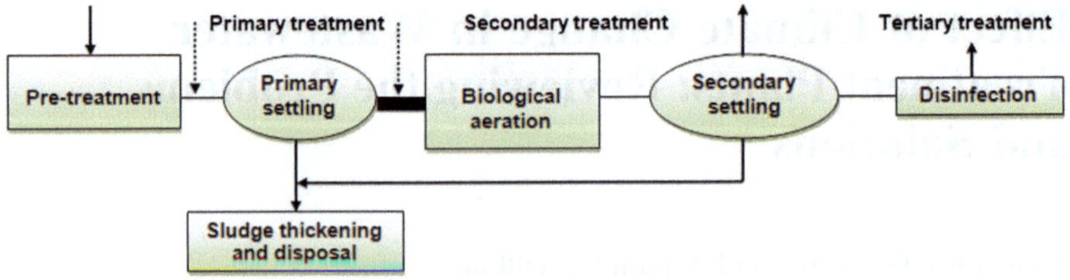

Fig. 1 Conceptual diagram of a typical activated sludge process

A conventional municipal WWTP includes the following main stages of processing:

(i) *Pretreatment*: removal of solid with relatively large diameters (e.g. >1 mm);
(ii) *Primary treatment*: removal of solids that settle relatively easily thereby aiming to reduce the concentration of particulates;
(iii) *Secondary treatment*: removal of biodegradable organic substances by biological processes (micro-organisms consume the organic content under aerobic or anaerobic conditions);
(iv) *Tertiary treatment and disinfection*: removal of residual solids and nutrients (N, P) and destruction of pathogenic micro-organisms;
(v) *Sludge disposal*: using landfills by composting or incineration (Zamboulis et al. 2003).

In the conventional municipal systems, wastewater after primary treatment, i.e. after removal of suspended solids, is treated by the activated sludge process (Fig. 1) consisting of an aeration tank followed by a secondary clarifier. Nevertheless, activated sludge is the most common and oldest biological process used for the treatment of municipal and industrial wastewater.

1.2 Climatic Events

Climate change is one of the main challenges for urban wastewater systems in future decades. Due to increasing concentrations of greenhouse gases (CHGs) in our atmosphere, temperatures are expected to rise between 2 and 5 °C globally by 2050.

Climate change is not just about changes in temperature. The whole water cycle is also affected. A warmer world means the atmosphere has the capacity to hold greater moisture. Consequently, there are changes in the amount of water vapour, rainfall, and circulation of water in the atmosphere. Rainfall has many characteristics, including amount, frequency, intensity, and type (Fig. 2) (EU WATCH 2014).

Climate change is affecting the hydrological cycle in various ways. Precipitation patterns are changing, snow and ice are melting, and atmospheric water vapour and evaporation are increasing. Evaporation increases because of surface heating and

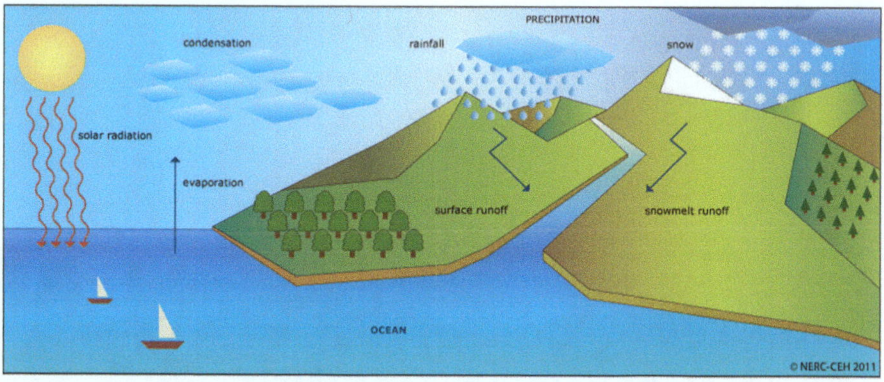

Fig. 2 Rainfall diagram (EU WATCH 2014)

with increased temperature the water-holding capacity of the atmosphere increases. As atmospheric moisture content directly affects precipitation, stronger rainfall events are expected with climate change (Trenberth 1999). Evaporation is an important process in the global water cycle. Solar radiation hits the surface of water or land and causes water to change state from liquid to gas. This is how water vapour enters the atmosphere: moisture in the atmosphere is linked to cloud formation and rainfall. Evaporation acts like an air conditioner for the surface because heat is used when water enters the atmosphere as moisture. However, at the same time, water vapour acts as a greenhouse gas by trapping radiation in the lower atmosphere. As temperature increases, so does the process of evaporation. In addition, the moisture-holding capacity of the atmosphere increases with temperature. For every 1 °C increase in global temperatures, there is a 7 % increase in the moisture-holding capacity of the atmosphere. Therefore, more moisture in the atmosphere ultimately leads to changes in rainfall patterns (EU WATCH 2014).

The wastewater industry is beginning to address the challenges posed by climate change, including regulatory burdens, pressure to reduce emissions, and the challenge of adapting to a changing climate (WEFTEC 2008).

The range of challenges related to climate change and cities in regard to the water supply and wastewater treatment sector is very wide; depending on geography, economics, administrative capacity, and demography. Many of the challenges are general, and some are more specific to particular cities (Major et al. 2011).

1.3 Greenhouse Gases (GHGs)

The global carbon cycle involves billions of tons of carbon in the form of CO_2 (Fig. 3). Carbon dioxide is absorbed by oceans and living biomass and is emitted to the atmosphere annually through natural processes. When in equilibrium, carbon movement among these various reservoirs is roughly balanced.

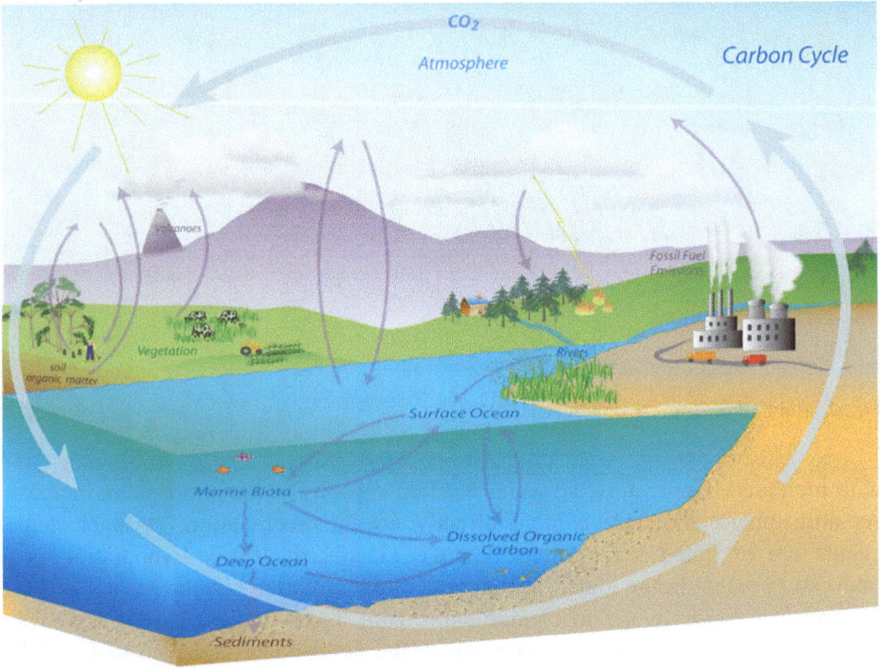

Fig. 3 The global carbon cycle (EPA 2012)

The concentration of CO_2 in the atmosphere has increased from a pre-industrial value of about 280 parts per million (ppm) to 379 ppm in 2005. Most scenarios of future emissions of CO_2 involve increases of CO_2. In 2004, 26.9 billion metric tons of CO_2 were emitted, and 33.9 billion metric tons are projected to be emitted in 2015. By 2030, 42.9 metric tons of CO_2 emissions are projected (EPA 2012).

Greenhouse Gases (GHGs): Gases that trap heat in the atmosphere are called CHGs. CO_2 is the principal greenhouse gas (Fig. 4), but other gases can have the same heat-trapping effect. Some of these other CHGs, however, have a much stronger greenhouse or heat-trapping effect than CO_2. For example, methane is 21 times more potent a greenhouse gas than CO_2.

Different GHGs have different atmospheric lifetimes, and therefore, actions to reduce emissions will take time to affect gas reduction in the atmosphere. The principal, human-generated CHGs that enter the atmosphere are the following:

Carbon Dioxide (CO_2): carbon dioxide enters the atmosphere through the burning of fossil fuels (oil, natural gas, and coal), production and transport of coal, natural gas, and oil.

Methane: methane emissions also result from livestock and other agricultural practices and by the decay of organic waste in municipal solid waste landfills and anaerobic WWTPs. CH_4 is a greenhouse gas approximately 21 times more potent than CO_2 and has an atmospheric lifespan of roughly 12 years.

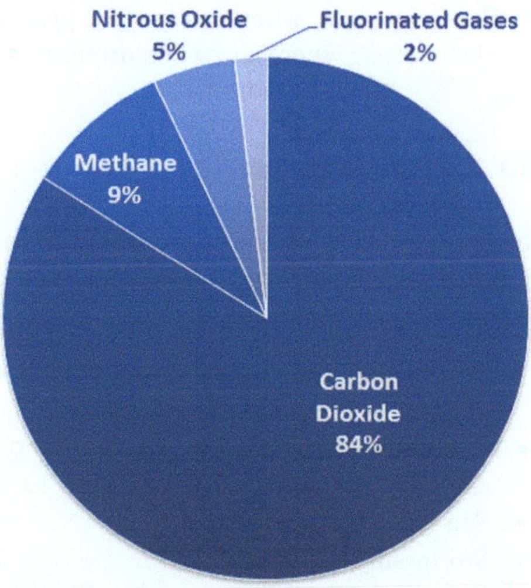

Fig. 4 Total emissions in 2011 = 6,702 million metric tons of CO_2 equivalent (EPA 2009)

Nitrous Oxide (N_2O): nitrous oxide is emitted during agricultural and industrial activities, as well as during the combustion of fossil fuels and solid waste. Nitrous oxide is also emitted from WWTPs during the nitrification and denitrification (NDN) processes. N_2O is 310 times more potent as a greenhouse gas than CO_2 and has an atmospheric lifespan of 120 years.

Fluorinated Gases: hydrofluorocarbons (HFCs), perfluorocarbons (PFCs), and sulphur hexafluoride (SF6) are synthetic powerful CHGs that are emitted from a variety of industrial processes. Fluorinated gases are sometimes used as substitutes for ozone-depleting substances (i.e. CFCs, HCFCs, and halons). These gases are typically emitted in smaller quantities but because they are potent CHGs, they are sometimes referred to as high global warming potential gases (High GWP gases). HFCs are 140–11,700 times more potent than CO_2 and have atmospheric lifespans of 1 to 260 years. Most commercially used HFCs remain in the atmosphere for less than 15 years. PFCs are 6,500–9,200 times more potent than CO_2 and have an atmospheric lifespan of several thousand years. Sulphur hexafluoride is 23,900 times more potent a greenhouse gas than CO_2 and is extremely long lived with very few sinks (EPA 2009).

2 WWT Affected by Climate Change

2.1 Associated Problems

The treatment, distribution, and disposal of wastewater as well as reuse of wastewater are subject to the effects of climate change through increased energy costs and through increases in the volumes of wastewater and storm water entering treatment facilities in areas where and when precipitation increases, and through increased needs for reuse where and when droughts become more prevalent (Major et al. 2011).

2.1.1 Climatic Events

In consideration of the wastewater infrastructure and baseline climate data, the following climate factors were identified as being particularly significant:

- Rainfall (intensity–frequency relationships, annual, and seasonal totals) Wastewater infrastructure is affected by rainfall storm events and, to a lesser degree, by the total annual rainfall.
- Snowfall (predicted increasing temperatures in future years for all months estimate that snowfall is expected to decrease).
- Sea-level elevation.
- Storm surge.
- Rain on snow events (another flood generation mechanism).
- Extreme temperatures (low and high).
- Drought conditions.
- Wind speed (extremes and gusts).
- Frost (freeze-thaw cycles).
- Ice (Kerr Wood 2008).

Wastewater systems are potentially affected by increased rainfall intensity, aggravating flooding, and combined sewer overflows affecting WWTP efficiency. The temperature increase will raise the likelihood of sewer corrosion and odour problems. The sea-level rise can decrease the hydraulic capacity of downstream sewers and increase salt water intrusion. Another challenge is caused by extreme rain events when combined sewer (wastewater and storm water) overflows can reach urban rivers. Combined sewer overflows need to be limited to protect the surface water quality (PREPARED 2010).

Temperature: Biological wastewater treatment is very much influenced by climate. Temperature plays a decisive role in some treatment processes, especially the natural-based and non-mechanised ones. *Warm temperatures* decrease land requirements, enhance conversion processes, increase removal efficiencies, and make utilisation of some treatment processes feasible. Some treatment processes, such as anaerobic reactors, may be utilised in diluted wastewater, such as domestic sewage, only in warm climate areas. Other processes, such as stabilisation ponds, may be applied in lower temperature regions, but occupying much larger areas and being subjected to a decrease in performance during winter. Other processes, such as activated sludge and aerobic biofilm reactors, are less dependent on temperature, as a result of higher technological input and mechanisation levels (Von Sperling 2005).

Warmer temperatures can also indirectly cause more severe weather; exacerbated by urban heat islands, which could in turn result in additional convective thunderstorms, hail, cyclonic events (i.e. tornadoes, cyclones, and hurricanes), and higher winds that may exceed the design capacity of the infrastructure (Major et al. 2011).

Fig. 5 Psittalia wastewater treatment plant, Athens, Greece (Psittalia 2013)

Sea levels: With the onset of rising sea levels, many water utilities have become threatened by flooding, which can have multiple negative consequences. Flooded wastewater facilities have the potential to release untreated waste into the ecosystem, thus causing significant damage to the environment and people alike. If the wastewater facility suffered structural damage, it may have to release untreated waste for an extended period of time until the plant can be fixed. Flood damage would be costly to wastewater municipalities, both in terms of financial loss and in the threat to public health. Careful advance planning to prepare for the consequences of sea-level rise and flooding is essential (Blumenau et al. 2011).

Particular effects of increased flooding on wastewater are as follows:

- Storm increase creates increased flooding, which can be harmful to infrastructure when WWTPs are built in coastal areas (Fig. 5).
- Sea levels are expected to rise in some areas by 2050, endangering the location of many plants.
- Rising downstream water levels may make pumping effluent a requirement, increasing energy need (Danas et al. 2012).

The impact of sea-level rise on WWT facilities will depend on both the degree and rate of the rise. It is expected that only facilities that lie near tidally influenced water bodies will be affected in the near future. According to projections developed by the United Nations International Panel on Climate Change (IPCC), the planet could experience a mean sea-level rise in a range of 18 to 58 cm during the twenty-first century excluding future rapid dynamical changes in ice flow (Mote et al. 2008).

As an example, the University of Washington's Climate Impacts Group (UW CIG) completed a regional analysis of sea-level rise for the Washington State major tidally influenced water bodies. The results of the UW CIG analysis indicate that the

Table 1 Puget Sound sea-level rise scenarios (Mote et al. 2008)

Scenario	Predicted sea-level rise	
	2050 (cm)	2100 (cm)
Very-low sea-level rise—low probability–low impact	7.6	15.2
Medium sea-level rise	15.2	33.0
Very-high sea-level rise—low probability–high impact	55.9	12.0

medium scenario expected in the Puget Sound is a sea-level rise of 15.2 cm by 2050 and of 33.0 cm by 2100 (Table 1).

Low-probability scenarios were estimated for sea-level rise at the low- and high-impact extremes. For the "very-low" scenario, sea-level rise is predicted to be less than 3 in. by 2050 and less than 6 in. by 2100. For the "very-high" scenario, sea-level rise is predicted to be more than 22 in. by 2050 and more than 50 in. by 2100 (Mote et al. 2008).

Rainfall: Increased frequency and intensity of rainfall is one of the most immediate effects of global warming and is already apparent in stream flow records from several previous decades. The expectation is that more severe storms will produce more severe flooding. This will inevitably result in additional water pollution from a large variety of sources. Chief among these are wastewater treatment, storage, and conveyance systems (Cromwell et al. 2007).

Storms: Increased tropical storm intensities will have negative effects on water resources (Fig. 6). More intense tropical storms can damage some infrastructures, because of increased flooding, which can overwhelm water infrastructure and cause pollutants to directly enter waterways and contaminate water supplies (EPA 2012).

Since they affect local tides, the frequency and intensity of storm events must be considered in the analysis of sea-level rise and its impact on WWTPs. Generally, intense storm events occur less frequently than smaller storms. Zervas (2005)

Fig. 6 Flooded wastewater treatment plants (EPA 2012)

analysed the response of extreme tide levels (from storms with a return frequency of 100 years) to long-term sea-level rise at various coastal stations.

Impacts on water pollution: According to EPA, for the most part, WWTPs and combined sewer overflow control programs have been designed on the basis of the historic hydrologic record, taking no account of prospective changes in flow conditions due to climate change. As a result, it is conceivable that water suppliers will face a continually increased influent challenge from sewage overflows producing high concentrations of Giardia, Cryptosporidium, and coliforms (Cromwell et al. 2007).

In the future, wastewater reuse and desalination will possibly become important sources of water supply in semi-arid and arid regions. An increase in wastewater treatment in both developed and developing countries is expected in the future, but point-source discharges of nutrients, heavy metals, and organic substances are likely to increase in developing countries. In both developed and developing countries, emissions of organic micropollutants (e.g. endocrine substances) on both surface waters and groundwater may increase, given that the production and consumption of chemicals, with the exception of a few highly toxic substances, is likely to increase. Several of these pollutants are not removed by current wastewater treatment technology. Changes in water quality may be caused by the impact of sea-level rise on storm water drainage operations and sewage disposal in coastal areas (Bates et al. 2008).

In addition, more frequent heavy rainfall events will overload the capacity of sewer systems and water and WWTPs. An increased occurrence of low flows will lead to decreased contaminant dilution capacity meaning higher pollutant concentrations, including pathogens. In areas with overall decreased runoff (e.g. in many semi-arid areas), water quality will be even worse (Bates et al. 2008).

2.1.2 Wastewater Infrastructure and Design Issues

Extreme rain events will result in higher risks associated with flooding and impact on the water and wastewater infrastructures (PREPARED 2010).

Engineers and owners are currently planning and designing improvements to water, wastewater, and storm water infrastructure all over the world. An engineer designs systems to meet the objectives of the owners, users, and other stakeholders within the constraints of the system. These constraints range from the physical composition of the land to the climate of the regulatory system. The engineer's main goal is to design a system that meets the objectives set forth by the project team. In addition, they seek to reduce the risk or frequency of failure to a certain extent, either socially or legally, such as designing a WWTP to be operational during a 100-year flood event (O'Neill 2010).

Most wastewater infrastructures consist of transmission facilities, treatment facilities, and discharge bodies. The impacts on the wastewater infrastructure can be categorised as follows:

- Impacts indirectly associated with climate change such as the decrease in water usage associated with water conservation.
- Impacts on infrastructure directly associated with climate change.

Indirect impacts: Climate change is primarily to blame for the water infrastructure. Reduced water usage decreases water that flows into the wastewater transmission and treatment systems of the community. This means that it decreases the overall water volume but not the waste load. The increased wastewater strength results in increased sewer cleaning and increased system corrosion. Wastewater viscosity will increase and system flushing will not occur easily. WWTPs will have the same contaminant loading rates at a reduced water volume creating potential hydraulic and corrosion issues that will need to be evaluated.

Direct impacts: Climate change is mainly responsible for the increase in frequency of intense rainfall. Rainfall infiltrates into sewer systems through cracks, poorly constructed or corroded manholes, and direct connections. Sewers are not hydraulically designed to convey large quantities of inflow. This causes the sewer to become hydraulically overloaded during intense rainfall and allows raw sewage to flow into receiving waters and homes as it escapes the sewer system. This is called a sanitary sewer overflow (SSO). As intense rainfall increases, SSO will also increase having an impact on receiving waters and system users. As engineers work with communities to combat SSO, they should contact the local climatologist to determine the overall impact of climate change on the current hydrology and the predicted change so as to determine appropriate design rainfall occurrences.

Another direct impact relates to the decrease in base flow of the waters, which receive the WWTP effluent. Climate change in drought-prone areas is likely to reduce the stream and river base flows. Base flow is used to determine the effluent parameters required by the WWTP, and therefore, as the base flow decreases, effluent requirements will become more stringent and may require the treatment plants to install additional treatment facilities to meet those requirements. Engineers should determine changes in climatic parameters with the local climatologist as they are designing wastewater treatment facilities and working with regulating agencies to determine the predicted impacts including any decrease in stream or river base flow predicted (O'Neill 2010).

2.1.3 Affected Processes in a WWTP

A typical activated sludge process is illustrated in Fig. 1, and the main processes in a WWTP, affected by the climate change, are listed below:

- *Sedimentation*
  - Warm wastewater increases the bacterial reaction rate, which reduces the density of settled sludge.
  - Inflow wastewater will be more dense, and consequently, experiments need to be carried out.

- *Biological Aeration of warm wastewater*
  - Increased BOD.
  - Activated sludge aeration, systems operating at high temperatures, support nitrification.
- *Processing of waste sludge*
  - Waste activated sludge must be thickened for efficient and effective digestion.
- *Stabilisation Ponds*
  - Pros: reliable treatment and minimal operation/maintenance.
  - Cons: land demand, infrastructure, sealed bottoms to prevent groundwater contamination, and potential emission of foul odours.
- *Chlorination* (Danas et al. 2012)

2.2 Applied Solutions

Wastewater infrastructure: Growing evidence indicates that the water sector will not only be affected by climate change, but its impact will continue through floods, droughts, or extreme rainfall events. Water resources will change in quantity, quality and water, storm water and wastewater facilities infrastructures will face greater risk of damage caused by storms, floods, and droughts. The effect of the climate change will be evidenced by difficulties in the operation of disrupted services and increased cost of water and wastewater. Governments, urban planners, and water managers must therefore re-examine the development processes for municipal water and wastewater services, and adapt strategies to incorporate climate change into infrastructure design, capital investment projects, service provision planning, operation, and maintenance.

Wastewater systems, built under historical design parameters, such as minimum flow levels or storm water capacity, will become obsolete and instead reconstruction or rehabilitation may become necessary. Reduced flow in receiving water meeting ambient standards after dilution of effluent from WWTPs may become increasingly difficult and result in a need for increased treatment standards.

Generally, the effects of climate change will require that water and wastewater service providers perform more frequent technical maintenance, undertake unscheduled rehabilitation, and in some cases, scale down operations at their facilities. Therefore, this will reduce the service to their clients. All of these imply additional utility costs. The utility may reduce additional expenditure through the implementation of improved planning, monitoring and maintenance, pass on the cost to consumers, let parts of the system deteriorate, provide lower service levels, or a combination of the above (Danilenko et al. 2010).

The underlying assumption for the design of storm water and combined wastewater systems is that the precipitation and flooding events in the past can be used to predict events in the future. Climate change therefore challenges wastewater system design. If the intensity of precipitation increases and the capacity of the wastewater systems are too low, more floods can be expected. In a separate system, this means surface flooding but in combined systems, basements can also be flooded, and more flow released through combined system overflows (CSO), which can lead to environmental problems in the receiving waters. Increased precipitation can lead to a higher ground water level, causing extra stress on the system with greater infiltration. Also, a higher ground water level can decrease the initial infiltration capacity of the drainage area and thus cause more runoff in the wastewater system. Higher precipitation intensity can wash pollutants off the surface more quickly, thus increasing the concentration of pollutants in the runoff (Berggren et al. 2007).

Adaptive capacity: The degree to which a municipality is able to deal with the impacts of climate change is often referred to as adaptive capacity. According to the Intergovernmental Panel on Climate Change (a global scientific body set up by the World Meteorological Organization and the United Nations Environment Programme), adaptation is defined as "the ability of a system to adjust to climate change (including climate variability and extremes) to moderate potential damages, to take advantage of opportunities, or to cope with the consequences" (IPCC 2007).

New York City is a prime example of a highly developed water supply and wastewater treatment system with an advanced adaptation planning process. Using a multistep adaptation assessment process, wide ranging adaptation studies are under way, including a study of the impacts of sea-level rise on the system, and a study of reservoir operations using future climate scenarios in reservoir modelling. Potential adaptations include operating system changes, flood walls for WPCPs, relocation of facilities, improved drainage, and enhanced water quality treatment (Major et al. 2011).

Assessment tool: To consider the effects of a facility flooding into the surrounding community, an impact assessment tool was created. Suitable tools were developed in wastewater facilities. The crucial factor considered for wastewater facilities is the ratio between the average flow rate and the design flow rate of the plant. This ratio measures how close to maximum capacity a wastewater treatment facility operates. A facility that operates close to maximum capacity will be less able to handle an increase in inflow, possibly caused by a storm or flood, than a facility which does not. Facilities that have an average flow rate of up to 50 % of their design capacities were rated as low impact. Facilities with an average flow rate above 50 % and up to 70 % of their design capacities were rated as medium impact. Facilities with an average flow rate above 70 % of their design capacities were rated as high impact (Blumenau et al. 2011).

Monitoring of wastewater treatment plants: During the last decade, continuous and long-term monitoring of urban wastewater infrastructure has increasingly been applied to wastewater treatment plans. This is due to the availability of reliable and affordable sensors, data communication, and data handling capacity, resulting in the availability of large data sets, describing the performance of wastewater infrastructure over a long time period. Combining this monitoring and meteorological data, it is possible to study the impact of anticipated climate changes by identifying periods of time in the data set resembling weather conditions representative of expected future climatic situations. This "data mining" could result in identifying relevant processes to be taken into account in a model-based analysis of climate change impacts (Langeveld et al. 2013).

Wastewater operation monitoring provides changes in volumes and composition of wastewater, brakes, and clogs in wastewater collection network, adequacy of existing technology to composition of wastewater, and wastewater treatment effluent and sludge (Danilenko et al. 2010).

The development of innovative monitoring of combined sewer overflows and an early warning system for faecal contamination in recreational waters will allow wastewater utilities to be better prepared and to respond faster to any contamination due to combined sewer overflows and uncontrolled runoff caused by more frequent and heavier rainfall (PREPARED 2010).

Vulnerability analysis: Water utilities across the country have initiated research efforts to investigate their vulnerability to climate change processes. Such efforts attempt to obtain a better analytical assessment of the possibility that current water resource development and facility plans could be disrupted by near-term (20–50 year) manifestations of climate change processes. This initial centre on vulnerability is a good means of identifying prior issues related to climate change and lays the foundation for follow-up actions. Two alternative approaches to vulnerability analysis have been articulated: "top-down" and "bottom-up". Many initial vulnerability analyses have been related to water resource and facilities planning. However, direct impact on water utility facilities from flooding due to more intense rainfall activity or sea-level rise is another obvious priority to be analysed.

Some of these efforts have engaged climate models (referred to as GCMs—general circulation models) to attempt to make climate change forecasting the forefront of water supply planning. This has been labelled, the "top-down" approach to vulnerability analysis. The major drawback of this approach lies in the current level of analytical resolution of the GCMs. Almost two dozen recognised GCMs are consistent in projecting increasing global mean temperature, but across a range of variability that is also partly the product of various GHG "emission scenarios", reflecting alternative global assumptions about the future path of economic growth and efficacy of GHG controls.

In contrast, a "bottom-up" approach to vulnerability analysis has also been articulated as a recommended path for utilities to follow in investigating impacts on climate change. The central idea of this approach is that utilities can work with their own water resource planning models to assess the vulnerability of their 20–50 year supply plans to climate change. The "bottom-up" analysis enables a utility to test the robustness of current plans relating to changes in key climate-related variables limited to one or several models, without trying to undertake new climate modelling work.

There is a considerable body in climate research that has been devoted to the understanding of similarities and differences between the predictions of alternative climate models, focusing on their comparative consistency, and overall accuracy versus precision. This type of comparative and interpretive analysis of climate modelling across the range of GCMs may be an important research priority to water utilities, which enable continuing improvements in the "bottom-up" approach to vulnerability analysis (Cromwell et al. 2007).

Membrane treatment processes: Many water suppliers in over-constrained settings have also turned to energy-intensive membrane treatment processes to enable desalination of water sources and reuse of highly treated wastewater effluent. These processes attempt to overcome any deterioration in the reliability of normal sources of supply by making it possible to meet part of the demand from sources abundant under most climate change scenarios (i.e. yields from water reuse and desalt supply options are drought-resistant). If these technologies fill a gap or hide vulnerability produced by climate change processes, in such a way so as to enable a broader scope for optimisation across the entire portfolio, they can play a critical role in improving the overall optimisation (Cromwell et al. 2007).

3 Emissions from WWTPs

3.1 Associated Problems

3.1.1 Background

By and large, the quantity of collected and treated wastewater is increasing in many countries in order to maintain and improve the quality of potable water as well as for other public health and environmental protection benefits. Concurrently, GHG emissions from wastewater will decrease in accordance with future increase in wastewater collection and treatment (Bates et al. 2008).

The study of gaseous emission, climate change, and air pollution is committed to physicochemical identification, inventories, measurements, and assessment methods, as well as quantitative study of the actual anthropogenic sources and its direct contributions. The causes provoked by human activities include the following:

(i) Emissions from wastewater discharges
(ii) Sewage collection and transportation
(iii) Wastewater treatment plants (WWTPs)
(iv) Associated activities

The level of uncertainty in the "carbon profile" of the wastewater industry is unacceptable in the emerging business environment of carbon pricing, and managerial commitments to "zero carbon emission". Methane and nitrous oxide emissions in particular have much higher global warming potentials than carbon dioxide (Foley et al. 2008).

3.1.2 GHG Emissions from Wastewater

Waste management and treatment activities are sources of greenhouse gas emissions (Fig. 7). In 2011, landfills were the third largest source of US anthropogenic CH_4 emissions, accounting for 17.5 % of total US CH4 emissions. Additionally, wastewater treatment accounts for 16.7 % of waste emissions, 2.8 % of US CH_4 emissions, and 1.5 % of N_2O emissions. Emissions of CH_4 and N_2O from composting grew from 1990 to 2011, and resulted in emissions of 3.3 Tg CO_2 Eq. in 2011 (EPA 2013).

Some significant trends in US emissions from wastewater treatment indicate that from 1990 to 2011, CH_4 and N_2O emissions increased by 0.2 Tg CO_2 Eq. (1.6 %) and 1.7 Tg CO_2 Eq. (49.7 %), respectively (Table 2) (EPA 2013).

During the wastewater treatment, GHGs including carbon dioxide (CO_2) from aerobic (oxidation processes), methane (CH_4) from anaerobic processes, and nitrous oxide (N_2O) associated with nitrification/denitrification (NDN) processes as an intermediate product can be emitted to the atmosphere (Fig. 8). Table 3 displays the expected GHG emissions that occur during the processes in a WWTP (Moore 2008).

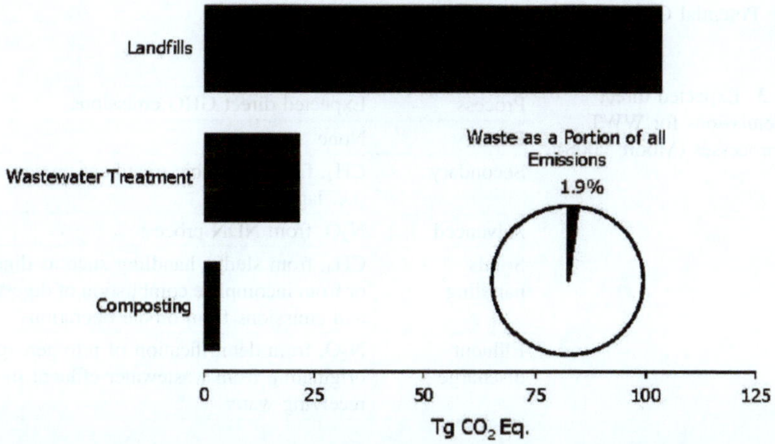

Fig. 7 Waste activities generated emissions of 127.7 Tg CO_2 Eq., or 1.9 % of total US greenhouse gas emissions (EPA 2013)

Table 2 Emissions from wastewater treatment (Tg CO_2 Eq.) (EPA 2013)

Gas	1990	2005	2007	2008	2009	2010	2011
CH_4 (Total waste emissions)	**164.0**	**130.5**	**129.8**	**129.8**	**131.9**	**131.4**	**124.7**
Wastewater treatment	15.9	16.5	16.7	16.6	16.6	16.6	16.4
N_2O (Total waste emissions)	**3.8**	**6.4**	**6.5**	**6.7**	**6.8**	**6.7**	**6.8**
Wastewater treatment	3.5	4.7	4.8	4.8	4.9	5.0	5.1

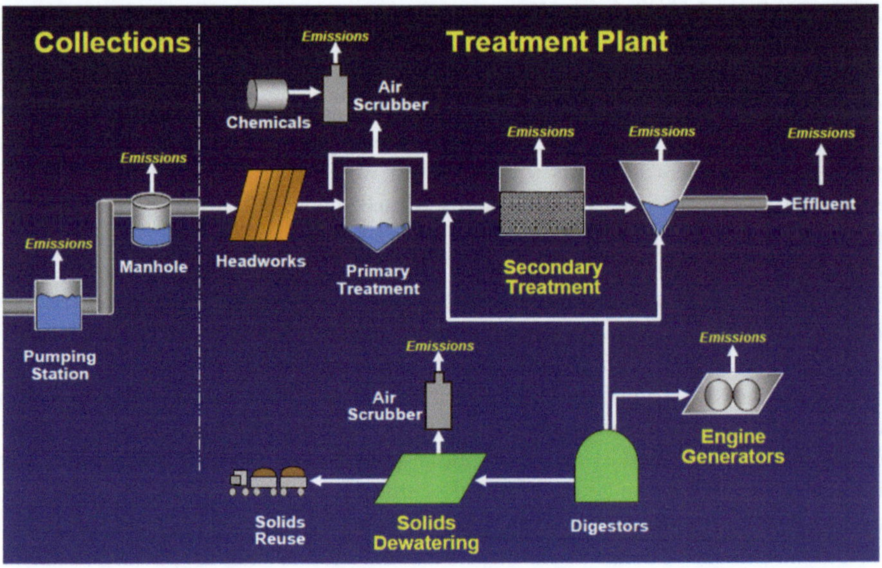

Fig. 8 Potential GHG emission sources (Moore 2008)

Table 3 Expected direct GHG emissions for WWT plant processes (Moore 2008)

Process	Expected direct GHG emissions
Primary	None
Secondary	CH_4, from anaerobic treatment processes (i.e. lagoons)
Advanced	N_2O, from NDN process
Solids handling	CH_4, from sludge handling such as digestion or from incomplete combustion of digester gas and emissions from offsite operations
Effluent discharge	N_2O, from denitrification of nitrogen species originating from wastewater effluent in receiving water

Worldwide CH_4 from wastewater accounted for more than 523 $MtCO_2eq$ in 2000. Wastewater is the fifth largest source of anthropogenic CH_4 emissions, contributing approximately 9 % of total global CH_4 emissions in 2000. India, China, the USA, and Indonesia combined, accounting for 49 % of the world's CH_4 emissions from wastewater. Global CH_4 emissions from wastewater are expected to grow by approximately 20 % between 2005 and 2020. Wastewater is also a significant source of nitrous oxide (N_2O). Worldwide, N_2O emissions from wastewater accounted for approximately 91 $MtCO_2eq$ in 2000. Wastewater as a source is the sixth largest contributor to N_2O emissions, accounting for approximately 3 % of N_2O emissions from all sources. Indonesia, the USA, India, and China accounted for approximately 50 % of total N_2O emissions from domestic wastewater in 2000.

Global N_2O emissions from wastewater are expected to grow by approximately 13 % between 2005 and 2020 (Scheehle 2012).

Methane emitted during wastewater transport, treatment, and disposal, including from wastewater sludge, amounts to 3 to 19 % of global anthropogenic methane emissions (IPCC 1996). Globally, the major sources of greenhouse gas, nitrous oxide (N_2O), are human sewage and wastewater treatment (IPCC 2007). Methane emissions from wastewater are expected to increase by about 50 % in the next few decades and N_2O emissions by 25 %. Thus, one of the most direct ways to mitigate greenhouse gas emissions is through improvements in the collection and management of urban wastewaters, using technologies most appropriate to the economies and settings involved (IPCC 2007). Technologies already exist for reducing, and perhaps reversing, these emissions growth rates.

To estimate the GHG emissions of the WWTPs in a comparable way, the considered emissions have to be listed (Snip 2010):

1. CO_2 and N_2O emissions at biotreatment, endogenous respiration, and BOD oxidation
2. Nitrification CO_2 credit and nitrogen removal
3. Energy use of the plant, for aeration, mixing and pumping, which leads to CO_2 emissions
4. Sludge digestion, biogas CH_4, and CO_2
5. Sludge disposal, truck emissions trip to reuse/disposal site, CO_2 emissions mineralisation
6. Power credit by use of biogas
7. GHG emissions from chemical use

To provide further detail, a WWTP is schematically displayed in Fig. 9. The different boxes show the treatment processes. The GHGs that can be released during the treatment processes are given in the circles. The numbers in the figure correspond with the numbers in the list of boundaries (Snip 2010).

Municipal sewage treatment plants play an important role in the abatement of water pollution, but they also produce a large amount of gaseous emissions into the atmosphere. The discharge of large volumes of fugitive gases containing low levels of chemical constituents may still lead to an excessive contribution to air pollution. Most centralised wastewater treatment methods consist of a combination of biological processes (activated sludge reactors, trickling filters, anaerobic digesters, etc.) promoting biodegradation of organic matters by micro-organism and the production of anthropogenic CH_4, and N_2O gaseous emissions.

Methane (CH_4) production is a direct result of anaerobic decomposition of the organic matter currently in sewers. The methanogenesis or CH_4 production rate depends primarily on the concentration of the degradable organic material in wastewater measured by biochemical oxygen demand (BOD_5) and chemical oxygen demand (COD). The main environmental factors that influence methane production include retention time, pH, temperature, presence of sulphate reducing bacteria, and methanogens (Guisasola et al. 2008; Listowski et al. 2011).

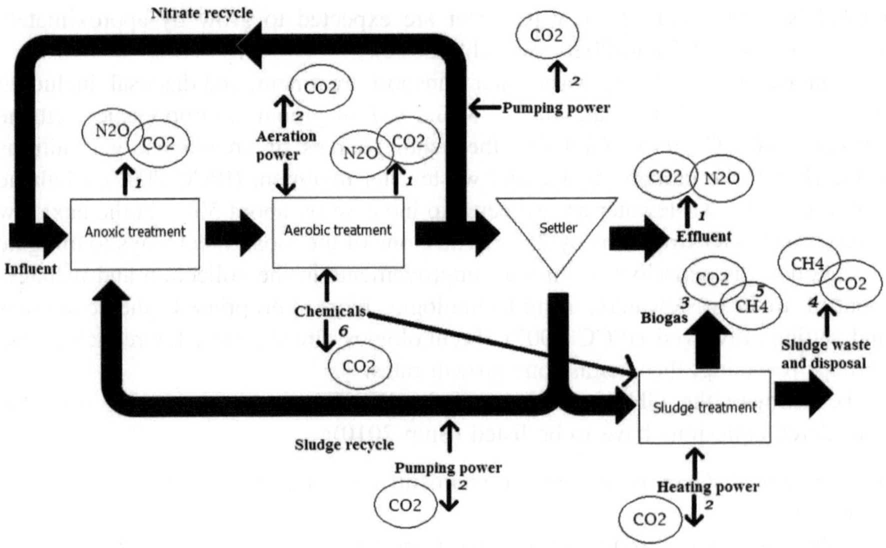

Fig. 9 Estimated greenhouse gas emissions of a wastewater treatment plant (Snip 2010)

CH_4 generation occurs as organic matter undergoes decomposition in anaerobic conditions. However, CH_4 generation varies widely depending on waste management techniques. Specifically engineered environments can increase the CH_4 generation rates. The quantity of CH_4 generated can be expressed in terms of several key activity and emissions factors (Scheehle 2012):

Domestic Wastewater

$$CH_4 \text{ Generation} = (\text{POP}) * (\text{BOD}) * (\text{PAD}) * (CH_4\text{P}) \tag{1}$$

where:

POP total population
BOD production of BOD per capita per year
PAD percentage of BOD anaerobically digested per year
CH_4P CH_4 generation potential per kg of BOD

Industrial Wastewater

$$CH_4 \text{ Generation} = (\text{IP}) * (\text{COD}) * (\text{PAD}) * (CH_4\text{P}) \tag{2}$$

where:

IP industry production
COD production of COD per unit of output
PAD percentage of COD anaerobically digested per year
CH_4P CH_4 generation potential per kg of COD

Nitrous oxide (N_2O) and nitric oxide (NO) production are associated with the breakdown of nitrogen components common in wastewater, e.g. protein and urea. Biological nutrient removal (BNR) processes have the ability to transform the ammonia and organic nitrogen compounds into nitrogen gas, which can be released into the Earth's atmosphere. The two-phase process involves nitrifying bacteria (Nitrosomonas) that oxidise ammonia to create nitrate (aerobic phase) while denitrifying bacteria reduces nitrate, turning it into nitrogen gas, which is then released to the atmosphere (anoxic phase). N_2O and NO can be released during both of these processes; however, it is mainly associated with denitrification (Park et al. 2000; Listowski et al. 2011). Denitrification is a four-step process, as heterotrophic bacteria can use nitrate, nitrite, nitric oxide, and nitrous oxide as an electron acceptor. The denitrification follows the four steps according to:

$$NO_3^- \rightarrow NO_2^- \rightarrow NO \rightarrow N_2O \rightarrow N_2$$

As can be seen, N_2O is an intermediary in this process. Thus, N_2O can be produced and released into the atmosphere due to incomplete denitrification (Snip 2010). The aerobic treatment process produces relatively small emissions, whereas in anaerobic processes, emissions can be increased by 50–80 % (Park et al. 2000; Listowski et al. 2011). N_2O production is typically estimated using an activity factor of annual per capita protein consumption (kilograms per year). However, it has been suggested that this factor alone underestimates the actual amount of protein entering wastewater treatment systems (Scheehle 2012).

Carbon dioxide (CO_2) production is attributed to two main factors: the treatment process and the electricity consumption. During the anaerobic process, the BOD_5 of wastewater is either incorporated into biomass or converted into CO_2 and CH_4. A fraction of biomass is further converted into CO_2 and CH_4 via endogenous respiration. Short cycle or natural sources of atmospheric CO_2 which cycle from plants to animals and to humans as part of the natural carbon cycle and food chain do not contribute to global warming. Photosynthesis produces short-cycle CO_2, removes an equal mass of CO_2 from the atmosphere that returns during respiration or wastewater treatment. Digestion processes, either aerobic or anaerobic, also only emit short-cycle CO_2 (Listowski et al. 2011).

The hydrogen sulphide (H_2S) gas evolves from the anaerobic decomposition of organic matter or from the reduction of mineral sulphites and sulphates. H_2S gas mixed with the sewage gases (CH_4 + CO_2) is highly corrosive to sewer pipelines, manholes, concrete junction chambers, and mechanical and electrical equipment (Guisasola et al. 2008; Listowski et al. 2011).

Volatile organic compound (VOC) emission occurs during the entire wastewater cycle. A significant fraction of VOCs is released into the atmosphere by gas–liquid mass transfer. VOC production during wastewater transportation in sewers occurs during turbulent flow and air exchange between ambient atmosphere and wastewater. The transfer rate of emission is affected by the physicochemical properties of chemicals, fluid and flow characteristics. There is a growing concern that several

VOCs present in wastewater; especially industrial effluents, find their way into the atmosphere. In particular, VOCs such as benzene, chloroform, ethyl benzene, toluene, *m*-xylene, and *o*-xylene are found in refinery and petrochemical wastewater in significant amounts as well as in many municipal wastewater (Bhattacharya 1989; Al-Muzaini et al. 1991; Listowski et al. 2011).

3.2 Applied Solutions

As climate change is a major concern, alternatives should reduce greenhouse gas emissions, making anaerobic treatment a more attractive component of novel approaches to treatment processes.

Vulnerability climate assessment: The definition of an assessment framework, proposed by Listowski et al. in 2011 for gaseous emission from urban wastewater systems (Fig. 10), appears necessary to the development of future adaptation strategies and knowledge to manage emissions from the wastewater cycle.

This framework should be developed to interact with the adaptive responses to address emission sources, infrastructure, the pathways for gaseous emissions and their concentrations, mitigation capabilities, and technologies. The main tasks for assessment framework incorporate several areas including:

- Understanding the emission generation processes (spatial, temporal, physical, and biochemical) with the key motivation issues including pathways for gaseous emissions and concentrations.

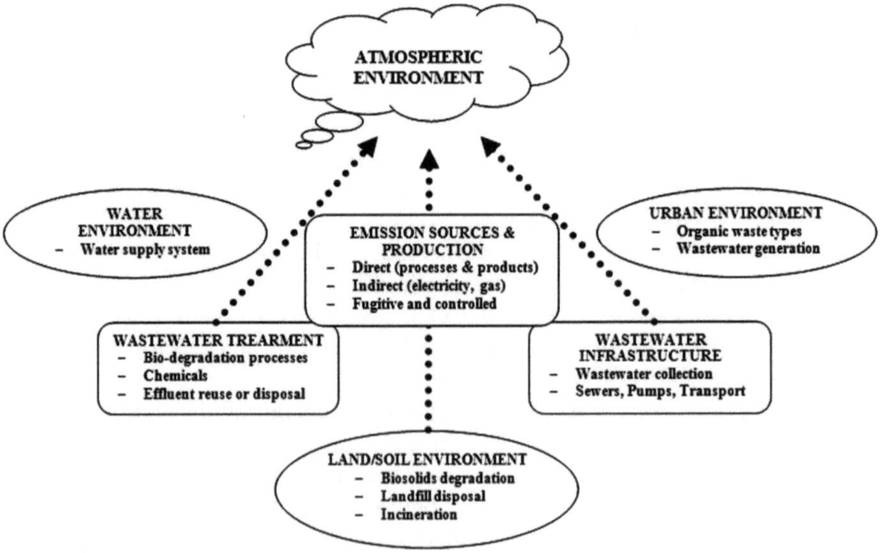

Fig. 10 GHG emission assessment frameworks from wastewater (Listowski et al. 2011)

- Identifying the appropriate and reliable parameters as a basis for the adaptation of the strongly variable combined wastewater flow to the actual treatment capacity.
- Establishing credible methods of obtaining data and information from defined emission sources.
- Quantifying and predicting gaseous emissions. Among the broad diversity of wastewater sector, the analysis of gaseous emissions could be assessed in two essential categories.
  - Direct emission linked with wastewater sources and activities that promote fugitive gaseous emission in relation to the physical and biochemical processes characteristic in wastewater and its by-products during the wastewater cycle.
  - Indirect emission; energy use associated with wastewater transportation, pumping, various treatment processes, effluent disposal, residual management, etc.

The main factors in this regard are the use of biological wastewater treatment, aerobic or anaerobic treatment technology, sludge processing, and also the electricity used. While assessment of emissions in relation to energy consumption (CO_2 equivalent) is relatively straightforward, quantifying direct fugitive emissions ("fugitive" emissions of nitrous oxide and methane from WWTP operations since both of these gases are major greenhouse contributors) from wastewater systems is an area of uncertainty for the industry, with less developed and less reliable methodologies. The diffused emissions include substances such as CH_4, CO_2, VOCs, NO_x, CO, mercury, cadmium and lead, hydrogen sulphide (H_2S), ammonia (NH_3), and sulphur dioxide (SO_2), which have an adverse effect on air quality, the environment, and public health (Listowski et al. 2011).

Methane as a fuel: The methane emission in relation to the anaerobic digestion of primary and secondary sludge counts for about three quarters of WWTP overall methane emission and causes a slightly larger greenhouse gas footprint than the carbon dioxide emission avoided by using the resulting biogas for energy generation.

Methane emissions can be significantly reduced by better handling of the ventilation air of sludge treatment facilities; one way to valorise the residual methane produced in the buffer tank is to use the ventilation air from the tank as combustion air in the gas engines of the cogeneration plant. The methane concentration in the ventilation air could of course be increased by using less fresh air for ventilation. This would result in less diluted methane streams, but then the ventilation system should be adapted to handle methane concentrations that exceed the lower explosive limit of methane in air, which is 4.4 % (Daelman et al. 2012).

Recovering the energy to provide heat and electricity for the WWTP process can offset significant fossil fuel-related GHG emissions. In general, intuitively sustainable practices for biosolids (energy recovery, recycling nutrients, and organic matter) reduce GHG emissions. Besides, the methane emitted into the atmosphere not only contributes to the greenhouse gas footprint of a WWTP, but also implies a waste of energy since the methane emitted from the unit processes related to the anaerobic digestion (7 to 2 % of the produced methane) could potentially be used as a fuel for the cogeneration plant. Although biogas production from waste sludge may be a sustainable technology from the energy aspect, it has no benefits in this case over fossil fuel-derived energy regarding greenhouse gas emissions. Nonetheless, it should be emphasised that the emission of methane is not intrinsic for anaerobic digestion, but that better design and good housekeeping may lead to drastic mitigation of the emission (Daelman et al. 2012).

4 Conclusion

The various problems associated with climate change and WWT operation and the solutions that can be applied to deal with them are summarised in this chapter. It is estimated that climate change has a dual effect on WWTPs.

The processes occurring in a WWTP are subsequently affected by climate change; more extreme weather events and earlier snowmelt runoff will lead to more untreated sewer overflows, increased flooding, etc. Most wastewater infrastructure consists of transmission facilities, treatment facilities, and discharge bodies. The impacts to the wastewater infrastructure can be categorised as:

- Impacts indirectly associated with climate change such as the decrease in water usage associated with water conservation.
- Impacts directly associated with climate change on the infrastructure.

The limitation of climate change effects on WWT processes can be achieved by applying an impact assessment tool, by the monitoring of WWTPs, and by using vulnerability as a good means of identifying the priority issues of a utility in relation to climate change.

In the other hand, we have the WWT contribution to climate change itself, as during the wastewater treatment, GHGs including carbon dioxide (CO_2) from aerobic (oxidation processes), methane (CH_4) from anaerobic processes, and nitrous oxide (N_2O) associated with NDN processes as an intermediate product, can be emitted to the atmosphere.

The development of future adaptation strategies and knowledge to manage emissions from the wastewater cycle appears necessary, and the vulnerability climate assessment should be developed to interact with the adaptive responses that could address emission sources. Furthermore, recovering the energy to provide heat and electricity for the WWTP process by using the resulting biogas from the anaerobic digestion of sludge can offset significant fossil fuel-related GHG emissions (CH_4 etc.).

References

Al-Muzaini S, Khordagui H, Hamoda MF (1991) Assessment and treatability of volatile organic compound emissions in proposed Shuaiba industrial wastewater treatment facility. Kuwait Institute for Scientific Research (KISR), Report No. KISR 4000, Kuwait

Bates BC, Kundzewicz ZW, Wu S, Palutikof JP (eds) (2008) Climate change and water. IPCC technical paper VI, Geneva

Berggren K (2007) Urban drainage and climate change—Impact assessment. Licentiate Thesis, Luleå University of Technology, Luleå, Sweden

Bhattacharya SK (1989) Removal and fate of RCRA and CERCLA toxic organics pollutants in wastewater treatment. US Environment Protection Agency, Report EPA-440/1- 82/303, US EPA, Washington, DC

Blumenau A, Brooks C, Finn E, Turner A (2011) Effects of sea level rise on water treatment & wastewater treatment facilities. 41-JPH B113. Date: 14 Oct 2011

Cromwell JE, Smith JB, Raucher RS (2007) Implications of climate change for urban water utilities. Association of Metropolitan Water Agencies (AMWA), Washington, DC

Daelman MRJ, van Voorthuizen EM, van Dongen UGJM, Volcke EIP, van Loosdrecht MCM (2012) Methane emission during municipal wastewater treatment. Water Res 46:3657–3670

Danas K, Kurdi B, Stark M, Mutlaq A (2012) Climate change effects on waste water treatment. CEE Jordan Group Presentation, Jordan, 18 Sep 2012

Danilenko A, Dickson E, Jacobsen M (2010) Climate change and urban water utilities: challenges & opportunities. Water Working Notes, Water Sector Board of the Sustainable Development Network of the World Bank Group, Washington, DC

EPA (2009) Overview of greenhouse gases. http://www.epa.gov/climatechange/ghgemissions/gases.html. Accessed 5 Jan 2014

EPA (2012) Watershed Academy Web. The effect of climate change on water resources and programs (PDF file adapted). http://www.epa.gov/watertrain. Accessed 5 Jan 2014

EPA (2013) Inventory of US Greenhouse Gas Emissions and Sinks: 1990–2011. EPA 430-R-13-001, 12 Apr 2013, Washington, DC, USA

Foley J, Lant PA, Donlon P (2008) Fugitive greenhouse gas emissions from wastewater systems. Water 38(2):18–23

Guisasola A, de Haas D, Keller J, Yuan Z (2008) Methane formation in sewer systems. Water Res 42:1421–1430

Intergovernmental Panel on Climate Change (IPCC) (1996) IPCC guidelines for national greenhouse gas inventories: reference manual. National Physical Laboratory, New Delhi, pp 6–15

Intergovernmental Panel on Climate Change (IPCC) (2007) Climate change 2007: impacts, adaptation and vulnerability. Contribution of Working Group II to the fourth assessment report of the IPCC. Cambridge University Press, Cambridge

Kerr Wood Leidal Associates Ltd (KWL) (2008) Vulnerability of Vancouver sewerage area infrastructure to climate change. File No. 251.219. Final report, March 2008

Langeveld JG, Schilperoort RPS, Weijers SR (2013) Climate change and urban wastewater infrastructure: there is more to explore. J Hydrol 476:112–119

Listowski A, Ngo HH, Guo WS, Vigneswaran S, Shin HS, Moon H (2011) Greenhouse gas (GHG) emissions from urban wastewater system: future assessment framework and methodology. J Water Sustain 1:113–125

Major DC, Omojola A, Dettinger M, Hanson RT, Sanchez-Rodriguez R (2011) Climate change, water, and wastewater in cities. In: Rosenzweig C, Solecki WD, Hammer SA, Mehrotra S (eds) Climate change and cities: first assessment report of the urban climate change research network. Cambridge University Press, Cambridge, pp 113–143

Moore MD (2008) Wastewater treatment impacts from AB 32 and climate change. Paper presented at the workshop: understanding AB 32: intent and impact statewide summit for municipal leaders on climate change mitigation and adaptation, San Diego, California, 4 Sept 2008

Mote P, Petersen A, Reeder S, Shipman H, Whitely Binder L (2008) Sea level rise in the coastal waters of Washington State. Report prepared by the Climate Impacts Group, University of Washington, Seattle, Washington, and the Washington Department of Ecology, Lacey, Washington

O'Neill II JA (2010) Climate change's impact on the design of water, wastewater, and stormwater infrastructure. In: Proceedings of AGU hydrology days 2010, Colorado States University, Colorado, 22–24 March 2010

Park KY, Inamori Y, Mizuochi M, Ahn KH (2000) Emission and control of nitrous oxide from a biological wastewater treatment system with intermittent aeration. J Biosci Bioeng 90 (3):247–252

PREPARED Enabling Change (2010). www.prepared-fp7.eu. Accessed 4 March 2014

Psittalia Wastewater Treatment Plant (2013). In: Athens Water Supply and Sewerage Company (EYDAP SA) web-site. http://www.eydap.gr/index.asp?a_id=205 Accessed 6 Jan 2014

Scheehle EA (2012) US EPA. http://www.epa.gov/climatechange/Downloads/EPAactivities/GM_SectionIII_Waste.pdf. Accessed 5 March 2014

Snip L (2010) Quantifying the greenhouse gas emissions of wastewater treatment plants. Thesis Systems and Control. Wageningen University, Agrotechnology and Food Sciences, Wageningen, The Netherlands

Trenberth KE (1999) Conceptual framework for changes of extremes of the hydrological cycle with climate change. Clim Change 42:327–339

Von Sperling M, de Lemos Chernicharo CA (2005) Biological wastewater treatment in warm climate regions. IWA Publishing, Padstow

Water Environmental Federation (WEFTEC 08) Workshop (2008) Wastewater treatment in tomorrow's climate change-driven world. Summary of pre-conference workshop, Chicago, USA, 18–22 Oct 2008

WATer and Global Change (WHATCH). http://www.waterandclimatechange.eu/. Accessed 4 March 2014

Zamboulis D, Zouboulis AI, Matis KA, Mavros P (2003) Introduction to chemical technology. Publications Aristotle University of Thessaloniki, Thessaloniki

Zervas CE (2005) Response of extreme storm tide levels to long-term sea level change. Oceans. In: Proceedings of MTS/IEEE, vol 3, pp 2501–2506

Part II
Adaptation

Managing Hydropower Under Climate Change in the Mekong Tributaries

Thanapon Piman, Tom A. Cochrane, Mauricio E. Arias, Nguyen D. Dat and Ornanong Vonnarart

Abstract The Mekong Basin is threatened by accelerated hydropower development and extreme events from climate change. The transboundary Srepok, Sesan, and Sekong (3S) basins contribute the largest discharge of Mekong River's tributaries, providing critical ecosystem services to the Tonle Sap and the Mekong delta downstream, including sediments, biodiversity, and fish production. This study aims to assess the potential impact of climate change and hydropower development scenarios on flow patterns and hydropower production in the 3S through multi-general circulation models (GCMs), hydrological simulations, and reservoir operation models. Full hydropower development coupled with energy-focused operations will increase dry season flows by 96 % and reduce wet season flows by 25 % at the basin outlet as compared to historical baseline conditions. Climate change is likely to decrease dry season flows by 6–24 %, but projections of wet season and annual flows using different climate change scenarios and GCMs are relatively uncertain. Energy production in the 3S is not likely to be affected substantially by climate-driven changes in flows; only minor changes resulting from either A2 and B2 climate change scenarios and different GCMs. Predicted climate change,

T. Piman (✉)
Climate Change and Adaptation Initiative, Mekong River Commission, Vientiane
Lao PDR
e-mail: gamekung2@yahoo.com

T. Piman · T.A. Cochrane · M.E. Arias
Department of Civil and Natural Resources Engineering, University of Canterbury,
Christchurch, New Zealand
e-mail: tom.cochrane@canterbury.ac.nz

M.E. Arias
e-mail: mauricio.eduardo.arias@gmail.com

N.D. Dat · O. Vonnarart
Information and Knowledge Management, Mekong River Commission,
Phnom Penh, Cambodia
e-mail: dat@mrcmekong.org

O. Vonnarart
e-mail: ornanong@mrcmekong.org

S. Shrestha et al. (eds.), *Managing Water Resources under Climate Uncertainty*,
Springer Water, DOI 10.1007/978-3-319-10467-6_11

however, will result in significant changes in the magnitude and frequency of extreme flood events, which will undoubtedly impact on future dam design and operation rules. Coordination of hydropower operations within the 3S basin will be critical to maximise development benefits within the basin and reduce negative environmental impacts at the local, national, and transboundary levels.

Keywords Mekong · Climate change · Hydropower · Southeast Asia · Reservoir operations

1 Introduction

The hydrology of the Mekong River is dominated by a monsoon climate, which forms a distinct dry/wet seasonality driving ecosystem productivity and livelihoods in the Lower Mekong floodplains. Productivity and development are often significantly affected by extreme events in terms of abrupt changes in rainfall and flow patterns over the Mekong region. For example, extreme droughts and floods that occurred in 2010 and 2011, respectively, caused severe property damage and loss of livestock and human lives (Kongthong 2011). The changing climate is increasing the frequency of extreme events (Delgado et al. 2012; Räsänen et al. 2013), affecting agricultural production, management of water resources, and the rich biodiversity and ecosystem productivity of the basin strongly related to the hydrological cycle and flood pulse. The strong relationship between the monsoon climate and hydrology in the Mekong has been well studied in recent years (Adamson et al. 2009; Delgado et al. 2010, 2012; Räsänen et al. 2013). For instance, Räsänen and Kummu (2013), Delgado et al. (2012) reported that the variability in Mekong's hydrology was strongly influenced by El Niño Southern Oscillation (ENSO), particularly in the decay year of ENSO events. These findings are important to improve climate prediction and disaster prevention. Building knowledge of potential climate change and its potential impact on water resources and ecosystems at regional, national, and sub-basin levels is also important to develop social and economic climate change adaptation plans for the Mekong countries (MRC and ICEM 2009).

A few large-scale studies have been conducted on the impact of climate change in the Mekong Basin on hydrology and water resources management based on projected climate change data of the greenhouse gas emissions scenarios from the IPCC (2000) through global climate models (GCMs) and corresponding downscaled regional predictions. The IPCC emission scenarios cover a wide range of the main demographic, economic, and technological driving forces of greenhouse gases (GHG) and sulphur emissions. The A1FI emission scenario has the highest possible emissions during a projected period from 2000 to 2050 followed by A2, A1B, A1T, B1, and B2 emissions scenarios, respectively (IPCC 2007). A summary of the large-scale studies carried out in the Mekong Basin is shown in Table 1. The results

Table 1 Summary of the impact of climate change on temperature, rainfall, and flow from large-scale studies over the Mekong Basin

Study	Emissions scenario	GCM	Temperature	Rainfall	Flow
Eastham et al. (2008)	A1b	NCAR_CCSM3 MIUB_ECHO MICRO3_2_MEDRES MICRO3_2_MEHIRES INV_ECHAM4 GISS_AOM CSRIO_MK3 CNRM_CM3 CCCMA_CGCM3_1_T63 CCCMA_CGCM3_1 BCCR_BCM2	Mean: +0.68 to 0.81 °C	Mean annual +0.5 to 36 %	Mean total annual for the whole basin: −8 to 90 %
Västilä et al. (2010)	A2	MPI_ECHAM4	Mean annual maximum: +1.0 to +2.0 °C	Mean Annual: +4 %	Mean annual at Kratie (Cambodia) +4.0 %
Hoanh et al. (2010)	A2 and B2	MPI_ECHAM4	A2-Mean annual maximum +0.70 °C (2.5 %)	A2-Mean annual +77 mm (5.3 %)	A2-Mean annual for the whole basin: +7.6 to 13.5 %
Mainuddin et al. (2010)			B2-Mean annual maximum: +0.80 °C (2.8 %)	B2-Mean annual: +47 mm (3.2 %)	B2-Mean annual for the whole basin: +4.1 to 9.0 %
Lauri et al. (2012)	A1b and B1	CCCMA_CGCM3.1 CNRM_CM3 GISS_AOM MPI_ECHAM5 NCAR_CCSM3	A1b-Mean annual maximum: +0.93 to 1.65 °C	A1b-Mean annual: −2.5 to 8.6 %	A1b-Mean annual at Kratie: −10.6 to +13.4 %
			B1-Mean annual maximum: +0.68 to 1.58 °C (2.5 %)	B1-Mean annual: +1.2 to 5.8 %	B1-Mean annual at Kratie: −6.9 to +8.1 %

indicate that temperature is very likely to increase in the future; however, the direction of changes in rainfall and river flows is uncertain and varies depending on the GCM and downscaling method chosen (Kingston et al. 2011; Lauri et al. 2012). Most GCMs predict an increase in annual rainfall and flows, but some GCMs actually show an opposite trend (Eastham et al. 2008; Kingston et al. 2011; Lauri et al. 2012). Recent studies from Hoanh et al. (2010), MRC (2011b), Lauri et al. (2012), Piman et al. (2013) have assessed basin-wide impacts from the combined effect of hydropower development and climate change on water flows in the Mekong. These previous studies, however, did not adequately capture the fine scale of hydropower operations in Mekong tributaries. Moreover, predicted changes were spatially and temporally different across the basin. Therefore, detailed studies in key Mekong sub-basins are necessary to improve understanding of the coupled effects of hydropower and climate change in the Mekong. Furthermore, there are no previous published studies on how climate change could affect hydropower operations and electricity generation in the Mekong. Because of alterations to extreme events, climate change has the potential to affect reservoir operations and hydropower generation. Hence, it is necessary to consider climate change impacts as early in the planning and implementing stages as possible (Schaefli et al. 2007; Minville et al. 2010).

Of all tributary contributions to the Mekong, the Sekong, Sesan, and Srepok (3S) rivers within Cambodia, Lao PDR, and Vietnam contribute 17–20 % of the total annual flow of the Mekong mainstream (an average of 2,886 m^3/s). This makes the 3S the most influential tributary to the Mekong in terms of water flow. Thus, climate-driven alterations in the 3S basin could have a significant effect on the overall of Mekong's discharge. Projected climate change impacts from the A2 and B2 emission scenarios (using only one GCM) for example have predicted an increase in water stress during the dry season in the Sesan River Basin (Ty et al. 2012).

The 3S basin is also one of the Mekong's most important tributary systems due to its contribution of aquatic biodiversity, ecosystem services, and fisheries to the Tonle Sap Lake and the Mekong Delta, particularly with regard to sediments, flooding patterns, and fish diversity (ADB 2010; Ziv et al. 2012; Arias et al. 2014b). Apart from the future effects of climate change, the natural hydrological flows in the 3S basin are under threat by ongoing hydropower development. Hydropower development capacity has been reached in Vietnam, and it is rapidly increasing in Lao PDR. As of December 2011, there were 9 operating dams, 11 projects under construction, and 21 other planned projects with a total installed capacity of 6,400 MW and a total active storage of 26,328 × 106 m^3 (Piman et al. 2013). This total active storage is about 30 % of the twenty-year average annual flow at the outlet of the 3S (86,987 million m^3). The redistribution of water from the wet to the dry season by dam operations in order to maximise energy production will undoubtedly change flow regimes (lower wet season flows and flood peaks and higher dry season flows). Patterns of flow alterations caused by hydropower development in the 3S follow a similar trend to cumulative alterations throughout the Mekong Basin (ADB 2004; WB 2004; MRC 2011b; Lauri et al. 2012). Thus, studies of hydropower and climate change in the 3S offer good representation of the situation in the greater Mekong Basin.

The main objectives of this study are to determine how projected climate change scenarios affect the flow regime in the 3S basin compared to flow alterations induced by hydropower development. Furthermore, the combined scenarios of climate change and hydropower development were investigated to determine the cumulative impacts on flows within the 3S basin. The final aim of the study is to quantify the uncertainty of climate change projections using different GCMs and downscaling methods on predictions of flow and hydropower energy production.

2 Methodology

The impact of climate change and hydropower development on river flows and hydropower production in the 3S was examined through hydrological and reservoir simulation models. Observed daily climatic data from the 1986 to 2005 hydrological years, which start on 1 June of the corresponding calendar year and end on 31 May of the following calendar year, were used as the baseline climate condition. The dry season period in the 3S spans from December to May, while the wet season period spans from June to November. Future climate change was represented with 40 years of projected daily climatic data (2010–2049) for the A2 and B2 greenhouse gas emission scenarios obtained from the downscaled MPI_ECHAM4 GCM (IPCC 2007). The A2 scenario assumes high global GHG emissions, while the B2 scenario assumes low GHG emissions. These two scenarios were chosen because they encompass a wide range of global emissions in the next 40 years (IPCC 2007). Moreover, three other GCMs (CCCMA_CGCM3.1, MPI_ECHAM5, and NCAR_CCSM3) were selected to obtain projected daily climatic data of the A2 emission scenario to investigate the uncertainty in global climate change projections on hydropower operations.

The climate data sets were input to the Soil and Water Assessment Tool (SWAT) model (Winchell et al. 2009) for simulating daily water flows at numerous points along the 3S river network. The output flows from SWAT were fed into the HEC-ResSim model (USACE 2000) to simulate regulated flows and power production for a range of hydropower development scenarios under the observed baseline climate and projected A2 and B2 emission scenarios from four GCMs. Changes in rainfall, flow patterns, and hydropower production from climate change and hydropower development scenarios were compared against the baseline scenario without hydropower power projects.

2.1 Climate Change Simulations

The spatial resolution of GCM outputs is too large for detailed modelling at the basin scale; thus, downscaling of the GCM data was required. All downscaled climate data were obtained from the Mekong River Commission (MRC). The downscaling process was carried out based on two approaches. Firstly, climate data from the A2 and B2 emissions scenarios from the MPI_ECHAM4 GCM (original resolution of 2.8 by 2.8°; Roeckner et al. 1996) were downscaled dynamically to 0.2 by 0.2° using the PRECIS regional climate model (Jones et al. 2004; Hoanh et al. 2010; Västilä et al. 2010). Secondly, the projected climate data of the A2 emissions scenario from the other three GCMs, including CCCMA_CGCM3.1 (original resolution of 3.75 by 3.75°), MPI_ECHAM5 (1.9 by 1.9°), and NCAR_CCSM3 (1.4 by 1.4°), were downscaled using the delta change method (Diaz-Nieto and Wilby 2005; Choi et al. 2009). In this method, change factors on a monthly basis between a baseline period (1981–2005) and future climate data were calculated and the difference applied to the observed historical data. The downscaled climate data were corrected for bias of daily maximum and minimum temperature and daily precipitation data by comparing computations with observed results to remove spikes and inconsistencies in the GCM outputs (Hoanh et al. 2010; Lauri et al. 2012).

2.2 Hydrological Simulation

The observed climate data from 1986 to 2005 and the predicted climate data of the A2 and B2 scenarios from 2010 to 2049 were used in the SWAT model to simulate water flow. The model allows for daily and long-term water yield simulations (Winchell et al. 2009). The model divides the 3S landscape into 118 sub-basins to calculate water yields to the main channel from each sub-basin and then route water through the channel network to the basin's outlet. The SWAT model was selected because it was calibrated (1986–2000) and verified (2001–2005) with observed flow data by the MRC for use in the Mekong Basin (MRC 2011a).

Spatial information of the 3S sub-basins (sub-basin delineation, land use/land cover properties, and soil properties) for hydrological modelling was generated from information in the MRC data catalogue, including satellite imagery, a 50-m-grid-cell-size-resolution digital elevation model (DEM) generated from topographic maps (scales 1:50,000 and 1:100,000), 2003 land use/land cover maps, and soil maps. Six climate stations were selected to obtain weather data including evaporation, humidity, wind speed, and solar radiation. Rainfall data were obtained from 35 stations throughout the sub-basins, and measured flow was obtained from six stations in the 3S Rivers for calibration and validation of the model parameters (MRC 2011a).

Future flow alterations may be caused not only by climate change but also by changes in land use/land cover. This important aspect has already been investigated in the Srepok by Ty et al. (2012) and should definitely be the subject of future research. The present study, however, focuses on climate change and hydropower, and land use/land cover is assumed to remain constant in all scenarios.

2.3 Reservoir Simulation

Computed flows of each sub-basin from the SWAT model were then used as inputs to the model HEC-ResSim at current and proposed dam sites (Fig. 1) to simulate regulated flows and power production from dam operations. The HEC-ResSim model, developed by the US Army Corps of Engineers (USACE), is a well-established and widely used model for simulating reservoir systems within basins (Cochrane et al. 2010; Minville et al. 2010; Piman et al. 2013). Physical characteristics, including dam height and width, spillways, and power plant release capacities, were defined for each dam (Table 2). This information is available in the MRC hydropower database (MRC 2009a). The Muskingum-Cunge method was selected for channel routing in which channel cross section geometry was defined by eight points (USACE 2000).

2.4 Reservoir Operation Rules

Monthly reservoir operation rules were used in the reservoir simulations of the Hec-ResSim model. Average monthly water level data in the reservoir were used to calculate target levels in the operation of existing. The information on how reservoirs might be operated for proposed dams is normally absent, so a linear optimisation model (Wurbs 1994) was used to determine how the reservoir will be operated during each month. The input data for the linear optimisation model include monthly inflows to the dam from the SWAT model, active storage of the reservoir, plant characteristics, reservoir volume—elevation relationship, tail water level, installed capacity, and design discharge of the plants. The objective functions for model optimisation were to (i) maximise annual energy generation, (ii) avoid running the reservoir dry before the end of the dry season, and (iii) minimise the risk of excess spillage due to the reservoir being full before the end of the wet season. The first objective was met by designing an upper rule curve that balances the gains in energy production resulting from operating at high reservoir levels and the losses of energy resulting from spilled water. The second and third objectives were met by designing a lower rule curve to the fraction of live storage that must be maintained. The monthly operation rule was then calculated by linear interpolation between upper and lower rule curves.

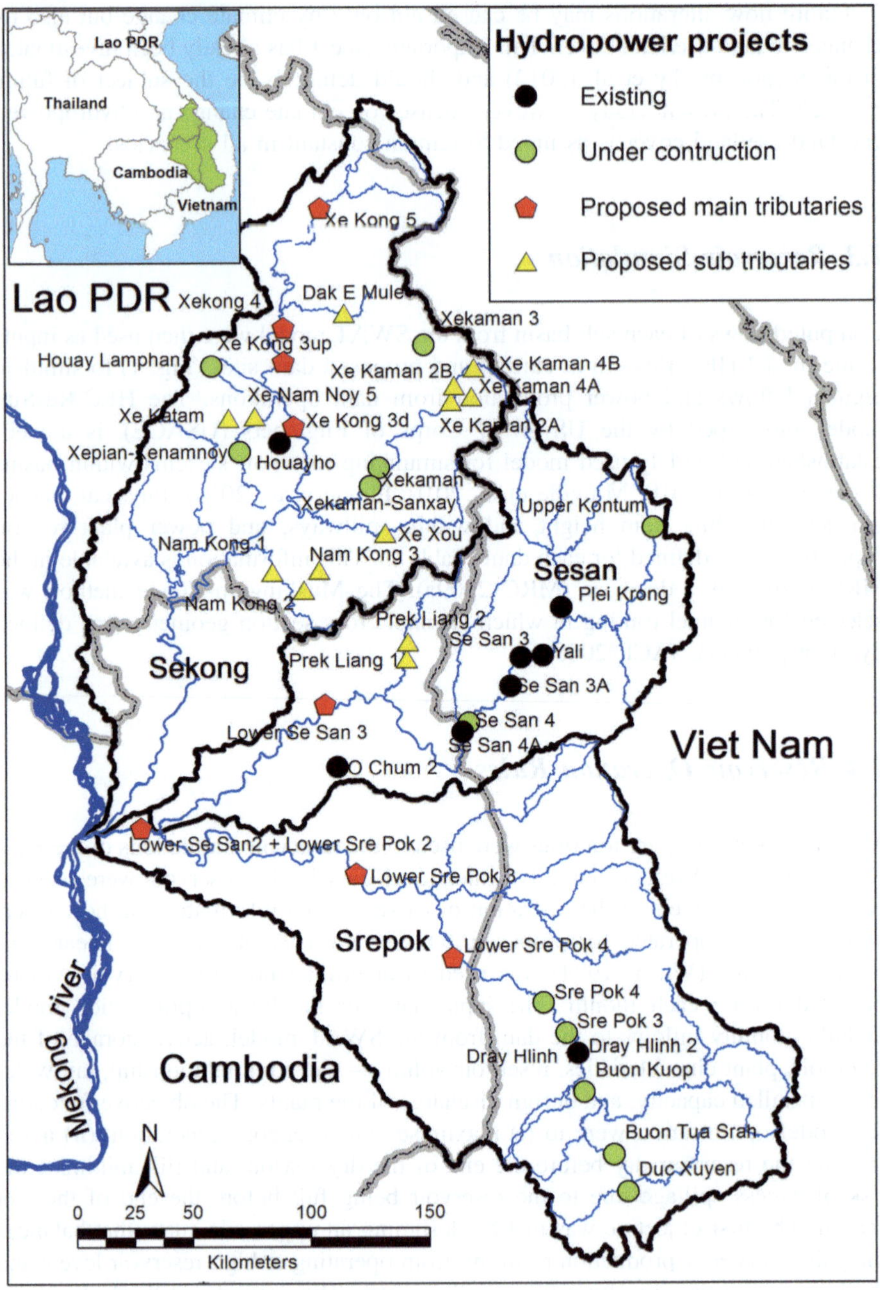

Fig. 1 Hydropower projects in the 3S basin

Table 2 Characteristics of the hydropower projects in the 3S basin

No.	Name	Status	Catchment area (km^2)	Full supply level (m.msl)	Low supply level (m.msl)	Live storage (mcm)	Designed discharge (m^3/s)	Designed head (m)	Installed capacity (MW)	Scenario	
										DF	AD
Lao PDR											
1	Hauayho	E	191.7	883.0	860.0	649.0	23.0	748.3	150.0	X	X
2	Xekaman 3	UC	712.0	960.0	925.0	108.5	62.5	477.7	250.0	X	X
3	Xekaman	UC	3,580.0	230.0	218.0	1,683.0	336.6	99.0	290.0	X	X
4	Xekaman_Sanxay	UC	3,740.0	123.0	122.0	7.1	378.0	12.2	32.0	X	X
5	Xekaman_Xenamnoy	UC	522.0	786.6	760.0	885.0	70.0	642.0	390.0	X	X
6	Houay Lamphan	UC	140.0	840.0	800.0	128.2	18.5	536.4	60.0	X	X
7	Xe Kong 3 up	PMT	5,882.0	160.0	155.0	95.1	460.0	33.7	144.6		X
8	Xe Kong 3 d	PMT	9,700.0	117.0	111.0	168.4	568.0	17.2	91.1		X
9	Xekong 4	PMT	5,400.0	290.0	270.0	3,100.0	240.0	140.0	300.0		X
10	Xe Kong 5	PMT	2,615.0	500.0	470.0	1,355.5	146.0	188.1	248.0		X
11	Dak E Mule	PST	127.0	780.0	756.0	154.0	27.4	433.8	105.0		X
12	Xe Kama an 2A	PST	1,970.0	280.0	275.0	3.7	155.0	48.6	64.0		X
13	Xe kam an 2B	PST	1,740.0	370.0	340.0	216.8	90.0	78.8	100.0		X
14	Xe Kam am 4A	PST	265.0	860.0	840.0	16.5	26.0	423.6	96.0		X

(continued)

Table 2 (continued)

No.	Name	Status	Catchment area (km^2)	Full supply level (m.msl)	Low supply level (m.msl)	Live storage (mcm)	Designed discharge (m^3/s)	Designed head (m)	Installed capacity (MW)	Scenario	
										DF	AD
15	Xe Kam am 4B	pst	192.0	865.0	850.0	21.2	18.4	459.1	74.0		X
16	Xe Katam	PST	263.0	910.0	890.0	115.0	16.0	450.0	60.8		X
17	Xe Nam Noy 5	PST	60.2	800.0	780.0	8.8	3.9	572.3	20.0		X
18	XE Xou	PST	1,273.0	180.0	160.0	1,714.0	131.3	51.8	63.4		X
19	Nam Kong 1	PST	1,250.0	320.0	287.0	505.0	44.5	186.0	75.0		X
20	Nam Kong 2	PST	860.0	460.0	437.0	139.0	76.5	106.5	74.0		X
21	Nam Kong 3	PST	650.0	540.0	520.0	298.6	37.6	80.0	25.0		X
Cambodia											
22	O Chum 2	E	44.7	254.0	251.0	0.1	3.8	32.6	1.0	Not modelled	
23	Lower Se San2 +	PMT	49,200	75.0	74.0	379.4	2,119.2	26.2	480.0		X
	Lower Sre Pok 2										
24	Lower Se San 3	PMT	15,600.0	150.0	147.0	3,120.0	500.0	58.5	243.0		X
25	Lower Sre Pok 3	PMT	26,200.0	125.0	118.0	5,310.0	775.0	31.5	204.0		X
26	Lower Sre Pok 4	PMT	1,300.0	190.0	185.0	2,700.0	327.0	52.2	143.0		X
27	Prek Liang 1	PST	883.0	330.0	310.0	110.0		153.0	35.0		X

(continued)

Table 2 (continued)

No.	Name	Status	Catchment area (km²)	Full supply level (m.msl)	Low supply level (m.msl)	Live storage (mcm)	Designed discharge (m³/s)	Designed head (m)	Installed capacity (MW)	Scenario	
										DF	AD
							27.2				
28	Prek Liang 2	PST	595.0	515.0	496.0	180.0	17.7	168.0	25.0		X
Viet Nam											
29	Plei Krongf	E	3,216.0	570.0	537.0	948.0	367.6	31.0	100.0	X	X
30	Y ail	E	7,455.0	515.0	490.0	779.0	424.0	190.0	720.0	X	X
31	Se San 3	E	7,788.0	304.5	303.2	3.8	486.0	60.5	260.0	X	X
32	Se San 3A	E	8,084.0	239.0	238.5	4.0	500.0	21.55	96.0	X	X
33	Dray Hlinh 1	E	8,880.0	302.0	299.0	1.2	94.9	15.0	12.0	X	X
34	Dray Hlinh 2	E	8,880.0	302.0	299.0	1.5	101.0	18.5	16.0	X	X
35	Se San 4A	E	9,368.0	155.2	150.0	705	Reregulating dam (no power plant)			X	X
36	Se San 4	UC	9,326.0	215.0	210.0	264.2	719.0	56.0	360.0	X	X
37	Upper Kontum	UC	350.0	1,170.0	1,146.0	122.7	30.5	904.1	250.0	X	X
38	Buon Kontum	UC	2,930.0	487.5	465.0	522.6	204.9	46.5	86.0	X	X
39	Buon Kuop	UC	7,980.0	412.0	409.0	14.7	316.0	98.5	280.0	X	X
40	Sre Pok 3	UC	9,410.0	272.0	268.0	62.6	412.8	60.0	220.0	X	X
41	Sre Pok 4	UC	9,568.0	207.0	204.0	10.1	168.9	17.1	70.0	X	X
42	Due Xuyen	PMT	1,100.0	560.0	551.0	413.4	81.0	71.0	49.0	X	X

Note E Existing, *UC* Under construction, *PMT* Proposed on main tributaries, *PST* Proposed on sub-tributaries

2.5 Scenarios

The degree of change in flows and energy production from climate change and hydropower development in the 3S basin was investigated using seven scenarios (Table 3). A list of hydropower projects included in each scenario is provided in Table 2. The definite future scenario consists of 8 existing and 11 projects under construction (total 19 projects), while the all dam scenarios comprise 8 existing, 11 under construction, and 22 proposed projects (total 41 projects as at December 2010). The effect of climate change and hydropower development on flows was assessed against baseline scenario flows (1986–2005). The computed results from the A2 and B2 emission scenarios (2010–2049) were compared to determine the impact of different climate change projections on rainfall patterns, flows, and energy generation. Multi-GCMs simulations were carried out under the A2 emission scenario in order to examine uncertainty in the GCMs and downscaling methods. The combination of climate change and hydropower development scenarios was not merely used to investigate the impact of climate change on hydropower production, but also to assess the cumulative impact of climate change and hydropower development and how hydropower operations would cope with future climate projections.

Table 3 Description of scenarios

Scenario	GCM	Abbreviation	Simulation period	No. of dams
Baseline without dams				
Baseline	No	ND-BL	1986–2005	No
Climate change without dam				
A2 emission scenario	MPI_ECHAM4	ND-A2-ECHAM4	2010–2049	No
	CCCMA_CGCM3.1	ND-A2-CGCM3.1	2010–2049	No
	MPI_ECHAM5	ND-A2-ECHAM5	2010–2049	No
	NCAR_CCSM3	ND-A2-CCSM3	2010–2049	No
B2 emission scenario	MPI_ECHAM4	ND-B2-ECHAM4	2010–2049	No
Hydropower development with baseline climate				
Definite Future	No	DF-BL	1986–2005	19
All dams	No	AD-BL	1986–2005	41
Hydropower development with climate change				
All dams with A2 emission scenario	MPI_ECHAM4	AD-A2-ECHAM4	2010–2049	41
All dams with B2 emission scenario	MPI_ECHAM4	AD-B2-ECHAM4	2010–2049	41

3 Results and Discussion

3.1 Climate Change Impacts on Rainfall

Cumulative seasonal and monthly average rainfall under the baseline and climate change scenarios over the 3S basin are reported and compared in Table 4 and Fig. 2. Firstly, the impacts of different emission scenarios (A2 and B2) from the MPI_ECHAM4 on rainfall patterns were examined. The average cumulative dry seasonal rainfall (December–May) of the A2 and B2 scenarios for 40 years (2010–2049) were 406.9 and 383.7 mm, respectively, corresponding to a reduction of 6.5 and 11.8 % from the mean of observed baseline scenario data (435.2 mm). Changes to average cumulative wet seasonal rainfall (June to November) and average annual rainfall of both climate change scenarios were less evident than during the dry season. In the wet season, the rainfall was predicted to increase by only 0.9 % for the A2 scenario and decrease by 1.5 % for the B2 scenario. The average annual rainfall reduction ranges from 0.7 to 3.7 % for both climate change scenarios. Overall, climate change has little impact on average monthly rainfall patterns as illustrated in Fig. 2. The maximum and minimum monthly rainfalls of the A2 and B2 scenarios occur in August and February, respectively, which are similar to the baseline scenario, but the monthly rainfall patterns of the A2 scenario are closer to the baseline patterns than the B2 scenario. The average monthly rainfall of both climate change scenarios in dry season months is less than the baseline scenario; the B2 scenario, particularly, results in lower flows in March and April. The average monthly rainfall in the early wet season months (June–August) of the B2 scenario is lower than the A2 and baseline scenarios by 10–30 mm, while

Table 4 Comparison of annual and seasonal mean area rainfall over the 3S basin for the baseline, and the A2 and B2 climate change scenarios

Scenario	Wet season (mm)	Relative change from the baseline (%)	Dry season (mm)	Relative change from the baseline (%)	Annual (mm)	Relative change from the baseline (%)
ND-BL	1,579		435.2		2,014	
ND-A2-ECHAM4	1,594	0.9	406.9	−6.5	2,000	−0.7
ND-A2-CGCM3.1	1,580	0.1	341.6	−21.5	1,922	−4.6
ND-A2-ECHAM5	1,639	3.8	350.8	−19.4	1,990	−1.2
ND-A2-CCSM3	1,600	1.3	373.8	−14.1	1,973	−2.0
ND-B2-ECHAM4	1,555	−1.5	383.7	−11.8	1,939	−3.7

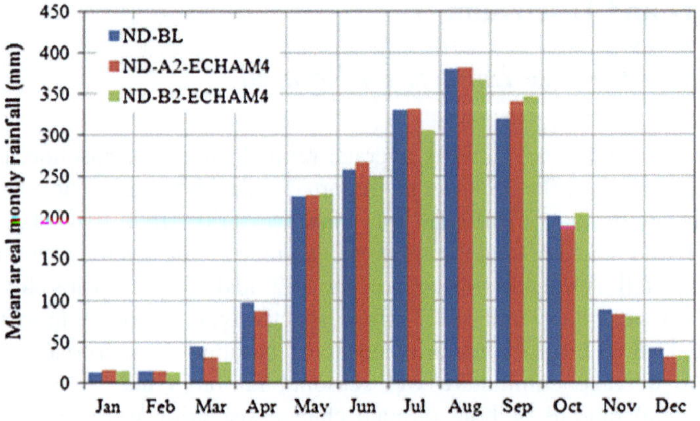

Fig. 2 Comparison of average monthly rainfall over the 3S basin between the A2 and B2 emission scenarios from the MPI_ECHAM4 and the baseline scenario

for late wet season months (September–October), the average monthly rainfall of the B2 scenario is higher than the A2 and baseline scenarios by 5–25 mm.

The effect of climate change on seasonal rainfall in 3S sub-basins was assessed via spatial analysis. ArcGIS spatial analyst tools were used to calculate cumulative rainfall in the wet and dry seasons from the predicted daily rainfall of A2 and B2 scenarios in each sub-basin. The computed cumulative rainfall in the wet and dry seasons was then averaged and compared with the baseline scenario to calculate the percentage change as presented in Fig. 3. The A2 scenario is projected to decrease the cumulative rainfall in the dry season by up to 20 % (relative to the baseline scenario) in most sub-basins except for sub-basins, which are located in the upstream and the outlet of the 3S. The cumulative dry seasonal rainfall is estimated to increase 20 % or less (Fig. 3a) in these areas because the upstream of the 3S basin is a mountainous region receiving more rainfall compared to the lowland region downstream. Only a few sub-basins have decreased or increased in cumulative dry seasonal rainfall by more than 40 % compared to the baseline scenario. The impact of the A2 scenario on the cumulative wet seasonal rainfall (June to November) is to increase rainfall by up to 20 % in most sub-basins in the Sesan Basin, upstream and downstream of the Srepok basin, and the middle part of the Sekong basin. For the remaining sub-basins, the cumulative rainfall in the wet season is estimated to decrease by less than 20 % (Fig. 3b). The B2 scenario results in a decrease of cumulative rainfall during the dry season in most sub-basins, particularly in the Srepok basin where the degree of change is more than the A2 scenario (Fig. 3c). Nearly half of the Srepok sub-basins experience cumulative dry seasonal rainfall alterations from −20 to −40 %. The impact of the B2 scenario on cumulative rainfall during the wet season is comparable to the A2 scenario in the Srepok and Sesan basins. In the Sekong basin, however, the number of sub-basins in which rainfall was predicted to increase is noticeably smaller in the B2 than in the A2 scenario (Fig. 3d).

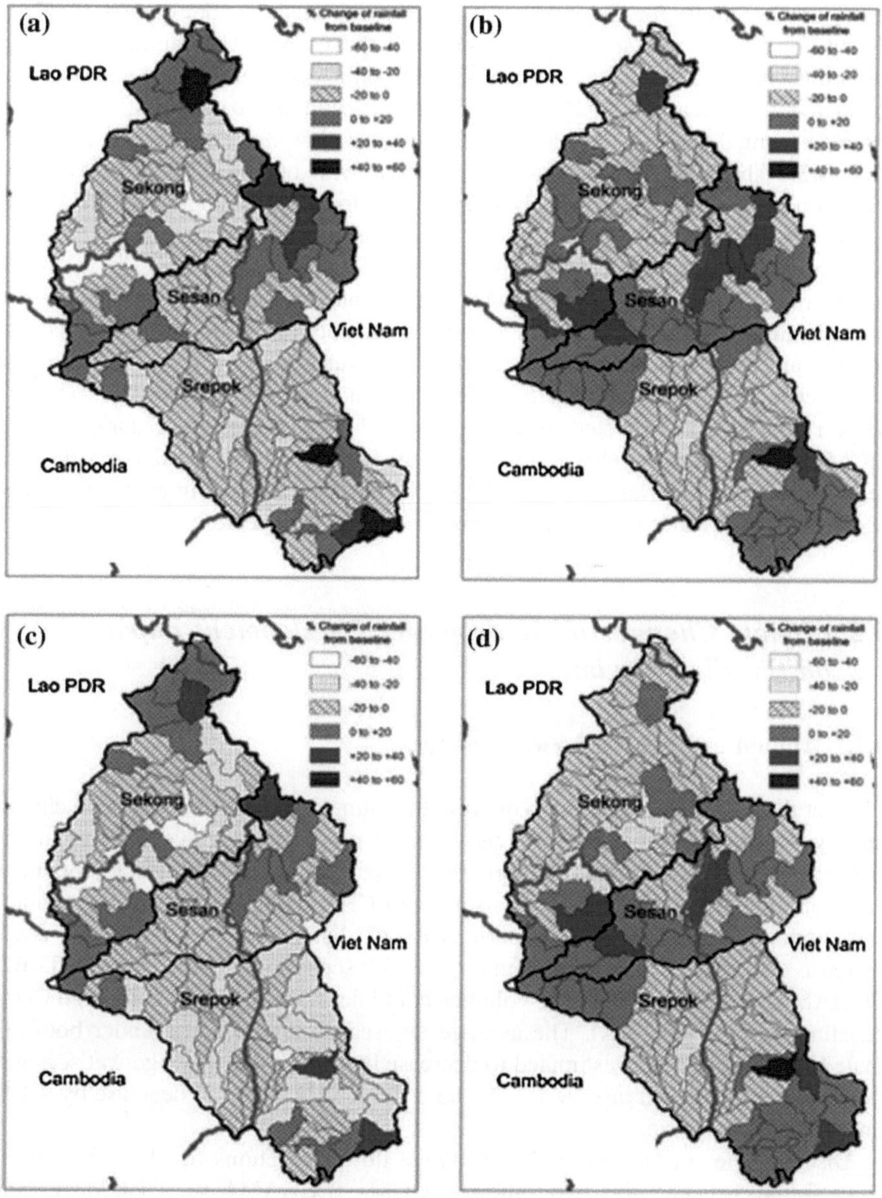

Fig. 3 Percentage change (%) of seasonal rainfall between the A2 and B2 emission scenarios from the MPI_ECHAM4 and the baseline scenario: **a** A2 dry seasonal rainfall, **b** A2 wet seasonal rainfall, **c** B2 dry seasonal rainfall, and **d** B2 wet seasonal rainfall

The uncertainty between GCMs and downscaling methods on rainfall projections in the A2 scenario for the 3S basin was investigated and presented in Table 4. In general, the direction of change (+ or −) in annual and seasonal rainfall prediction is similar for all GCMs and downscaling methods, but the magnitude of change during the dry season varies between the GCMs. The rainfall in the dry season is projected to decrease by 6.5–21.5 %, while the rainfall in the wet season is predicted to increase by 0.1–3.8 % compared with the baseline scenario. The average annual rainfall decreased by only 0.7–4.6 %. The results from the CGCM3.1 show the largest change compared with other GCMs. However, the direction of the projected monthly rainfall was found to be more uncertain, with GCMs predicting a mixture of higher and lower rainfall values compared to the baseline scenario for the same months.

Overall, climate change simulations suggest small impacts on the wet season and annual rainfall at the basin scale, while the impacts on monthly and dry season rainfall are more prominent, particularly at the sub-basin level. The B2 scenario could change rainfall patterns to a larger extent than the A2 scenario. The dry season rainfall was predicted to decrease for all climate change scenarios and all GCMs. The decrease of dry seasonal rainfall will impact on the dry season flow regime and increase water stress, which will have an adverse impact on wetland ecosystems and water users, particularly the agricultural sector.

3.2 Climate Change and Hydropower Development Impact on the Flow Regime

3.2.1 Annual and Seasonal Flow Changes

Flow at the 3S outlet was used to analyse the cumulative impact of climate change and hydropower development over the basin. A comparison of average annual and seasonal flow volumes for all scenarios is presented in Table 5. Comparison between different climate scenarios using the ECHAM4 shows that average annual flow volumes from the A2 scenario without dams (ND-A2-ECHAM4) could increase by only 0.2 %, while under the B2 scenario without dams (ND-B2-ECHAM4), average annual flow volumes could decrease by 4.6 % compared to the baseline scenario (ND-BL). The average dry season flow volume under both climate change scenarios is estimated to decrease by 6–7 %. The average wet seasonal flow volume could increase by 1.5 % under the A2 scenario and decrease by 4.1 % under the B2 scenario.

Discrepancies in the seasonal and annual flow projections for the A2 scenario were found between the two different GCMs (ECHAM4 vs. CGCM3.1) and downscaling methods (dynamic vs. statistical). All GCMs, based on both dynamic and statistic downscaling, show that dry season flows decrease by 6.0–23.6 %, while wet season flows increase by 0.6–7.5 % compared with the baseline flows. However, the direction of net change (increase or decrease) in average annual flows is unclear. Average annual flows from the ECHAM4, ECHAM5, and CCSM3 runs were predicted to increase by 0.2–3.4 %, while the CGCM3.1 runs result in a decrease of 3.6 % when compared with the baseline scenario.

Table 5 Comparison of average annual and seasonal flows for different scenarios at the 3S outlet

Scenario	Average total wet season volume (km^3)	Relative change from the baseline (%)	Average total dry season volume (km^3)	Relative change from the baseline (%)	Average total annual volume (km^3)	Relative change from the baseline (%)
Baseline						
ND-BL	74.3		15.5		89.8	
Impact of climate change						
ND-A2-ECHAM4	75.4	1.5	14.6	−6.0	90.0	0.2
ND-A2-CGCM3.1	74.8	0.6	11.8	−23.6	86.6	−3.6
ND-A2-ECHAM5	79.9	7.5	12.9	−16.6	92.8	3.4
ND-A2-CCSM3	77.0	3.6	13.8	−10.9	90.8	1.1
ND-B2-ECHAM4	71.2	−4.1	14.5	−6.6	85.7	−4.6
Impact of hydropower development						
DF-BL	67.3	−9.4	19.3	24.8	86.7	−3.5
AD-BL	55.7	−25.0	30.3	95.7	86.1	−4.2
Impact of climate change and hydropower development						
AD-A2-ECHAM4	57.0	−23.3	29.4	89.5	86.4	−3.8
AD-B2-ECHAM4	53.0	−28.7	29.2	88.3	82.1	−8.5

The definite future hydropower development, operating to maximise energy production, under the baseline climate scenario (DF-BL) results in an increase of dry season flow by 24.8 % and a decrease of wet season flow by 9.4 % (Table 5). These changes mainly result from the high level of hydropower development (existing and under construction) in the Vietnam highlands, particularly Yali, Plei Krong, and Buon Tua Srah dams (Fig. 1), which have large active storages (>500 million m^3). The eventual construction of all hydropower in the 3S basin under baseline climate condition (AD-BL) shows substantial alterations in average seasonal flow during both wet and dry seasons. The average dry seasonal flow nearly doubles (+95.7 %), while the average wet seasonal flow volume decreases by 25.0 %. The large additional alterations in seasonal flows in this scenario are due to the large storage dams in the main channels of the 3S (Fig. 1), particularly the proposed hydropower projects in Lao PDR (upstream of the Sekong River) and in Cambodia (downstream of Srepok and Sesan Rivers). The total active storage in the DF-BL scenario is only 6,203 million m^3, while the total active storage in the basin of the AD-BL scenario would increase to 26,328 million m^3 (325 % increase from the DF-BL scenario). This total active storage is about 30 % of the average annual flow at the outlet of the 3S basin (86,987 million m^3).

The combined effect of climate change and full hydropower development scenarios (AD-A2-ECHAM4 and AD-B2-ECHAM4) decrease wet season flows by 23–29 % and increasing dry seasonal flows by 29 % compared to the ND-BL scenario. These results clearly show that hydropower development has a much greater impact on downstream seasonal flows than climate change.

Monthly Flow Changes: Average monthly flow patterns from climate change scenarios with multi-GCM runs, hydropower development scenarios, combined climate change, and hydropower development scenarios were compared to the baseline scenario at the 3S outlet (Fig. 4). The climate change scenarios without dams using ECHAM4 (ND-A2-ECHAM4 and ND-B2-ECHAM4) show a decrease in monthly flows during the early wet season (June to August) and an increase in flows during September to November (Fig. 4a) compared to the baseline scenario (ND-BL), showing a clear shift of the peak flows by about one month. Similar to the alterations in rainfall amounts (Table 4), B2 scenario impacts on river flows are more pronounced than impacts from the A2 scenario. The average monthly flow patterns in the dry season of both climate change scenarios were close to the baseline flows except for May, when the average monthly flow was predicted to decrease.

Comparison of average monthly flows from the A2 IPCC scenario with varying GCMs and downscaling methods is presented in Fig. 4b. Results show that there is more variation in monthly flow predictions between GCMs than between IPCC scenarios A2 and B2. Overall, there are similarities in the direction and magnitude of changes, particularly from January to June. The difference varies from −23.4 to

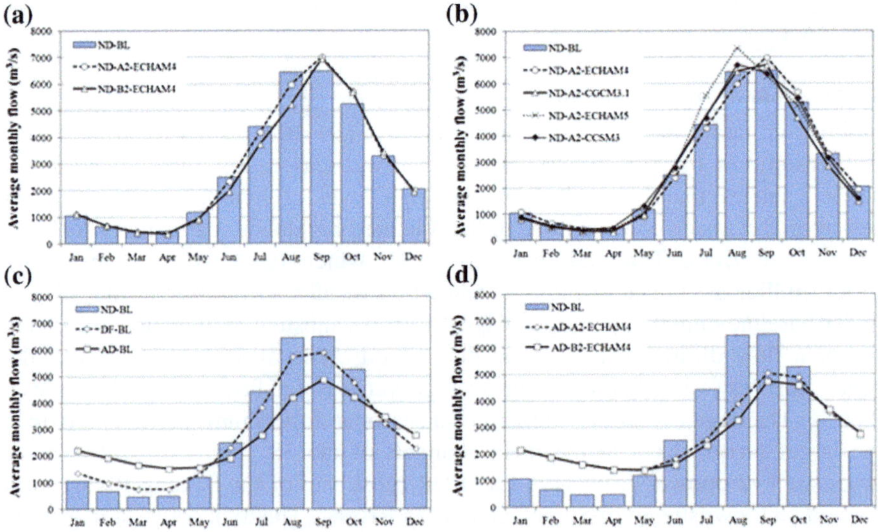

Fig. 4 Comparison of monthly flows between the baseline scenarios at the 3S outlet: **a** climate change scenarios, **b** the A2 emission scenario with different GCMs and downscaling methods, **c** hydropower development scenarios, and **d** combination of hydropower development with climate change

25.7 % compared to the baseline scenario. Monthly flows were projected to decrease during November to April for most GCMs, ranging from 3.7 % (in the month of November) to 32.1 % (in the month of March). Peak flows were computed from the ECHAM5 and CCSM3 to occur in August, while the ECHAM4 and CGCM3.1 projected the peak flows to occur in September. Findings at the sub-basin level are comparable with the basin-wide studies from Kingston et al. (2011), Lauri et al. (2012), which show that the variation in predicted flows between individual GCMs is relative high and as significant as the variation between IPCC greenhouse gas emission scenarios.

Hydropower development on its own without climate change impact could drastically modify the average monthly flow distribution (Fig. 4c). Overall, both hydropower development levels (DF and AD) decrease wet season flows and increase dry season flows. The operation of the DF-BL and AD-BL scenarios reduces the magnitude of the average monthly peak flow in September by 600 and 1,700 m^3/s, respectively. The scenarios of full hydropower development with climate change (AD-A2-ECHAM4 and AD-B2-ECHAM4) yield monthly flow alterations of similar magnitude as the AD-BL (Fig. 4d). Full hydropower development operations to maximise energy would overwhelm potential changes in predicted average monthly flows caused by climate change.

3.2.2 Daily Flow Changes

The average, standard deviation, and average annual maximum and minimum of daily flows were calculated at the 3S outlet to the Mekong River (Table 6). Results show that climate change or hydropower development scenarios decrease average daily flows by less than 5 %, whereas the combined full hydropower development with the B2 climate change scenario (AD-B2-ECHAM4) decreases average daily flows by 8.5 %. The A2 and B2 scenarios increase the standard deviation of daily flows by 12.7 and 8.5 %, respectively. Climate change scenarios increase the average annual maximum daily flows up to 19.7 % (3,582 m^3/s) compared with the baseline scenario and decrease the average annual minimum daily flows by up to 18.1 % (52 m^3/s). These results show that projected climate change will increase the range of daily flow variations and extreme events (as described in the next section). The increase in daily flow variability may impact on day-to-day water resource operations such as water supply and irrigation, as well as aquatic and wetland communities, which are sensitive to daily changes in flow patterns.

Conversely, in the hydropower development scenarios, DF-BL and AD-BL, standard deviation of daily flow decreases by 15.9 and 45.6 %, respectively (Table 6). The average annual maximum flows decreased by 51.6 % (9,360 m^3/s) from the baseline scenario when all hydropower projects were active in the 3S basin, while the average annual minimum flows increased by 2.7 times (767 m^3/s). The operation of all hydropower projects under climate change scenarios (AD-A2-ECHAM4 and AD-B2-ECHAM4) results in a net reduction in daily flow variability. For example, the standard deviation of daily flow under the A2 scenario

Table 6 Comparison of average daily flows and variations for different scenarios at the 3S outlet

Scenario	Average daily flow (m^3/s)	Relative change from the baseline (%)	Standard deviation of daily flow (m^3/s)	Relative change from the baseline (%)	Average annual max daily flow (m^3/s)	Relative change from the baseline (%)	Average annual min daily flow (m^3/s)	Relative change from the baseline (%)
Baseline								
ND-BL	2,846		2,915		18,156		287	
Impact of climate change								
ND-A2-ECHAM4	2,851	0.2	3,284	12.7	21,738	19.7	266	−7.3
ND-B2-ECHAM4	2,715	−4.6	3,162	8.5	21,259	17.1	235	−18.1
Impact of hydropower development								
DF-BL	2,747	−3.5	2,452	−15.9	15,171	−16.4	520	81.2
AD-BL	2,727	−4.2	1,584	−45.6	8,796	−51.6	1,054	267.5
Impact of climate change and hydropower development								
AD-A2-ECHAM4	2,737	−3.8	1,890	−35.2	10,295	−43.3	880	207.0
AD-B2-ECHAM4	2,603	−8.5	1,827	−37.3	9,663	−46.8	828	188.6

without dams (ND-A2-ECHAM4) was 3,284 m^3/s, while the standard deviation of daily flow under the A2 with hydropower development scenario (AD-A2-ECHAM4) decreased to 1,890 m^3/s. These findings show that climate change could result in increased daily flow extremes, but hydropower development and operation would dampen this effect to some extent. The net reduction in flow variation caused by hydropower development, even under different climatic change scenarios, is of great concern because it could impact on habitats and ecological productivity in downstream floodplains (Arias et al. 2014a).

3.2.3 Extreme Events

Scatter plots of joint distribution of the annual peak flow and the annual total wet season volume were developed to understand the effect of climate change and hydropower scenarios on annual peak flood events (Fig. 5a, c). The average annual peak flow and the average annual total wet season volume of the ND-BL scenario were calculated and used for comparison with climate change and hydropower development scenarios. The squares in Fig. 5a, c indicate one (1σ) and two standard (2σ) deviations from the annual peak flow and the annual total wet season volume above and below their respective averages. Events inside the smallest frame represent those within one standard deviation from the mean and were defined as "normal", while events outside the one and two standard deviation frames were defined as significant and extreme flood events, respectively. The 40 annual peak events (2010–2049) from the A2 and B2 scenarios without dams using ECHAM4 (ND-A2-ECHAM4 and ND-B2-ECHAM4) were plotted against 20 annual peaks (1986–2005) of the ND-BL scenario in Fig. 5a. In general, the flood events of those three scenarios were uniformly scattered. There were two significant (1996 and 2000) and one extreme flood event (2005) in the ND-BL scenario. In contrast, the A2 scenario has six significant and three extreme flood events occurring. The operation of all hydropower projects under baseline climate (AD-BL) and under climate change (AD-A2-ECHAM4 and AD-B2-ECHAM4) conditions substantially decreases the magnitude of the annual peak flow and increases the number of years with dryer wet seasons (Fig. 5c). These results show that hydropower development could reduce the frequency and magnitude of extreme flood events downstream, even beyond the range of natural variability. Further research should study dam operations under extreme flood events to ensure safe passage of flows under such extremes enhanced by climate change.

Scatter plots of the joint distribution of annual minimum flow and the annual total dry season volume were used to understand the impact of climate change and hydropower on annual drought events (Fig. 5b, d). Events inside the smallest frame represent those within one standard deviation from the mean and were defined as "normal", while events outside the one and two standard deviation frames were defined as significant and extreme drought events, respectively. In the ND-BL scenario, only one event could be pin pointed as a significant drought (2004; Fig. 5b). In contrast, the number of significant droughts in the B2 scenario increases

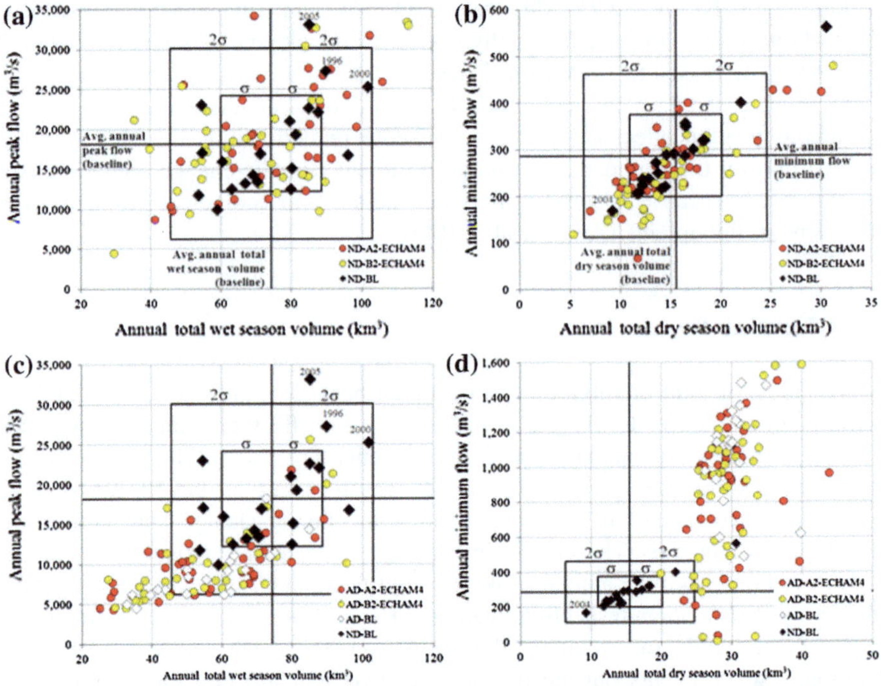

Fig. 5 Scatter plots of the joint distribution of the annual peak flow and the annual total wet season volume (**a** and **c**) and scatter plots of the joint distribution of the annual minimum flow and the annual total dry season volume (**b** and **d**) of climate scenarios without dams (**a** and **b**) and all hydropower development with and without climate change scenarios (**c** and **d**) at the 3S outlet. The squares indicate one (1σ) and two (2σ) standard deviations for each variable above and below their respective means

to 14 and one extreme drought is also present. There are fewer drought events in the A2 scenario, with only five significant droughts and one extreme drought. Under the hydropower development scenario, the dry season was wetter and the minimum flows increase, resulting in no extreme or even significant droughts. Nevertheless, the combined effect of climate change and hydropower operations causes several annual minimum flow events to occur significantly beyond the normal range for the ND-BL (Fig. 5d). These results show that climate change-enhanced droughts could affect hydropower operations; hence, seasonal operational rules may need frequent adjustment in order to meet electricity demands in years of low flow.

3.2.4 Climate Change Impacts on Electricity Generation

In the definite future scenario (DF-BL), the hydropower installed capacity will be 3,642 MW, and this could eventually increase to 6,363 MW when all hydropower projects are developed (AD-BL). The average energy production for the DF-BL

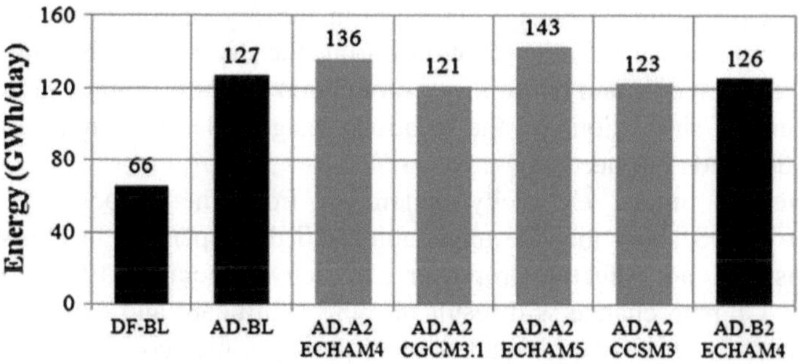

Fig. 6 Energy production from hydropower development scenarios with and without climate change impact

scenario under a maximum energy production rule is 66 GWh/day, whereas the average energy production for the AD-BL scenario is 127 GWh/day, representing an increase of 92 % from DF-BL (Fig. 6). The average energy production under the A2 with all dam scenarios using multi-GCM runs varies from 121 to 143 GWh/day. The ECHAM5 and CGCM3.1 generated the highest and lowest energy production, respectively. The energy production of the A2 scenario (average from the four GCMs) is 130.8 GWh/day, which represents only an increase of 3 % from the AD-BL scenario. Energy production under the B2 scenario using the ECHAM4 (126 GWh/day) is similar to the energy production under baseline climate conditions. In short, these annual estimates suggest that climate change may have a low effect on hydropower electricity generation in the 3S.

4 Conclusion

Full hydropower development in the 3S basin will increase flows at the basin outlet in the dry season by over 95 % and reduce flows in the wet season by 25 % from baseline conditions. When addressing climate change, different greenhouse gas emission scenarios, GCMs, and downscaling methods were considered. All climate change simulations agreed on the direction of rainfall and flow changes in the dry season (a reduction from baseline scenario); the magnitudes of alterations, however, were different. Changes in rainfall patterns are more prominent, particularly at the sub-basin level. The impact of climate change on the wet season and the annual flows varies depending on IPCC climate scenarios, GCMs, or downscaling methods. Therefore, it is important to always incorporate these factors of uncertainty through improved spatial and temporal climate change modelling across the basin for future adaptation planning.

Climate change impact on flows is low compared to hydropower development. The combined effect of climate change and hydropower development shows that flow alterations from full hydropower power operation in the 3S basin are predominant over climate change. Increase in the dry season flows from hydropower operations may benefit irrigation during the dry season, but the timing of irrigation requirements and hydropower production needs may not overlap. Further information and simulations on the locations, magnitude, and timing of potential irrigation demands are needed.

Estimates of annual electricity generation from the proposed hydropower scheme in the 3S show that the uncertainty and discrepancies in climate change simulations will not affect hydropower electricity generation to any significant magnitude. Climate change will result in more significant and extreme flood and drought events in terms of magnitude and frequency, which could affect dam design and operation rules. Further research should study how hydropower operations can be adapted to ensure proper production under future extreme events. Coordination and cooperation among dam operators and countries will be important in reducing risks of negative impacts downstream at local, national, and transboundary levels and maximising overall benefits to the basin. Further work is needed to quantify changes in sediment flows and trapping, water quality, ecosystems, social vulnerability, and economic impacts.

Acknowledgments The authors wish to thank the Mekong River Commission for providing the databases used in this paper. Funding for this project was provided by the John D. and Catherine T. MacArthur Foundation through a project entitled "Critical Basin at Risk: Assessing and managing ecosystem pressures from development and climate change in the 3S basin".

References

Adamson PT, Rutherfurd ID, Peel MC, Conlan IA (2009) The hydrology of the Mekong River. The Mekong: biophysical environment of an international river basin. Elsevier, New York, pp 53–76

ADB (2004) Cumulative impact analysis and Nam Theun 2 contributions: final Report, Asian Development Bank, Manila

ADB (2010) Sesan, Srepok and Sekong River Basins development study in Kingdom of Cambodia, Lao People's Democratic Republic, and Socialist Republic of Vietnam, Asian Development Bank, Manila

Arias ME, Cochrane TA, Kummu M, Lauri H, Koponen J, Holtgrieve GW, Piman T (2014a) Impacts of hydropower and climate change on drivers of ecological productivity of Southeast Asia's most important wetland. Ecol Model 272:252–263

Arias ME, Piman T, Lauri H, Cochrane TA, Kummu M (2014b) Dams on Mekong tributaries as significant contributors of hydrological alterations to the Tonle Sap Floodplain in Cambodia. Hydrol Earth Syst Sci Dis 11:2177–2209

Choi W, Rasmussen PF, Moore AR, Kim SJ (2009) Simulating streamflow response to climate scenarios in Central Canada using a simple statistical downscaling method. Climate Res 40:89–102

Cochrane TA, Arias ME, Teasley RL, Killeen TJ (2010) Simulated changes in water flows of the Mekong River from potential dam development and operations on the Sesan and Srepok tributaries. Proceedings of the IWA world water congress and exhibition, vol 7. Montreal

Cronin R, Hamlin T (2012) Mekong turning point: shared river for a shared future. Stimson Center, Washington, DC, pp 45–46

Delgado JM, Apel H, Merz B (2010) Flood trends and variability in the Mekong River. Hydrol Earth Syst Sci 14:407–418

Delgado JM, Merz B, Apel H (2012) A climate-flood link for the lower Mekong River. Hydrol Earth Syst Sci 16(5):1533–1541

Diaz-Nieto J, Wilby RL (2005) A comparison of statistical downscaling and climate change factor methods: impacts on low flows in the River Thames, UK. Clim Change 69:245–268

Eastham J, Mpelasoka F, Mainuddin M, Ticehurst C, Dyce P, Hodgson G, Ali R, Kirby M (2008) Mekong River Basin water resources assessment: impacts of climate change, CSIRO, Water for a Healthy Country National Research Flagship report, Australia

Hoanh, CT, Jirayoot, K, Lacombe, G, and Srinetr, V (2010) Impacts of climate change and development on Mekong flow regimes: First assessment—2009. Technical Paper No. 29, Mekong River Commission, Vientiane, Lao PDR

IPCC (2000) Special report on emissions scenarios: a special report of working group III, Intergovernmental Panel on Climate Change. Cambridge University Press, Cambridge

IPCC (2007) Climate change (2007) synthesis report: an assessment of the intergovernmental panel on climate change, Intergovernmental Panel on Climate Change, Cambridge University Press, Cambridge

Jones RG, Noguer M, Hassell DC, Hudson D, Wilson SS, Jenkins GJ, Mitchell J.F.B. (2004) Generating high resolution climate change scenarios using PRECIS, Met Office Hadley Centre, Exeter

Kingston DG, Thompson JR, Kite G (2011) Uncertainty in climate change projections of discharge for the Mekong River Basin. Hydrol Earth Syst Sci 15:1459–1471

Kongthong O (2011) Drought in Mekong River and vulnerability of livelihood in Chiang Khan District, Lower Mekong Basin, Thailand. Master Thesis, Lund University, Sweden

Lauri H, de Moel H, Ward PJ, Räsänen TA, Keskinen M, Kummu M (2012) Future changes in Mekong River hydrology: impact of climate change and reservoir operation on discharge. Hydrol Earth Syst Sci 16(12):4603–4619

Mainuddin M, Hoanh CT, Jirayoot K, Halls AS, Kirby M, Lacombe G, Srinetr V (2010) Adaptation options to reduce the vulnerability of Mekong water resources, food security and the environment to impacts of development and climate change, CSIRO, Water for a Healthy Country National Research Flagship report, Australia

Minville M, Brissette F, Leconte R (2010) Impacts and uncertainty of climate change on water resource management of the Peribonka River System (Canada). J Water Res Plann Manage ASCE 136(3):376–383

MRC and ICEM (2009) Climate change adaptation in the Lower Mekong Basin Countries: Regional synthesis report, Mekong River Commission, Vientiane, Lao PDR and International Centre for Environmental Management, Hanoi, Vietnam

MRC (2011a) Application of MRC modelling tools in the 3S basin, Mekong River Commission, Phnom Penh, Cambodia

MRC (2011b) Assessment of Basin-wide Development Scenarios: Main Report, Basin Development Plan Programme, Mekong River Commission, Vientiane, Lao PDR

Piman T, Cochrane TA, Arias ME, Green A, Dat ND (2013a) Assessment of flow changes from hydropower development and operations in Sekong, Sesan and Srepok Rivers of the Mekong Basin. J Water Res Plann Manage 139:723–732

Piman T, Lennaerts T, Southalack P (2013b) Assessment of hydrological changes in the lower Mekong basin from basin-wide development scenarios. Hydrolo Process 27:2115–2125

Räsänen TA, Kummu M (2013) Spatiotemporal influences of ENSO on precipitation and flood pulse in the Mekong River Basin. J Hydrol 476:154–168

Räsänen TA, Lehr C, Mellin I, Ward PJ, Kummu M (2013) Palaeoclimatological perspective on river basin hydrometeorology: case of the Mekong Basin. Hydrol Earth Syst Sci 17:2069–2081

Roeckner E, Arpe K, Bengtsson L, Christoph M, Claussen M, Dümenil L, Esch M, Giorgetta M, Schlese U, Schulzweida U (1996) Report No.218: The atmospheric general circulation model ECHAM-4-model description and simulation of present-day climate, Max-Planck Institute for Meteorology, Hamburg

Schaefli B, Hingray B, Musy A (2007) Climate change and hydropower production in the Swiss Alps: quantification of potential impacts and related modelling uncertainties. Hydrol Earth Syst Sci 11(3):1191–1205

Ty TV, Sunada K, Ichikawa Y, Oishi S (2012) Scenario-based impact assessment of land use/ cover and climate changes on water resources and demand: A case study in the Srepok River Basin, Vietnam-Cambodia. Water Resour Manage 26:1387–1407

USACE (2000) Hydrologic modeling system HEC-HMS: Technical reference manual, U.S. Army Corps of Engineers, Institute for Water Resources, Hydrologic Engineering Center, Davis, California

Västilä K, Kummu M, Sangmanee C, Chinvanno S (2010) Modelling climate change impacts on the flood pulse in the Lower Mekong floodplains. J. Water Climate Change, IWA, pp 67–86

WB (2004) Modelled observations on development scenarios in the Lower Mekong Basin. Mekong Regional Water Resources Assistance Strategy, World Bank, USA

Winchell M, Srinivasan R, Di Luzio M, Arnold J (2009) ArcSWAT interface for SWAT2009: User's guide, Soil & Water Research Laboratory, Temple, Texas

Wurbs RA (1994) Computer models for water-resources planning and management, Texas A&M University, Texas, pp 143–150

Ziv G, Baran E, Nam S, Rodriguez-Iturbe I, Levin SA (2012) Trading-off fish biodiversity, food security, and hydropower in the Mekong River Basin. Proc Natl Acad Sci 109:5609–5614

Managing Water Resources Under Climate Uncertainty: Opportunities and Challenges for Cambodia

Thun Vathana

Abstract Despite rich water resources, Cambodia faces numerous challenges in managing and maximising the use of these for its population and for the country as the whole. The typical challenge is too much water in the wet season, leading to widespread flooding, and lack of water in the dry season, causing drought. Cambodia is one of the most natural disaster-prone countries in East Asia, and climate change has already been observed there. This article gives an overview of the links between climate and water, as well as the effects of climate change on water resources. Potential ways of managing water resources from the climate change aspect are discussed. An overview of the importance of climate information in water resource management provides an insight into the role climatology plays as a means of predicting and managing future water resources. The Regional Climate Outlook Forum (RCOF) is a process that can be employed within water resource management. A Hydrological Outlook Forum can be developed based on the processes and outcomes of a RCOF.

1 Introduction

Cambodia is located in the south western part of the Indochina peninsula, bordering Lao PDR and Thailand to the north, Vietnam to the east and southeast, and Thailand and the Gulf of Thailand to the south and southwest (Fig. 1). Roughly flat and square in shape, the country has two dominant features:

- The Tonle Sap Lake (the Great Lake), the largest freshwater lake in Southeast Asia, is located in the centre of the country. It is one of the most intensively fished freshwater areas in the world, with more than 200,000 tons of fish harvested each year. Cambodia's fish production reached 730,000 tons in 2013, according to Nao

T. Vathana (✉)
Prekleap National College of Agriculture, National Road 6A,
Chroy Changva, Phnom Penh, Cambodia
e-mail: thunvathana@yahoo.com

S. Shrestha et al. (eds.), *Managing Water Resources under Climate Uncertainty*,
Springer Water, DOI 10.1007/978-3-319-10467-6_12

Fig. 1 Map of Cambodia

Thouk, Director of the Department of Fisheries of the Ministry of Agriculture, Forestry, and Fisheries (Hor 2014). This included 550,000 tons of fresh fish, 100,000 tons of shellfish, and 80,000 tons of varieties bred through aquaculture.

- The Mekong River, the twelfth-longest river in the world, cuts through Cambodia from north to south. This is also one of the most intensely fished freshwater areas in the world. "The 2.6 million tons of fish caught annually in the Mekong basin represents seven times more than the catches of the North American inland fisheries sector and more than 10 times the entire (inland) fish catch in Australia" (IRIN 2008).

Located in a tropical region with all the natural features this entails, Cambodia is rich in water resources. The average annual discharge of the Mekong River entering Cambodia is estimated at 324.45 km^3. Other inflows to the Mekong–Tonle Sap system from outside the country include 1.19 km^3 from Thailand and 29.9 km^3 from Vietnam.[1] This abundance of water resources allows the country to utilise and depend on these resources for agriculture, fishery, transport, and other economic activities.

---

[1] http://www.fao.org/nr/water/aquastat/countries_regions/cambodia/index.stm.

Cambodia is an agricultural economy, and the majority of the population lives in rural areas and engages in that sector, which absorbed approximately 56 % of the total employed labour force in 2011. Agricultural values were even higher in 1998, at 78 %. Cambodia's most important agricultural products are crops (mainly rice). Even with relatively strong growth in recent years, the sector remains volatile, for a number of reasons, the lack of effective water resource management in particular. A Khmer saying links farming with water: "Do farming with water, do war with rice". That is, effective irrigation and water resource management are crucial to support agricultural growth.

Beyond this, water is a source of food, especially fish, which represents one of the major sources of both protein and income for Cambodians. Freshwater fish is one of the country's most traded commodities. From the Tonle Sap, from where more than 200,000 tons of fish are harvested each year, fish is marketed across the country, with the largest domestic trade route going to Phnom Penh (van Zalinge et al. 2000). Apart from in the Tonle Sap Lake, Cambodians also catch fish in other smaller lakes and rivers, especially the Mekong River, making the total inland fish catch in the country approximately 400,000 tons/year.

Despite its rich water resources, Cambodia faces numerous challenges in managing and maximising the use of these for its population and for the country as a whole. Cambodia has a wet season and a dry season that are relatively equal in length; the typical challenge is of (1) too much water in the wet season, leading to widespread flooding, and (2) lack of water in the dry season, causing drought, often in the latter part of the season. Floods and droughts can happen in the same year and sometimes even in the same season, albeit in different parts of the country.

The lack of water results in high competition and/or even serious conflicts in the agricultural sector, given multiple claims on water as a shared resource. Problems arise mostly among rice farmers, who need water to save their paddies in times of drought. Meanwhile, when there is too much water, upstream farmers discharge it and flood farmland downstream, causing further conflicts. Generally speaking, both lack of water and too much water can give rise to conflicts, requiring intervention and negotiation by various levels of authority for resolution.

Finally, water is necessary for health, sanitation, and agriculture and thus plays an important role in poverty reduction in Cambodia and elsewhere. However, maintaining a sufficient quantity of good-quality water is increasingly difficult, owing to climate change and increasing demand, arising as a result of population increase, urbanisation, and industrialisation.

This article explores how water resources can be managed from a climatological perspective. Its objectives are as follows:

- To develop an understanding of the impact of climate change on water resources in Cambodia.
- To explore how climate information can be used in water resource management.
- To develop a concept for a project on water resource management from a climatological perspective, called the Hydrological Outlook Forum.

The article is structured in four sections. Section 2 introduces Cambodia and its current water situation. Section 3 gives an overview of the links between climate and water. It reviews the hydrological cycle and water resource issues, as well as the effects of climate change on water resources. The last part of Sect. 3 describes potential ways of managing water resources from the climate change aspect. Section 4 discusses the importance of climate information in water resource management. It provides an insight into the role climatology plays as a means of predicting and managing future water resources. This section identifies the Regional Climate Outlook Forum (RCOF) as a process that can be employed within water resource management. Section 5 discusses how a Hydrological Outlook Forum can be developed based on the processes and outcomes of a Regional Climate Outlook Forum.

2 Understanding the Cycle of Water: Water Resource Issues

Nature builds block of water resources everywhere to ensure the continuity of lives on Earth. Water resource availability is affected in terms of quality and quantity when climatological factors such as temperature and precipitation change with space and time. Like other countries, Cambodia's climate has varied and changed over time, by region and by seasonality. Figure 2 shows the mean historical monthly temperature and rainfall for Cambodia during the time period 1960–2009.

Based on data from the Climatic Research Unit of the University of East Anglia,[2] average total annual rainfall in Cambodia is 1,838 mm. Monthly average rainfall reaches a peak in October, at 308.5 mm, and gradually goes down to the lowest level in February, at 15.4 mm. It then starts to go up from March, jumping up in May to 187.8 mm. The change in the availability of water resources by season obviously has the capacity to have a fundamental impact on all segments of society and can sometimes have disastrous consequences that harm agriculture and other sectors.

The vast majority of the Earth's water resources is made up of saline water in oceans, which accounts for 97.5 %. The other 2.5 % consists of freshwater found in icecaps and glaciers (1.85 %), groundwater (0.64 %), and lakes, rivers, soil moisture and atmosphere (0.01 %) (Cech 2003). Although oceans make up the largest proportion of saline water resources, they also represent the largest source of evaporation, which contributes to precipitation creation in the hydrologic cycle of water resources. According to Cech (2003), ocean evaporation provides approximately 90 % of the Earth's precipitation.

[2] http://sdwebx.worldbank.org/climateportal/index.cfm?page=country_historical_climate&ThisRegion=Asia&ThisCCode=KHM accessed 19 March 2014.

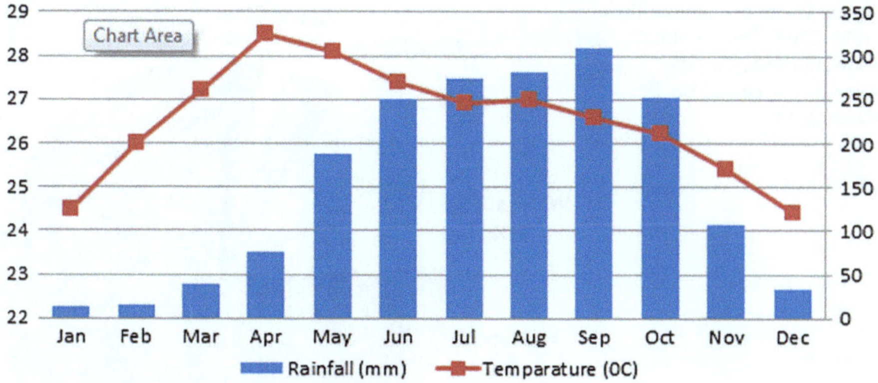

Fig. 2 Average monthly temperature and rainfall for Cambodia, 1960–2009. *Source* http://sdwebx.worldbank.org/climateportal/index.cfm?page=country_historical_climate&ThisRegion=Asia&ThisCCode=KHM accessed 19 March

In Cambodia, about 86 % of the territory is included in the Mekong River Basin, with the remaining 14 % draining directly into the Gulf of Thailand. On average, 471 km^3/year flows out of the country in the Mekong channels and tributaries to Vietnam. Cambodia's total renewable water resources are estimated at 476.110 km^3/year. Average rainfall in Cambodia in recent years is estimated at 1,463 mm, varying between about 1,000 to 4,700 mm; mean annual evaporation varies from 1,000 to 2,300 mm/year.[3]

Although water resources are generated and regenerated through the hydrologic cycle, water availability is not distributed homogenously across different spatial locations of the world. More often, the amount of precipitation varies by region. According to the UN Educational Scientific and Cultural Organization (UNESCO 2009), Latin America is the most water-rich region, with about a third of global run-off. Asia is next, with a quarter of global run-off, followed by the countries of the Organisation for Economic Co-operation and Development (mainly North America, Western Europe, and Australia), with the combined amount of 20 %, Sub-Saharan Africa and Eastern Europe, and the Caucasus and Central Asia each with about 10 %. The Middle East and North Africa is the most water-limited region, with only 1 % each.

Over the past decades, global economic development, population growth, and technological modernisation have led to an increase in the number and severity of problems in the water resources sector. Population and economic growth means energy consumption is on the increase. Water availability is therefore constrained by the increased use of these resources for various sectors such as agriculture, household consumption, electricity and gas, and manufacturing and mining (Fig. 3).

[3] http://www.eoearth.org/view/view/156934/ accessed 19 March 2014.

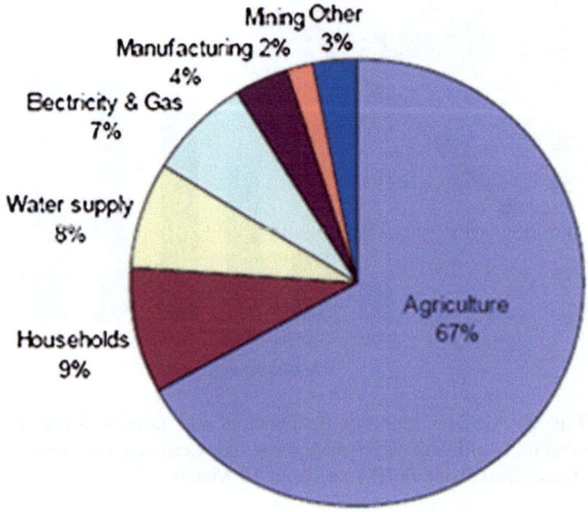

Fig. 3 Water use around the world, 2005. *Source* http://www.climate.org/topics/water.html accessed on 19 March 2014

Without sufficient food, malnutrition is obviously a possibility. According to Fig. 3, agriculture is the largest water user, accounting for 67 % of all water use. Although agricultural irrigation produces return flows and seepage to the river basin, a large amount of water is consumed by evapotranspiration. The comprehensive assessment of water management in agriculture estimates that the total amount of water losses through evaporation in crop production is presently at 7,130 km^3 (IWMI 2007). The amount of evaporation in crop production will increase to 12,000–13,500 km^3 by 2050, corresponding with an increase of 70–90 % in the next four decades. This estimated figure suggests the future requirement of water for irrigation will be on the increase in order to ensure sufficient food production. Such rising water demand will put more pressure on water, particularly in regions with water scarcity.

Besides agriculture, there has been widespread hydropower dam construction, particularly since the middle of the twentieth century, when improvements in engineering and construction skills, hydrologic analysis, and technology made it more possible to build dams safely (Gleick 1998). According to Gleick (1993), global hydropower production increased by more than 20 % during the 1980s. He argues that the development of new hydropower facilities has slowed greatly in industrialised nations as the best sites have been developed and the environmental costs of further construction are rising. The greatest amount of hydropower development is now occurring in those regions that have seen little development. During the 1980s, hydroelectric production increased by 50 % in Asia, for example.

Hydropower dams provide many benefits in terms of economic growth, food production, surface water recreation enhancement, and storing water during the wet season for later use during dry periods; they also lead to much disadvantage, including urban sprawl, loss of wild habitats, and destruction of river corridors. As Cech (2003: 149–150) puts it, "Dams and reservoirs have enhanced the health and economic prosperity of citizens around the world. However, dam construction comes with a price: altering natural and human environments. Reduced flows, degraded water quality, and impacts on migrating fish are among serious problems caused by dams".

As such, the increase in dam building in recent decades is likely to induce conflict between upstream and downstream between dam owners and local communities, whose livelihoods depend on natural resources, including the quality of water generated by the river system. Much has been written about the environmental impacts of dams (e.g. Cech 2003; Cernea 2004; Covich 1993; Gleick 1993, 1998; Goldsmith and Hildyard 1984; McCully 1996; Postel and Richter 2003).

What is described above is important for Cambodia to study in an effort to find ways to prepare for changes that affect water resources. Cambodian agriculture is the country's most important sector as it contributes significantly to economic development and produces food for the increasing population. Even though Cambodia does not have such data available, agriculture consumes most of the country's water.

The most important issue in water resources in Cambodia may be that there is too much water in the wet season and too little in the dry season, as we have seen. Hydropower projects across the country are providing energy for Cambodia; the most energy-hungry nation in the region. Apart from the provision of energy, hydro projects are designed to mitigate floods and reduce drought in the neighbouring areas. Hydro project reservoirs are supposed to be sources of fish and other water resources, but are also resorts for tourists, mainly domestic. However, negative aspects include flooding, displacement of local communities from their land, affecting their livelihoods and customs, and loss of forests and natural methods of fish movement.

3 Effects of Climate Change on Water Resources and Adaptation

3.1 Effect of Climate Change on Water Resources

As precipitation is the ultimate source of freshwater, any change in its levels will affect the availability and condition of water on Earth. Since climate is an important determinant factor affecting the condition of precipitation, climate change will affect water resources. Increased temperatures are expected to cause evaporation and possible changes in mean rainfall, rainfall intensity, and rainfall seasonality that would have an impact on soil moisture, stream flow and groundwater recharge, and the occurrence of floods and droughts (Yun et al. 2005).

One of the main issues in the discussion of global warming and climate change impact on water resources is related to where and when precipitation is taking place. Although increased precipitation has generally been observed to occur globally over the past decades, distribution is uneven. In Cambodia, some regions are experiencing increased precipitation, whereas others are seeing the opposite. Generally, global projections of changes in the total amount of precipitation indicate that increases are likely in the tropics and at high latitudes, whereas decreases are likely in the subtropics, especially along its polar edge.[4] The Intergovernmental Panel on Climate Change (IPCC) also reports that, over the twentieth century, precipitation has mostly increased over land in high northern latitudes; decreases have dominated from 10°S to 30°N since the 1970s (Bates et al. 2008).

Deforestation occurring at a rapid rate in the past two decades in Cambodia is believed to contribute to a high level of water evaporation, leaving small ponds and rivers dry. The consequences of drying up rivers and ponds include the extinction of some water resources, especially large fish and vegetation.

Although it is logical to argue that regions with increased precipitation will receive more water and may benefit greatly from global warming and climate change, there are uncertainties involving the intensity and duration of rainfall occurring, and unexpected intense rainfall could increase the risks of flooding. The IPCC projects that the frequency of heavy precipitation events will be very likely to increase over most areas during the twenty-first century, with consequences for the risk of rain-generated floods (Bates et al. 2008). Although floods generate soil fertility and increase fishery production, they can damage agricultural crops, livestock, property, and infrastructure as well as claiming human lives. The World Water Development Report 3 (UNESCO 2009) emphasises that floods represent one of its extreme high-impact water-related events. For instance, floods in Europe in 1997 and 2002 and floods in China in 1996 claimed a combined cost of US$26 billion in material damage. In Cambodia, the Disaster Loss and Damage Database shows that, for the period from 1996 to early 2014, floods took 1,091 lives and claimed almost 12 million victims. Drought saw 2.5 million victims over the same period of time.

Higher temperatures mean increased evaporation rates, which can put pressure on water bodies, such as lessening the amount of groundwater and of surface water from rivers, lakes, and wetlands. The greatest deficits are expected to occur in the dry season, leading to decreased soil moisture levels and more frequent and severe agricultural drought. More frequent and severe droughts arising from climate change will have serious management implications for water resource users. Cambodian agricultural producers are particularly vulnerable, since this sector consumes the largest amount of water compared with all other water-using sectors. Drought occurs when precipitation is significantly below the usual minimum recorded level of approximately 1,000 mm, causing serious hydrological imbalances that adversely affect land resource production systems. In Cambodia, prolonged drought during the wet season may cause destruction of rain-fed agricultural crops if there is no supplementary irrigation.

A study on irrigation in Cambodia shows that estimates of the extra yield produced as a result of irrigation, when measured in terms of rice production in the wet season, are very low: for an increase of 1 % in the amount of water used, rice yield

[4] http://www.climate.org/topics/water.html accessed 19 March 2010.

increases by only 0.06 % in the wet season (Naag et al. 2011). The same study indicates that production in the dry season is not generally feasible without irrigation. Therefore, there is a real need to keep water available for dry season farming in order to increase the productivity of land and water.

Natural ponds and rivers aside, irrigation management is still a great challenge in Cambodia. Schemes are often jointly funded by the government and external donors, with in-kind contributions (such as land and labour) from project beneficiaries (Naag et al. 2011). The present governance system, however, is challenged by a lack of effective feedback mechanisms and coordination among different levels of government (ibid.).

As drought involves the greatest fall in the level of rainfall below normal, it can be seen to cause deterioration in the quality of freshwater bodies, since changes in water resources may affect the chemical composition of groundwater and surface water in rivers and lakes. Increased temperatures may also result in warmer water, causing fisheries and aquatic life to die. This has a serious impact for local populations whose livelihoods depend on fisheries. Lowering the water level in rivers also makes navigation impossible and therefore limits low-cost alternatives for transportation and the exchange of goods between regions and countries.

Another effect of climate change on water resources is a rise in sea levels, causing salinity intrusion into the river system, leading to a decrease of freshwater availability for humanity and ecosystems in coastal areas. This issue has been noted by farmers living along Cambodia's coastline.

The foregoing discussion shows that climate change has a great impact on the availability and quality of water resources. Uncertainty relating to future water availability and quality is greatly dependent on precipitation. Other factors limiting water availability and quality, which this paper has not fully discussed, are human factors such as population growth, changes in land use, expansion of industrial areas, and changes in lifestyle owing to economic growth all of which result in increased water consumption and lower water quality as a consequence of pollution (see Bates et al. 2008; UNESCO 2003).

3.2 Climate Uncertainty in Cambodia

Cambodia is one of the most natural disaster-prone countries in East Asia, and climate change has already been observed in the country in terms of more serious lightning and storms, floods and droughts, rising temperatures, and changing rainfall patterns. Some studies predict that the Cambodian climate will see an increase in extreme weather events. The average temperature in the country has increased by about 0.8 °C from 1960 to the present, and the frequency of unusually hot days and nights has increased (DCA and CA 2011).

Rainfall is expected to increase more in the lowlands than in the highlands, with precipitation and flooding increasing predominantly in the central agricultural plains, which are already vulnerable to flooding and drought (Ministry of Environment 2002). Floods and droughts occurring in many provinces in Cambodia have destroyed agriculture significantly and led to a reduction in the national income and a massive loss of income to farmers, as well as the social impact. Recent extreme climate events such as La Niña and El Niño have already resulted in more severe and frequent flooding and droughts. In the past 10 years, the economic cost of these disasters has been evaluated at US$214 million.[5]

It is useful to establish a mechanism to monitor water resources affected by climate change so as to provide more concrete data for analysis and policymaking. Water resources are heavily dependent on precipitation. Precipitation can be understood and forecast by climatologists, who can help inform water users so they can plan future investment activities, for instance in the selection of crop types resistant to drought during the dry season.

3.3 Adaptation

3.3.1 What Can Cambodia do to Manage Water Resources Affected by Climate Change?

Even with limited data and information on water resources affected by climate change, general observation shows that Cambodia faces a two-pronged dilemma involving floods and droughts in the same years. Therefore, there is a need for a mechanism for better water management in order to reduce the impact of floods and droughts. The question is: What mechanism? Two potentially useful mechanisms are described below:

- *Building more water gates to regulate water flow to the Mekong and the sea*: The fresh water level in Cambodia increases during the wet season for two reasons. Firstly, there is an increase in rainwater, and secondly, the volume of water flowing downstream from China, Myanmar, Thailand, and Lao PDR rises. The highest water level is generally seen in October; this starts to recede in November, with the lowest level seen around April. Cambodia has a limited number of water gates to regulate the flow of water, meaning water levels go up and down very quickly.

When the rain starts in May and June, it would be useful to retain water and prevent it flowing into the Mekong and the sea. This would mean there is water

[5] http://go.worldbank.org/SYXPJ6M450 accessed 17 March 2014.

available for agriculture, albeit not enough for the whole year. As rainfall intensity increases from August, water should be released into the Mekong and the sea. This should continue until October, when rainfall density reaches its peak. This release in August means there will be space to receive more water when the rain starts to intensify in September and October. This will help reduce the impact of potential flooding usually seen in October. In October, water in lakes, ponds, canals, and reservoirs should be maintained at optimal levels and all water gates must be closed to reserve water for the dry season.

- *Rehabilitation of existing canals and ponds*: Cambodia is a relatively flat country, with many canals, ponds, and lakes. These provide many benefits, including an attractive environment, habitats for wildlife and native birds, water for agriculture and native species, and a source of refill for underground water.

Moreover, ponds and lakes help control water run-off in times of rain, so can help reduce the impact of floods and reserve water for use in the dry season, preventing drought. They are of benefit not only for vegetation, livestock, and fish but also for people living in the area, as they can enjoy the landscape and fishing. This is particularly true for ponds and lakes close to cities. Therefore, it is good to rehabilitate or create new ponds in communities, not only for the present but also as a strategy to prepare for future changes in water systems brought about by climate change.

Strengthening community forestry: Cambodia currently has approximately half a million hectares of community forestry located in 400 different locations, out of which only 100 have been registered at the ministry level. The rest has also been registered but below the ministry level, and they are at risk of being taken by the government for development. Therefore, there is a need to ensure that all are registered at the ministry level. The validity of registered community forestry is approximately 15 years, while that of land concession is several times longer. Community people see the validity of community forestry is short and want to have longer validity. Longer validity is necessary to ensure community members' participation and their interest.

Creating more parks in and around towns and cities: Cambodia does not have many parks in and around cities and towns. More than that, the country has lost some of its parks due to new development. It is high time for Cambodia to take this issue into consideration seriously. Creation of more small and medium parks in and around cities and towns will benefit not only the residents but also preparedness for the climate change.

Reforestation is also important for Cambodia: forests are a source of income for people and the country as a whole, as they help ensure clear air and water for the benefit of mankind. Forestland also helps to retain water, absorb rain, refill underground water, and works as a windbreaker to slow down storms and reduce soil erosion and run-off.

4 Importance of Climate Information in Water Resource Management

4.1 Climatology as a Means of Predicting and Managing Future Water Resources

Cambodia is not yet advanced in terms of climatology, but highlighting its importance will help the country to make advances in the future. Climatology is the study of the weather and its changes over long periods of time. It focuses on how changes in climate occur and how those may affect future conditions. Climatology can therefore be considered a way to study climatic conditions and to make predictions for the management of future natural resources or warning about possible future hazards, etc.

Climate is concerned with temperature, precipitation, and wind. Changes in one or all of these parameters will affect the availability of freshwater on the Earth. Since the uncertainty of climates may be predictable, climate prediction and information will provide useful means to help water resource managers and policymakers work in advance to prepare for and prevent various disasters and hazards.

Climatology may be of benefit to hydrologists to enable them to understand precipitation and temperature at different times and spatial scales. For instance, a seasonal climate forecast predicts how rainfall or temperature in a coming season is likely to be different from the average weather calculated over a long period of time, usually 30 years (Dialogue on Water and Climate n.d.). Climate forecasts are generally given in terms of the probability that rainfall or temperature will be either below normal, normal, or above normal over certain regions and time periods.

As shown in the above example, climate forecast information can be used to inform people in the region, particularly farmers, of the likelihood of rainfall so they can take decisions for the future, such as whether to grow less water-intensive crops or to supplement their crops with irrigation. It can also help government agencies or developers to take decisions to prepare farmers and provide them with irrigation facilities and crop seeds resistant to drought.

Since rainfall may occur at any rate and at any time, this forecast technique has limitations: it does not say when and in what quantity rainfall is likely to occur over the forecasting period. However, it can be used as a means to foresee the general picture of the availability of rainfall during the whole forecasting season.

Forecasting of temperatures and rainfall can lead to predictions relating to soil moisture and river flows. As discussed in the previous section, freshwater availability is dependent largely on the amount of precipitation. Other determinant factors include temperature and soil moisture in basins, which directly influence the evaporation rate, run-off, and seepage of water.

4.2 Climate Information Service

The foregoing discussion has already briefly shown the general role of climatology in water resource management. We now explore how climate prediction and dissemination of information are being implemented in order to link the climate services product into the hydrological aspect of water resource management.

4.2.1 What is Climate Information?

The term "climate information", and what it constitutes and the purpose it will serve, is rather unclear at present. In fact, there are varying degrees and different types of climate information. According to the World Meteorological Organization (WMO), it may include historical data, analyses and assessments, forecasts, predictions, outlooks, advisories, warnings, model outputs, model data, climate projections and scenarios, and climate monitoring products and can be in the form of text, maps, charts, trend analyses, graphs, tables, GIS overlays, photographs, satellite imagery, etc.

Climate prediction is perhaps one of the most important tools. According to WMO, it includes forecasts, outlooks, and predictions at monthly, seasonal, inter-annual, decadal, and multi-decadal scales. This range includes long-range forecasts (monthly outlooks, three-month or 90-day outlooks, and seasonal outlooks) and climate forecasting (including climate variability prediction and climate prediction beyond 2 years).

Climate information can be used to serve the individual needs of the many climate-sensitive sectors. At the World Climate Conference 3 organised by WMO in Switzerland in September 2009, WMO (2009b) stressed the importance and need for climate information in nine climate-sensitive sectors: human health; sustainable energy; water; transport; tourism; biodiversity and natural resource management; sustainable city; land degradation, agriculture, and food security; and oceans and coasts. Climate information must be generated, regenerated, produced, and reproduced to meet the needs of each specific sector. Information about the climate can be used to help in making decisions on the urgency and desirability of adaptive measures in all sectors.

4.2.2 Who Should be Involved in Climate Service?

Climate service operates at three levels: national, regional, and global. At national level, responsible agencies are placed under the authority of different ministries or institutes, which vary from country to country. Usually, climate service falls under the responsibility of the National Climate Service or National Climate Centre within the National Meteorological Service (NMS) or National Meteorological and Hydrological Service (NMHS). In some countries, climate functions are mandated to other national entities, including government agencies, universities or research institutes, in addition to the NMHS (WMO 2009a). In Cambodia, the service is within the Department of Meteorology of the Ministry of Water Resources and Meteorology, which carries out climate studies, conducts climate predictions and projections, and develops and provides climate services.

In addition to the above national climate agencies, other regional and global centres are being developed to back them up and promote regional and global cooperation in providing climate services. Supported by WMO, regional climate centres are developed as "centres of excellence that assist WMO members in a given region to deliver better climate services and products including regional long-range forecasts that support regional and national climate activities, and thereby strengthen the capacity of WMO members in a given region to deliver better climate service to national users" (WMO n.d.). At global level, WMO also supports the establishment of the Global Producing Centres (GPCs) as "operational centres producing long-range forecasts of global large-scale fields of temperature, precipitation, and other major climate variables (WMO 2009a)". Table 1 provides a brief summary of the responsibilities of national, regional, and global centres.

4.2.3 Role of the Regional Climate Outlook Forum

At the current time, climate information services are still in need of improvement. Over the past 15 years, WMO and NMHS regional institutions and other international organisations have jointly initiated and established an innovative process known as the RCOF. This has the aim of providing consensus-based early warning seasonal climate information to reduce climate-related risks and support sustainable development efforts at a regional scale.

The RCOF concept has been implemented in many regions worldwide, including South America, Central America, Asia, and the Pacific islands (WMO n.d.). Although its scope may differ from region to region, the core process is similar and can be generalised as follows (ibid.):

- Meetings of regional and international climate experts to develop a consensus on the regional climate outlook, typically in a probabilistic form.
- The forum proper, which involves both climate scientists and representatives from user sectors, for identification of impacts and implications, and the formulation of response strategies.
- A training workshop on seasonal climate prediction to strengthen the capacity of national and regional climate scientists.
- Special outreach sessions involving media experts, to develop effective communications strategies.

After the climate outlook has been formulated at a regional level, national forums are organised to develop detailed national-scale climate outlooks and risk information, including warnings for communication to decision-makers and the public.

RCOF has been working towards providing climate information for various sectors, including agriculture, human health, disaster management, and water

Table 1 Responsibilities of national, regional, and global centres

Category	Responsibilities
National climate centres	• Develop climate services at the national level for various sectors
	• Monitor, conduct climate watch, and issue weather warnings and climate advisories to support national early warning systems and disaster risk reduction activities and programmes
	• Exchange climate data and operational products with regional and global centres
	• Link to the regional and global centres
	• Downscale global and regional climate information, including diagnostic (present and past) and prognostic (future) information at various timescales
Regional climate centres	• Downscale, interpret, and assess relevant prediction products from global centres
	• Monitor regional climate variability and extremes
	• Implement and conduct climate watches
	• Develop quality-controlled regional climate datasets for temperatures (minimum, maximum, and mean) and for total precipitation, rainfall, and snowfall
	• Share regional and sub-regional products and information
	• Downscale climate change scenarios
Global climate centres	• Produce long-range forecasts of global large-scale fields of temperature, precipitation, and other major climate variables
	• Provide calibrated probability information for 2-m temperature over land, sea surface temperature, and precipitation

Source adapted from WMO (2009a)

resource management. The climate outlook provides probabilistic information on rainfall or temperature that can be used to generate other climate-related sectoral outlooks. For instance, in the health sector, a malaria outlook for the Greater Horn of Africa region has been developed based on RCOF products. As with the climate outlook, this malaria outlook is developed to forecast the probability of occurrence of vector-borne diseases in the next one to three months. Based on this, interventions for malaria prevention can be initiated for the regions where vulnerabilities are projected to occur (Ogallo et al. 2008).

5 Moving to a Hydrological Outlook

5.1 Background and General Concept

As we have seen, climate outlooks established over the past century have evolved from just the provision of information on the amount of rainfall and air temperature to more useful data for the further development of other outlooks, such as on malaria and food security. Hydrological outlooks have also been developed and

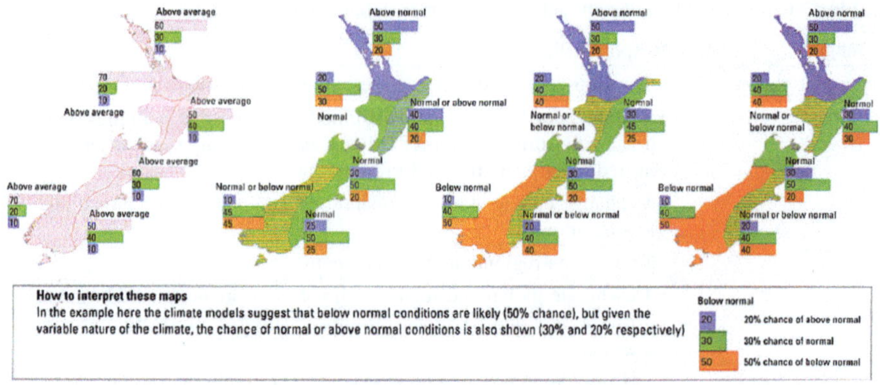

Fig. 4 Hydrological outlook based on climate information. The *map* shows May to July 2008 tercile probability predictions for (*left to right*) air temperature, rainfall, soil moisture, and river flow for six New Zealand regions

implemented, for instance, in New Zealand (Fig. 4). Good practice on moving to refine climate outlook information for hydrological prediction is being explored, documented, and extended for implementation elsewhere in the world.

As stressed earlier, some NMHS produce climate information and predictions for up to three months in advance for their country. The information typically produced includes status of rainfall and temperatures for the immediate past, and prediction of the likelihood of rainfall and of the temperature. This kind of information is useful for NMHS to further utilise for the hydrological prediction of river systems, such as predictions on the availability of water and river flows. Such information can then be used for irrigation planning, water resource management, hydropower operation, and navigation and warnings of the likelihood of floods and droughts.

Beside good climate data and predictions, including rainfall and temperatures, good baseline hydrological data are required to predict future hydrological flows as well as availability of water in basins. This includes data on soil moisture status, mean river flows, groundwater, and evaporation. In addition, hydrological prediction should take into account water use in the basin, influencing the availability of water and river flow, especially cumulative reservoirs for hydropower and irrigation, which take up most water and thereby reduce the availability of water in the river system.

Based on climate outlook information (such as temperature and rainfall prediction data), together with baseline hydrological data, forecasting of soil moisture and river flow can be generated. Figure 4 presents the example of a climate-based hydrological outlook implemented in New Zealand. It shows the four sets of predictions carried out. The first two maps represent the climate prediction information used to further predict soil moisture and river flow. As discussed in Sect. 2, availability of water and river flow is initially determined by rainfall and evaporation rate. Soil moisture also influences river flow because of the soil's capacity to hold rainwater and because of seepage into river channels. If the soil condition is too dry, a great deal of rain water will be absorbed into it, meaning less or no seepage to the riverbed. As such, a prediction of soil moisture is required before it is possible to predict river flow. As Fig. 2 shows, in the regions where rainfall and soil moisture are predicted to be high, river flow is also likely to be above normal. However, note that each region demarcated on the map is not necessarily drawn according to river basin boundaries; other factors apart from rainfall need to be taken into account. For example, it is important to understand the river networks criss-crossing the regions, as the level of water in a specific area can be affected by the water level in other areas (upstream). Other factors such as land use and hydropower dams must also be considered.

Although it is simple to outline how climate information theoretically fits hydrological prediction, integration of the two disciplines has not always been easy, as they do not blend the two sciences together under one roof in hydro-climatological prediction. Refinement of climate data for hydrological prediction needs to be carried out through a close working collaboration between the two groups of scientists, climatologists and hydrologists. In this sense, a forum that paves the way for the two disciplines to work together needs to be designed and established. Cambodia must strengthen the capacity of its own scientists and work towards building climate data to enable better prediction.

5.2 *Justification for Project Intervention*

The idea of establishing a Hydrological Outlook Forum is to take the climate information services produced under RCOF one step further to predict hydrological patterns so water resources can be further managed along the baseline of a river basin under the theme of climate change.

The rationale behind the establishment of the Hydrological Outlook Forum is the close interaction and relationship of climate and water. The availability of freshwater on the Earth is influenced through the hydrologic cycle, which is largely driven by various weather and climatic conditions, including air temperature, precipitation, and wind.

Since global warming and climate change induce changes in water availability over space and time in many parts of the globe, a new approach for water resources management needs to be explored to tackle the complexity of the hydrological system under climate change. To simplify water management under climate change, a river basin approach is required so hydrological assessment and prediction can be easily implemented.

Since water is not static, a river basin is viewed not just as a geographical area but also as a unit in which strong relations exist between the different elements: land and water, groundwater and surface water, quantity and quality, and upstream and downstream (Mostert 1999). These interrelations turn river basins from a geographical area into a coherent system (Lundqvist et al. 1985) of interacting and interdependent elements. For instance, an increase in irrigated agriculture and use of pesticides upstream can decrease the quantity and quality of the water available downstream. Mostert (1999) argues that sustainable development can be ensured only if these interrelations are taken into account in the management of the natural environment, and this makes the notion of the river basin so important. This means the river basin needs to be treated as the unit for water resource management.

As mentioned earlier, water availability is determined through the hydrological cycle, which is largely driven and influenced by the complexity of the climate system in the atmosphere. As such, any climatic changes in the atmosphere induce a change in water availability in the river basin, whether too much or too little. Given that the hydrological cycle is closely linked to the climate system, hydrological forecasts for water resource management in the river basin will certainly be based on the climate information provided by climatologists. In this way, hydrologists and climatologists must work together to identify necessary data and information.

Over the course of nearly 15 years, WMO has assisted in the establishment of RCOF to bring regional climatologists to meet every year to develop a consensus on the regional climate outlook and produce information on various climatic predictions to serve and promote agriculture and health services. The Hydrological Outlook Forum will predict water availability or hydrological events in the short term (seasonal) and long term (decadal) in the river basin. It will make use of the existing RCOF concept, which is considered to have laid the groundwork for its foundation.

5.3 Establishment of the Hydrological Outlook Forum: The Way Forward

As climate outlook products are the main inputs for a hydrological outlook, the Hydrological Outlook Forum should initially target one of the regions where RCOF has already taken place. This will make the process of establishing the Hydrological Outlook Forum easier and more likely to succeed. Diffusion of the concept can be implemented once the practices have been simplified. In this sense, the design of the Hydrological Outlook Forum should be formulated jointly by both types of technical expert (climatologists and hydrologists), under the lead and facilitation of hydro-climatological experts with experience in climate-related hydrological outlook. To kick off the Hydrological Outlook Forum, an initial technical workshop should be held to discuss the integration of climate information into hydrological outlook.

The Hydrological Outlook Forum should be established in the Greater Mekong Sub-region to share information between countries in the upper and lower Mekong regions. The forum will allow downstream countries such as Cambodia and Vietnam to benefit from past, present, and future hydrological information for better understanding of risks and opportunities. The forum will also evolve a culture of working together to manage risks and coordinate policies in various climate-sensitive sectors, such as agriculture, irrigation, and sanitation health.

References

Bates B, Kundzewicz ZW, Wu S (2008) Climate change and water. IPCC, Geneva

Cech TV (2003) Principles of water resources: history, development, management and policy. Wiley, New York

Cernea MM (2004) Social impacts and social risks in hydropower programs: pre-emptive planning and counter-risk measures. Keynote address, session on social aspects of hydropower development at the UN symposium on hydropower and sustainable development, Beijing, 27–29 Oct

Covich AP (1993) Water and ecosystems. in: Gleick PH (ed) Water in crisis: a guide to the world's fresh water resources. Oxford University Press, New York

DCA (DanChurchAid) and CA (Christian Aid) (2011) Climate change and disaster management policy mapping and analysis in Cambodia. DCA and CA, Phnom Penh

Gleick PH (1993) Water in crisis: a guide to the world's fresh water resources. Oxford University Press, New York

Gleick PH (1998) The World's water 1998–1999: the biennial report on fresh water resources. Island Press, Washington DC

Goldsmith E, Hildyard N (1984) The social and environmental effects of large dams. volume 1: overview. Wadebridge Ecological Centre, Wadebridge

Hor K (2014) Fish exports up 10 per cent from 2012. The Phnom Penh Post, 3 Jan

IRIN (Integrated Regional Information Networks) (2008) Cambodia: freshwater fish resources under increasing pressure. 22 Aug: http://www.irinnews.org/report/79937/cambodia-freshwater-fish-resources-under-increasing-pressure

IWMI (International Water Management Institute) (2007) Water for food water for life: a comprehensive assessment of water management in agriculture. Earthscan, London

Lundqvist J, Lohm U, Falkenmark M (eds) (1985) Strategies for river basin management; environmental integration of land and water in a river basin. D. Reidel Publishing Company, Dordrecht

McCully P (1996) Silenced rivers: the ecology and politics of large dams. Zed Books, London

Ministry of Environment (2002) Cambodian first national communication under the UN framework convention on climate change. Ministry of Environment, Phnom Penh

Mostert E (ed) (1999) River basin management and planning: institutional structures, approaches and results in five european countries and six international basins. research report 10. Delft University of Technology, Delft

Naag P, Khiev D, Hirsch P, Whitehead I (2011) Understanding stakeholders roles, perceptions and constraints for effective irrigation and catchment management and development. Working paper 54, CDRI, Phnom Penh

Ogallo L, Bessemoulin P, Ceron J, Mason S, Connor SJ (2008) Adapting to climate variability and change: the climate outlook forum process. WMO Bull 57(2):93–102

Postel S, Richter B (2003) River for live: managing water for people and nature. Island Press, Washington DC

UNESCO (United Nations Educational, Scientific and Cultural Organization) (2003) Water for people water for life. UNESCO, Paris

UNESCO (United Nations Educational, Scientific and Cultural Organization) (2009) World water development report 3: water in a changing world. Earthscan, Paris

Van Zalinge, N, T Nao, ST Touch, and L Deap (2000) Where there is water, there is fish? Cambodian fisheries issues in a mekong river basin perspective. In: Ahmed M, Hirsch P (eds) Common property in the mekong: issues of sustainability and subsistence. ICLARM, Manila

WMO (World Meteorological Organization) (2009a) Draft global framework: concept note ver. 3.4, 11.03.2009. WMO, Geneva

WMO (World Meteorological Organization) (2009b) Report of the world climate conference 3: better climate information for a better future. WMO, Geneva
WMO (World Meteorological Organization) (2011) RCC flyer, http://www.wmo.int/pages/prog/wcp/wcasp/documents/RCC_flyer_October2011_en.pdf. Accessed 19 Mar 2014
Yun W, Kim H, Cho K (2005) Climate services towards water supply and distribution. Technical conference: climate as a resource, Beijing, 1–2 Nov

Transboundary Water Resources Management in the Context of Global Environmental Change: The Case of Bhutan Himalaya

Om N. Katel, Dietrich Schmidt-Vogt and Ngawang Dendup

Abstract The transboundary Himalayan Rivers flowing through Bhutan to India and Bangladesh constitute an enormous asset for economic development in a region which contains the largest number of poor people in the world. However, the rapid retreat of Himalayan glaciers has made South Asia vulnerable to a variety of water-related natural hazards and disasters such as floods, landslides, and glacial lake outbursts. As a result, the region is increasingly experiencing water-related stress such as a decline in freshwater supplies for drinking and agriculture purposes. International cooperation in transboundary water resources management could help to reduce such impacts. The objectives of this paper are to (a) illustrate water resources and their management as strategic cooperation for regional development, (b) appraise how cooperation between concerned countries can be successful in using water resources as an engine for economic growth, and (c) elaborate on climate change and its impacts on the lives and livelihoods of people in Bhutan and beyond. This paper also presents recommendations for possible areas of cooperation with a focus on water resources management in the Ganges–Brahmaputra–Meghna basin.

O.N. Katel (✉)
College of Natural Resources, Royal University of Bhutan, Lobesa, Punakha, Bhutan
e-mail: katelombhutan@yahoo.com

D. Schmidt-Vogt
Centre of Mountain Ecosystem Studies, Kunming Institute of Botany, Chinese Academy of Sciences, Kunming, China
e-mail: schmidt-vogt@mail.kib.ac.cn

D. Schmidt-Vogt
World Agro-Forestry Center, East Asia Node, Kunming, China

N. Dendup
Department of Economics, Sherubtse College, Royal University of Bhutan, Kanglung, Bhutan
e-mail: ngawangdendup@sherubtse.edu.bt

S. Shrestha et al. (eds.), *Managing Water Resources under Climate Uncertainty*, Springer Water, DOI 10.1007/978-3-319-10467-6_13

Keywords Climate change · Hydropower development · Transboundary cooperation · Water resources management

1 Introduction

The transboundary Himalayan Rivers flowing through Bhutan to India and Bangladesh are potentially a huge asset for economic development in a region where the largest number of poor people in the world live. All the rivers flowing through Bhutan drain into the River Brahmaputra, one of the ten largest rivers in South Asia, which links the Himalaya and the plains in India and Bangladesh. The Brahmaputra basin is connected to four countries, namely Bangladesh, Bhutan, China, and India, and is the second largest basin in South Asia covering an area of 651,335 km^2 with a population density of 182 persons per square kilometre. The largest river basin in South Asia, the Ganges river basin, covers 1,016,124 km^2 with a population density of 401 persons per square kilometre and is also the most populated river basin in the world (ICIMOD). These basins, though constituting some of the most fertile areas in the world in terms of agricultural productivity, also contain the highest concentrations of poor people, especially the Brahmaputra River Basin. This is a clear indication that people have not fully benefitted from the available resources such as water and fertile land. The benefit of these resources can be shared more equally by development through better regional and local cooperation mechanisms. However, limitations may be imposed by recent threats such as global warming and climate change.

Countries linked by transboundary rivers can use water resources for poverty alleviation and for accelerating economic development. This can be achieved through hydropower generation and by trading electricity with other nations as in the case of the Bhutan–India cooperation (Nakayama and Maekawa 2013). Other immediate benefits can be certified emission reduction (CER) credits from a clean development mechanism based on the Kyoto Protocol. An example from Bhutan is the Dagachu hydropower project (ADB 2011).

The U.S. Energy Information Administration (2013) has projected that the demand for energy in non-OECD countries will increase by 90 % and a significant share of this increase will be required for development in India and China. South Asian countries can benefit from developing transboundary water resources by stimulating economic growth through revenue generation and by reducing emissions of carbon dioxide by fulfilling their energy needs with hydropower instead of coal and fossil fuel (Chattopadhay and Fernando 2011). In this sense, transboundary water resources management can also contribute in mitigating the impacts of climate change.

The Greater Himalayan Region (GHR) covers approximately 7 million km^2 and is also known as the Water Tower of Asia (Xu 2007). GHR is the source of 10 of the largest rivers in Asia, and among these rivers, the Brahmaputra and Ganges in India, also known as Meghna in Bangladesh, are the most important water resources and engines of growth for countries such as Bhutan, India, and Nepal (Xu et al. 2007, 2009; Biswas 2011; Bisht 2012). One example of transboundary water resources management is the Bhutan–India cooperation. In this connection, Bhutan benefits by exporting electricity generated from its rivers to India, and India benefits from higher industrial and economic growth boosted by cheap and available energy supplies from Bhutan. Bhutan's major perennial water resources are fed by permanent glaciers and associated glacial lakes, seasonal southwest monsoon rains, and significant forest cover. Because of the location of these rivers in deep narrow valleys, the hydropower potential in Bhutan is estimated to be 30,000 Mega Watt (MW), of which 27,000 MW is identified as feasible (Berkoff 2003; Biswas 2011; Bisht 2012; Dhakal and Jenkins 2013). Currently, Bhutan is gearing towards ten new hydropower projects and with many more in the pipeline.

Considering the transboundary nature of the major rivers in Bhutan, the Bhutanese government decided in the early 1980s to cooperate closely with India. These water-based developments have increased Bhutan's per capita GDP from being the lowest in south Asian in 1980 to being the highest in the Ganga–Brahmaputra–Meghna (GBM) region in 2008 at US$1,932.8. Given this trend of economic development, Bhutan would have the highest per capita GDP in entire South Asia by the year 2020 (Biswas 2011; Bisht 2012). However, what is unique about the current form of electricity trading between Bhutan and India is that it is arranged through bilateral agreements between the governments of two countries, which is not the case between the other South Asian countries. Judging from the case of Bhutan and India, it can be argued that other countries of South Asia could benefit from similar cooperation in terms of socio-economic development by designing a regulatory framework for such cooperation in transboundary water resources management (Chattopadhay and Fernando 2011).

However, there are also potential risks related to stress on hydrological regimes in the Himalayas due to significant variability in climate such as annual precipitation and temperature patterns, affecting the rate of glacier retreat and other effects through disruption of water-related ecosystem services. Although Bhutan's contribution to negative factors affecting climate change is negative, the country is not going to be spared the impact of climate change on its vulnerable water resources. This could have enormous implications not only for Bhutan but also for other countries in the region in terms of poverty alleviation, environmental sustainability, climate change adaptation, and even political stability. Therefore, international cooperation in transboundary water resources management is important and could be a key to reducing such impacts.

Since most of the Himalayan rivers are transboundary in nature, involving water resources management, climate change as a driver of environmental change has become a priority issue for disaster reduction and holistic water management at the catchment level. This could be dealt with by regional economic cooperation such as cross-border economic exchange and trade and infrastructure development with a central focus on water. The development of mechanisms for using water resources in a more efficient and effective manner is likely to play an important role in providing substantial benefit to people of the countries concerned and a platform for regional stability.

The objectives of this paper are to examine the following: (a) the water resources and the potential of these resources for economic development in Bhutan; (b) the potential impacts of climate change and variability for water resources and the risks of the livelihoods of millions of people living in the region; and (c) the importance of transboundary cooperation in water resources management in the region. This paper is based on a literature review related to impacts of climate change and its effect on water resources in Bhutan and the need for regional cooperation.

2 Location of Bhutan

Bhutan has a total land area of 38,394 km^2, inhabited by about 700,000 people, and located in the Eastern Himalayas or eastern Hindu Kush Himalaya (HKH). Bhutan is a landlocked country bordering in the east, south, and west by India and in the north by China (Fig. 1). In the west, Bhutan is separated from Nepal by the Indian state of

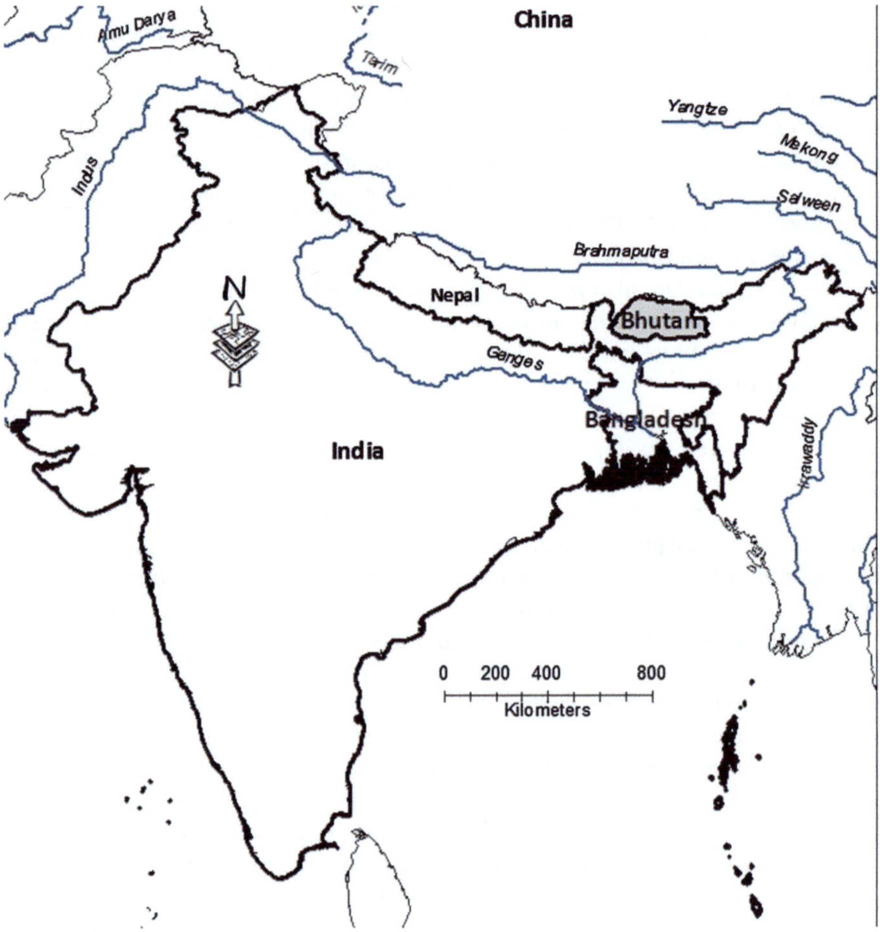

Fig. 1 Location of Bhutan (*Data source* GIS laboratory, College of Natural Resources; Map generated by Om Katel)

Sikkim and in the south from Bangladesh by the Indian state of Assam and West Bengal. Southern Bhutan is characterised by a subtropical climate, northern Bhutan belongs to the sub-alpine Himalaya. The elevation ranges from 150 m in the south at the Indian border to more than 7,000 m in the north on the border with China.

Bhutan climatic features differ due to the great variation in elevation, high mountains, and narrow deep valleys. Precipitation is affected by the summer monsoon, which brings about 60–69 % of the region's rainfall and northeast winds in winter. Bhutan is divided into three lateral zones from south to north, namely the southern foothills, the inner Himalaya, and the great Himalaya with 20 % of the northern region under perpetual snow. The inner Himalaya of Bhutan is the largest physiographic region where broadleaved pine and mixed forests are commonly found. This zone is characterised by broad valleys and forested hillsides located from 1,100 to 3,000 m above sea level. The southern foothills are characterised by a humid climate and receive the highest rainfall of all the regions in Bhutan.

Due to rising elevations from south to north, temperature also varies from south to north with decreasing trends. Most of the central part of the country experiences a cool, almost year round temperate climate, while the south climate is hot and humid with temperatures ranging between 15 and 30 °C. Precipitation also varies widely; in the north, the high mountains receive as much as 40 mm annually, which is mostly located in the rain shadow, the Inner Himalayas receives around 1,000 mm, and the humid south up to 7,800 m (Ohsawa 1987).

3 Water Resources in Bhutan

Bhutan is a landlocked country, and most of its water resources are contained in rivers. It has four major river basins, namely Amochu, Wangchu, Punatsangchu, and Drangmechu. The Drangmechu River Basin comprises the Mangdechu, Bumthangchu, Kurichu, Drangmechu, and Gongrichu Rivers. Among these, Drangmechu and Kurichu are transboundary. The Kurichu flows from Southeast China and the Drangmechu flows from Arunachal Pradesh in India. In Western Bhutan, Amochu originating from China is the only transboundary river, while all other rivers originate within Bhutan and are mostly fed by glaciers in the northern Bhutan. All the rivers originating and passing through Bhutan drain into the Brahmaputra River (Fig. 2).

These rivers play an important role in Bhutan's economic, social, and cultural geography and have enormous potential for hydropower development. This is because river valleys at higher elevations, i.e. within the Greater Himalaya, are broader than the valleys in the Inner Himalayas and in the southern foothills. As the rivers pass through the Inner Himalayas towards the southern foothills, the valleys become narrower and steeper, providing higher potential for hydropower development (Fig. 3).

The rivers flowing from northern part of Bhutan have steep longitudinal gradients and narrow valleys, which open up occasionally where small tracts of flat

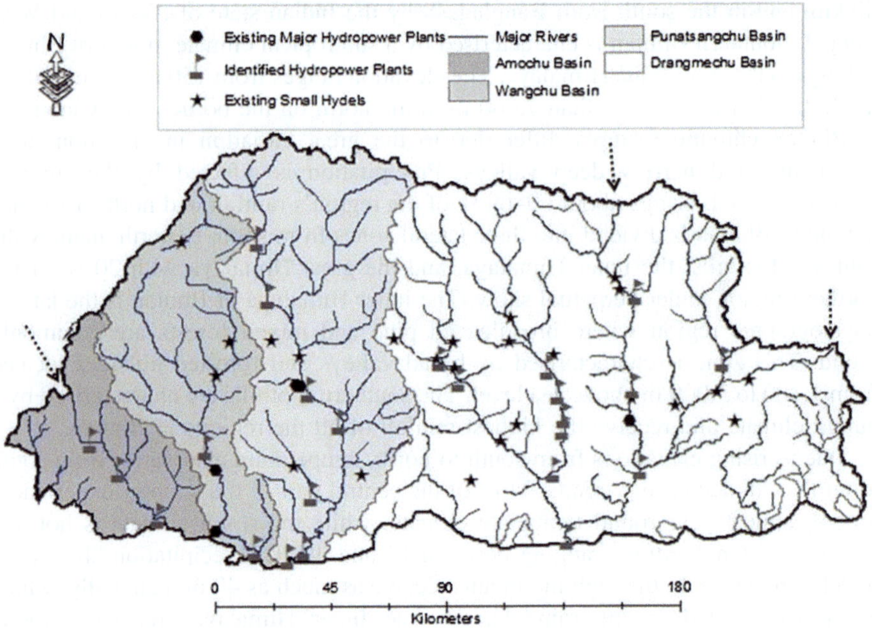

Fig. 2 River basins in Bhutan (*Data source* GIS laboratory, College of Natural Resources: Map generated by Om Katel). Transboundary rivers (marked with *arrows*)

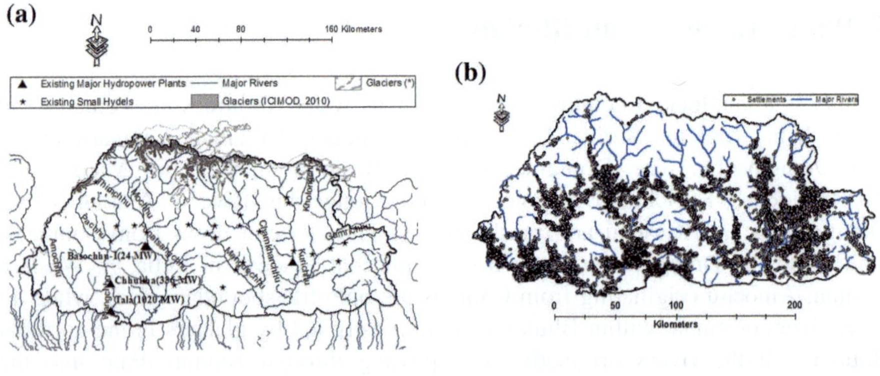

Fig. 3 Drainage patterns of major rivers and hydropower plants in Bhutan ([Glacier (*) data is downloaded from freegisdata.rtwilson.com, while ICIMOD, 2010 data is downloaded from ICIMOD database (www.icimod.org) from Bhutan glacier records for the year 2010 (ICIMOD, 2014)]; *Data source* GIS laboratory, College of Natural Resources: Map generated by Om Katel)

land can be used for cultivation. However, due to the alteration of dry and rainy seasons, flow of rivers also varies seasonally. Rivers carry large volumes of flow and sediment during monsoon season but low volumes during the dry season this is because of the limited base flow from insufficient groundwater recharge. The base flow during the dry season depends on snow melt at high altitudes and the source of

Table 1 Summary of glaciers and glacial lakes in Bhutan

S. No	Sub-basins of rivers	Glaciers			Glacial lakes	
		Numbers	Area (km^2)	Ice reserves (km^3)	Numbers	Area (km^2)
1	Amo chu	–	–	–	71	1.83
2	Ha chu	–	–	–	53	1.83
3	Pa chu	21	40.51	3.22	94	1.82
4	Thimphu chu	15	8.41	0.33	74	2.82
5	Mo chu	118	169.55	11.34	380	9.78
6	Pho chu	154	333.56	31.87	549	23.49
7	Dang (Tang) chu	–	–	–	51	1.81
8	Mangde chu	140	146.69	11.92	521	17.59
9	Chamkhar chu	94	104.1	8.11	557	21.03
10	Kuri chu	51	87.62	6.48	179	11.07
11	Drangme chu	25	38.54	2.26	126	5.82
12	Nyera ama chu	–	–	–	9	0.08
13	Northern Basin	59	387.73	51.72	10	7.81
Total		**677**	**1316.71**	**127.25**	**2674**	**106.78**

Source Chhopel et al. (2011) (Climate Summit for living Himalaya document)

the river water is mainly by melting of glaciers either flowing directly or through formation of glacial lakes. There are about 2,674 glacial lakes (Table 1) located in the remote high-altitude alpine areas. Outburst from these lakes has occasionally resulted in enormous flash floods causing damage to downstream areas.

4 Water Availability and Demand in Bhutan

Bhutan has abundant water resources with average annual flows of 73,000 million m^3 per year (Table 2), which at 109,000 m^3 is within the range of highest per capita mean annual flow availability of water (Chhopel et al. 2011; ICIMOD 2011a, b). Gross consumptive demand of 22 million m^3 has been estimated for 2002, and the demand is forecasted to grow to 516 million m^3 by 2012 and to 541 million m^3 by 2022 in Bhutan (WRMP 2003) (see Table 3). Non-consumptive water demand exists in the form of hydropower demand. It was estimated at 6,700 million m^3 in 2002; it is forecasts to grow exponentially to 26,900 million m^3 by 2022, in addition to the upcoming and potential hydropower projects.

It is important to note that the consumptive and non-consumptive demand for water is increasing over time. However, one of the visible impacts of climate change that Bhutan has witnessed during last couple of years is a drying-up of water

Table 2 Water resources in Bhutan

S. No	Characteristics of national features	Values
1	Country land area	38,394 km^2
2	Total population	634,982
3	Long-term mean annual flow for entire country	2325 m^3/s = 73,000 million m^3/year
4	Per capita mean annual flow availability	109,000 m^3
5	Minimum 7 days flow of 10 year period	427 m^3/s = 13,500 million m^3/year
6	Per capita minimum flow availability	21,207 m^3

Source Chophel et al. (2011) (Climate Summit for living Himalaya document)

Table 3 Gross national consumptive and non-consumptive demand in Bhutan

S. No	Demand category	2002 (million m^3/year)	2012 (million m^3/year)	2022 (million m^3/year)
1	Municipal demand	10	19	37
2	Irrigation demand	393	472	472
3	Rural demand	11	15	20
4	Industrial demand	0.6	0.9	1.5
5	Livestock demand	7.5	8.8	10.2
6	Sum of maximum consumptive demand	422	516	541
7	Sum of water supply demands	29.1	43.7	68.7
8	Add on irrigation demand	15		26
9	Non-consumptive hydropower demand	6,700	16,600	26,900

Source Water resource management plan report (WRMP: 2003)

sources and change in precipitation and temperature patterns (SNC 2011). The increase in demand for water during the next decade in the one hand, and the depletion of water sources on the other, could create an acute shortage of fresh drinking water and water for agriculture. At the same time, change in rainfall and temperature patterns could pose a significant threat to the food security of farmers who depend on seasonal rainfall for subsistence agriculture. Climate change is therefore a cause of worry to people living both in the mountains and in the plains. It is thus, essential to initiate water resources management both within and beyond the national boundary. One of such initiatives is the payment for ecosystem services which links people living in the watershed areas and in the downstream areas. It is hypothesized that designing and implementing policy instruments such as payment for ecosystem services would promote conservation and protection of water resources.

In Bhutan, forest cover has increased from 64.4 % in 1995 to 70.5 % in 2010 (RGOB 2012; SNC 2011). Shrub cover has also increased from 8.1 to 10.4 %. Since maintaining forest cover is considered important to conserve water resources and maintain water quality and quantity, Bhutan restricted the commercial sale of forest products such as timber since the 1970s and started developing hydropower projects in the 1980s with the understanding that retaining forests would protect water sources to sustain hydropower projects in the long run.

5 Water Resources Management as Strategic Cooperation for Regional Development

The fact that most of the rivers in South Asia originate in the Himalaya provide a great opportunity for South Asian countries to initiate transboundary water resources management and cooperation. Biswas (2011) argues that the success of transboundary agreements and cooperation can be jeopardised by hostilities between these nations, caused by "historical rivalries, political mistrust, asymmetric power relationships, increasing nationalism, and short-term requirements of national political parties as compared to long-term national interests".

The expansion of economies in the South Asian region has led to a substantial increase in energy demand providing a good opportunity to build transboundary cooperation and enhance hydropower trading between countries. In future, the demand for hydropower is expected to increase; for instance, in the next two to three decades, India's requirement alone is projected to increase by more than 140 %, whereas the total global requirement in the same period is expected to increase by 55 % (Singh 2013). The growing energy needs in India can be met from the hydropower generated from the rivers in Bhutan and Nepal, which are substantially underdeveloped (Nexant study, 2004 cited in Singh 2013). South Asian river basin, particularly Brahmaputra-Ganges river basin, has good resource potential and tremendous scope for energy co-operation, which can be harnessed to address energy security concerns in the region (Srivastava and Misra 2007). However, regional energy generation and transboundary trading requires greater cooperation within South Asia for which an example can be taken from the Bhutan–India cooperation. The cooperation started in the 1970s and is generally considered to be a success.

According to some literature, one of the most important prerequisites for successful international cooperation on transboundary river basin management is mutual trust, which involves political and technical cooperation (Biswas 2011; Bisht 2012). Technical cooperation usually precedes political cooperation, which requires a strategic approach, linking upstream and downstream stakeholders with water resources management to take care of conflicting interests of the stakeholders where necessary (Bisht 2012). It should be noted that political will, mutual trust, and confidence developed between India and Bhutan have helped the two countries to achieve consensus for developing such cooperation for mutual benefit. The cost of the project was agreed on the basis of 60 % grant and 40 % loan, which follows a very different path compared to other countries such as Bangladesh, India, or Nepal (Biswas 2011). Thus, this is a unique case in South Asia in terms of water resources development promoted as an engine of economic growth.

Trust is important as managing resources means managing resource users, which can be a complex undertaking that links political stability, trust, and mutual benefits. Bhutan has benefitted from its water resources because it was able to negotiate with India through transboundary power trading and investment mechanisms, while this has failed in the case of Nepal. The failure of power trading between India and Nepal has been attributed to India exercising monopsony power over water resources (Dhakal and Jenkins 2013) rather than initiating agreements from which both countries could benefit. Nepal has substantial water resources, four times as much as that of Bhutan, and could therefore use these rich resources to benefit that country and at the same time help to mitigate carbon emissions by substituting fossil-based electricity production in India.

6 Power Cooperation Between Bhutan and India

Bhutan initiated its first development plan for hydropower in the 1980s, realising that water is one of its main natural resources, which if developed might help to lift Bhutan from its position of the lowest per capita income country in South Asia (Biswas 2011). However, since Bhutan lacked technology and investment capital to develop its water resources, the Bhutan–India cooperation was viewed as a necessity. The Bhutan–India cooperation is based by interest-driven mutual benefits. The main approach to this cooperation is integrative, synergising both countries domestically and bilaterally (Bisht 2012).

The demand for electricity in India has increased due to accelerated economic development and industrialisation, while Bhutan needs revenue to finance its development activities using its rich water resources. While developing its hydropower infrastructure and exporting power to India, Bhutan has also provided land, timber, and fuel wood for projects; without imposing taxes on construction materials. On the other hand, the domestic transmission infrastructure was developed with Indian assistance and Bhutan's domestic grid connection with India. Thus, the Bhutan–India cooperation has at least two positive effects. Firstly, the development of infrastructure for power transmission in Bhutan with the help of India has also formed the basis for other related infrastructure development. Secondly, the revenue earning benefits Bhutan in financing its own socio-economic development and at the same time enabling India to meet its energy demand for economic development (Dhakal and Jenkins 2013).

Hydropower is one of the most important projects in Bhutan in terms of current and future socio-economic development. It is also used as a foreign policy tool to promote cooperation with neighbouring countries such as India and Bangladesh. The history of the introduction of electricity in Bhutan dates back to 1960s when diesel generators was used to generate electricity. In 1967, Bhutan started importing electricity from India and in 1989, the Chukha Hydropower project was commissioned. Since then, Bhutan started exporting electricity to India (Bisht 2012). The construction of the 336 MW project in Chukha was agreed by both countries on the basis that a 60 % grant and 40 % loan be provided by India for the estimated cost of the project INR 2,540 million (1 USD = 45 INR). Since India supported the planning, construction, and management of the project, Bhutan agreed to sell the excess electricity from the project to India at a mutually agreed rate (Biswas 2011).

The Chukha hydropower project was significant as a foundation for the future transboundary cooperation. About 75 % of the total power generated from the Chukha hydropower project was exported to India. The example of Chukha demonstrated that hydropower could significantly contribute to the national revenue of Bhutan. Therefore, two other hydropower projects at Kurichu and Tala, with the power generation capacity of 60 and 1,020 MW, were installed with a similar memorandum of understanding as in the Chukha hydropower project (Bisth 2012).

Electricity contributes about 16 % to the total energy supply of Bhutan and constitutes 23 % of its total GDP (RMA 2013a, b). Of the total electricity generated, 99 % is from hydropower. Only 5 % (1,480 MW) of Bhutan's hydropower resources of 23,760 MW has been harnessed so far (Bisht 2012). Bhutan's bilateral relationship with India is strengthened by trading in hydropower, which also forms the basis for Bhutan to generate as much as 10,000 MW of power by 2020. Currently, about 80 % of Bhutan's total trade is with India and a further increase in the export of electricity to the Indian grid accounts for a sizeable portion of trade indicating that hydropower is the key feature of India–Bhutan relations and is expected to be so in the future.

With the completion of the Chukha, Kurichhu, and Tala hydropower projects and the successful completion of the first phase of power cooperation, the two countries have embarked on a second phase. In 2008, "The Bhutanese cabinet endorsed the Sustainable Hydro Development Policy, which delineated measures to invite private sector participation and foreign investment to develop hydropower resources in Bhutan" (Bisht 2012) extending the opportunity for more investment in hydropower projects. In 2009, Bhutan and India, agreed to achieve a target of 10,000 MW of power generation by 2020, which according to Bisht (2012) means blocking any participation of other countries in the hydropower sector in Bhutan.

7 Existing Hydropower, Potentials, and Hydropower Plants of the Future

The total hydropower potential of Bhutan is 30,000 (MW) with an economically feasible potential of 27,000 MW (Biswas 2011; Bisht 2012; Dhakal and Jenkins 2013). So far, Bhutan has developed only four major hydropower stations in the country, namely Basochu (BHP-64 MW); Chukha (CHP-330 MW), Kurichu (KHP-60 MW), and Tala (THP-1020) (Fig. 3) (BEA 2010). This accounts only about 5 %

Table 4 Ongoing and upcoming hydropower projects in Bhutan

S. No	Projects	Capacity MW	Start date	Completion date	Mode
1	Punatsangchhu-I	1,200	2009	2015	Bilateral
2	Mangdechhu	720	2010	2017	Bilateral
3	Punatsangchhu-II	990	2010	2019	Bilateral
4	Sunkosh Reservoir	4,060	2011	2020	Bilateral
5	Kuri-Gongri	1,800	2012	2020	Bilateral
6	Amochhu Reservoir	620	2012	2018	Bilateral
7	Kholongchhu	650	2012	2018	JV*
8	Chamkharchhu-I	670	2012	2018	JV
9	Wangchhu	600	2012	2018	JV
10	Bunakha Reservoir	180	2012	2018	JV
11	Nikachhu	208	2012	2017	DGPC PPP*
12	Khomachhu	327	2014	2017	DGPC PPP
13	Rotpashong	918	2012	2019	DGPC PPP
14	Gamri	102	2013	2017	DGPC PPP
15	Dagachhu	114	2009	2013	DGPC PPP
16	Nyera Ama Chu	473	2016	2021	IPP*
Total		**13,632**			

Source http://www.drukgreen.bt/index.php/projects-updates, downloaded on January 01, 2013
*JV** Joint Venture, *DGPC PPP** Druk Green Power Corporation Private Public Partnership, *IPP** Independent Power Production

of the potentially feasible power production in Bhutan. Since the beginning of the power trade between India and Bhutan, the two countries agreed to expand their collaboration to other hydropower developments in Bhutan in 2009 (Table 4) with the goal that about 10,000 MW power should be generated by the year 2020. Thus, ten projects were selected by both countries (SHDP 2008). Six projects will be developed bilaterally by the two governments; the other four will be taken up as joint ventures (JV) between government corporations of the two countries (Table 4). The projects developed under the bilateral agreements will be handed over to the Druk Green Power Corporation (DGPC) after two years from the date of completion. In addition to this, DGPC will be developing the remaining projects on its own (Table 4).

8 Water Resources as an Engine of Economic Growth

In Bhutan, energy has been the primary focus for revenue generation and financing development. As a result of hydropower development, the GDP has increased over the 5-year plan period (1980–1985) from BTN. 1,204.8 million to BTN. 1,674.5 million (1 USD = 12.61), which is a compound annual growth rate of 6.8 % (RGOB 1987). When all hydropower projects (small and major) are completed, Bhutan can earn a revenue of more than US$100 million annually provided India (the buyer) revises the power tariff periodically (Bisht 2012). For Bhutan's population, the amount earned from hydropower can provide a substantial income, contributing significantly to infrastructure and other social service developments. The revenue that has so far been generated by the sale of power to India has been used for developmental activities and for paying the salaries of the employees in various sectors of the civil service in addition to repaying debts.

In terms of access to electricity, about 60 % of the rural households in Bhutan have had power since 2012, an increase of about 20 % from 2003. Currently, hydropower contributes as much as 45 % to the country's revenue and is one of the major sources of revenue to the Bhutanese economy (Singh 2013). Bhutan is suitable for hydropower generation as there are swift flowing rivers with an altitudinal variation of about 150 m in the south to about 7,000 m in the north over a span of 150 km. After the government's decision to exploit its water resources for the production of electricity, the economic scenario for Bhutan has changed significantly. Cheap electricity may in future attract foreign investors in electricity intensive industries to Bhutan, assuming that other factors necessary for establishing industries in Bhutan are favourable; Bhutan would benefit in terms of employment and tax revenues from these industries. Developing the hydropower industry in Bhutan is a key to solving economic problems such as unemployment and the acute shortage of foreign exchange.

9 Hydropower Generation and Its Relationship with the Use of Forest Resources

The guiding principle of the national forestry policy of Bhutan is to maintain a minimum of 60 % of the total land area under forest cover into perpetuity. This policy has played a major role in maintaining the large forest cover of Bhutan (RGOB 1974). Currently, the forest cover is recorded as 80.9 % with 70.5 % under tree cover and cultivated agricultural area at 2.93 % (RNR 2013). Due to this policy, logging operations were nationalised after 5 years (in 1979) and extensive areas were designated as protected areas in the 1980s. Degraded forested areas were improved by implementing forestation programmes. Among the main aims of the national forestry policy is to reduce illegal logging and protect water catchment and watershed areas in Bhutan, for the benefit of developing hydropower as a cornerstone of Bhutan's economic development. The unit cost of hydropower generation has steadily declined since the Chukha plant was first constructed because of the greater economic scale of production and emergence of more efficient operations and management. One of the reasons for efficient production of hydropower is the improvement of forested areas reducing soil erosion and its negative effect on hydropower plant turbines.

Erosion is one of the main factors affecting water quality in the Himalayan region. Ives and Messerli (1989) argue that soil erosion in the Himalayas probably is higher than in most other major mountain systems. The higher rate of erosion is because of high annual total precipitation that occurs in an area of high relief, vulnerable terrain, and high seismic incidence (ibid.). Furthermore, Tejwani (1987) shows that erosion and stream flow in the Brahmaputra lowlands is also changing because of the significant soil deposition eroded from high Himalayas. The soil erosion in high Himalayas is either due to forest degradation or due to increasing intensity of use of land for farming on the slopy lands. Thus, soil conservation and watershed management practices are extremely important not only for the livelihoods and security of upstream communities but also for downstream communities and long-term sustenance of the hydropower plants.

The forest degradation can be reduced by expanding electrical network as is evident in Bhutan. For instance, expansion of the electrical network in Bhutan has reduced the use of fuel wood and diesel import from India. About 91 % of the residential sector energy demand is used to and is still met by biomass mainly in the form of fuel wood. The total annual fuel wood consumption in the residential sector is estimated at 0.54 million tons, which is 0.85 tons per capita per year for the country and 1.19 tons per capita per year for the rural areas (Palit and Garud 2010; Kuenselonline 2013). The average fuel wood consumption in Bhutan has decreased from 1.27 to 1.19 tons per capita per year for the rural areas although the current consumption is still high (Palit and Garud 2010). However, it is expected to decrease further with more access to electricity in future. In 2005, the fuel wood consumption in Bhutan was 725,000 tons accounting for 57.7 % of the total primary energy supply mix and the highest per capita fuel wood consumption in the world (DOE 2009).

10 Climate Change Impacts and Risks

The Himalaya represents one of the most complex landscapes in the world to be affected by climate change. Run-off from glaciers, snow, and ice, feeding into downstream flows, can be linked to the impact of climate change. Melting of the glaciers can, on the one hand, increase river discharge; while on the other hand, water supply can be low during the dry season. Xu et al. (2009) note that the greatest loss in water availability will be observed in rivers such as the Brahmaputra. In the Himalayan region, including the Brahmaputra River Basin, hazards such as landslides, debris flow, and flash floods are projected to increase significantly in the upland areas (300–3,000 m) as a result of climate variability and change in addition to loss of water availability in some months with fluctuation of water discharge (ibid.).

The changing water resource is associated mainly to changing temperature in turn affecting precipitation. The average annual mean of global warming is projected to be about 3° by the 2050s and 5° by the 2080s (Xu et al. 2009). However, the warming in the Himalayas has been much greater than the global average of 0.74 °C over the last 100 years (IPCC 2007). In case of the Indian subcontinent and in Tibetan plateau, temperature is projected to rise between 3.5 and 5.5 °C by 2100 (IPCC 2007). This trend could have tremendous implications for Bhutan and the millions of poor people living in the river basin regions who are dependent on natural resources for their livelihoods.

Increasing temperature also mean melting of glaciers in the Himalayas. In Bhutan himalaya, the rate of glacial retreat is found to be higher than any other regions (Karma et al. 2003; Ageta et al. 2006; Bajracharya et al. 2007). Bhutan has about 677 glaciers and 2,794 glacial lakes (NEC 2011) out of which 25 glacial lakes are potentially at risk (Richardson and Reynolds 2000) of Glacial Lake Outburst Floods (GLOFs). A study of approximately 66 glaciers showed that the retreat rate was 8.1 % from 1963 to 1993 and the shrinkage of smaller glaciers was found to be faster than the larger glaciers, which contribute to faster expansion of the glacial lakes (Karma et al. 2003). Even the debris-covered glaciers are found to retreat faster in Bhutan Himalaya than in other Himalayan regions because of the location at the near end of the eastern edge of the Himalayan range, which receives heavier precipitation than other Himalayan regions with stronger influence of the summer monsoon season (Ageta et al. 2000). This could increase the risk of GLOFs despite some variation in the distribution of glacial lakes across elevations (Table 5). GLOFs constitute major hazards in the Himalayas, and in Bhutan and Nepal, the formation of glacial lakes is more common, which may lead to more occurrences of GLOFs (Richardson and Reynolds 2000). GLOF is a "commonly used term to describe catastrophic bursts from proglacial moraine damned lakes" (Richardson and Reynolds 2000). GLOFs occur as a result of glacial melts and water draining rapidly from the lakes and also as a result of the collapse or over topping of ice dams formed by the glacier itself (ICIMOD 2011a, b; UNDP 2012).

Table 5 Statistics of high-altitude wetlands in Bhutan

Wetland type	Numbers	Area subtotal (m^2)	Average area (m^2)	Largest lake (m^2)	Smallest lake (m^2)
Supra-snow lake	110	52,327.0	475.7	4,758.8	36.2
Supra-glacial lake	495	28,554,801.3	57,686.5	1,517,436.4	133.6
Glacial lake	637	23,230,604.6	36,468.8	878,311.5	114.7
Lake	1,722	49,973,272.8	29,020.5	868,048.9	34.6
Marsh	63	497,334.4	7,894.2	63,811.1	126.1

Source Inventory of high-altitude wetlands in Bhutan (Sherub and Norbu n.d.)

The change in mountain glaciers is one of the best indicators of climate change (Oerlemans 1994) which may also mean increasing risk of GLOFs. Studies reveal that glaciers are melting rapidly resulting in the formation of new glacial lakes and an increase in the number of GLOFs (Watanabe et al. 1994; Fujita et al. 2001; Bajracharya and Mool 2009). This shows that there has been a significant impact on the high mountain glacial environment and most of the lakes in high elevations are formed during the second half of the twentieth century in response to warming temperatures (Mool et al. 2001; Bajracharya and Mool 2009). Such lakes are hazardous to communities and infrastructure downstream because of their significant destruction potential. According to ICIMOD (2009), there are about 8,790 glacial lakes within the selected parts of the Hindu Kush-Himalayas, of which about 204 of the glacial lakes are considered to be potentially dangerous and may burst out anytime leading to GLOFs. In the eastern Himalayan region, 35 GLOFs have been reported to have occurred within Bhutan, China, and Nepal during the twentieth century.

In Nepal Himalaya, some of the studied glaciers showed that in the years between 1970 and 2000, the loss of glacier area has increased together with the number of lakes formed by glacier melt (Bajracharya et al. 2006). There has been at least one GLOF event occurring every 3 to 10 years in the region resulting in loss of life and destruction of houses, bridges, forests, fields, roads, and other infrastructure. In Bhutan, glaciers covered about 10 % of the total surface area in the 1980s and are important source of river water but sometimes glacial melts and movement also blocks water flow and creates lakes, which may ultimately burst as a GLOFs (UNDP 2012). Similarly, GLOFs occurred in Bhutan in the year 1957, 1960, 1968, and 1994, devastating lives and property in the downstream areas (Komori 2008; SNC 2011; UNDP 2012).

In Bhutan, GLOFs have been a concern since most of the rivers in Bhutan are fed by glaciers in the northern part of Bhutan. GLOFs can damage agricultural fields, lives and livelihoods, and critical infrastructure, such as hydropower dams in downstream areas (Figs. 3, 4). For instance, the Thorthomi glacier poses a risk to the valleys of Central Bhutan and in particular could damage the infrastructure of projects such as the Basochhu and Punatsangchu hydropower projects.

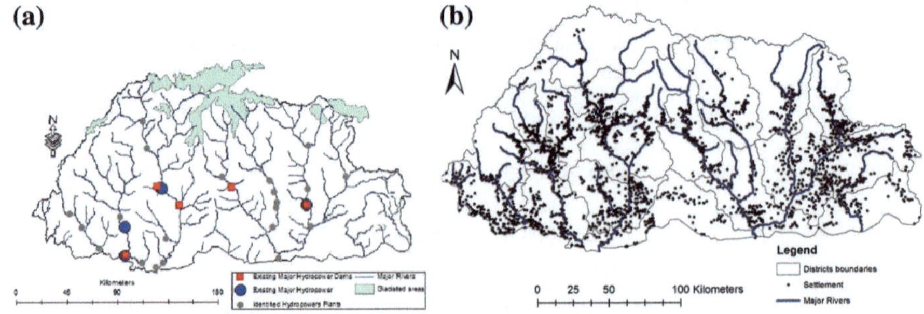

Fig. 4 Location of settlements (*Data source* GIS laboratory, College of Natural Resources)

The effects of GLOFs also increase sediment load. The sediment load is estimated to be significantly higher in rivers flowing through deep valleys in the Himalayan region and could be further augmented by GLOFs with an adverse effect on turbines, reducing their efficiency and lifespan, which can in turn affect electricity generation in the long run. The retreat of glaciers also cause fluctuation in the discharge of river waters and may create a shortage of water for electricity generation as a result of volume in the storage capacity of the glaciers. Furthermore, fluctuating river water significantly affects the lives and livelihoods of the millions of people living in the river basins. Such events will not only affect poor people but also the wider communities as this region is fertile and supplies food crops for people living within and beyond the region. Therefore, policy must target communities and associated institutions within and beyond the political boundaries providing a platform to stakeholders to build adaptive capacities towards climate-related disasters.

11 Effects of a Changing Climate on Ecosystems and Livelihoods

The temperature and precipitation in the Bhutan Himalaya have been found to be increasing steadily over the past three decades (SNC 2011). This could have a tremendous impact on the Himalayan ecology. Although there is evidence to show specific impacts of climate change in Bhutan, evidence from the Tibetan plateau and the Indian Himalayas, highlights the urgency of the problem. One of the projected impacts of climate change is the shifting of species upwards with the possible danger of species extinction (Becker et al. 2007 cited in Xu et al. 2009). As a result of a shift in species from lower to higher elevations, the emergence of invasive species and weeds from lower elevations is very likely (McCarty 2001).

The impacts of climate variability and change may potentially affect sectors such as water resources, forests, biodiversity, agriculture, and human health. Changing climates may affect the drinking water with change in flow regimes reducing water availability to wildlife as well as humans. In case of Bhutan, about 79 % of the Bhutanese population are subsistence farmers (RGOB 2012), therefore, changing patterns of temperature and precipitation would mean a risk to farmers' food security and threat to their livelihoods. Rising temperatures may also increase the incidences of tropical diseases and heat stress in the higher elevation and this also means the loss of farmers' productivity both in farming input and output affecting government revenues while providing assistance to farmers.

12 Need for Regional Cooperation

Water resources management in mountain areas is essential for the sustainable development not only the Himalayas but also the downstream plains in Bangladesh and India. Energy is fundamental for economic growth, and access to reliable energy sources at reasonable cost is a critical factor for sustainable economic development. Therefore, cooperation on transboundary water resources between countries such as Bangladesh, Bhutan, and India is important. The cooperation on water resources management not only provides a conducive environment in terms of local livelihoods and management of risks during natural disasters but also substantial benefits for economic development at least in South Asia. For instance, Nepal has a potential of about 82,000 MW of which about 44,000 MW is estimated to be viable, while Bhutan has a potential of about 30,000 MW of which 27,000 MW may be viable (Berkoff 2003; Dhakal and Jenkins 2013) and exporting these power hungry countries like India and Bangladesh can be a win win situation both for sellers and as well as for the buyers.

Water resources management, however, reflects both opportiunities and threats demanding regional level co-operation. The opportunities is that generating energy in the Himalayan region like in Bhutan would provide substantial benefits for economic development. Generating energy has the potential to meet the growing demand for electricity in the region, for instance, by lowering carbon emissions and also providing a means of regional cooperation through transboundary water resources management. India's annual electricity requirement roughly doubled between 1990 (28,539 MW) and 2000 (57,070 MW) with a projected requirement of approximately 228,310 MW in the year 2020, an increase of 102,740 MW per year (Chattopadhyay and Fernando 2011). Talmiz (2006) cited in Dhakal and Jenkins (2013), writes that India's requirement for power generation capacity will be 778,000 MW by the year 2032. Currently, India has an installed capacity of 375,000 MW (Singh 2013).

On the other hand, the threat is that there have been growing uncertainties regarding the investment in hydropower projects due to increasing variability in precipitation and temperature affecting sediment load, water shortages in the dry season, and probability of occurrence of natural disasters such as GLOFs and landslides. The power grid may also collapse as a result of natural disasters. Therefore, attention should be given to disaster prevention and relief management strategies including the prevention of grid collapse. However, since Bhutan lacks the technology and capacity to deal with such problems, regional cooperation is required and should not be constrained by bilateralism discouraging the participation of third party(ies).

13 Conclusion

It can be concluded that uncertain environmental changes will have a serious impact, causing increased water scarcity and GLOF-related problems. This could provide a window of opportunity for building cooperation at basin and regional levels and would not only help to address the water or disaster management issues but also security and development needs. Thus, to mitigate and adapt to the uncertain and adverse environmental impacts, the associated basin countries must share a long-term vision and cooperative action to manage water resources targeting communities and infrastructure development. Management of water should focus on energy generation and use, agriculture, and environment and health through which countries linked together by transboundary water resources can benefit significantly, particularly in transferring water-related technologies, developing water storage structures, disaster management, and improving information dissemination systems. One of the areas for requirring immediate response could be the development of disaster management strategy that may arise from GLOF and erratic monsoons which could be done by sharing information, better research collaboration covering structural and non-structural projects of flood preventive measures by improving governance systems of water resources at transboundary level.

Forests and forestry are integral parts of mountain community economies with extensive ecosystem services to the larger society. Understanding the changing patterns of ecosystem services such as those provided by forests and their valuation in mountainous areas could be the basis to link upstream and downstream communities. This should not only limit to local level and within a country but should be at the transboundary and at a regional level. This is because transboundary water resources management is an international issue and, therefore, cooperation is needed between different government institutions: International NGOs and all other international stakeholders linking national government sectors, (coordination among the different sectors) experts, and the concerned stakeholders. Floods are the main natural disasters in the Himalayan region influencing poverty, so technical advances in flood forecasting and management offer an opportunity to cooperate beyond the national boundaries.

While Bhutan and India share a common concerns on the impact of climate change, focusing on policy areas such as coordination on adaptation measures and supporting policies towards mitigation in mountainous areas such as Bhutan could benefit both the neighbours (Bhutan and India) including Bangladesh in the long term. The initial focus could be on how to ensure safe provision of good quality ecosystem services such as water quality and quantity. Linking ecosystem services and impact of climate change would help better understand the approaches that could help alleviate poverty. Furthermore, a comprehensive understanding of climate change and its impact on water resources would help design policies related to climate change response choices. Addressing issues such as hydrological regime, management systems, water equity with particular emphasis on improved understanding of upstream—downstream linkages is essential and urgent.

One of the constraints in improving transboundary water resources management appears to have been constrained by not been able to link the water issues to multidisciplinary issues. This is because transboundary water resources management encompasses various disciplines such as politics, culture, management, engineering and others which could be the only possible way to achieve cooperation between river basin member countries. The first step could be linking academic Institutions of member countries countries and initiating quality research to identify the indicators of the drivers of change. To understand what policies are requirred and how much to invest can be estimated by mapping the vulnerable areas towards changing climate. therefore, development of scientific programmes for climate change and variability monitoring is highly recommended. This is because credible up-to-date scientific knowledge linking mitigation and adaptation strategy is essential to develop climate change response policies. The Himalayas are not well represented in global climate models mainly because of the coarse resolution. Therefore, constructing a regional climate model with a higher resolution than a global one is highly recommended, which may be useful for evaluating climate variability and change at different spatial and temporal scales.

References

ADB (2011) Case story on cross-border power export from Dagachhu hydropower development. Green Power Development Project, Bhutan

Ageta Y, Iwata S, Yabuki H, Naito N, Sakai A, Narama C, Karma (2000) Expansion of glacier lakes in recent decades in the Bhutan Himalayas. Debris-covered glaciers, IAHS publ. no. 264, 165–175

Ageta Y, Matsuda Y, Suzuki R, Karma T, Yabuki H (2006) Glacier shrinkages and climate conditions around Jichu Dramo glacier in the Bhutan. Bull Glaciol Res 23:51–61

Bajracharya SR, Mool P (2009) Glaciers, glacial lakes and glacial lake outburst floods in the Mount Everest region, Nepal. Ann Glaciol 50(53):81–86

Bajracharya SR, Mool PK, Shrestha BR (2006) The impact of global warming on the glaciers of the Himalaya. International symposium on geo-disasters, infrastructure management and protection of world heritage sites, 25–26 Nov 2006. Nepal Engineering College, Ehime College and National Society for Earthquake Technology Nepal, proceedings, pp 231–242

Bajracharya, SR, Mool, PK, Shrestha, BR (2007) Impact of climate change on Himalayan glaciers and glacial lakes: case studies on GLOF and associated hazards in Nepal and Bhutan. http://books.icimod.org

Berkoff J (2003) Hydropower in Bhutan and Nepal: why the difference? World Econ 4(3):121–142

Bhutan Electricity Authority (BEA) (2010) Druk Green Power Corporation Limited, Tariff Review Report, Thimphu Bhutan

Bhutan RNR Statistics (RNR) (2013) RNR statistical coordination section. Policy and Planning Division, Ministry of Agriculture and Forests, Royal Government of Bhutan

Bisht M (2012) Bhutan–India power cooperation: benefits beyond bilateralism. Strateg Anal 36 (5):787–803

Biswas AK (2011) Cooperation or conflict in trans-boundary water management: case study of South Asia. Hydrol Sci J 56(4):662–670

Chattopadhyay D, Fernando PN (2011) Cross-border power trading in South Asia: it's time to raise the game. Electr J 24(9):1040–6190

Chen S, Parnetta JC, Duda AM (2013) Towards a new paradigm for trans-boundary water governance: implementing regional frameworks through local actions. Ocean Coast Manag 85:244–256

Chhopel GK, Dulal IR, Chhophel K, Wangchuk C, Rinzin U, Dupchu K, Kunzang, Tse-ring K, Tshethar K, Dorji U, Wangchuk T, Lhamo T, Nidup J (2011) Securing the Natural Freshwater Systems of the Bhutan Himalayas: climate Change and Adaptation measures on Water Resources in Bhutan. A climate summit for a living Himalaya. Available from http://sa.indiaenvironmentportal.org.in/files/file/Water_Paper_Bhutan.pdf

Department of Energy (DOE) (2009) Overview of energy policies of Bhutan. Ministry of Economic Affairs, Thimphu, Bhutan

Dhakal DNS, Jenkins GP (2013) Risk sharing in hydropower development: case study of the Chukha Hydel Project, Bhutan

Fujita K, Kadota T, Rana B, Shrestha RB, Ageta Y (2001) Shrinkage of Glacier AX010 in Shorong region, Nepal Himalayas in the 1990s. Glaciological Res 18:51–54

Geographical Information Systems (GIS) laboratory (n.d.) College of natural resources. Royal University of Bhutan, Lobesa, Punakha, Bhutan

International Centre for Integrated Mountain Development (ICIMOD) (2009) Climate change in the Himalayas. Information sheet no. 3/09. Kathmandu, Nepal

International Centre for Integrated Mountain Development (ICIMOD) (2011a) Glacial lakes and Glacial lake outburst floods in Nepal. ISBN: 978 92 9115 194 3

International Centre for Integrated Mountain Development (ICIMOD) (2011b) Securing the natural freshwater systems of the Bhutan Himalayas: climate change and adaptation measures on water resources in Bhutan. A climate summit for a living Himalaya. Available from http://sa.indiaenvironmentportal.org.in/files/file/Water_Paper_Bhutan.pdf

International Centre for Integrated Mountain Development (ICIMOD) (n.d.). www.icimod.org. Accessed 5 Jan 2014

IPCC (2007) Climate change 2007: impacts, adaptation and vulnerability. Contribution of working group II to the fourth assessment report of the intergovernmental panel on climate, Cambridge University Press, Cambridge, p 976

Ives JD, Messerli B (1989) The Himalayan Dilemma: reconciling development and conservation. United Nations University. ISBN 0-203-26465-7

Karma T, Ageta Y, Naito N, Iwata S, Yabuki H (2003) Glacier distribution in the Himalayas and glacier shrinkage from +30-to +33-in the Bhutan Himalayas. Bull Glaciol Res 20:29–40

Komori J (2008) Recent expansions of glacial lakes in the Bhutan Himalayas. Quatern Int 184:177–186

Kuenselonline (2013) Fuelwood consumption in decline. http://www.kuenselonline.com/fuelwoodconsumption-in-decline/#.UtNhf9IW2o0. Accessed 13 Jan 2014

McCarty JP (2001) Ecological consequences of recent climate change. Conserv Biol 15:320–331

Mool PK, Bajracharya SR, Joshi SP (2001) Inventory of glaciers, glacial lakes and glacial lake outburst floods: monitoring and early warning systems in the Hindu Kush–Himalayan region, Nepal. Kathmandu, International Centre for Integrated Mountain Development with United Nations Environment Programme/Regional Resource Centre for Asia and the Pacific

Nakayama M, Maekawa M (2013) Economic benefits and security implications of trading hydropower through transboundary power grids in Asia. Int J Water Resour Dev 29 (4):501–513

National Environment Commission (NEC) (2011) Second national commission to the UNFCCC. Royal Government of Bhutan, Thimphu, Bhutan

Oerlemans J (1994) Quantifying global warming from the retreat of glaciers. Science 264 (5156):243–245

Ohsawa M (ed) (1987) Life zone ecology of the Bhutan Himalaya. Laboratory of Ecology, Chiba University, Japan

Palit D, Garud S (2010) Energy consumption in the residential sector in the Himalayan kingdom of Bhutan. General 58:34–36

Richardson SD, Reynolds JM (2000) An overview of glacial hazards in the Himalayas. Quatern Int 65–66:31–47
Royal Government of Bhutan (RGOB) (1974) National Forest Policy. Department of Forestry, Ministry of Trade, Industry and Forests, Thimphu, Bhutan
Royal Government of Bhutan (RGOB) (1987) Fifth five year plan document. Gross National Happiness Commission, Thimphu, Bhutan
Royal Government of Bhutan (RGOB) (2012) Statistical year book. National Statistics Bureau, Thimphu, Bhutan
Royal Monetary Authority (RMA) (2013a) Annual Report. http://www.rma.org.bt/annualreporttp.jsp. Accessed 14 Jan 2014
Royal Monetary Authority (RMA) (2013b) National Budget financial year 2013–2014. Ministry of Finance, Thimphu, Bhutan
Second National Communication (SNC) (2011) Second national communication to the UNFCCC. National Environment Commission, Royal Government of Bhutan
Sherub WN, Norbu N (n.d.) The wetlands sky high: mapping wetlands in Bhutan, Ugyen Wangchuck Institute of Conservation and Environment
Singh BK (2013) South Asia energy security: challenges and opportunities. Energy Policy 63:458–468
Srivastava L, Misra N (2007) Promoting regional energy co-operation in South Asia. Energy Policy 35:3360–3368
Sustainable Hydropower Development Policy (SHDP) (2008) Bhutan Sustainable Hydropower Development Policy, Ministry of Economic Affairs, Thimphu, Bhutan
Tejwani KG (1987) Sedimentation of reservoirs in the Himalayan region. Mt Res Dev 7(3):323–7
UNDP (2012) Glacial lake outburst flood, reducing risks and ensuring preparedness. The report on the international conference, Thimphu, Bhutan
US Energy Information Administration (2013) Today in energy. Available from http://www.eia.gov/todayinenergy/detail.cfm?id=10611. Accessed 14 Jan 2014
Watanabe T, Ives JD, Hammond JE (1994) Rapid growth of a glacial lake in Khumbu Himal, Himalaya: prospects for a catastrophic flood. Mt Res Dev 14(4):329–340
Water Resources Management Plan (WRMP) and Update of the Power System Master plan, Bhutan (2003) (vol A–D3). Department of Energy, Ministry of Trade and Industry, Thimphu, Bhutan
www.icimod.org. Accessed 8 Jan 2014
www.drukgreen.bt/index.php/projects-updates. Accessed 10 Jan 2014
www.kuenselonline.com/fuelwood-consumption-in-decline/#.UtNhf9IW2o0. Accessed 13 Jan 2014
Xu J (2007) The highlands: a shared water tower in a changing climate and changing Asia. Working paper. Published by Nagarjuna Publication (P) Ltd. ISBN 978-99946-853-6-3
Xu JC, Shrestha AB, Vaidya R, Eriksson M, Hewit K (2007) The melting Himalayas: regional challenges and local impacts of climate change on mountain ecosystems and livelihoods. Technical paper. International Center for Integrated Mountain Development, Kathmandu, Nepal
Xu J, Grumbine RE, Shrestha A, Eriksson M, Yang X, Wang Y, Wilkes A (2009) The melting Himalayas: cascading effects of climate change on water, biodiversity and livelihoods. Conserv Biol 23(3):520–530

Addressing Climate Change Impacts Through Community-Level Water Management: Case Studies from the South-Eastern Coastal Region of India

Ronjon Chakrabarti, Ni Luh Made Ashanapuri Hertz, Sibylle Kabisch, Alice Reil and Till Wolf

Abstract With regional variations, climate change has significant direct and indirect impacts on the water sector in coastal south-east India. In most cases, it is a combination of changes in different climate signals such as temperature and precipitation as well as non-climatic factors that exacerbate water-related problems for rural communities. Overall, the data from 18 villages show how climate change affects the most basic needs of vulnerable coastal communities, namely water, occupation and health. Thus, adapting to water management is one of the major challenges for rural development in the region. In many communities, significant changes in local water resources have been observed during the last decades that can be linked to changing climatic conditions indicating risks for water management in the future. Combining scientific and local perspectives, the major climate-related challenges for the water sector can be prioritised as growing water scarcity, decrease in water quality and the risk of increased floods. Following vulnerability and needs assessment at community level, the AdaptCap project together with communities, developed and implemented a set of adaptation measures to address the most pressing challenges faced by the water sector: Mitigating water scarcity by

R. Chakrabarti (✉)
Adelphi Research, Eco-India Project Office, School of Water Resources Engineering (SWRE), Jadavpur University, Kolkata, 700032 West Bengal, India
e-mail: chakrabarti@adelphi.de

N.L.M.A. Hertz · S. Kabisch · A. Reil · T. Wolf
Adelphi Research Gemeinnützige GmbH, Berlin, Germany
e-mail: hertz@adelphi.de

S. Kabisch
e-mail: kabisch@adelphi.de

A. Reil
e-mail: reil@adelphi.de

T. Wolf
e-mail: till.wolf@posteo.de

S. Shrestha et al. (eds.), *Managing Water Resources under Climate Uncertainty*, Springer Water, DOI 10.1007/978-3-319-10467-6_14

introducing effective irrigation technologies in Poovula (a) Doruvu and rainwater harvesting for irrigation and domestic water usage in Motumala; (b) Improving *water quality* in the drinking water supply in Desarajupalli and providing a sanitation infrastructure in Govupetta; (c) Reducing *flood risks* from sea water and backwater through the construction of bunds in Kaduvetti and Perumalpettai.

Keywords Climate change · Climate impacts · Water sector · Participatory IWRM · Community adaptation · AdaptCap

1 Introduction

Communities along India's east coast have been struggling for a long time with poor infrastructural and socio-economic situations. Key challenges for the region include insufficient availability of water, improper drainage systems, lack of sanitation facilities, unsuitable solid waste management systems and a limited number of employment opportunities (goi 2008). With the projected climatically induced change in rainfall patterns, increased droughts and higher intensity of cyclones, rural coastal communities in India are becoming even more vulnerable. Yet, measures taken to protect the population and their livelihoods against negative impacts from climate change currently still lag behind what is required, especially in the water sector (GoI 2013).

Adapting water management to a changing climate is one of the major challenges to rural development in the south-east coast of India. The major climatic factors causing these challenges are the increasing temperature and changing rainfall patterns relating to alterations in the monsoon trend.

The related key impacts manifesting along the east coast of India are water scarcity, due to the unavailability and/or inadequate quality of water, as well as floods. These impacts inevitably exacerbate non-climate-related social, economic and demographic stresses, thus limiting opportunities for the communities to make use of their main livelihoods such as agriculture, fisheries and livestock (Fig. 1).

This paper presents an assessment of climate change impacts in the east of India. The assessment is followed by the elaboration of a methodology based on community participation processes for addressing the aforementioned key impacts manifesting in the area. Furthermore, result examples taken up as part of the project "Strengthening Adaptation Capacities and Minimizing Risks of Vulnerable Coastal Communities in India" (AdaptCap 2011–2013) will be displayed.

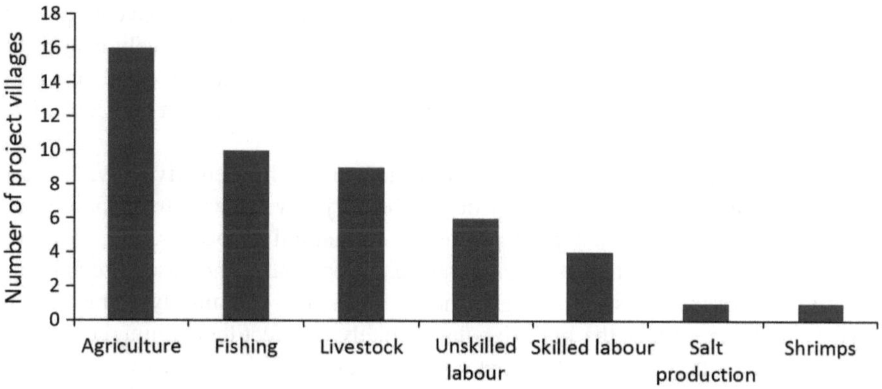

Fig. 1 Livelihoods in the project villages in India (*Source* AdaptCap. Accessed 14 Dec 2013a)

1.1 Climate Change Observations and Projections in India

Since 1970, the average temperature in India has risen by approximately 0.15–0.22 °C per decade, mainly due to intense warming in the decade 1998–2010. In particular, winters and post-monsoon seasons, which have warmed by around 0.80 °C during the last hundred years, have contributed to this trend (IMD 2009, p. 22). According to the Indian Network for Climate Change Assessment, this trend is very likely to continue. Simulations of rising temperatures for the 2030s indicate all-round warming throughout the Indian subcontinent. On the east coast, the annual surface air temperature is set to rise from 28.7 to 29.3 °C. The rise in temperature from the 1970s to the 2030s is of the order of 1.6–2.1 °C. The maximum increase in temperature in all simulations for March, April and May ranges from 1 to 3.3 °C (INCCA 2010, p. 41).

Although no definite trend can be observed for overall annual rainfall, considering extreme rainfall events, a significant increase in intensity during 1980 and 2009 was reported in the sectoral and regional analysis of the Indian Network for Climate Change Assessment (INCCA) for Indian coastal regions (ibid., pp. 34–35). Accordingly, rainfall on the east coast of India is projected to increase in the 2030s corresponding to the 1970s by 0.2–4.4 %. The maximum increase in rainfall is projected for March, April and May. The winter rainfall is projected to decrease (ibid., p. 37).

With regard to storm surges and cyclones, observations since 1961 have revealed that the frequency has decreased on a decadal scale. However, the intensity of cyclones is projected to be increasing (ibid., p. 44, 49). Compared to the west coast of India, the east coast is more prone to disasters linked to cyclones. The states Andhra Pradesh and Tamil Nadu are among those very vulnerable to cyclone disasters as shown by the past occurrence of cyclones Laila and Jal (2010), Thane (2011), Nilam (2012) and Helen (2013) (NCRMP 2014).

In the past, along the Indian coast, the sea level was observed to have an average rising trend of 1.3 mm/year for a period of 20 years which is relatively near to the global trend. It is generally expected that the expansion of sea water and hence sea level will continue to increase similar to observations from the past (INCCA 2010, p. 132ff.).

Areas along the coast of the south-east of India are highly sensitive towards these climate-induced changes, particularly due to the very low capacity to adapt, i.e. lack of infrastructure, poverty, high dependency on natural resources, etc. With the projected sea level rise, increased temperature, unseasonable increased precipitation as well as more intense storms surges and cyclones, the community along the east coast of India live under the high risk that is likely to affect their living conditions.

Due to the many facets of climatic challenges faced by the east coast India, urgent action is called for to support the community in the area to adapt to all these changes. This support may manifest itself through an effective and efficient community-based climate change adaptation intervention. Nevertheless, questions remain: what are the most effective adaptation measures for each of the affected areas? To be effective, the measures must include an increase in the adaptive capacity of the coastal community of south-east India through increasing awareness and knowledge, provision of acceptable technologies, leverage of financing capacity, assuring sustainable operation, coordinating relevant stakeholders and, lastly but most importantly, building a sense of ownership in the infrastructure.

2 Background, Methods and Materials

2.1 The AdaptCap Project

Although there is a consensus on the general regional climate change impact, the actual technical implementation of measures for adaptation on the local level lacks effectiveness and best practice cases for replication are rare. The following elaborations showcase results of in-depth vulnerability and needs assessment and participatory project development and implementation approaches. Thus, sustainable climate change adaptation measures in the field of integrated water resource management are presented for replication in communities with similar challenges and profiles. The elaborations are primarily based on the findings of the AdaptCap project in India.

The AdaptCap project started in early 2011 and finished at the end of 2013 with the aim of increasing the knowledge, planning and adaptation capacities of coastal communities concerning climate change mitigation, and disaster risk reduction in Andhra Pradesh and Tamil Nadu, India (Fig. 2). Ultimately, the AdaptCap project supported the Millennium Development Goals 1 and 7 of lasting poverty reduction and to ensure environmental sustainability as well as achieving the targets under the Indian National Action Plan on Climate Change. To address these goals, the specific objectives of the project were designed and carried out as follows:

Fig. 2 Target Areas (*Source* AdaptCap. Accessed 14 Dec 2013b)

- To create and implement measures for climate change adaptation linked to mitigation and disaster risk reduction in coastal communities, and support local authorities in addressing related challenges.
- To develop and carry out pilot initiatives on adaptation and mitigation in rural communities (e.g. by improving local infrastructure).
- To improve capacities and decision-making skills of local bodies and communities on adaptation and mitigation and provide advisory services on disaster risk management.
- To create public awareness and improve regional and global visibility of the project and stimulate networking.

Through the implementation of the AdaptCap project, it was expected that the affected areas located in Tamil Nadu and Andhra Pradesh, in the south-east coast of India, would be able to receive the most appropriate intervention in adapting to the changing climate. Furthermore, based on the project design, the AdaptCap project was believed to be able to serve the needs of increased awareness and knowledge in the community, providing appropriate technologies on the ground, and most importantly the assurance for a sustainable operation even at the end of the project.

2.2 *The AdaptCap Approach: A Methodology for Developing, Selecting and Implementing Local Adaptation Projects*

As part of the AdaptCap project, a methodology that ultimately aimed at establishing sustainable adaptation measures in the target communities was prepared (Fig. 3):

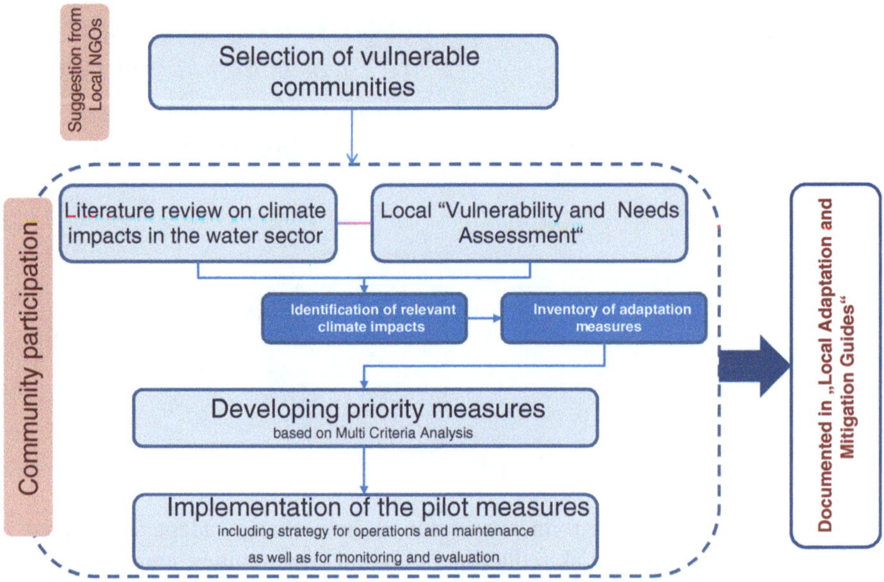

Fig. 3 AdaptCap approach for addressing climate change impacts

- Following a selection of communities, the identification of relevant climate change impacts was approached from two sides:
  (1) A literature review and
  (2) A vulnerability and needs assessment carried out within each of the communities.
- The identification of the relevant impacts also included an impact ranking for each community.
- An inventory of adaptation measures related to different impacts was then set up, followed by a selection of measures based on multi-criteria analysis.
- Finally, the measures were only implemented in the community once a strategy for operation and maintenance as well as monitoring and evaluation had been set up.
- For each community, the findings were collected in a "Local Adaptation and Mitigation Guide" serving as a living reference document for the community and the NGOs working there. These guides are hybrid documents that include background information and theory on climate change adaptation, mitigation and disaster risk reduction, as well as (interim) results from the consecutive AdaptCap project steps.

2.3 *Identification of Relevant Climate Change Impacts in the Water Sector*

The changes in the different climate signals are today already affecting India's population and environment. Future changes and associated risks will further increase the pressure to adapt to changing climatic conditions as highlighted by studies assessing climate impacts and observations from local communities. This particularly holds true for the water sector in the study region as detailed below.

2.3.1 Literature Review of Climate Change Impacts in the Water Sector

Climate change is impacting on the natural ecosystems and is expected to have a substantial adverse effect in India, mainly for water management in agricultural sector on which 58 % of the population depends for their livelihood. Climatic changes will cause increased frequency of extreme events such as floods and droughts. The sea level rise will increase risks for India's long coast line and habitation with implications for food and water security (GoI 2013, pp. 256–57).

Studies assessing the climate signals discussed above show that a changing climate will have an impact on the regional water cycle of coastal India as a whole (INCCA 2010). At the same time, local water cycles will be affected in terms of quantity and quality. Ranuzzi and Srivastava show that increased temperatures cause an intensification of the water cycle with more extreme variations in weather events and longer-lasting droughts. Furthermore, the expected temperature increase is likely to exacerbate drought conditions during subnormal rainfall years. Regarding the study region, large areas in Andhra Pradesh and some areas in Tamil Nadu are already experiencing recurrent drought with several regions currently experiencing water deficits (Ranuzzi and Srivastava 2012, p. 8).

In terms of precipitation, the recharge of groundwater is directly affected by changing climatic conditions. Groundwater is showing a significant decline in parts of Andhra Pradesh and Tamil Nadu. The agricultural irrigation sector uses 83 % of the captured water in these regions and is therefore expected to suffer most from shortages. Projections for 2050 of run-off in India's river basins indicate that the overall quantity of surface run-off is likely to show a general decrease. Despite the rise in overall annual precipitation, increased evapotranspiration and seasonal effects could have negative effects on river run-off, probably leading to drought (INCCA 2010, pp. 119–120).

Coastal and delta communities will face an increased risk of flooding during cyclones. For all Indian coastal regions, an increase of flooding magnitudes from 10 to 30 % in the 1970s to 2030s is projected (ibid., p. 126). Salinity ingress is already a major problem in the coastal regions of Tamil Nadu. With its possible impact on the hydrological cycle, climate change can further exacerbate the problem of saline water intrusion in the coastal and island aquifers (UNICEF et al. 2013, p.15).

Lastly, rising sea levels and subsequent sea water intrusion into coastal and island aquifers is causing a deterioration in drinking water quantity and quality, leading to the salinisation of agricultural land (INCCA 2010, p. 115). Low-lying coastal areas on the east coast of India are already partially inundated by backwaters, and sea level rise would worsen this impact (ibid., p. 53). The INCCA report concludes that "increase in the salinity of water due to sea-level rise, increase in the intensity of cyclones and storm surges, leading to a rise in waterborne diseases and scarcity of potable water may be the major cause of morbidity in the coastal regions of Tamil Nadu and Andhra Pradesh in the 2030s" (ibid., p. 137).

2.3.2 Vulnerability and Needs Assessment at Community Level

Adaptation activities do not come with a one-size-fits-all solution, but rather require location-specific adaptation approaches that are community based, integrated and innovative and which simultaneously address climate impacts, livelihood improvements and environmental sustainability (cf. DEFRA 2010, p. 4). Therefore, alongside the detailed literature review on the climate change impact, a participatory climate vulnerability and local needs assessment was also carried out within the AdaptCap project. Vulnerability and needs were assessed with a participatory bottom-up approach (with top-down input on climate change information) for each village.

The assessment started with the compilation of a general profile of each community. Next to general socio-economic and geographic information, queries on existing infrastructure with respect to water management included information on water sources and supply, water treatment of the water source and household treatment, waste water discharge and treatment, irrigation sources and techniques. In parallel, the general needs of the communities were assessed to get an initial overview of the concerns of the people at the target location. This was also essential to identify the general, oftentimes pressing needs of the communities and to find synergies with climate change adaptation to ensure that the projects were actually wanted.

The vulnerability and needs assessment activity was carried out with support from the local partner non-governmental organisations (NGOs). They were given training on the topics to further apply the participatory rural appraisal tools in the communities. Large group meetings were complemented with focus group discussions (e.g. according to livelihood as in fishermen or farmers and/or according to gender). Additionally, community task forces were formed in the project areas of Andhra Pradesh and Tamil Nadu to support local climate vulnerability and needs assessments. These task forces helped to institutionalise the project by increasing the ownership and willingness of key decision makers. Several town Panchayats signed a formal memorandum of understanding on the goals and steps with the project partners.

The next step was identifying past, current and future climate change impacts on the different livelihoods based on the observation of local communities. In many communities, significant changes in local water resources observed during the past few decades may be linked to changing climatic conditions, indicating risks for water management in the future.

Table 1 Local water problems linked to climate change

Desarajupalli	Scarcity of drinking water, existing drinking water sources are contaminated; climate change puts additional stress on the available water resources
Govupetta	Scarcity of drinking water, drinking water source is contaminated; lack of sanitation further pollutes the drinking water source; climate change puts additional stress on the available water resources and increases risks of vector-borne diseases
Kaduvetti	Agriculture highly vulnerable to flooding during cyclones; salt water intrusion through backwater channels
Motumala	Intense surface water run-off and lack of water for irrigation; output from agriculture reduced due to drought and shorter, unseasonable, intense rainfall and resulting floods
Perumalpettai	Flooding of habitation by sea water in the case of storms and cyclones
Poovula Doruvu	Decreasing rainwater catchment and groundwater stores; water scarcity for irrigation; limited extraction options

According to locals, changing rainfall patterns and increased surface water run-off have led to decreased groundwater recharge and shorter water storage periods for water reservoirs. For example, wells providing freshwater in different villages throughout the year in the past dried up in summer or provided only saline water during recent years. At the same time, water reservoirs in the catchment areas dry up earlier in the summer, storing less water than a decade ago. Freshwater is only available after a rain period and if enough groundwater is being recharged. The decrease in groundwater recharge has in some cases led to salt water intrusion from the sea into the aquifer. Furthermore, the reduced water flow of rivers is in some regions not enough to infiltrate and retain a groundwater level, which can prevent the sea water from intrusion. Table 1 provides examples of local water sector problems linked to climatic changes for six coastal villages (their locations are shown Fig. 4):

Although not all of the water-related problems can directly be linked to climate change, the likelihood of the occurrence of weather-related extreme events affecting the water sector is increasing with climate change. Furthermore, the data suggest that certain impacts could not possibly be attributed merely to a single climate signal but also to various non-climatic environmental, demographic and technological factors. Changes in groundwater level can be linked to increasing temperatures and changing rainfall patterns and the availability of new electric groundwater extraction pumps. Non-climatic factors can amplify climate change hazards and vice versa. Thus, they need to be taken into account thoroughly when discussing climate impacts and designing adaptation measures.

Accordingly, discussions reveal that most of the time the combination of changes in different climate signals as well as non-climatic factors exacerbates water-related problems for rural communities in India. Overall, the data demonstrate how climate change affects the most basic needs of vulnerable coastal communities, namely occupation, water and health.

Fig. 4 Location of the selected Adapcapt pilots (*Background Source* OpenStreetMaps)

2.3.3 Assessment of Climate-Related Challenges for the Water Sector

As elaborated above, there were two different starting points for developing adaptation approaches in the water sector as part of the AdaptCap project: scientific findings on climate projections and potential future climate impacts on the one hand and discussions with local communities regarding their observations and perceptions of weather-related events and climate change on the other. The links between the different climate signals and their direct and indirect impacts on the water sector were discussed in a series of workshops with local stakeholders. In combining scientific and local perspectives, the major climate-related challenges for the communities could be prioritised and grouped as summarised in Table 2:

Table 2 Combined scientific and local perspective of climate-related challenges for the water sector

Growing water scarcity	Decrease/slow groundwater recharge due to altered hydrological cycles with shorter intense rainfalls and faster run-off
	Salt water intrusion (salinisation) due to drawdown of the groundwater table
	Reduction in capacity of surface water sources due to siltation caused by erosion from stronger rain events
	Increased evaporation due to rising temperatures and longer dry periods
	Deterioration of natural water resources due to decreased recharge by freshwater making them unusable for human consumption
	Often, water scarcity is also linked to the anthropogenic impacts of excessive groundwater extraction due to increasing water demands, which may result in higher water requirements due to increased heat and longer dry periods or droughts
Deterioration of water quality	Surface water bodies receiving reduced freshwater recharge can lessen diluted pollution loads from industries, domestic effluents or agricultural run-offs. On the other hand, an increase in the rainfall intensity is expected to cause a higher solid suspension and introduce pollutants (e.g. pesticides) into the water due to soil erosion
	Erosion and sedimentation potentially increases the level of suspended solids in the water and hence make it unsuitable for drinking, washing and irrigation; the water will bury aquatic fauna, preventing the photosynthesis process and affecting the fish population
	Sedimentation will also decrease surface water body capacities and result in decreased contaminant dilution capacity and thus higher pollutant concentrations
	It has further been observed over recent decades that climate-related warming changes water species composition, organism abundance and productivity
Flooding risk	Increased flooding risks are linked to a variety of climatic and non-climatic processes, e.g. coastal erosion and sea level rise but also a change in rainfall patterns with shorter more intense rain events in combination with decreased capacities of surface water bodies
	The increasing trend of flood recently is also driven by land use change and sealing of settlement areas not allowing infiltration
	Flooding is affecting both, quantity and quality of available water resources

2.4 Identification of Relevant Climate Change Adaptation Measures

Based on these challenges, initial concepts and approaches for adaptation, mitigation and disaster risk reduction were developed in a participatory manner with the communities, village leaders, local authorities and city officials. This provided an inventory of potential adaptation interventions for each community with water management being the most prominent topic. In principle, the potential measures should be oriented towards at least one of the three main strategies, which entail avoiding, reducing and sharing the impact damage.

A general inventory of adaptation measures can serve as a sound knowledge base for the project stakeholders. In Fig. 5 below, a matrix-type inventory was set up to link different types of adaptation measures to different climate change impacts. A specific inventory measure was established for each project community.

Matrix for Identification of measures which address climate change impacts (1 = address impact, 0 = has no relevance in addressing the impact)		Water problems	Scarcity of water	Scarcity of irrigation water	Scarcity of drinking water	Scarcity of water for livestock	Decreasing ground water table	Contamination of drinking water	Salination of surface water	Salination of ground water	Piped water supply water becomes saline
	Potential adaptation measures										
Water	**Improve water resource (management) / water reservoir**										
	Improve rain water harvesting		1	1	1	1	1	1	1	1	1
	Renovation of pond / tank / reservoir (e.g. deepening/desilting/bunds...)		1	1	1	1	1	1	1	1	1
	Water reservoir quality conservation systems		1	0	1	1	0	1	1	1	1
Water	**Improve water extraction**										
	Cleaning / renovating / maintaining existing well		1	1	1	1	0	1	1	0	0
	Drilling / deepening of well		1	1	1	1	0	0	0	0	0
	Improve water pumping systems (extraction)		1	1	1	1	0	0	0	0	0
Water	**Water treatment / purification systems**										
	Surface water treatment system		1	0	1	1	1	1	0	1	0
	Groundwater treatment systems		1	0	1	1	0	1	0	0	0
	Desalination system		0	0	1	0	0	0	0	0	0
	Well desinfection		0	0	1	0	0	0	0	0	0
Water	**Improve water storage and supply**										
	Construct ground level storage water tank		1	0	1	0	1	0	0	0	0
	Improve drinking water supply network		0	0	1	0	0	0	0	0	0
	Improve water pumping systems (distribution)		0	0	1	0	0	0	0	0	0
Water	**Improve demand side water management (general / usage / consumption practices)**										
	Technical water saving measures		1	1	1	1	1	0	1	1	0
	Implement technical less water consuming practices; e.g. system of rice intensification (SRI)		1	1	1	1	1	0	1	1	0
Water	**Improve waste water / sewage / drainage / canal system**										
	Improve / renovate sewage system / desilt drainage channels		0	0	0	0	0	1	0	0	0
	Construct sewage systems / canals		0	0	0	0	0	1	0	0	0
Water	**Improve waste water treatment**										
	Provision of house hold soak pits		0	0	0	0	0	1	0	0	0
	Cultivation of wetlands / sedimentation ponds		0	0	0	0	0	1	0	0	0
	Waste water treatment plant		0	0	0	0	0	1	0	0	0
Water	**Improvement of flood protection**										
	Construction / strengthening of coastal belt / sea bund		0	0	0	0	0	1	1	0	1
	Heighten / strengthen river / canal bund / revetment		0	0	0	0	0	1	1	0	0
	Heighten / strengthen bunds to protect fields		0	0	0	0	0	0	0	0	0
	Heighten / strengthen bunds to protect salt pans / salt forms		0	0	0	0	0	0	0	0	0
	Construct / renovate / desilt storm water drains / flood water diversion channe		0	0	0	0	0	1	1	1	1
	Taking other technical measure to protect water system infrastructure		0	0	0	0	0	1	1	1	1

Fig. 5 Example of Inventory Matrix for the AdaptCap Project (for complete version, see http://www.hrdp-net.in/live/hrdpmp/hrdpmaster/hrdp-asem/content/e199/e11168/e41814/e41817/AdaptCap_ruralVNA_print_Impacts_measures.pdf)

2.5 Project Selection Criteria

After developing an inventory with adaptation options, the measures need to be prioritised to make sure that the resources available are used efficiently for the most urgent adaptation measures. Following the preselection of alternatives, selected adaptation measures can be described, evaluated and compared in more detail. Defining appropriate evaluation criteria is essential for the assessment and prioritisation of adaptation options. Deciding on these criteria in a participatory manner with all key parties involved ensures that the selected options are then carried out appropriately.

Proceeding from the long list of options, three measures were preselected for each project community and inserted into an "evaluation format". Multi-criteria analysis was chosen as a quick and appropriate method for assessing and comparing adaptation options. A two-level evaluation approach was then applied:

At the first level, the pilots were evaluated locally in the given evaluation format using the criteria:

- *General eligibility* (i.e. climate change relevance, approval from community and other relevant stakeholders and contribution to sustainable development)
- *Sectoral eligibility* (i.e. linking vulnerability reduction to a specific sector, e.g. coastal zones, integrated water resource management, and drinking water as well as description of measure; support documents)
- *Cost and time requirements* (i.e. capital expenditure, operational expenditure, construction time and preferred suppliers/PPP partners)

At a second level, a third-party evaluation with evaluation criteria followed. Among others, the key criteria given in the third-party evaluation were as follows: vulnerability reduction, acceptance/feasibility/costs and possible side effects. These criteria were also agreed upon in advance with the key stakeholders.

The communities took part in the selection process due to their awareness of local weather and climate-related problems and knowledge of possible negative impacts on major livelihoods. Following a series of training programmes on climate change and adaptation in the villages by the local NGOs, the communities were involved in the pilot development and assessment project.

A series of discussions were held with the task forces, and technical support has been obtained from other stakeholders such as the Rural Water Supply Department, Irrigation Department, and Revenue, Irrigation and Fisheries Departments. Officials from these departments visited the proposed pilot sites and provided technical support to develop technical specifications.

2.6 Implementation of the Selected Measures: Ensuring Sustainability

An implementation strategy was developed with the communities wherein the roles of the contractors, local authorities and technical experts were clearly discussed and defined. Strategies were also developed for obtaining approval from various departments, material procurement, community contribution, quality check-ups by community-based organisations, mobilisation of cash and in-kind contribution from the communities, together with technical support. Detailed terms of reference (ToRs) including technical and contractual specifications for the suppliers were developed. The European Union procurement procedures were followed, and transparency was maintained for the selection of suppliers and awarding of contracts. Community resolutions were submitted with each of the measures to increase sustainability and ownership of the pilots by the community.

To ensure longevity of the installations, operations and maintenance (O&M) training was implemented for the local partner NGOs in each village and specifically for each technology. The training included technical background information, operation and maintenance procedures and organisation, and an exemplary O&M plan for one pilot village. O&M plans were developed for each community after the workshop and included in the respective adaptation and mitigation documentation for each village.

3 Results and Discussion: Water Management Case Studies from the Indian East Coast

Following the process of development, 18 efficient adaptation measures were designed to meet the elaborate criteria for addressing the needs of local communities to cope with climate-related challenges. All pilots have been implemented through the AdaptCap project from 2011 to 2013 in Andhra Pradesh and Tamil Nadu and have been monitored in their successful operation in 2013. Many AdaptCap community interventions resulted in improved water management activities ranging from projects addressing safe drinking water and availability of clean water sources for domestic and agricultural purposes to the provision of sanitation and protection against sea water intrusion.

In this section, six successful water management projects are described. These pilot projects are proven to increase the adaptive capacity of the coastal communities in dealing with the changing climate while at the same time safeguarding their prime livelihood. In total, six examples are presented for addressing three main impacts of climate change on the water sector in the region as follows:

1. *Water scarcity* is addressed by
   (a) Rainwater harvesting for irrigation and domestic water usage in Motumala helping farmers to keep their farm irrigated at any time of the year by capturing the run-off rainwater during intense rain.
   (b) Effective irrigation technologies in Poovula Doruvu for agricultural production helping to save water and to irrigate more crops sustaining the livelihood of the community.
2. *Water quality* deterioration is addressed by
   (a) Improvement of the drinking water supply in Desarajupalli by removing pathogens from the open well water. The installation of the water treatment unit not only improved the quality of water but also ensured its constant supply;
   (b) Provision for the sanitation infrastructure in Govupetta prevents the pollution of surface water from open defecation practices. The measure protects the drinking water catchment area as well as reduces the risks of waterborne diseases.
3. *Flooding* is addressed by
   (a) Preventing inflow of saline water in Kaduvetti with the construction of a backwater bund protecting agricultural land from salinisation. The measure helps to improve land productivity and increase village crop yield.
   (b) Protection from cyclones and flooding of sea water in Perumalpettai by coastal bund construction to protect lives, belongings and the local economy of the community.

3.1 Water Scarcity

Climate change-related phenomena such as increasingly heavier rains within a shorter time span or unseasonable rains are putting agricultural resources under pressure. The AdaptCap project contributed to climate-proofing local livelihoods, particularly in the cross-cutting water–agricultural sector, by implementing adaptation measures to support continuous irrigation throughout the year, independent of changing seasons.

3.1.1 New Water Sources for the Local Agriculture in the Village of Motumala, Andhra Pradesh

Located in the Prakasam District of the State of Andhra Pradesh, Motumala has a population of 4,300. Water is particularly crucial to sustain the livelihood of the villagers as 325 families are employed in the agricultural sector. The villagers mainly use groundwater for their daily domestic needs. Meanwhile, surface water from ponds is utilised for agricultural and shrimp production activities.

During the last few years, drought as well as flash floods caused by shorter, unseasonable, and more intense rainfall impacted agricultural productivity. Such changes could be related to climate-induced changing rainfall patterns as well as heavy precipitation events, and these are projected to increase. Despite the scarcity of water for agriculture, the demand for irrigation rose due to higher temperatures and longer drought periods. The 600-acre government-built irrigation tank which was supposed to supply 14 villages in the region had only enough water capacity to irrigate the cultivated fields during the dry season, from January to February. Meanwhile, in the rainy season, rainwater run-off flowed through a drainage canal past the tank and was directly discharged into the sea.

Through a participative vulnerability and needs assessment in Motumala, growing water scarcity, particularly for irrigation, was identified as a major risk to the most important economic activities in the village. Based on the fact that water was scarce during drought periods but went to waste during the rainy season, farmers proposed capturing rainwater, which could be used as a freshwater source to irrigate the fields. This would make water continuously available for irrigation.

To harvest the rainwater that runs through the drainage canal in Motumala's agriculture fields during the monsoon period, members of the community dug a pond in the irrigation canal, which is 150–200 m away from the irrigation tank. The pond fills up during heavy downpours and pumps and lifts the water into the irrigation tank. The rainwater harvested can be used to irrigate 200 acres of land and water an additional 400 acres during times critical to paddy cultivation. Figure 6 illustrated the process of rainwateer harvesting and its distribution into the fields.

As a village famous for its vegetables, Motumala is now able to use the fresh rainwater to protect standing crops during drought as well as during erratic monsoons. The overall cost of this measure was partly shared by the community. This investment allowed the community to increase their rainwater storage capacity by approximately 150,000 m^3 per year and thus have constant water available for irrigation. This resulted in a higher crop yield of up to 140 tons per year. The measure

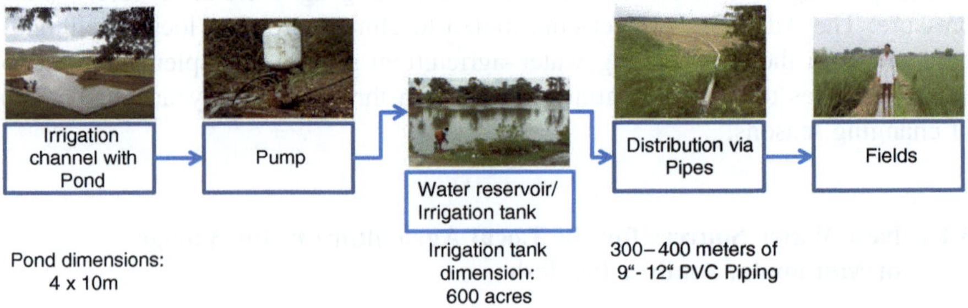

Fig. 6 Technical scheme of implemented measure in assuring sufficient irrigation storage (*Source* Irrigation channel, pump and field: Flickr doc; water reservoir and distribution pipe: Wikipedia). Accessed 22 Dec, 2013

has helped over 200 families in the area to safeguard their livelihoods. The AdaptCap project activities contributed to sustaining and climate proofing the agriculture of Motumala.

Box 1 Potential of Harvesting Rainwater for Irrigation
Changing rainfall patterns increase the surface run-off due to short intense rain events. They quickly and directly run off into the sea without having enough time to recharge the freshwater sources. Thus, there is a huge potential for harvesting rainwater, which can compensate for the decreased water resource recharge.

3.1.2 Saving Water and Energy with an Efficient Irrigation System in the Village of Poovula Doruvu, Andhra Pradesh

Poovula Doruvu of the Annagaripalem Panchayat, Kavali (rural) Mandal, is home to 554 farmers. Most families earn their income from floriculture and kitchen gardens. Each family cultivates flowers, vegetables, paddy and groundnuts in their 0.15–0.25 acres of land. In addition, 12 women's self-help groups manage the agricultural labour and farming.

In the past, rainwater catchment and groundwater storage capacities have decreased and the usable acreage around the village became smaller and less productive. In the face of a projected climate change, uncertainties regarding the future water supply are increasing. Prior the intervention of the AdaptCap project, villagers extracted groundwater using hand pumps or low-powered well motors to irrigate their small gardens. To water the plants, they made use of short hoses or buckets. This technique was inefficient as it consumed high amounts of water and energy, especially when energy supply was scarce in the summer months. Water scarcity and an energy crisis led to low yields and insufficient income for the farmers. In order to safeguard the villagers' main source of income, the introduction of sprinkler irrigation scheme was agreed as the first priority. This measure was expected to save water and energy while simultaneously increasing agricultural productivity in all of the 60 acres of village land.

When the pilot was implemented, 35 sprinkler kits were distributed. Each sprinkler kit included an electric pump, a sprinkler set including pipes of 6 m length, quick action couplings, six sets of sprinkler couplers, sprinkler nozzles, riser pipes as well as enough hose for about an acre of land. One sprinkler was used jointly by two families. The scheme is described in Fig. 7.

This adaptation measure successfully stabilised horticulture productivity and community income and even made it possible to use new land for cultivation. It has made it possible to irrigate additional new fields and broaden the crop base to include maize, gram, sweet potatoes and other crops.

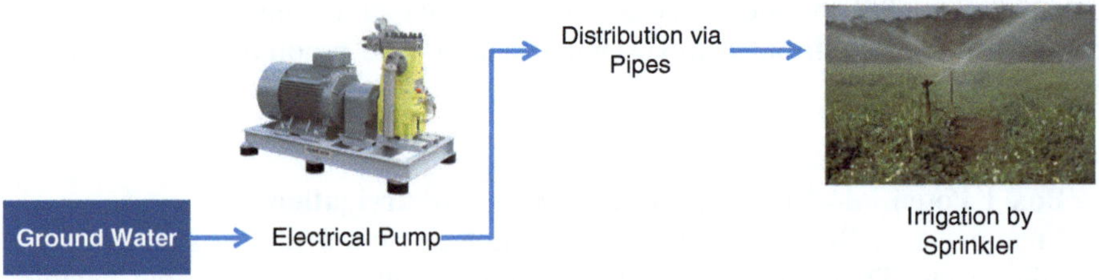

Fig. 7 Sprinkler irrigation scheme (*Sources* Electrical pump: Wikipedia; Irrigation by Sprinkler: Flickr doc). Accessed 20 Dec 2013

With technical assistance and a sprinkler sharing plan developed together with local partner NGOs, the distributed sprinkler kits decreased the use of energy and water while improving productivity of the land. The sprinklers now irrigate 60 acres of sandy land and have helped to improve the employment situation as well as food security in the village. This particularly benefits marginalised social groups, such as Dalits.[1] The sprinkler systems decrease the burden on human labour as well as water resources. Based on estimates, each sprinkler unit saves approximately 13.44 m^3 water per day, i.e. 15–25 % less than before. With 35 units operating 200 working days a year, the pilot project will save a total of 94,080 m^3 annually. In relation to energy, one sprinkler unit saves 5 kWh energy per day, meaning 35,000 kWh of energy is saved per year. This project helped prepare the villagers of Poovula Doruvu for future water shortages while contributing to a sustainable development in a changing climate.

Box 2: 2/3 of the Cultivated Land in India is Suffering from Water Scarcity

While the pilot project helped prepare the local population of Poovula Doruvu for changing rainfall patterns and increasing temperatures, it is also in line with the sustainable agriculture strategy of the Indian Government. India's National Action Plan on Climate Change stated that 2/3 of the cultivated land is suffering from water scarcity, and thus, the *efficient use of water emerged as one crucial issue to be addressed* (GoI 2010, p. 23).

3.2 Deterioration of Water Quality

With a changing climate and current water consumption habits, safe drinking water sources become scarcer. Villagers are forced to consume contaminated well water or purchase expensive filtered water during the dry summer. Additionally, communities

[1] Group of people considered as belonging to the lowest cast in India.

are not encouraged to change their polluting habits, which can worsen the quality of local water resources (e.g. practice of open defecation, which contaminates the catchment area of the local water source).

3.2.1 Making a Community's Water Supply Safe and Reliable in the Village of Desaraju Palli, Andhra Pradesh

Desaraju Palli is inhabited by 650 people. It is located in the State of Andhra Pradesh approximately 16 km from the coast of the Bay of Bengal. The village is equipped with one groundwater well, two water tank reservoirs fed by the river and two canal-filled ponds. They were the main water sources for the villagers' daily life. These sources were not only scarce but also highly contaminated. The piped water for Desaraju Palli arrived irregularly (once every three days in the dry months) and was contaminated with bacteria. The rural water department pumped water to the village directly from the Gundlakamma River to the Alluru reservoir. It was distributed to the villagers without being treated beforehand. The community was accustomed to using this tank water for domestic purposes. Due to water quality issues, villagers had been purchasing water cans for drinking for the last two years. Those who could not afford bottled water were forced to drink untreated and highly contaminated well water supplied from the canal-filled pond reservoir. This inevitably resulted in waterborne diseases occurring in the community.

Improving the drinking water situation was prioritised during the community meetings in 2011. The selected intervention was implemented by installing a low-carbon, climate-resilient water filtration plant near the village well. The filtration unit operates by pumping the water from reservoir (pond) to the filtration unit using a solar pump. A combination of roughing filter and slow sand filter is used to treat the turbidity and bacteria contamination. The filtered water was stored in an elevated tank from where it is directly distributed (Fig. 8). Additionally, the water is tested locally with a water quality testing kit to control the final quality. Depending on the initial water quality and performance of the system, the water is disinfected at the end.

This pilot project addresses both climate change adaptation and mitigation by providing a clean, climate-resilient water supply supported by solar pumps. The technology is resilient to climate change impacts on energy supply as the pumps gathering and distributing the water run on photovoltaic power, which produces 10 kWh per day. Furthermore, the measure provides an alternative source for a clean and safe water supply on a daily basis. This, in turn, improves the community's health by mitigating waterborne diseases. Approximately 650 people are now supplied with safe water. Moreover, this intervention has resulted in an increased quality and constant availability of 15 l of clean water per day per capita (10 KLD[2]) as well as the extra 1.5 h per day, which they previously had to spend collecting water.

[2] Kilo Litres per Day.

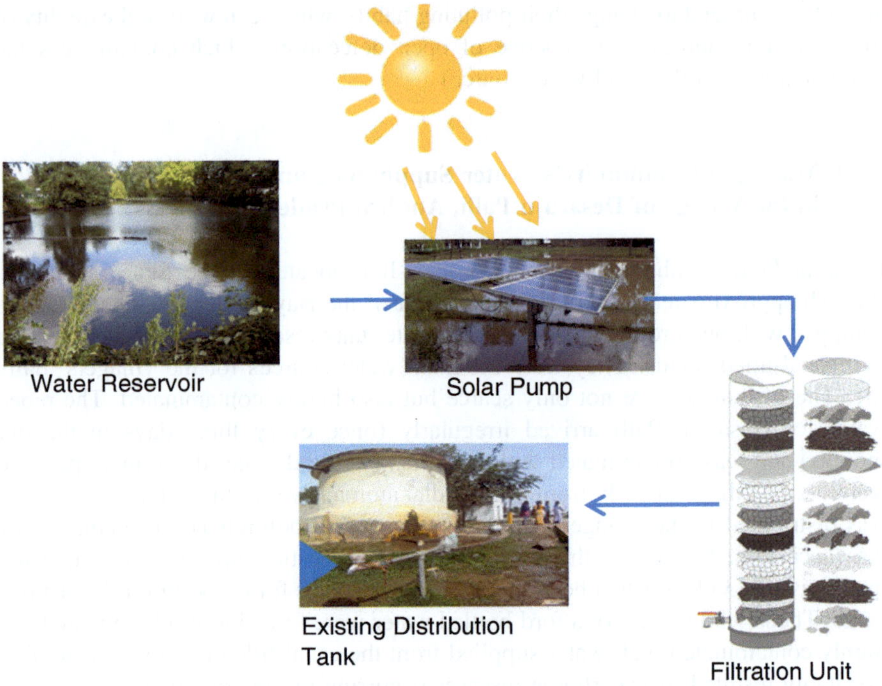

Fig. 8 Technical scheme of the implemented water filtration plant (*Sources* Water tank: Geograph doc.; Solar pump and Filtration unit: Wikipedia; Distribution tank: AdaptCap (GOI 2013b). http://www.geograph.org.uk/photo/2605532. Accessed 20 Dec 2013c

Box 3 Any Water Treatment System Needs Proper Operation and Maintenance

Filtration systems require regular O&M activities to assure that the water treatment plant continuously operates without any interruptions. Systems with simple O&M activities had to be identified so that the community could take ownership of the treatment plant. In Desaraju Palli, the community willingly took over the responsibility of operating and regularly maintaining the filtration unit, while the local Panchayat takes care of the distribution system. In general, 95 % of the O&M activities can be conducted by the community members themselves. Only maintenance of the solar pump and its electric set-up is done by external service providers.

3.2.2 Water Resource Protection and Improving Sanitary Infrastructure at Govupeta, Andhra Pradesh

Govupeta is a fishing village located in the district of Visakhapatnam. The village is a home to 345 people. Insufficient sanitation infrastructure was available in the village. Government community bathrooms in this district were not maintained and had instead been used as cattle sheds due to preference for open defecation. Prior to the AdaptCap intervention, the village women used a field owned by the Municipal Corporation as the communal toilet, while men used the beach area. The community faced sanitation-related health problems. Health risks associated with deficient sanitation could increase with climate change contributing to vector-borne diseases. The catchment area of the water supply for the adjoining villages was also affected and led to deteriorating water quality.

Through the participative process of AdaptCap, village women leaders and others pinpointed the urgent need for individual toilets like their Chinna Uppada neighbours to the north. Responding to the needs of the community, the installation of latrines was the selected and prioritised adaptation measure.

Dual pit individual latrines were installed throughout the village according to a land use plan agreed upon by the pilot project committee and families using them (Fig. 9). In total, 80 toilets were provided by the project with the community covering 45 % of the costs. Each latrine is shared among two families (12 people).

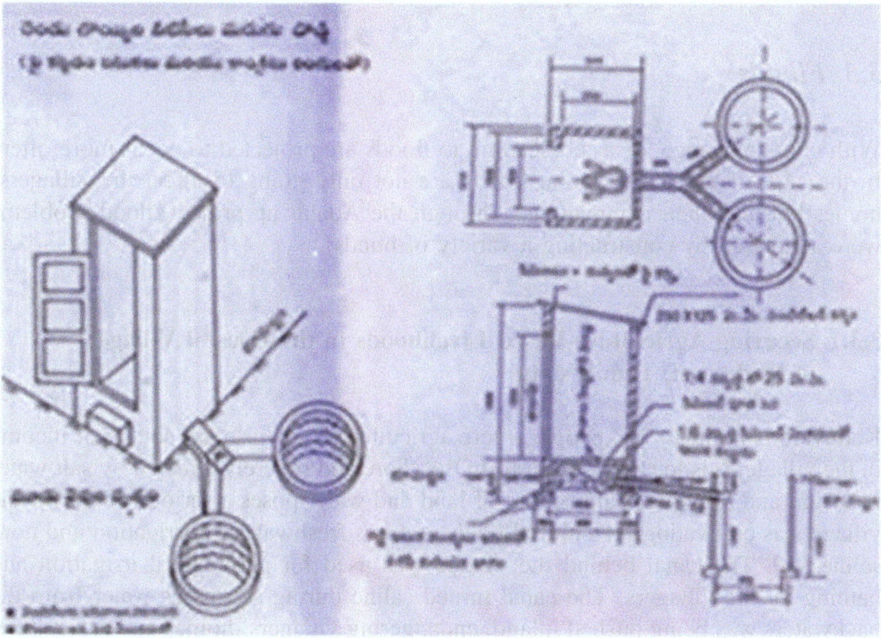

Fig. 9 Technical drawing of dual pit latrines distributed to the community (*Source* AdaptCap)

This intervention addresses the sanitation needs of the community. It prevents unsafe open defecation for women and men and avoids the contamination of water sources within the village as well as in the downstream village of Chukkavani Palem. The installed latrines help to prevent diseases from spreading and thereby avert the negative health effects of climate change. Accordingly, health risks for the users and especially women and children are reduced. Furthermore, the presence of the new latrines has resulted in the availability of an additional 25 acres of land, which was formerly used for open defecation. This increased acreage also ensures a safe catchment area.

Box 4 Adapting to Climate Change by Improving Sanitation
Health risks associated with deficient sanitation are aggravated by climate change. As part of their vision for 2030 on making water supply and sanitation resilient to climate change, the World Health Organization (WHO) emphasised that sanitary toilets are crucial to avoid an overburdening pollution load in the water bodies (WHO 2009). With an increasing temperature and reduced recharge of surface water bodies, the water is less diluted, which leads to the deterioration of the overall water quality. Providing adequate sanitary infrastructure can help to safeguard the catchment area and reduce the probability of waterborne diseases occurring from domestic effluent contaminating the water.

3.3 Floods

With climate change, disasters leading to floods are projected to occur more often. In the case of east coast India, floods are not only going to affect the villagers' homes but also their occupations. Through the AdaptCap project, flood problems were addressed by constructing a variety of bunds.

3.3.1 Securing Agriculture-Based Livelihoods in the Coastal Village of Kaduvetti, Tamil Nadu

Kaduvetti is home to 273 people where agriculture is the primary source of income in the village. It is located 2.5 km from the shore and severely affected by salt water intrusion and flooding. Salinisation of land and water poses a serious threat for the villagers as cultivating agricultural fields requires freshwater for irrigation and non-saline soil. The canal behind the village was used for agricultural irrigation and bathing by the villagers. The canal turned saline during storms as water from the backwaters was being pushed inland, endangering villagers' homes and agriculture. However, during recent years, the canal became brackish for longer periods of time, putting all agricultural activities of the village at risk. The agricultural office gave interest-free loans of fertiliser to farmers, but these loans were contingent upon annual harvest returns. If harvests were lost, the village would not be able to repay loans and secure another loan for the next season, which meant a serious risk of loss of income.

With increasing temperatures and more frequent and intense floods, climate change is expected to contribute to drawing saline groundwater at least six feet below the surface. This would make shallow rainwater-fed irrigation ponds the only source of domestic water for the community. But even these sources were flooded by brackish water from the canal.

Based on the participatory selection process, village leaders called for a ca. 1-km-long bund along the backwater shore to protect their fields from floods and salt water intrusion. The height of the existing bund was increased to 1.3 m and the width to 3 m. New soil was deposited for the foundation. Overall, 5,200 m^3 of soil, gravel and sand were used to construct the bund, guaranteeing high stability. Moreover, planting works have been carried out to prevent erosion of the bund. The road gradient was re-established.

The construction of the canal bund helped keep it from flooding homes, fields and freshwater supplies. It provides the people of Kaduvetti with more secure water sources, allowing the village to produce at least one good harvest per growing season.

The measure was financed jointly by the community and the AdaptCap project. It significantly helps safeguard the people of Kaduvetti and their livelihoods by strengthening local capacities to cope with current and potential climate-related impacts such as storm surges and flooding. In total, 60 families and 76 acres of agricultural land are now being protected. The previously barren land can be desalinated and recultivated, giving the farming families the chance to sustain their lives. Moreover, 37 tons more groundnuts are estimated to be harvested annually. After harvesting the groundnuts, the land can then be used for paddy cultivation, which makes it possible to harvest a second crop. The bund can also be used as a road and prevents homes from being destroyed during floods.

Box 5 Combating Land Salinisation

According to the Central Soil Salinity Research Institute, India, 13,231 ha of land in Tamil Nadu are now affected by salinisation (GoI 2013). The construction of a wing-wall river bund protects not only homes and water sources from the inflow of salty water but also agricultural land. With the help of the bund 76 acres of previously uncultivable saline land could be recovered, increasing crop production. Even though the measure only improved the situation in a small area of land in Tamil Nadu, this adaptation option could be replicated rapidly and cost-efficiently in other areas.

3.3.2 Protecting the Livelihoods of the Fishing Communities of Perumalpettai Village, Tamil Nadu

Perumalpettai is a fishing and agricultural village located in the Nagapattinam District of the State of Tamil Nadu, India. The village is located along the low-lying coastal area of the Bay of Bengal, 500 m from the coast. It is inhabited by 390 families with a total population of 1,500. The village was rebuilt after the Tsunami in 2014 in a traditional manner with additional new structures and is still used by fishermen and their families. The original Perumalpettai village, only a few metres from the coast, as well as the post-tsunami village 500 m from the coast, could both be flooded during storm surges. With climate change, cyclones are projected to become more intense, thus posing a higher risk of flood disaster for the village. Moreover, fishing boats and tied nets from the fishermen line up on a narrow strip of the beach and may easily be destroyed by the floods. The lack of infrastructure in the village, exacerbated by the effects of changing climatic conditions, puts the properties and livelihoods of the villagers at an alarming rate risk.

A joint community decision prioritised the construction of a bund as an adaptation measure to protect the villagers from suffering severe losses during storm surges and cyclones. Consequently, a coastal bund was constructed just behind the dunes over a length of 700 m. It was connected to the existing bund and became part of the dune structure on the north side of the village. To ensure particularly high stability, the bund was constructed at a height of 4 m and a width of 6 m, planted with undergrowth on the slopes to prevent erosion. The bund was built using stiff clay, hard red earth and gravelly soil for the foundations.

The constructed bund is now protecting over 1,500 inhabitants, village property and fishing equipment from future storm surges and flooding. It also safeguards ground and surface water from intruding sea water. With the constructed bund, 25 acres of arable land is protected from floods and salt water. In addition, 60 families and a school building are also directly protected. The bund allows for agricultural plots further inland contributing to the efforts of a sustainable development in Perumalpettai under a changing climate.

The participation of the community and the endorsement of the local authority made this pilot successful. The project's concept and preparation assured the necessary political buy-in. The combination of two adaptation measures increased the adaptive capacity—the bund construction and the digging of a pond, while at the same time an additional freshwater source was made available.

Box 6 Cyclones Along the East Coast of India

In December 2011, cyclone Thane hit the south-east coast of India. Reportedly, more than 40 people were killed as the disaster took place and damages to local infrastructure were estimated to exceed that of the tsunami in 2004 (NDTV 2011). As the cyclone passed, an increasing occurrence of storm surges continues. With the growing knowledge of society regarding climate change as well as the toll of the impact, the coastal bund was constructed with the support of the community. A joint contribution from the villagers amounted to 20,000 INR and helped safeguard the properties of villagers including fishing equipment which is worth up to 500,000 INR.

4 Conclusion

During the final project phase, a number of the lessons learnt were identified as being essential for the success of the project.

Firstly, the AdaptCap project demonstrated how local perceptions can lead to a better understanding of local climate-related vulnerabilities and facilitate discussions about climate change adaptation within the communities. Indigenous knowledge is essential in developing sound community-based adaptation measures, not only in connection with different perceptions of weather-related changes but also in connection with the local water cycle impact. If communities develop and present their own conclusions, this supports joint discussions of climate-related hazards and adaptation measures and helps to overcome barriers in understanding and acting upon climate change projections and uncertainties.

Secondly, it became clear that the start phase of the project on the ground, assessing the climate-related vulnerabilities and needs of the communities, has to receive special attention to ensure that actual problems are addressed. The focus should be on applying a transparent and participatory approach and investing sufficient time to carefully identify climate-related risks and development needs jointly with local stakeholders. Only then can communities develop an interest to support the implementation of the project to create a sense of belonging, sustaining the long-term impact of the project.

Furthermore, during the development and selection of adaptation pilot projects, it is important to encourage community members to identify short-, medium- and long-term responses to climate change that connect to existing knowledge, activities and organisational structures to aid risk management. The criteria for selecting projects and implementation should be predefined with key stakeholders and include sustainability, life cycle costs and feasibility as well as maintenance aspects. Clear responsibilities for operation and maintenance are to be agreed upon and documented within the communities before pilot projects are implemented. A sustainable business model, for the long-term availability of funds, needs to be secured. For the long-term success of the operation and maintenance of adaptation measures, it is important to raise awareness of potential maintenance problems and solutions and to strengthen local knowledge for the operation of the respective adaptation technologies.

Finally, single adaptation projects at the local level, such as the six pilots described above, will only have a limited effect on the overall resilience of the population in the region. However, the integration of climate change adaptation approaches at multiple levels and enhanced dialogue between different actors can strengthen overall adaptive capacities to cope with climate change. Therefore, one objective should be to integrate plans on local adaptation with existing plans on for example disaster risks reduction (e.g. District and Village Disaster Management Plans) as well as, in the long-run, adaptation strategies developed at State and district levels (e.g. State Action Plans on Climate Change) to further foster coherence and effectiveness. Applying the "climate lens" when planning infrastructure investment and systematically climate proofing any development efforts will be key to protect people and their livelihoods from future climate-induced risks.

The information on vulnerability and needs gathered at the local level provides an idea of the climate-related risks and adaptation measures most frequently identified. Local adaptation guidance and planning documents like the ones developed under the AdaptCap project can help to track the findings and decisions regarding climate change adaptation activities at the community level. They can also provide the basis for future project funding applications. The data gathered at the local level can help district-level officers not only to prioritise future adaptation actions in the water sector, but also to identify the most relevant offices for mainstreaming climate change into their existing activities. The pilot measures implemented can serve as a point of reference to roll out climate change adaptation activities on a larger scale. Support for similar measures could be included in district, state-specific and central-level funding schemes.

The AdaptCap project gathered a group of initiatives in order to serve as a climate change adaptation platform. This platform focuses on the upscaling and replication of successful pilot projects by identifying funding sources. The AdaptCap project also generated a projects Website[3] on which proposals for adaptation measures that require funding can be showcased and linked to donors also after the project period.

References

AdaptCap (2011–2013) Climate change adaptation project implemented by GIZ-ASEM, India. http://www.adaptcap.in. Accessed 13 Mar 2014

AdaptCap (2013a) Livelihoods at the project villages in India. Available at http://www.adaptcap.in/. Accessed 14 Dec 2013

AdaptCap (2013b) Target area map. Available at http://www.adaptcap.in/. Accessed 14 Dec 2013

AdaptCap (2013c) Distribution tank. Available via http://www.adaptcap.in/projects/desaraju-palli-adaptation-project/. Accessed 20 Dec 2013

Department for Environment, Food and Rural Affairs (DEFRA) Adapting to Climate Change Programme (2010) Measuring adaptation to climate change—a proposed approach (p 4). http://archive.defra.gov.uk/environment/climate/documents/100219-measuring-adapt.pdf. Accessed 22 Dec 2013

[3] www.adaptcap.in/projects.

Government of India (GoI), Prime Minister's Council on Climate Change (2008) National action plan on climate change (p 35). http://pmindia.nic.in/Pg01-52.pdf. Accessed 22 Dec 2013

Government of India (GoI), Ministry of Agriculture (2010) National mission for sustainable agriculture, (p 23). http://agricoop.nic.in/Climatechange/ccr/National%20Mission%20For%20Sustainable%20Agriculture-DRAFT-Sept-2010.pdf. Accessed 22 Dec 2013

Government of India (GoI), Central Soil Salinity Research Institute (2013a) Extent and distribution of salt affected soils in India. http://www.cssri.org/index.php?option=com_content&view=article&id=122&Itemid=126. Accessed 22 Dec 2013

Government of India (GoI), Ministry of Finance (2013b) Economic survey 2012–13. http://indiabudget.nic.in. Accessed 22 Dec 2013

Government of India (GoI), Ministry of Water Resources (2013c) Report of the Committee for Drafting of National Water Framework Law (p 3–4). http://mowr.gov.in/writereaddata/linkimages/nwfl1268291020.pdf. Accessed 22 Dec 2013

http://upload.wikimedia.org/wikipedia/commons/7/73/Hammelmann_Stationary_unit_with_electric_motor.jpg. Accessed 20 Dec 2013

Indian Meteorological Department (IMD), Ministry of Earth Sciences (2009) The warmest year since 1901. http://www.imd.gov.in/doc/warm2009.pdf. Accessed 22 Dec 2013

National Cyclone Risk Mitigation Project (NCRMP) (2014) Cyclones and their Impact in India. http://ncrmp.gov.in/ncrmp/Cyclone_Impact.html. Accessed 12 Mar 2014

New Delhi Television Limited (NDTV) (2011) Cyclone Thane to hit Tamil Nadu coast early on Friday morning. http://www.ndtv.com/article/tamil-nadu/cyclone-thane-to-hit-tamil-nadu-coast-early-on-friday-morning-161889. Accessed 22 Dec 2013

Ranuzzi A, Srivastava R (2012) Impact of climate change on agriculture and food security. In: ICRIER Policy Series no. 16. Indian Council for Research on International Economic Relations. http://www.icrier.org/pdf/Policy_Series_No_16.pdf. Accessed 22 Dec 2013

The Indian Network for Climate Change Assessment (INCCA), Ministry of Environment and Forest (2010) Climate change and India: a 4 × 4 assessment—A sectoral and regional analysis for 2030s. http://moef.nic.in/downloads/public-information/fin-rpt-incca.pdf. Accessed 22 Dec 2013

UNICEF et al 2013: UNICEF, FAO, SaciWATERs (2013) Water in India: situation and prospects. http://www.unicef.org/india/Final_Report.pdf. Accessed 22 Dec 2013

Wikipedia (2013) Water reservoir: free use via http://en.wikipedia.org/wiki/File:%E0%AE%95%E0%AF%81%E0%AE%B3%E0%AE%AE%E0%AF%8D.JPG. Accessed 22 Dec 2013

Wikimedia (2013a) Solar pump: free use via http://commons.wikimedia.org/wiki/File:Solar_Panels_in_Farm,_India.jpg. Accessed 20 Dec 2013

Wikimedia (2013b) Filtration unit: Free use via http://commons.wikimedia.org/wiki/File:Water_Filtration_Systems.png. Accessed 20 Dec 2013

World Health Organization (WHO) (2009). Vision 2030: the resilience of water supply and sanitation in the face of climate change. http://www.who.int/water_sanitation_health/vision_2030_9789241598422.pdf. Accessed 22 Dec 2013

www.flickr.com/photos/indiawaterportal/2639960064/. Accessed 22 Dec 2013

www.flickr.com/photos/steve-wilson/1230580583/. Accessed 22 Dec 2013

www.flickr.com/photos/cgiarclimate/8551245564/. Accessed 22 Dec 2013

www.flickr.com/photos/iita-media-library/5908565297/. Accessed 20 Dec 2013

www.geograph.org.uk/photo/2605532 Accessed 20 Dec 2013

[3] www.adaptcap.in/projects.

Climate Adaptation and Governance and Small-Scale Farmers' Vulnerability Through Artificial Glacier Technology: Experiences from the Cold Desert of Leh in North-West Himalaya, India

Falendra Kumar Sudan and Jennifer McKay

Abstract In the context of ongoing climate change, the critical factors for the livelihood and environmental sustainability of small farmers are the resilience and the capacity to cope using little technology, which needs to be strengthened to manage climate resilience. This paper uses several methods to examine the vulnerability and livelihood interaction between small farmers, especially how they cope and adapt to climate variability using artificial glacier technology. This case study can be used to draw lessons to support the climate adaptation strategies for small farmers and enhance their resilience to livelihood stresses. The study is based on the primary data and information collected from 675 households in 27 villages of Leh (Ladakh) in north-west Himalaya, India, and reveals that 61.18 % of the population is fully supporting their livelihoods from agriculture and allied activities. With increased irrigation potential due to the use of artificial glaciers, food security has been assured to 77.56 % of the poor smallholder households and health vulnerability has been reduced in 31 % of the households. The seasonal migration has declined in nearly two-thirds of smallholder households as a diversification mechanism, thereby improving their livelihood strategies. The use of tactical adaptations by small farmers in response to the persistent droughts such as the selling of livestock, expanding agricultural lands, and the use of relief cash and food has declined to 20.44, 24.74, and 63 % of smallholder households, respectively. However, these measures are unsustainable in the long term. How should the policy makers and other societal stakeholders act in this context? To address the small farmers' livelihood challenges in the context of climate change, the role of innovative technology is vital to improve climate adaptations and governance using a

F.K. Sudan (✉)
Department of Economics, University of Jammu, Jammu 180 006, Jammu and Kashmir, India
e-mail: fk_sud@rediffmail.com

F.K. Sudan · J. McKay
Centre for Comparative Water Policies and Laws, University of South Australia, Way Lee Building, City West Campus, GPO Box 2471, Adelaide, SA 5001, Australia
e-mail: jennifer.mckay@unisa.edu.au

S. Shrestha et al. (eds.), *Managing Water Resources under Climate Uncertainty*, Springer Water, DOI 10.1007/978-3-319-10467-6_15

multidisciplinary approach involving multilateral collaboration among different stakeholders. The presence of social entrepreneurs (the new actors on the adaptation scene) is necessary to bring forth the necessary measures. To this end, better linking of science and technology policies together with other policies should be encouraged. The social transfers through better health care, access to safe drinking water, better sanitary conditions, and improved standards of education and infrastructure are the win-win measures to enhance the community's adaptive capacity, which requires significant amounts of additional investment. Besides, developing the institutional mechanisms for specific adaptation interventions can be one of most effective ways of implementing a plan and achieving the expected outcome.

Keywords Climate change · Vulnerability · Governance · Adaptation · Artificial glacier technology · North-west Himalaya · India

1 Introduction

Climate change is expected to have major consequences in the developing countries (Oxfam 2007; Stern 2006; UNDP 2007; UNFCCC 2007; World Bank 2006) with the greatest negative impact on the poorest section of the population (CPRC 2008) who will be affected first and the most (IPCC 2007), pushing them into poverty traps by creating a vicious cycle of poverty (Ibarraran et al. 2009). Climate change could cause agricultural productivity in general to decline between 10 and 25 % by 2080, and the yield decline could be as much as 50 % in rain-fed agriculture (Cline 2007), which will clearly threaten the achievement of the Millennium Development Goals (MDGs) (UNDP 2007). Small-scale farmers are most vulnerable to the impacts of climate change on agriculture (Easterling et al. 2007) through increased likelihood of crop failure, increase in diseases and mortality of livestock, and/or forced sales of livestock at disadvantageous prices, increased livelihood insecurity resulting in asset sales, indebtedness, out-migration and dependency on food aid, and a downward spiral in human development indicators such as health and education (IPCC 2007), which will further aggravate the vulnerability already associated with subsistence production.

Comprehending the dynamics of vulnerability is as important as understanding the climate itself (Handmer et al. 1999). The vulnerability of a household or community is determined by the socio-economic and political factors (Blaikie et al. 1994). It is the '*ability or inability of individuals or social groupings to respond to, in the sense of cope with, recover from, or adapt to, any external stress placed on their livelihoods and well-being*' (Kelly and Adger 2000: 328). From a natural hazard perspective, vulnerability may be defined as the characteristics of a person, or group in terms of their capacity to anticipate, cope with, resist, and recover from the impact of a natural hazard (Blaikie et al. 1994). In the climate change context, vulnerability is used as an integrative measure of the threats to a system

(IPCC 2001; Kelly and Adger 2000). Therefore, vulnerability is a function of the character, magnitude and rate of climate variation, and adaptation (IPCC 2001). Adaptation is among the key building blocks for a strengthened response to climate change. Considerable attention has been paid to identifying the characteristics that influence the ability to adapt (Smit and Pilifosova 2001) due to the existence of differential vulnerabilities (Bohle et al. 1994). IPCC (2001: 15) recognises that '*even within regions, impacts, adaptive capacity, and vulnerability will vary*'. The vulnerability often highlights the significance of poverty and inequality or differential resource access (Adger and Kelly 1999). The small-scale farmers use complex adaptation processes. Most adaptation practices serve multiple purposes and are strongly interrelated (Smit and Skinner 2002; Adger et al. 2007). Thus, adaptation is an iterative, dynamic, multiscale, and multiactor process (Osbahr et al. 2008). The adaptation practice needs to take into account the linkages between actors and levels (Smit and Skinner 2002). Therefore, adaptation is highly context sensitive. The adaptation policies do not have to start from scratch. People have been managing (or failing to manage) the climatic vulnerabilities for centuries (Adger et al. 2003). Effective adaptation requires institutions to support the enabling policies and systems to ensure their effective coordination so that the policies and systems may target the poorest and most vulnerable and enhance organisational capacity in order to carry out adaptation. The community-based adaptation builds on existing technical knowledge and the coping strategies of the individuals and communities (Chatterjee et al. 2005). The social protection programmes such as direct cash transfer and disaster cash relief are likely to become a key vehicle for supporting adaptation in developing countries (Stern 2007).

The social dimension of climate change is no longer peripheral to science, technology, and innovation (STI). Indeed, STI is being mobilised to address the small-scale farmers' vulnerability and adaptation to climate change. The experiences from the cold desert of Leh (Ladakh) in the north-west Himalaya illustrate the potential of STI to address the challenges of climate change and vulnerability of small-scale farmers through the use of artificial glacier[1] techniques. In Leh, melting water from glaciers has been the only source of irrigation for 80 % of the small-scale farmers. During the last four decades, the region has observed decreased and untimely snowfall and retreating glaciers due to climate change, causing declining irrigation and agricultural productivity of the smallholders, domestic water use,

[1] Artificial glacier is a simple water harvesting technique suited for high-altitude cold deserts that are totally dependent on glaciers. Glacier melts water at different altitudes and is diverted to a shaded area of the hill, facing the north side, where winter sun is blocked by a ridge or a mountain slope. At the start of winter (or November), the diverted water is made to flow onto the sloping hill face, through appropriately designed distribution channels/outlets. At regular intervals, stone embankments are built, which impede the flow of water, making shallow pools. In the distributing chambers, 1.5″ diameter G pipes are installed after every 5 ft for proper distribution of water. Water flows in small quantities with less velocity through the G pipes and freezes instantly. The process of ice formation continues for 3 to 4 winter months, and a huge reserve of ice accumulates on the mountain slope, aptly termed as an 'artificial glacier' (Leh Nutrition Project, www.lehnutritionproject.org).

ecological purpose, and increasing livelihood vulnerability. Due to the short summer season, the small-scale farmers are able to cultivate only one crop per year, which needs to be sown in April or May. If it is not sown at this time, the crop cannot fully mature, resulting in low yield. However, at that time of the year, water in streams is not sufficient. This is because natural glaciers are located at a higher altitude and further away from villages and start to melt in June, which is too late for sowing. Keeping the above conditions in mind, the small-scale farmers devised a unique technique of water harvesting to augment irrigation, called ‘artificial glaciers’. These provide an intricate network of water channels and dams along the upper slope of a valley located closer to the villages and at a lower altitude than the natural glaciers. These glaciers start to melt much earlier and are supplemented with additional irrigation to address the small-scale farmers’ vulnerability and adaptation to climate change, improve their livelihoods, and for a host of other purposes. Therefore, the issues of vulnerability, adaptive capacity, and adaptation strategy need to be understood in a local context in the communities and regions where people live. Leh (Ladakh) in the north-west Himalaya provides a case study for exploring the ways in which the adaptation to climate change is taking place on a community scale using the artificial glacier technology by the adaptive and resilient communities. With the above backdrop, an attempt has been made to analyse the rural poor households’ vulnerability and adaptation practices for climate change using artificial glacier technology and to draw lessons focusing on vulnerability-–livelihood interactions in the cold desert of Leh (Ladakh) in the north-west Himalaya, India.

2 Data and Methodology

The study is based on the data and information collected as part of the ‘Baseline Survey of Minority Concentration Districts of India—Leh District, Jammu and Kashmir’ and funded by the Ministry of Minority Affairs, and the government of India during August 2007 to September 2008 and has been confined to rural areas in the Leh district of Jammu and Kashmir State. The sample has been taken from 25 households, each from 27 villages for a detailed survey, thus making a total of 675 households. The team strategies including the participatory rural appraisal (PRA) technique have been used to collect the data and information. Both unfocused and focused observation techniques have been used. The unfocused initial observations have been used to become increasingly familiar with the inside world so as to refine and focus subsequent observation and data collection. All the observations have been recorded on site, and any misunderstandings have been corrected. In all the focus group observations, 10–12 stakeholders have been recruited from different settings. The highly formal interviews have been conducted using the structured interview schedules. The in-depth interviews have also been conducted to elicit the opinions of the stakeholders with the extensive knowledge of the phenomenon under study. The content analysis technique has been used to analyse the data and information qualitatively and quantitatively (using descriptive statistics). The content analysis technique has been supplemented by the use of the code and label field notes, sorting, shifting, constructing, and reconstructing these materials.

The small-scale farmers' vulnerability is understood by knowing how the communities, natural systems, institutional structures, and social relationships are affected by climate variability. Vulnerability is the physical exposure to extreme events and adverse outcomes as well as failure of entitlement to resources. It is the susceptibility to damage from an environmental extreme and relative inability to recover from that damage. Vulnerability is a chronic state of survival rather than merely an outcome of environmental extremes. The present study has used the case study approach for measuring the vulnerability index. An attempt has been made to find those who are vulnerable and the factors in the study area for future policy intervention. Both qualitative and quantitative aspects of vulnerability determinants were used to arrive at a vulnerability index, which provide a baseline to determine the extent of vulnerability after some interval. Therefore, vulnerability has been assessed in its dynamic form.

The available standard research tools such as questionnaire, participatory rural appraisal (PRA), social and resource mapping, timelines, focused group discussions (FGDs), individual interviews, and transect walks were used to collect the data and information. Three main parameters, namely patterns of vulnerability, the nature of ecosystems, and the status of communication, were used to capture the context of study area. The patterns of vulnerability were assessed by (i) livelihood analysis using existing livelihood options and household occupations; (ii) uninformed migration by analysing trends and patterns of migration, individual household migration, and preferred destinations of migration; (iii) food insecurity by analysing monthly food deficiency and seasonable food habits; (iv) hazards and vulnerable groups identified by major hazards, affected areas and households and vulnerable groups; (v) level of services and opportunities by analysing major services and their status, levels of access, and satisfaction; (vi) physical infrastructure by types of infrastructure and their status, including transportation, agricultural inputs, electricity, and hospitals; and (vii) social capital and belief systems by existing community-based organisations and their focuses, social norms and local traditions, and local perceptions of hazards.

For communication parameters, the variables used were as follows: (i) existing means of communication and trends; (ii) information needs; (iii) existing information on available weather forecasts and interdepartmental coordination in disseminating information; and (iv) access to information, gaps, and limitations.

Similarly, an ecosystem parameter was analysed using variables such as (i) existing ecosystems; (ii) changing patterns of availability and use of natural resources; and (iii) impacts of disasters and livelihood practices on ecosystems using the techniques of resource mapping, focused group discussions, and small group meetings.

The vulnerability index assessment was carried out using the following processes: mapping exposure to climate hazards, understanding sensitivity, and assessing adaptive capacity. The challenge was to understand who is vulnerable and vulnerable to what. An indicator was needed to encapsulate the complex reality into a single, quantifiable or measurable comparative construct. The development of a quantitative measure of vulnerability was not an easy task, and the process has its own limitations. Therefore, qualitative dimension of vulnerability assessments such as socio-ecological change and coping and adaptive mechanisms were taken into considerations.

The data and information on the indicators of vulnerability are available only at the macrolevel and not at the local level, where planning and projects take place. Therefore, the case study has developed a quantitative vulnerability index at microlevel to understand the susceptibility to suffer damage and the ability to recover from it. The non-material aspects of vulnerability, institutions, and attitudes have also been taken into consideration. It also elicits local-level perception on vulnerabilities and capacities to develop an integrated approach focusing on both the physical and social dimensions of vulnerability from an interdisciplinary perspective.

Both the quantitative and qualitative analyses have been used to assess the symptoms and causes of vulnerability. Vulnerability assessment has a dynamic aspect, which not only tracks vulnerability reductions but assesses the types of interventions useful for reducing vulnerability. Therefore, it provides a baseline for future monitoring and learning processes and is useful for climate adaptation policy frameworks. The qualitative analysis was carried out to link the vulnerability analysis with existing livelihood strategies, ecosystem services, and physical and social infrastructures including information and communication systems for identifying interventions, which could strengthen ability to adapt and build resilience.

The process of vulnerability assessment is discussed in the following paragraphs. Firstly, a situation analysis was carried out to identify external threats such as risks, hazards, and climate variability to analyse the vulnerable and how they are coping/adapting. The changes in land use, access to water resources, debt, hunger, credit, and access to and control of natural and social resources were also assessed. This was followed by an in-depth analysis of the differing causes of vulnerability and to identify differential vulnerability at the community level. Lastly, assessment of capacities to cope or adapt by different individual/communities in a given social, institutional, and governance context was made to understand rights and entitlements, social networks, and physical infrastructures.

Vulnerability assessments have been carried out using a simple tool with multiple weighted indicators using equal weight of 0.20, and 30 variables used were classified into five groups of indicators as follows:

(i) demographic indicators (D): family size, dependency, sex ratio, unproductive family members below 15 years of age, number of family members in the age group 15 to 24 years, number of family members above 60 years of age, and number of family members below 5 years of age;
(ii) social indicators (S): social status, access to public distribution system and food security, age at marriage, access to educational facilities, educational attainment and skills' possessions, access to health care, level of awareness,

access to or affiliation with political parties, access to or affiliation with community organisations, and membership in a self-help group;
(iii) economic indicators (E): work participation, dependent on traditional sources of livelihood, occupational mobility (capacity to diversify), net assets, and income and expenditure;
(iv) access to infrastructure (I): housing facility; access to safe drinking water; access to sanitation facilities; and access to health and education infrastructure; and
(v) access to communication and transport (C): access to a communication network such as mobiles and TVs, and access to private transport.

The vulnerability index (VI) combines components into a composite index structure as

$$\mathrm{VI} = (\mathrm{dD} + \mathrm{sS} + \mathrm{eE} + \mathrm{iI} + \mathrm{cC})/(\mathrm{d} + \mathrm{s} + \mathrm{e} + \mathrm{i} + \mathrm{c}) \quad (1)$$

where D, S, E, I, and C are the subcomponents, and d, s, e, i, and c are weighting factors (subjective).

On the basis of the aggregate score of all the variables, the households have been ranked on the vulnerability scale as follows: (i) if >0.80, gravely vulnerable, (ii) if 0.65–0.79, very highly vulnerable, (iii) if 0.50–0.64, highly vulnerable, (iv) if 0.25–0.49, moderately vulnerable, and (v) if <0.25, less vulnerable.

3 Overview of the Region

Ladakh is situated in one of the most important water towers of the Himalayas, within the upper reaches of the Indus watershed. The catchment area of the Indus system in India is about 321,289 km^2 (out of a total 1,165,500 km^2), and the average annual run-off is about 73 km^3 in India. The utilisable surface water is about 46 km^3, and estimated replenishable groundwater is about 26.5 km^3. There are 29 major dams in the Indus river system of which 10 are in India. Ladakh is a high-altitude, arid mountain region characterised by climatic and seasonal extremes and challenging terrain. Temperatures in winter drop to −30 °C. In summer, maximum temperature is around 25 °C. Air moisture content is very low. The high-altitude desert supports a unique ecosystem. Being mountainous with arctic desert conditions and scanty rainfall, irrigation depends on the glaciers, which give birth to a number of rivulets. However, efforts are made to bring more barren land under cultivation by constructing irrigation canals on the Indus, Shayok, and other tributaries (Fig. 1).

Ladakh is a semi-autonomous region, comprising the Leh and Kargil districts, of approximately 97,000 km^2 in the state of Jammu and Kashmir. The main source of irrigation in Ladakh is surface water, with approximately 10,190 ha of land around the tributaries to the Indus. The irrigation system is managed carefully through the

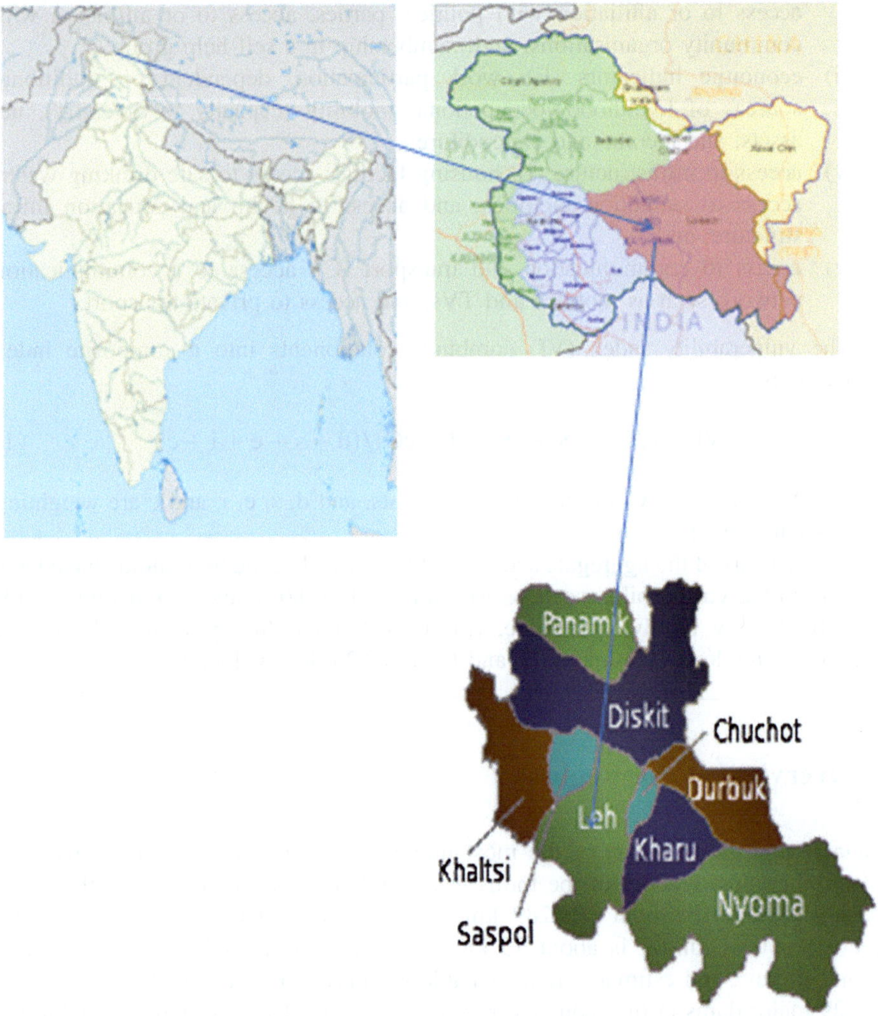

Fig. 1 Location map of Leh (Ladakh)

effective use of seasonal run-off from melting snow and ice at high altitudes using small hand-built mud canals. However, the irrigation water from melting glaciers is available from the beginning of June, which causes huge water stress on crop cultivation. Climate change in the region has also been putting additional stress on water requirements for agriculture in the early months of sowing. In order to improve the agricultural practices and performance in the district, the small-scale farmers with the support of the Leh Nutrition Project devised the new technique of artificial glacier by snow harvesting in the winter months to be used for irrigation purpose in the summer months. The snow water harvesting using artificial glaciers is an innovative technology, which provides water during the critical first months of the growing season (April–May). Due to topography, the use of groundwater for irrigation is negligible. The agricultural economy is characterised by seasonal migration, temporary communities, and nomadic subsistence farming. Besides agriculture, horticulture plays a major role in supplementing the income of the farmers and has assumed great importance in recent years. The main horticultural productions are apricot and apple, but in some parts of Khaltsi block, almond and grapes are also grown. The fruit produced is marketed in Leh town among other places and supplied to defence forces stationed in the region through cooperative marketing societies. The rearing of livestock is a very crucial and core economic activity in the district. It is adopted as a subsidiary occupation by the majority of the rural population. The nomadic population depends exclusively on sheep and goat rearing for their livelihoods, and the district is famous for Pashmina wool.

The Leh district is industrially backward. It is remote and inaccessible and also lacks basic infrastructure facilities. Handloom and handicrafts are another important income-generating activities in the district. Cooperative societies occupy an important place in the economic life of the people, and it has helped them diversify their economic activities. Currently, more than two-thirds of the households are members of cooperative societies. The cooperative public distribution system controls three-quarters of consumer business in essential commodities and marketing of agricultural produce and cent per cent fertiliser distribution. The generation and distribution of power continues to be one of the most problematic areas of development in the district. The major source of power in the district is Stakna Hydel Project. For various reasons, it only generates power for approximately 7 months a year. However, solar energy and diesel energy are used for electrification. It is significant to note that cent per cent villages are electrified in the district. Leh district is connected to the Block Headquarters through a network of roads. The Boarder Roads Organisation (BRO) maintains most of the highways connecting the Block Headquarters, and the Public Works Department (PWD) maintains a road length of 1,060 km.

4 Results and Discussion

4.1 Vulnerability Assessment

Small-scale farmers face many climate risks due to regular and persistent drought, deepening poverty, chronic food insecurity, malnutrition, and high risk of illness. Climate vulnerability is mainly experienced locally, even though it is affected by age, culture, resource tenure regimes, and gender and is determined by the local institutional, environmental, political, and market context. Therefore, on the basis of climate-induced livelihood challenges being faced by the small-scale farmers, it is possible to identify locally appropriate adaptation measures to increase their resilience.

4.1.1 Demographic Indicators

The average household size is 5.29 persons, the lowest being Muslim (5.06) and the highest Buddhist (5.33). The overall dependency is very small (0.46), which is higher among Muslims (0.60) than Buddhist (0.44). The gender ratio is favourable (1,026), which is comparatively high for Buddhist (1,038) and low for Muslims (937). The high gender ratio in Buddhist households reflects a comparatively higher female status in the community. The low gender ratio among Muslims is due to the predominance of traditional practices and prejudices against women, preference for sons, and dowry practices in the community. All this indicates lower female empowerment in the community compared to Buddhist.

Less than one-fifth of the population is in the unproductive age group of below 15 years. Buddhist households have comparatively less population in the child age group (17.84 %) than Muslim households (20.85 %). Therefore, educational needs of the Muslim households are higher than those of Buddhist households. Gender inequity in children ratio is sharp. About 18 % of the population is in the age group of 15–24 years. The concentration of Muslim households is comparatively less in this youthful age group (15.85 %: 10.71 % male and 21.33 % female). Thus, more females are found in the age group 15–24 years. This is true for both the communities. Buddhist households have a higher concentration of older age groups, which implies relatively more longevity in the community. The life expectancy is comparatively low in the Muslim community, and 8.96 % of the population is found in the age group above 60 years. The lower life expectancy reflects exposure to more vulnerability from climate hazards and higher risk incidence of climate-induced morbidity and mortality. The increased concentration of the population in children reveals higher climate-induced health risks in future. Similarly, a high youth proportion poses a greater challenge for labour markets due to climate change-induced demography.

4.1.2 Social Indicators

More than one-third of the households (36.86 %) are living below the poverty line (BPL) and are economically highly vulnerable. Nearly, 29.85 % had BPL ration cards, and of them, one-half are using the BPL Public Distribution System (PDS) facility. The huge difference in those falling under the BPL category and holding BPL ration cards and receiving benefits from the PDS is a very serious issue. With the increased irrigation potential due to the use of artificial glaciers, food security is assured for 77.56 % of households. A food crisis is a common consequence of climate risk as poor households have very little food, stores, or savings. Small-scale farmers whose livelihoods are based on agriculture and a daily wage often find themselves in a dire predicament, not just during and immediately after a drought, but for months afterwards.

Marriage before the age of 18 years is a reality for many young girls (20 %). Parents encourage the marriage of their daughters, while they are still children with the hope that it will benefit them both financially and socially, while also relieving financial burdens on the family. The social and economic effect of climate change does not provide a sense of security to girls with little education and poor vocational training, reinforcing the gendered nature of poverty and vulnerability.

Educational facilities for girls are inadequate, which is likely to make them more vulnerable in future, specifically in coping with climate hazards. Given the poor educational facilities for girls, the gender differential in literacy is high. Female literacy stood at 62.24 % among Buddhists and 77.20 % among Muslims. Nearly 69 % of children are enroled in government-run educational institutions and 6.57 % of them are never enroled, which makes them highly vulnerable to climate hazards. The drop out is one-fifth as children are forced to do domestic and paid labour outside the home due to poverty. About one-third of the population is educated to high school level and above. Due to lower educational attainment, both vocational and technical, youths have poorer prospects in the labour market, which is likely to increase their poverty status and vulnerability in future climate events if suitable measures are not initiated by the government.

Health accessibility is very inadequate, and all villages are devoid of health facilities. To avail themselves of such facilities, villagers have to travel to district headquarters at a mean distance of 33–42 km, creating appalling health vulnerability. However, increased agricultural productivity due to the use of artificial glacier techniques has resulted in higher income and improved livelihood conditions, helping to reduce climate change-induced health vulnerability in nearly 31 % of poor households. Dependence on government hospitals for medical treatment is significantly higher compared to private hospitals. Due to poverty and economic vulnerability, most of the poor households are unable to access private hospitals for medical treatment.

4.1.3 Economic Indicators

Work participation is significantly high (76.14 %), and gender differentials are noticeable. Agriculture is the major land-based source of livelihood. The majority of the households are small-scale farmers with landholdings of less than 1 ha. Due to the cold desert and adverse climatic conditions, a large proportion (61.18 %) of the poor households derive their livelihoods from agriculture and allied activities, and this is responsible for their poverty and vulnerability. Regular drought, lack of input, or pest attack adversely affect crop productivity. Poor small-scale farmers are forced to take loans at high interest to meet household expenses. However, in the case of farmers who are using irrigation water from artificial glaciers, farm production and productivity have improved significantly, causing seasonal migration to decline in more than two-thirds of farming households. In the past, this was one of the most effective livelihood diversification mechanisms to cope with the adverse effects of climate hazards. Small-scale farmers were vulnerable locally due to the lack of a diversified agricultural system and income source. Most small-scale farmers were heavily dependent on agricultural labour to generate household income. Therefore, irrigation water from artificial glaciers has helped to diversify the farming economy of the smallholders in the context of climate change. It has helped in the production of crops adaptable to climate risk and reduced the impact of climate change. These strategies have not only improved the local farming economy but also ensured local food security.

Seasonal migration is an important alternative to local livelihood options and a popular and convenient way of adapting to climate hazards. Nearly two-thirds of small-scale farmers do not need to resort to migration after the use of irrigation water from artificial glacier technology. Although migration is clearly an effective coping mechanism, it adversely impacts on the women and children who are left behind. It increases their workload, forcing them not only to engage in household activities but also to carry out farming and other activities. This adversely affects their health. With men away for long periods of time, there is the feminised impact of climate hazards. Both pull and push factors are at work, and climate change impact exacerbates the push factor. The perceived or constructed rigidity of certain traditional occupations, such as rearing yaks, also hampers the livelihood diversification of small herders. Their ancestors have been doing this work for generations, and they possess no other skills. There are both psychological and real skill-based barriers to occupational mobility, influencing vulnerability levels. Due to declining stocks and increasing climate risks, 75.26 % of small herders prefer to diversify their traditional livelihood strategy.

Assets are identified as the primary factor in determining vulnerability, involving 'resilience' through the use of assets and entitlements to manage hardship. Vulnerability is, therefore, closely linked to asset ownership. The more assets people have, the less vulnerable they are, and the greater the erosion of assets, the greater their insecurity. Landlessness is very high among households of the Leh district. About 43.65 % of the Buddhist households are landless. The average size of landholding is small (1.2 ha) compared to the national average of 1.37 ha. Landlessness, poor quality of land, vagaries of weather, consequent persistent drought and desertification, and the small amount of land possessed by poor households not only reduce their livelihood options but also make them highly vulnerable by working on low wage levels, keeping the landless and small-scale farmers in a poverty trap. The per capita value of livestock owned by the households stood at INR 16,008. The quality of livestock also seems to be poor. The possession of more livestock provides rural households with milch animals, meat, and other products (depending upon the types of livestock owned and maintained) and helps to reduce the effects of drought. The mean value of productive and modern assets possessed by the households has increased significantly to INR 26,648 and INR 22,959, respectively, over the last few years due to the use of irrigation water from artificial glaciers. In addition, small-scale farmers also possess more improved agricultural tools and implements, which in turn improves agriculture production and productivity further assuring greater food security.

4.1.4 Access to Infrastructure

The majority of households (87 %) are living in semi-concrete and concrete houses, and 12.06 % are living in mud and thatched houses. More than two-thirds of the households are living in single-room accommodation, which invades their privacy. The condition of safe drinking water facilities is not satisfactory in the rural areas of the district. Nearly 61 and 31.26 % of households use drinking water from public sources and natural water sources, respectively. The dependence on unprotected natural sources of drinking water exposes them to climate change, induced health vulnerability and more. Half of the households are defecating in open fields, which are totally unhygienic, exposing them to frequent health hazards, with adverse outcomes in the case of climate hazards. Household drainage conditions are also unsatisfactory. All this causes increased vulnerability in poor households in normal times and more so in the case of climate hazards.

4.1.5 Access to Communication and Transport

Rural road accessibility and communication are much poorer in villages in the Leh district than in the villages of other districts in Jammu and Kashmir. Nearly two-thirds of villages have a bus stop, a post office, and a public telephone connection. Television and mobile network services are available in about 70 % of the villages. About 33 % of the households have poor transport and communication accessibility, which results in lesser dissemination of information. It obstructs the mobility caused by climatic events, making these already poor households more vulnerable.

4.1.6 Vulnerability Index

Vulnerability varies across communities. Buddhists are the most vulnerable in terms of all indicators: demographic, social, economic, access to infrastructure, and access to communication and transport. Vulnerability is highest in access to communication and transport across the communities. Demographic indicators present a comfortable situation due to better literacy, non-farm employment in the modern sector, and higher engagement of women in economic activities (see Table 1). Poor households face multiple short- and long-term problems, including droughts, landslides, heavy snowfalls, and avalanches. These have a direct impact on their livelihoods and survival. The existing vulnerability of poor households has increased due to (a) ineffective communication of early warnings; (b) poor sanitation, especially during disaster times; (c) regular droughts; (d) fragile ecosystem; (e) lack of livelihood diversification; and (f) lack of access to clean drinking water.

Table 1 Vulnerability index (out of a total score of 1.00)

Indicators	Muslim	Buddhist	Total
Demographic indicators	0.32	0.37	0.35
Social indicators	0.67	0.73	0.71
Economic indicators	0.71	0.74	0.73
Access to infrastructure	0.69	0.76	0.75
Access to communication and transport	0.83	0.89	0.87
Weighted average of all indicators	0.66	0.71	0.70

Note A five-point scale was constructed to measure the level of household vulnerability to selected variables. Values are as follows: >0.80: gravely vulnerable, 0.65–0.79: very highly vulnerable, 0.50–0.64: highly vulnerable, 0.25–0.49: moderately vulnerable, and <0.25: less vulnerable

5 Adaptation Measures

Vulnerability to climate change has necessitated the development of adaptation practices to cope with climate impacts. Adaptation refers to adjustments in livelihood capitals in response to actual or expected climatic events and their effects or impacts. Adaptation practices include reactive/responsive measures such as changing the pattern and timing of the cropping system, emergency response to disasters, and migration, and proactive/anticipatory adaptation such as crop and livelihood diversification, climate forecasting, community-based disaster risk reduction, early-warning systems, insurance, water storage, and supplementary irrigation.

5.1 Community Participation

Traditionally, community members including small-scale farmers have participated in consultations and local institutional (such as Panchayat) meetings to screen environmental management measures for use in their communities and identify those acting as an adaptation to climate hazards, particularly drought. However, these do not always give everyone the right to be heard (McKay and Keremane 2006). Community members have also participated in categorising measures according to the primary type of livelihood capital targeted, evaluating benefits generated, identifying actors, assessing the resources and capacities needed for implementation, and identifying important uncertainties that might affect their performance.

5.2 Community Mobilisation

The artificial glaciers have been constructed near to villages. Therefore, their benefits are equally distributed among all the villagers. The whole population of each village located close to an artificial glacier has been the beneficiary of this innovative intervention. The maintenance and monitoring of the interventions are carried out collectively by the village community members. Before operationalising the construction of the artificial glaciers, the following steps are taken by the village communities as the main stakeholders. They know the available resources, dynamics, and constraints thoroughly. The first step is to mobilise the community for intensive discussions on issues relating to availability of water in the stream during the peak winter season, presence of shady areas along the course of stream, timing of sunrise and sunset, and village history regarding water availability. The villagers are also oriented towards the innovative technique of artificial glaciers and their role in irrigation management system.

5.3 Technical Considerations

The technical components are equally important and given due priority. Various aspects of technical components are the direction of the village, water availability in the stream during peak winters, location, etc. The villages where artificial glaciers have been constructed are south facing, i.e. on the south of the Indus River, so as to ensure proper formation during winter and its timely melting during the spring season. These artificial glaciers must be located as close as possible to the village so that they melt more quickly compared to the natural glacier and reach the adjoining village at the crucial time, i.e. the sowing period in April and May. Most of the artificial glacier projects in Ladakh have been funded by the government and the Sadbhavana, the philanthropic wing of the Indian army.

5.4 Additional Income from Allied Activities

The timely and adequate irrigation of water to the barley and wheat fields and some other crops has resulted in increased agricultural productivity and cash income for farmers. The availability of assured water in early spring time has helped the farmers to harvest double the crops compared to a traditional single harvest per annum, resulting in the generation of additional income. The additional water available through the use of the artificial glaciers has helped the villagers to increase the number of tree plantations, which are a major source of income as the twigs/branches and the main trunk are mainly used to construct the roof and wooden floor of the houses. The increased availability of water has also resulted in pasture development for the village, improving conditions for cattle rearing, thereby creating additional sources of income from dairy farming.

5.5 Promoting Vegetative Growth

The artificial glaciers have numerous environmental benefits, including diverted channels in shady areas to slow down the water, which helps to reduce surface run-off, thereby recharging underground aquifers and increasing the groundwater table. Water discharge from 'chumiks' (natural springs) has increased due to the increased water table. The total agricultural landholdings have also significantly increased causing an increase in the green belt cover. The artificial glacier techniques have also improved soil moisture conservation and humidity, which is responsible for creating conducive conditions for plantations and promoting vegetative growth in desert land. An increase in cattle population has increased the use of animal excrement as manure in agriculture rather than chemical fertilisers.

5.6 Restoring the Ecological Balance

The artificial glacier techniques have also restored the ecological balance by harnessing, conserving, and developing natural resources, i.e. land, water, and vegetative growth. Traditionally, the use and distribution of water resource has been a source of conflict in the villages of the Leh. The construction of artificial glaciers has made additional water available for agricultural and domestic use, thereby helping to reduce water disputes among neighbours and families. The agricultural operations have been more intensive than in the past, resulting in the decreased migration for those seeking new employment opportunities.

5.7 Planting of Trees

The artificial glacier technique has been a local community-driven adaptation measure taken in response to loss of moisture and scarcity of irrigation water in nearly one-third of the villages surveyed. In three-quarters of the villages, the planting of trees in arid land has been introduced in response to soil erosion and land degradation resulting from frequent drought.

5.8 Tactical Measures

The use of tactical adaptations by small-scale farmers in response to persistent drought such as the selling of livestock, expanding agricultural lands, and use of relief cash and food has declined to 20.44, 24.74, and 63 % of households, respectively. However, over time, small-scale farmers have discovered that this type of adaptation measure is unsustainable, since it does not address the root cause of the problem but only the symptoms. As a result, they would face a similar or worse situation in the future with much less capacity to cope.

5.9 Agricultural Operation and Management

The unsustainable nature of tactical adaptation has led the small-scale farmers to initiate measures focusing on structural changes in agricultural operation and management as a long-term coping strategy. Changes in land management have been made by small-scale farmers such as limiting crop areas (31.26 %), land rehabilitation (16.44 %), livestock improvement (21.92 %), agro-forestry (40.88 %), and replacing goats with sheep (6.37), creating less impact on the environment and the diversification of income sources (one-third) by practising different economic activities.

5.10 Improved Management of Water and Land

Other measures used by the small-scale farmers include improved management of water and land through water harvesting techniques (29.3 %), earth bunds and a terracing system (44.74 %), increasing recharge of the surface and groundwater reservoirs (54.96 %), and enhanced equity in access to water (42.07 %).

5.11 Consuming Less Food and Mortgage

Food deficiency is an enduring consequence of recurrent drought being faced by nearly 22.44 % of the small-scale farmers. During scarce months, the main coping strategy is to consume less food for survival. In adverse situations, poor households are forced to obtain credit at high interest from moneylenders by mortgaging their agricultural land or household assets. Both of these strategies, however, serve to perpetuate poverty. While all communities are engaged in them, such responses are gender-differentiated, with women more likely to be denied sustenance.

5.12 Change in Crop Varieties

Small-scale farmers are also using an early variety of barley (25.63 %), vegetable cultivation (49.03 %), water-resistant crops (42.52 %), early-maturing varieties of wheat (38.37 %), higher drought-resistant and/or pest-tolerant crop varieties

(48.74 %), and promoting interventions in emerging opportunities as a result of climate change by altering their livelihoods away from agro-ecosystems and implementing non-farm interventions (33.92 %).

5.13 Non-land Interventions

Non-land-based adaptation interventions have been used by small-scale farmers to adjust to various climate stresses and evolve new livelihood strategies, which include efforts to increase and diversify income and provide food security, ensure access to irrigation through community-based artificial glacier irrigation systems, reduce the fragility of the existing infrastructure by raising hand pumps, and rehabilitate ecosystems by improving drainage and using available surface water.

5.14 Migration

With the use of community-based artificial glacier irrigation systems, the incidence of poverty and vulnerability has declined. The consequent improved livelihood opportunities have resulted in a decline in migration as a survival strategy in 29.48 % of poor farming households. Nearly one-fifth migrated long term and the remainder on a short-term basis, and of those, one-quarter opted to migrate within the district or state and three-quarters migrated outside. The households who prefer to stay in native villages possess no skills and training for jobs outside the traditional occupations of agriculture and allied activities and have greater affinity with family and friends, seeking comparatively better livelihood options locally due to the use of community-based artificial glacier irrigation systems. Many of those who migrate come from the most depressed, remote, and isolated villages, where there is currently little potential for productive income generation. These are often also regions of dry land subsistence agriculture influenced by climate hazards, where incomes play a vital role in the sheer survival of poor households. Unsurprisingly, they are also the regions in which economic vulnerability is widely prevalent because of persistent backwardness and rising inequality.

6 Lessons Learnt

Climate change will have a significant impact on the small-scale farmers' livelihoods. The majority of small-scale farmers are no longer able to produce traditional crops due to desertification and purchase much of their food. They cannot afford capital investment (e.g. irrigation) to produce alternative crops. This affects small-scale farmers badly, who are often excluded from credit programmes due to lack of awareness and credit worthiness. Similarly, this is the case with poor women, who are denied credit due to lack of land title. Small-scale farmers living in remote and inaccessible areas are facing an increasing shift to monocultures, which constrain traditional methods for risk-spreading based on biodiversity. Their reliance on safety nets is increasingly strained due to political, ethnic, or gender-related marginalisation. Small-scale farmers are unable to achieve the collective action needed to sustainably manage common property resources. Local adaptation requires linkage with other activities and ongoing development programmes such as managing local natural resources. Local communities can advocate support for micro-issues effectively if those issues are regularly discussed and experiences shared among communities. The management of natural resources becomes sustainable only if a sense of ownership is inculcated. Government support in accessing agricultural input and equipment can help small-scale farmers reduce the risk of crop loss and other negative impacts associated with climate change.

6.1 Role of Technology and Innovation

Innovation in climate change mitigation technologies has accelerated since the late 1990s. Technology and innovation are increasingly perceived as essential for tackling the challenges of climate change. Solutions to societal challenges of climate change, food security, and vulnerability require innovations to create cooperative actions even though everyone cannot participate collectively in the decision-making process (see McKay and Keremane 2006). The need to invest in technology and innovation to help address these challenges and maximise their impact raises corresponding challenges in the policy context. Cooperation and collective action are necessary to sustain the benefits and minimise the cost of addressing livelihood challenges of climate variability. The targeted research and development (R&D) expenditure as well as policy measures, investment grants, and obligations can be a significant inducement to innovation in technologies. While addressing challenges of climate change requires a mix of policy instruments, their relationships and impacts on society need to be assessed carefully. Each policy instrument should address a specific market failure or barrier and foster a sustainable societal change based on technological innovation. Incentives to adopt appropriate innovative technologies can be created to alleviate vulnerability to climate change.

There is tremendous potential for science and technology to unleash innovation to address challenges of climate change through new entrepreneurial and policy experiments. There is urgent need to make the case for new policies to enable innovation to support the creation of shared social and economic values. Innovation in the twenty-first century differs from profit-oriented and nationally targeted models embraced in the last century. The likely challenges ahead for society in the future call for the construction of a new system of collaborative innovation to address the challenges of climate change. Thus, there is a need to find ways to foster innovation to generate social and collective values.

6.2 Collective Action of Public and Private Stakeholders

The livelihood challenges posed by climate change urgently call for new forms of collective action between public and private stakeholders in order to better integrate the social challenges into research and innovation. A new approach is necessary to solve problems where social and technological progress coevolves in order to generate social and public value. Most societal challenges are multidisciplinary in nature; therefore, to address livelihood challenges, the role of innovative technology is critical in a multidisciplinary approach involving multilateral collaboration among different stakeholders. The presence of social entrepreneurs; new participants on the innovation scene are necessary to bring social dimension in focus. To this end, a better linking science and technology policy together with other policies should be encouraged.

6.3 Addressing Opportunities and Synergies

There are opportunities and synergies to be exploited by better integrating challenges posed by climate change at the core of innovative activities. Social challenges have a strong mobilising effect, allowing the unprecedented gathering of competences and resources; beyond institutions, sectors and discipline boundaries. Thus, there is a need to create social awareness, involve various stakeholders, provide space for good future dialogue with stakeholders to interact and liaise, generate initial funding, maintain innovative systems, involve communities, cooperate with local public entities and non-governmental organisations (NGOs), extract good practice from model cases, and maintain and build networks to address livelihood challenges posed by climate change. To operationalise innovation activities, there is a need to link science and technology policy with livelihood policies, wherein the government has a dominant role.

6.4 Social Transfers and Social Protection

Social transfers and protection programmes are essential to achieve universal access to basic services and address the underlying causes of inequalities in well-being. Therefore, increased and more predictable central aid and assistance provide a window of opportunity to support local government to invest in social transfers to bring development benefits to those who have previously lacked them. Social transfers do not always work efficiently due to less transparent governance arrangements and leaking bucket phenomenon, which needs immediate attention from the appropriate authorities to improve performance and intended outcomes. Ongoing social transfers and social protection programmes have not paid enough attention to the long-term risks associated with climate change. Furthermore, current social protection programmes are insufficient to make a significant dent on the livelihoods of small-scale farmers. Climate change challenges the well-being of the poor small-scale farmers and increases the need for service provision. Social transfers represent a targeted means of addressing these challenges. Financial aid and assistance for poverty reduction and climate adaptation are both likely to rise significantly in the near future. Climate-induced poverty and vulnerability could be addressed also through this channel to support the climate adaptive capacity of the poor small-scale farmers. However, social transfers to support climate adaptive capacity require significant amounts of additional investment.

6.5 Institutional Mechanism

Developing an institutional mechanism for specific adaptation intervention can be one of the most effective ways of implementing a plan and achieving the expected outcome (Hughes and McKay 2010). Unless people have the capacity and are given the space to make decisions and to act, it will be difficult to achieve the expected outcomes. Local institutions play a major role in designing and implementing adaptation plans. Community-based organisations (CBOs) help facilitate the participation of vulnerable households, framing specific rules and regulations to guide people in performing certain assigned tasks, developing self-initiated decision-making processes to enhance a person's confidence to act, help everyone to work together in a transparent and equitable fashion, provide feedback and corrective mechanisms to enable people to share learning, and redesign and/or reorganise activities if required (McKay and Keremane 2006).

6.6 Revival of Ecosystem Services

Poor households suffer more from climate hazards and increased climate variability, which has led to degradation of fragile ecosystems. Given the limited capacity of ecosystems to sustain a degradation in the provision of goods and services, the adaptation measures used by poor small-scale farmers to revive ecosystem services include large-scale plantation, shifting to early-maturing varieties of crops or promoting higher drought-resistant and/or pest-tolerant crop varieties, altering livelihoods away from agro-ecosystems, and implementing non-farm interventions as a means to bridge livelihoods and income deficit at household levels. These strategies are likely to yield encouraging results in the building of resilience and adaptive capacity of small-scale farmers due to increased climate variability and uncertainty. These interventions need to be taken forward, enabling sustainable policies and ensuring their proper and effective implementation.

6.7 *Building Adaptive Capacity*

Building adaptive capacity involves providing the regulatory, institutional, managerial, and financial support needed for adaptation actions (Keremane and McKay 2006). The government has an important role in putting in place an effective policy and institutional framework, filling information and knowledge gaps, creating the right incentives, and allocating adequate public resources for adaptation. Improving adaptive capacity must be an urgent priority. Enhancing adaptive capacity has been high on the development agenda. Although uncertainty may make it difficult to fine-tune adaptation, win-win measures exist to address climate change and adopt good sustainable development practices. The long-term nature of climate change makes timing crucial to adaptation decisions. Despite the uncertainties, one of the best adaptation measures available may be to extend ongoing efforts towards sustainable development, as these are justifiable even without climate change. Better health care, access to safe drinking water, better sanitary conditions, and improved standards of education and infrastructure are measures that, while useful in their own right, will also enhance community adaptive capacity.

6.8 *Role of the State*

The government has a vital role to play in providing incentives and a policy framework for small-scale farmers and communities to adapt effectively to climate change and enhance their adaptive capacity. Adaptation decisions are largely decentralised. Some adaptations will have local public good characteristics and as such may be provided by the state, also called policy-driven adaptation. However, the majority of decisions will be taken by individuals, households, and communities with local benefits, known as autonomous adaptation. Since adaptation is a decentralised process, there is the question of how incentives can be provided to support it. Strengthening adaptive capacity efforts requires mainstreaming climate change adaptation into development planning. Adaptation must be considered not only as a technical solution focused on natural systems but also as an integral part of sustainable development and poverty reduction strategies.

References

Adger WN, Agrawala S, Mirza MMQ, Conde C, O'Brien K, Pulhin J, Pulwarty R, Smit B, Takahashi K (2007) Assessment of adaptation practices, options, constraints and capacity. In: Parry ML, Canziani OF, Palutikof JP, van der Linden PJ, Hanson CE (eds) Climate change 2007: impacts, adaptation and vulnerability. Contribution of working group II to the Fourth Assessment Report of the Intergovernmental Panel on Climate Change. Cambridge University Press, Cambridge

Adger WN, Kelly M (1999) Social vulnerability to climate change and the architecture of entitlements. Mitig Adapt Strat Glob Change 4:253–266

Adger WN, Huq S, Brown K, Hulme M (2003) Adaptation to climate change in the developing world. Prog Dev Stud 3(3):179–195

Blaikie P, Cannon T, Davis I, Wisner B (1994) At risk: natural hazards, people's vulnerability and disasters. Routledge, London

Bohle H, Downing T, Watts M (1994) Climate change and social vulnerability. Glob Environ Change 4(1):37–48

Chatterjee K, Chatterjee A, Das S (2005) Community adaptation to drought in Rajasthan. IDS Bull 36(4):33–52

CPRC (Chronic Poverty Research Centre) (2008) The chronic poverty report 2008–09: escaping poverty traps. Chronic Poverty Research Centre. Brooks World Poverty Institute, University of Manchester, Manchester

Cline WR (2007) Global warming and agriculture: new country estimates show developing countries face declines in agricultural productivity. Center for Global Development Brief (Sept 2007)

Easterling WE, Aggarwal PK, Batima P, Brander KM, Erda L, Howden SM, Kirilenko A, Morton J, Soussana J-F, Schmidhuber J, Tubiello FN (2007) Food, fibre and forest products. In: Parry ML, Canziani OF, Palutikof JP, van der Linden PJ, Hanson CE (eds) Climate Change 2007: impacts, adaptation and vulnerability. contribution of working group II to the fourth assessment report of the intergovernmental panel on climate change. Cambridge University Press, Cambridge, UK, pp 273–313

Handmer J, Dovers S, Downing TE (1999) Societal vulnerability to climate change and variability. Mitig Adapt Strat Glob Change 4(3 and 4):267–281

Hughes S, McKay JM (2010) The contribution of actors to achieving sustainability in Australia through water policy transitions. In: Huitema D, Meijerink S (eds) Water policy entrepreneurs: a research companion to water transitions around the globe. Edward Elgar, and IWA, Cheltenham, pp 175–194

Ibarraran MI, Ruth M, Ahmad S, London M (2009) Climate change and natural disasters: macroeconomic performance and distributional impacts. Environ Dev Sustain 11:549–569

IPCC (2001) Climate change 2001: impacts, adaptation and vulnerability. In: McCarthy JJ, Canziani OF, Leary NA, Dokken DJ, White KS (eds) Contribution of working group II to the third assessment report of the intergovernmental panel on climate change. Cambridge University Press, Cambridge

IPCC (2007) Climate change 2007: impacts, adaptation and vulnerability. In: Parry M, Canziani O, Palutikof J, vander Linden P, Hanson C (eds) Contribution of working group II to the fourth assessment report on climate change. Cambridge University Press, Cambridge

Kelly PM, Adger WN (2000) Theory and practice in assessing vulnerability to climate change and facilitating adaptation. Clim Change 47(4):325–352

Keremane G, McKay JM (2006) Self-created rules and conflict management processes: the case of water users' associations on Waghad canal in Maharashtra, India. Water Resour Dev 22 (4):543–559

McKay JM, Keremane G (2006) Farmers perception of self-created water management rules in a pioneer scheme: the Mula irrigation scheme India. Irrig Drainage Syst 20:1324–1540

Osbahr H, Twyman C, Adger WN, Thomas DSG (2008) Effective livelihood adaptation to climate change disturbance: scale dimensions of practice in Mozambique. Geoforum 39(6):1951–1964

Oxfam (2007) Adapting to climate change: what's needed in poor countries, and who should pay. Briefing Paper 104, Oxfam

Smit B, Skinner M (2002) Adaptation options in agriculture to climate change: a typology. Mitig Adapt Pract Policies Glob Change 7(1):85–114

Smit B, Pilifosova O (2001) Adaptation to climate change in the context of sustainable development and equity. In: McCarthy JJ, Canzianni OF, Leary NA, Dokken DJ, White KS (eds) Climate change 2001: impacts, adaptation, and vulnerability. Contribution of working

group II to the third assessment report of the intergovernmental panel on climate change. Cambridge University Press, Cambridge
Stern N (2006) The economics of climate change: the stern review. Cambridge University Press, Cambridge
Stern N (2007) The economics of climate change: the stern review. Cambridge University Press, Cambridge
UNDP (2007) Human development report 2007/2008: fighting climate change: human solidarity in a divided world'. Palgrave MacMillan, Basingstoke
UNFCCC (2007) Climate change: impacts, vulnerabilities and adaptation in developing countries. United Nations Framework Convention on Climate Change, Bonn
World Bank (2006) Clean energy and development: towards an investment framework. Prepared for the meeting of the World Bank—International Monetary Fund Development Committee. World Bank, Washington (23 April)

Governing Disaster Risk Reduction in Nepal

Dallas Blaney

Abstract How can a community cope with the risk of water-induced disaster when its government is either unable or unwilling to act? Communities typically rely on governments to coordinate disaster mitigation efforts and finance risk reduction initiatives; however, climate change may exceed the capabilities of this government-centered model. In the search for an alternative, this chapter considers the special case of Nepal. Already weakened by misguided development, extreme poverty, and a decade of conflict, Nepal must also cope with the rapidly increasing risk of water-induced disaster. To do so, this chapter argues in favour of expanding private sector participation in disaster risk governance. Support for this argument includes a brief review of the disaster literature, a summary of endogenous and exogenous threats, as well as an overview of recent international risk mitigation efforts. Following this, the focus returns to Nepal and the potential benefits of increasing private sector participation in its disaster management efforts. The paper concludes with recommendations for expanding the private sector role in disaster risk governance.

1 Risk, Natural Disaster, and Water-Induced Disaster

The concept of risk is fraught with multiple and often contradictory meanings: it can be something to avoid yet also a strategic opportunity; it can reference a sudden onset event or something latent and pernicious; it can be the product of endogenous or exogenous forces; and it can be the object of political struggle or the subject of actuarial science. It is peculiar as well that risk shares an inverse relationship with uncertainty, so that as the degree of uncertainty increases, risk decreases. For example, while it is highly probable that a meteor will strike the earth, it is nonetheless highly uncertain as to when and where this will happen. Consequently, we can say that the risk of this event is low. On the other hand, it is all but certain

D. Blaney (✉)
Department of Public and Environmental Affairs, University of Wisconsin-Green Bay, 2420 Nicolet Dr., Green Bay, WI 54311, USA
e-mail: blaneyd@uwgb.edu

S. Shrestha et al. (eds.), *Managing Water Resources under Climate Uncertainty*, Springer Water, DOI 10.1007/978-3-319-10467-6_16

that human activity has fundamentally transformed the chemical composition of the earth's atmosphere, whereby the risk of climate change is high (IPCC 2013). Nevertheless, great risks do not necessarily produce great actions. In the final analysis, risk is nothing more than an assessment of vulnerability. To borrow from Beck and Levy (2013, 19), risk simply refers to "a future that needs to be prevented". Real and imagined differences in the vulnerability to a dystopian future tend to complicate both the assessment of risk and the efforts to mitigate it.

To be vulnerable is to lack the ability to cope with risk. For example, an employee may be exposed to greater risk of losing her job during an economic recession; however, she can minimise her vulnerability by acquiring high-value skills and developing a strong performance record. Countries are like individuals in the sense that their degree of vulnerability is often a function of their level of preparedness (Buckle 2012). The best way to prepare for risk is to avoid it altogether; however, in the context of water-induced disaster, this option often involves costly and politically contentious strategies, including community relocation, large-scale redevelopment, and environmental restoration (FAO 2009). In the absence of effective mitigation, Bankoff et al. (2004) suggest that countries can reduce their vulnerability by expanding their population's social, economic, and cultural ability to cope with disasters. Comfort et al. (2010, 34) echo this sentiment, finding "to respond effectively it is necessary to coordinate across each nodal point in the decision-making network and practice subsidiarity by allocating decision-making authority to the lowest appropriate level". When a country is vulnerable, it means the actions taken to prepare for risk were insufficient, and therefore when an event occurs it is far more likely that the situation will rapidly deteriorate into an immanent disaster (Comfort et al. 2010).

A natural disaster is a sudden act of nature so devastating that it exceeds the affected community's ability to cope. These events have three defining features: speed, intensity, and human centeredness. Extractive institutions, misguided development, and other endogenous pressures can increase a community's susceptibility to exogenous shocks and reduce its capability to anticipate, resist, and recover (Bohle 2001). However, it is ultimately the speed and severity of an event that exposes these shortcomings, which is why we tend to differentiate between disasters like volcanoes, tsunamis, and earthquakes on the one hand, and invasive species and rising sea levels on the other.

In recent years, the risk of natural disasters has increased exponentially, thereby compounding the challenges and costs of preparing for disaster. Over the last three decades, there has been a threefold increase in the frequency of natural disasters and a sixfold increase in related economic damages (Bogardi 2006). In the last decade alone, natural disasters have killed as many as one million people and caused more than one trillion dollars in damage (UNDP 2013). The good news is that consistent investments in disaster risk reduction have led to a sharp decline in the loss of life; the bad news is that economic assets like roads, homes, and businesses are far more exposed to environmental shocks, which will likely increase the economic costs of natural disasters and make it more difficult for communities to cope with the damage (IPCC 2007; Habiba et al. 2013; Bankoff et al. 2004).

It is particularly difficult for communities to cope with the risks of water-induced disasters, which are the most common and perhaps complex of all natural disasters. In 2010, water-induced disasters accounted for 90 % of all natural hazards. In the same year, water-induced disasters claimed nearly 300,000 lives and cost as much as US$110 billion (WWAP 2012, 602). These disasters result from having either too much water or too little, producing either a flood or drought. Of the two general forms, floods can be further subdivided into several types. Landslide dam outburst floods (LDOF) occur when landslide debris temporarily blocks a river or stream, prompting water to build up and eventually overwhelm the dam. While LDOFs are unusual and therefore the risk is relatively low, their occurrence can nevertheless result in devastating outcomes. For example, in 1996, an LDOF in Larcha, Nepal, claimed 54 lives and destroyed 22 homes (COE-DMHA 2012). Glacial lake outburst floods (GLOF) occur when glacial run-off overwhelms natural dams or when a suspended block of ice falls into a glacial lake, producing a wave that weakens or overtops the sediment dam.

While these devastating events are frequent occurrences in Nepal, it is difficult to determine where and when a water-induced disaster will strike. GLOFs are always difficult to predict, particularly when the cause is falling ice or an avalanche. In 1985, one such ice-generated GLOF forced water over the moraine dam at Dig Tsho, emptying the small lake into the valley below. The flood water displaced the community of Khumbu for several months and destroyed a sorely needed and nearly completed Namche Hydropower Plant project (Vuichard and Zimmerman 1987). Damage estimates for this single event exceeded US$3 million. Of course, flooding is most commonly the result of rainfall, the precise timing and location of which is also difficult to predict. In 1993, brief but intense rainfall in the Bagmati River system compounded the effects of an unusually heavy monsoon weather system that year, triggering floods that claimed 1,300 lives, displaced over 73,000 families, and caused extensive economic losses. These are but a few examples of the severity and uncertainty that are intrinsic to water-induced disaster events. The challenge of avoiding rainfall events or at the very least minimising their effects is further complicated by several factors, including unexpected changes in rainfall intensity and the spatial distribution of rainfall, as well as topographic and surface conditions (WWAP 2012). Ultimately, the risk of water-induced disaster is a function of forces that are largely exogenous to human behaviour; however, the vulnerability to risk is largely a function of political and economic decisions. The next section explores these exogenous and endogenous sources of risk and vulnerability in greater detail.

2 Exogenous and Endogenous Drivers of Risk

Rapidly increasing risk of water-induced disaster is the consequence of exogenous and endogenous forces. Global climate change is undoubtedly the most powerful exogenous driver of risk. According to the Intergovenmental Panel on Climate Change (IPCC 2013), global climate change has been most clearly manifested in the global mean surface temperature, which increased by 0.9 °C between 1901 and 2012, with most of this warming taking place after 1951. While every decade since 1980 was warmer than all previous decades in the instrument record, the decade of the 2000s is the warmest of all. Thus far, most of this additional energy has been absorbed by the oceans; however, the IPCC has high confidence in its finding that annual mean surface warming is also underway.

This trend is most evident in the Northern Hemisphere, where warming since 2000 was so substantial that it reversed a 5,000-year cooling trend (IPCC 2013). The symptoms of this trend include decreased ice/snow cover and increased precipitation. The IPCC (2013) report has high confidence that snow cover has decreased in the Northern Hemisphere by 7 % between 1922 and 2012. Whereas satellite technology allows for accurate measurements of annual snowfall, changes in precipitation are far more challenging to measure and predict. Nevertheless, the IPCC (2013) has medium confidence that human activity has also caused changes in global precipitation patterns. The IPCC (2013) is virtually certain, however, that future changes in average precipitation will result in more frequent and prolonged droughts in some areas, whereas others will experience more frequent and more intense flooding.

While there are multiple endogenous drivers of risk, it is possible to trace many of these back to their economic and political sources. The economic drivers of risk ultimately stem from problems of value. Because water is finite, essential, and non-substitutable, attempts to assign water an economic value are made to compete with more established social and spiritual values (Postel 1997; Shiva 2002). This also means that the economic value of water, once established, rarely conveys the full cost of treatment and delivery (Saleth and Dinar 2004). While there has been some success in assigning an economic value to water from a lake, river, or aquifer, the value of rainwater has thus far been relegated to virtual status (Hoekstra 2003). These problems of valuation have obfuscated the environmental and social costs of production practices as they relate to water consumption, withdrawal, and treatment (Allan 2009). But more importantly, they have made existing patterns of production and consumption highly vulnerable to the risks of water-induced disasters. Because the value of water has been largely external to economic decision-making, many development decisions were made without regard for the changing reality of environmental conditions (WWAP 2012). Consequently, the costs of adaptation are massive and multifaceted. These costs include the cost of adjustment, the cost of adjusting to this adjustment, and the costs of failing to adjust (Kates 2009). While we know that every dollar invested in disaster preparedness saves seven dollars in the aftermath of a disaster, this fact alone has thus far failed to produce the kinds of investments that are needed to reduce the risk of a water-induced disaster (Abramovitz and Starke 2001).

Getting the value of water right may also prove essential for making water-induced disasters a priority on the political agenda. Governments are the largest sources of financing for water-induced disaster risk reduction efforts. In a recent survey of more than 130 countries, most acknowledged that the risk of water-induced disaster has increase over the last 20 years, yet only those countries with a high ranking on the human development index classified the importance of water for ecosystems as a high priority (UN WATER 2012). These findings highlight a critical flaw in the way that most policy-makers value water. The tendency is to value water only as an economic input. This approach is problematic from a disaster risk reduction perspective because it blinds decision-makers to the immense value water provides in sustaining ecosystem services. Foremost among these is the role that healthy forests, wetlands, and other ecosystems provide in mitigating disasters. Consequently, rising concerns about water-induced disaster have thus far failed to generate corresponding increases in international investments for risk reduction. According to one report (Kellett and Caravani 2013), between 1991 and 2010, the sum of all international aid spent on natural disasters was US $106.7 billion. Of this total, 87.3 % was spent on post-disaster activities, including reconstruction and emergency response. The remaining balance of US$13.5 billion was spent on disaster risk reduction efforts, making up just 0.04 % of all international aid over the last two decades.

3 International Disaster Management

Water-induced disasters are often discrete and local events with international implications. Floods in a headwater country may negatively affect downstream neighbours. Prolonged and severe droughts can reduce agricultural production and prompt price increases in the global food markets. Infrastructure damage can disrupt global supply and production networks. Prolonged or severe disasters can displace millions, generating spillover effects for neighbouring countries. Disasters can also result in the loss of critical habitat for migratory species or eutrophication, which may result in biodiversity loss in neighbouring countries. Related healthcare problems and the loss of livelihood for affected populations often produce spillover effects as well. Indeed, Habiba et al. (2013, v) point out that the international dimensions of disaster risk are so severe that "Unless we help vulnerable and poor nations, regions and cities prepare and adapt to current and future climate and disaster risks, we could see decades of development progress rolled back".

Ultimately, the risks these disasters pose is not just to the affected communities; rather, these disasters also pose a significant risk to the maintenance and stability of the entire international system. Indeed, this is precisely the rational offered by a recent Central Intelligence Agency-commissioned report on the threat climate change poses to US national security. In its conclusion, the study attributes at least part of the natural security threat to the risk that natural disasters pose to the stability of "globally integrated systems":

> It is prudent to expect that over the course of a decade some climate events — including single events, conjunctions of events occurring simultaneously or in sequence in particular locations, and events affecting globally integrated systems that provide for human well-being — will produce consequences that exceed the capacity of the affected societies or global system to manage and that have global security implications serious enough to compel international response (Steinbruner et al. 2013, 5).

Thus far, international responses to these threats have been fraught with contentious and dysfunctional decision-making processes. In a recent study of disaster resilience, Habiba et al. (2013) observed entrenched institutional resistance against any effort to integrate response approaches across communities of practice. The result is insufficient, temporary, and misdirected investments and disaster interventions. In response, the authors call for adopting a comprehensive approach to disaster risk management, which includes all of the following elements: coordinating institutions, risk identification and reduction, preparedness, financial and social protection, and resilient reconstruction. Although lingering concerns about scientific uncertainty and incomplete data are problematic, the report concludes that institutional challenges nevertheless remain the single greatest barrier to bringing about a durable and comprehensive approach to disaster risk reduction.

In 2005, the UN-sponsored World Conference on Disaster Risk Reduction endeavoured to address at least some of these institutional challenges when it launched the Hyogo Framework for Action. The Hyogo Framework builds on the lessons learnt from previous international disaster risk reduction efforts by marrying these foundational principles with the clear, measurable, time-bound objective setting principles established in the Millennium Development Goals process. The Framework contains just three broad goals, which are then divided into five targets or priorities for action, each of which contains a variety of quantifiable measures of progress. The goals include the following: (1) the integration of disaster risk reduction into sustainable development policies and plans; (2) the development and strengthening of institutions to enhance resilience to hazards; and (3) the systematic incorporation of risk reduction approaches into post-disaster programmes. Hyogo also set an ambitious deadline, calling upon the international community to meet the Framework objectives by 2015. To reach this goal, Hyogo prescribes an inclusive implementation process involving a wide variety of stakeholders, including nation states, civil society, the scientific community, and the private sector. Consequently, the Hyogo Framework offers a foundation for raising awareness and building the international cooperation needed to overcome the institutional challenges related to disaster risk reduction.

Unfortunately, the Hyogo Framework has thus far achieved only modest incremental reductions in disaster risk. A recent report by the UN Office of Disaster Risk Reduction (UNISDR 2013a, 7) credits Hyogo with producing more "thoughtful reflection on specific hazards and vulnerabilities", including a greater recognition of gender-specific vulnerabilities. However, on the subject of policy implementation, the report is more despondent, noting a strong correlation between the intensity of implementation efforts and national income, widespread resistance to community engagement, and subsequently low levels of stakeholder buy-in.

In his analysis of international or system-level disaster risk reduction efforts, Haase (2010) uncovered a host of additional international or system-level barriers to goal attainment. Foremost among these is the patchwork of international laws and policies on disaster reduction and response, which breeds uncertainty, waste, confusion, and costly litigation. A recent surge in the number of disaster management actors has added another layer of complexity to the immense challenge of disaster risk mitigation efforts. According to UNISDR (2013b), this assemblage of national and system-level challenges culminates in the most significant barrier of all: a lack of financial resources.

4 Valuing Disaster Risk Reduction in Nepal

Few countries face a greater risk of water-induced disaster with fewer financial resources than Nepal. Nepal is a landlocked country of 30 million, over half of whom live on less than $2 per day (World Bank 2013b). Landlocked countries face higher transportation costs and therefore tend to orient their economies to serving the needs of their immediate neighbours. Consequently, landlocked countries often find it more challenging to achieve and maintain robust levels of economic growth and, subsequently, face more significant economic challenges in preparing for the risks of climate change and other environmental pressures (Collier 2007). As of 2012, Nepal's gross domestic product was US$19 billion; however, remittances made up as much as 20 % of this sum (World Bank 2013d). During the fiscal year 2012–2013, Nepal had US$769 million in exports, which included iron and steel, carpets, textiles, plastics, and agricultural goods. However, imports soared to US$.5.6 billion for the same fiscal year, leaving the country a trade deficit of US$4.8 billion. Neighbouring India is Nepal's largest trade partner; however, exports to India increased by only 2.8 % in the last fiscal year, whereas imports from India increased by 22.6 % (HNS 2013). In 2012, this economic scenario endowed Nepal with a GDP per capita of US$700, which is a significant improvement over its previous US$490 in 2009 (World Bank 2013a). Nepal also suffers from a highly skewed distribution of wealth, earning a low score on the Gini index of 32.8 in 2010 (World Bank 2013c).

It is almost always the case that the communities that face the greatest risk of water-induced disaster are also the poorest. Poverty is not merely an economic condition but rather a more holistic signifier of well-being. The Human Development Index offers perhaps the best holistic measure of poverty, and on this measure, Nepal ranks near the bottom, coming in at 157 out of 186 countries (UNDP 2013). Contributing factors include a high maternal mortality ratio of 170, low adult literacy rate of 60.3, and gender inequality. These endogenous factors exacerbate the vulnerability communities already face from the rapidly growing threat of floods and droughts: insufficient education can translate into an insufficient awareness of risk; a high maternal mortality ratio and pronounced gender inequality tend to weaken the bonds that make a community resilient; and economic poverty leaves communities without the resources they need to resist, cope with, and recover from disasters.

The risk of water-induced disaster in Nepal is among the highest in the world. In its Fifth Assessment Report, the IPCC (2013) cites evidence of increases in the minimum and maximum temperatures in the Western Himalayas, a trend which seems to coincide with evidence of declines in snowfall and glacial mass. In a separate report, the World Bank (Schellnhuber et al. 2012) cited evidence that suggests the warming rate is highest in winter and that higher elevations are warming faster than low-lying areas. In his review of recent glacier studies of Himalayan, Inman (2010) argued that these warming trends have been manifested in widespread evidence of retreating and thinning glaciers, with most retreating by an average of 10–20 m per year. According to Gautam et al. (2013), these trends have coincided with an increase in precipitation across the entire Tibetan plateau. Bolch et al. (2012), however, attribute the rapid increase in flooding risks to the change in precipitation patterns as opposed to glacial melting. In heavily glaciated drainages, the authors expect peak discharge rates to remain largely unchanged for the foreseeable future; however, they also contend that discharge rates will likely decrease in less glaciated valleys. Ultimately, these projections rely upon potential climate-induced changes in monsoon intensity; yet current climate models lack the precision to project these changes with a high degree of certainty. It was on the basis of these findings and others that the IPCC concluded that observed changes in flood events worldwide are highest in places like Nepal, where spring run-off takes place earlier each year (IPCC 2013). These findings suggest it is a change in the type of precipitation and not just the amount that is behind the growing risk of water-induced disaster in Nepal.

As if to prove this point, recent findings indicate that shorter and warmer winters have coincided with more frequent heavy rainfall events, culminating in a growing number of catastrophic floods (IPCC 2013). Floods and landslides are the most deadly natural disasters in the country, accounting for an average 211 deaths per year between 1998 and 2008 (IFRC 2011). In 2013, Nepal endured the most severe monsoon floods in a decade after heavy rain in the western region flooded the Rapti River. This flood affected over 60,000 villagers, claimed more than 2,000 animals, and destroyed more than 10,000 tons of food stocks (Worldwatch 2013). Additionally, early monsoon rains triggered a number of flash floods and landslides. Because Nepal already suffers from persistent and severe food shortages, these disasters are difficult for the country to overcome (WFP 2013). Unfortunately, every indication points to a future filled with more frequent and severe flooding events. In fact, the IPCC has medium confidence that modern flood events in the region have already reached or surpassed historic floods in magnitude and/or frequency (IPCC 2013).

A multitude of endogenous trends increase Nepal's vulnerability to water-induced disaster. Vulnerability to disaster risk is a function of available resources as well as the capability of individuals and communities to successfully utilise these resources in a disaster scenario (Adger et al. 2003). Drivers of vulnerability can therefore include poverty, environmental degradation, weak or ineffective governance, severe social and ethnic cleavages, and the scarcity of resources. Nepal struggles with all of these. Widespread poverty and rapid urbanisation leave communities without the resources or the capability to avoid disasters. Environmental degradation exacerbates these vulnerabilities by increasing the likelihood that disaster events will be more frequent and severe (Renaud 2006). Healthy forests have an incredible capacity to mitigate the otherwise harmful effects of heavy precipitation. Unfortunately, deforestation is a serious problem in Nepal, where pressures from population growth, unmanaged settlement, and unemployment have greatly depleted the country's forests (FAO 2009). After a decade of conflict, Nepal has also struggled to create an effective system of governance. As of this writing, the country is without a working constitution. High turnover in key national government positions and weak local governance systems have been corrosive as well as eroding public trust and increasing its vulnerability to disaster events (Taylor et al. 2013; IRIN 2013). While the central government has achieved limited success in producing disaster risk reduction legislation, a host of structural political problems have made it difficult to implement these plans at the district and local levels (IFRC 2011).

5 The Case for Private Sector Participation

It is remarkable that Nepal has accomplished so much given its political and economic disadvantages. Since 2005, the government has developed a comprehensive plan for disaster risk reduction, enlisted robust support from the international community to assist in the creation and implementation of this plan, and established a central agency to coordinate these efforts. Nevertheless, persistent governance challenges have produced uncertainty, the erosion of public trust in government, and insufficient investments in disaster management initiatives.

In the absence of effective governance, expanding private sector participation in disaster mitigation efforts may be the best way to increase investments and secure community support for mitigation and adaptation initiatives. Habiba et al. (2013) argue that the key to improving disaster risk reduction is the creation of durable yet flexible financial programmes. They note that the success of a programme should not be measured by the size of the investment but rather by its ability to prompt a desirable and widespread change in behaviour. Among other things, this means that disaster reduction efforts only achieve their full potential if they are fully integrated into development and poverty reduction efforts, rather than functioning as add-ons. To achieve this goal, the authors support initiatives that endeavour to fully integrate strong risk identification procedures into decision-making processes, with the ultimate objective of producing complementary and coordinated disaster mitigation efforts across multiple sectors and levels of analysis. When viewed in this context, increasing private sector participation in disaster risk reduction strategies takes on greater meaning; it is not merely about avoiding the next catastrophe but rather part of a sustainable and resilient development strategy (Wisner et al. 2012).

The disaster risk reduction literature contains useful insights into the sequence of steps needed to launch such an ambitious project. The first of these is to develop a comprehensive risk assessment. Among other things, a comprehensive assessment should endeavour to map the "great variation of types of community vulnerabilities" (NRC 2006, 228), including environmental, political, economic, and social/human health vulnerabilities (Few et al. 2006). As Benson (2012) notes, many risk assessments overlook the economic benefits of disaster risk reduction efforts. Sustainable land use planning and resilient infrastructure are not just sound strategies for disaster mitigation efforts; they also promote economic growth as well as the health, well-being, and productivity of the community. These benefits are particularly evident in poor communities, which are almost always the most vulnerable to disaster events. The threat of climate change increases the cost of inaction. Comfort et al. (2010) remind us that when risk is either not recognised or the action taken to mitigate risk is insufficient, the situation can deteriorate rapidly.

Risk assessments are therefore a prerequisite for taking the next step in disaster mitigation: raising awareness. Creating greater awareness of disaster risk creates the occasion for a community to act to mitigate and adapt to this risk. It is not enough to produce a comprehensive risk assessment; it is also necessary to effectively communicate these risks to the community. Community strategies are context specific. To develop an effective strategy, it is necessary to have a thorough knowledge of the preferences, perceptions, and values of the target community. Because preferences, perceptions, and values are dynamic, an effective communication strategy must evolve in concert with these changes. The costs associated with developing a disaster risk communication strategy are therefore bound up in the costs of adjusting to a condition of increasing risk and uncertainty.

In rich countries, governments take responsibility for many if not all of these risk reduction functions; however, poor countries like Nepal often lack the capabilities to shoulder these burdens on their own. Whereas rich countries can rely on a broad and deep tax base, resource rents, or catastrophe bonds to produce a relatively stable and durable flow of disaster risk reduction financing, poor countries tend to rely on less predictable financial sources, including international aid, remittances, the sale of assets, and loans (Linnerooth-Bayer 2012). Unfortunately, pledges of aid often fail to materialise, remittances are highly vulnerable to global economic shocks, and asset sales and loans are not durable.

Rapidly increasing risk, inadequate financing, and exclusive decision-making practices have culminated in calls for business sector involvement in disaster risk reduction efforts. Traditionally, businesses externalised the cost of disaster management by passing on these costs to the public sector; however, the rapidly increasing risk of climate change means that traditional business and governance practices may well prove inadequate for dealing with the challenges of the future. To remain viable, businesses will need to internalise some of these disaster mitigation and adaptation costs. Interest in expanding the private sector footprint in disaster risk reduction efforts intensified in 2013, when this subject emerged as the topic of a plenary session in the biennial Global Platform for Disaster Risk Reduction. This development followed the publication of the UNISDR Global Assessment Report on Disaster Risk Reduction (2013b), which focused almost exclusively on the need for greater private sector engagement in disaster risk reduction efforts. In its report, UNISDR found that in most economies, private sector investment made up 70–85 % of overall investments. This finding led to the conclusion that private sector investment decisions largely determined the fate of disaster risk reduction efforts. In the plenary discussions, business leaders described their experiences in planning for and coping with disaster events. Discussants also highlighted the importance of effective governance and comprehensive risk assessments, as well as the immense business opportunities that exist in disaster-resilient investment. Among these investment opportunities were developing resilient construction methods, building resilient infrastructure, production and delivery systems, as well as environmental restoration work (UNISDR 2013a). The panellists also noted that disaster has only recently appeared as a subject of consideration in business risk assessments and that this practice has not yet been universally accepted across all business sectors. Audience members added that small- and medium-size firms often lack the resources or incentives to undertake such a comprehensive risk assessment, which means they also lack the resources to adequately prepare for disasters.

It remains to be seen whether businesses will play a greater role in shaping disaster risk reduction efforts at the state and local levels. In Nepal, an iron triangle of bureaucrats, foreign donors, and non-governmental organisations (NGOs) dominate disaster risk reduction efforts. Thus far, this arrangement has produced important policy instruments, including a National Strategy for Disaster Risk Management, as well as the Nepal Risk Reduction Consortium (NRRC). Established in 2009, the NRRC assists the government of Nepal in meeting its obligations under the Hyogo Framework for Action. Its members include the World Bank, the Asian Development Bank, the UN Development Programme, and several non-governmental organisations. Foremost among its objectives is the task of raising financial resources to support disaster risk reduction efforts in Nepal. Its organisational architecture consists of a Steering Committee tasked with developing the overarching vision and direction of NRRC work, and a Secretariat which provides the Steering Committee with technical and advisory support. NRRC efforts are organised into five issue areas or Flagship Programmes, which include school and hospital safety, emergency preparedness and response capacity, flood management in the Kosi River Basin, integrated community-based disaster risk reduction, and policy and institutional support for disaster risk management.

In their review of the NRRC, Taylor et al. (2013) identified several strengths and weaknesses. The authors note that in the absence of effective government, the NRRC has performed a critical role in governing disaster risk reduction efforts in Nepal. Specifically, the authors credit the NRRC with mobilising financial and technical resources, raising the national profile of disaster risk reduction and preparedness, and creating an institutional framework that allows actors with diverse and seemingly incommensurable interests to voice their concerns regarding disaster risk reduction proposals. Nevertheless, the authors note the need to expand NRRC efforts, particularly in the area of flood mitigation where there is a clear need to apply mitigation efforts across all river basins in Nepal. Additionally, the authors advised the NRRC to increase its efforts to draw upon a wider range of actors and increase its visibility and profile on the national stage.

The general sentiment of this report and others is that for all their merits, current risk reduction efforts have thus far failed to solve the complex problem of changing behaviour within the community. Insufficient community support has long been the undoing of disaster mitigation efforts in Nepal. For example, in the mid-1990s, Nepal received US$1 million in World Bank funding to build a sophisticated GLOF early warning system in the Tamakoshi Valley (Ives et al. 2010). While the system was sufficiently robust to withstand extreme climate conditions in the Valley, it nevertheless failed within 4 years of installation. An analysis of the project discovered that local community members had destroyed the system. It was determined that political uncertainty and the lack of community participation in the project design and implementation process were the principal flaw in the project design, noting "the people in the area thought the lake had been reduced to a safe level and lost interest in the warning system" (Ives et al. 2010).

While the government of Nepal has successfully developed a comprehensive risk assessment, it has been less successful in generating the investments and community support needed to implement its risk reduction strategy. The private sector can contribute to these efforts by taking steps to internalise the costs of disaster mitigation into their corporate risk assessments and developing an effective risk communication strategy that targets their suppliers and consumers. These activities are not about providing a public service; rather, the objective should be to protect corporate assets and ensure profitability. The first step is to develop a comprehensive risk assessment to determine the level of risk exposure throughout the supply, production, and distribution networks.

Fortunately, there are several tools that greatly reduce the costs of this risk assessment process. For example, The Water Footprint Network, The Nature Conservancy, World Wide Fund for Nature, and other organisations offer a robust water footprint (WF) assessment tool to help businesses quickly and accurately assess their exposure to water-induced risk (WFN 2013). WF analysis allows companies to develop product-specific profiles to identify and mitigate physical water-related risks throughout their supply networks. This tool is particularly effective in identifying vulnerabilities in agricultural supply chains (Chapagain and Tickner 2012). WF analysis can also be used to mitigate the financial and regulatory risks associated with poor water quality or increased water scarcity. Additionally, it allows companies to identify reputational risks that could erode their social license to operate (Orr 2009). Coca Cola recently used this tool to identify its exposure to water resource risk and took action to reduce the water requirements in its production process. In addition, Coca Cola is working with its suppliers to reduce their water footprint and it has shared these experiences at major international water resource conferences, including the 2012 World Water Forum. Chapagain and Tickner (2012) found that this type of corporate activity can be effective in raising the profile of water resource concerns. In the case of Coca Cola, WF analysis led to a shift in its corporate communications strategy to include a significant emphasis on water resource challenges. This is not to suggest that these activities constitute a comprehensive or even substantial shift in the governance of water-induced disaster; however, this anecdotal evidence may signify a growing awareness among corporations that involvement in risk mitigation is not just about good corporate citizenship, it is also good business.

Greater awareness of water resource vulnerability is only a first tentative step towards more substantial private sector engagement in disaster risk reduction decision-making processes. To address these threats, businesses must do more than talk; they must invest. The private sector needs to make substantial investments to protect its assets, but it will need to invest as well in the well-being of the larger community it serves. To this end, Nepal's National Strategy for Disaster Risk Management (MoHA 2009) recommends that the private sector should contribute to a national emergency response fund; however, there is a dire need for private sector investments in mitigation and adaptation efforts as well. Presently, the private sector is neither engaged in nor attracted to the work of disaster risk reduction in Nepal (MoHA 2009). Although it is sometimes suggested that increased interest in disaster risk reduction may open new business opportunities in the microcredit and insurance industries, these possibilities have thus far failed to live up to their promise.

One possibility for overcoming this problem is to expand the size of key decision-making bodies to include representatives of the business community. For example, it may be possible to expand the size of the NRRC Steering Committee by two seats to include representatives from large- and medium-sized firms in Nepal. Founded with just six members in 2009, the Consortium has since doubled in size (Taylor et al. 2013). Notably, the business community does not have any direct representation in this deliberative body. Inviting the business community to participate in the forum may help raise awareness within the business community, allow business leaders the opportunity to share their ideas and concerns, and potentially lead to enhanced buy-in from the business sector. Business representatives might be nominated to the NRRC Steering Committee by an established national business organisation, such as the Nepal Chamber of Commerce, with final confirmation resting with the permanent Steering Committee members. Business representatives would serve three-year appointments, during which time they would be obligated to report back to their national business organisation regarding NRRC proceedings and proposals.

This approach also promises to address several critical needs that were recently identified in a comprehensive evaluation and review of NRRC activities (Taylor et al. 2013). Given the uncertainty inherent to international sources of financial and technical support, increased buy-in by the business sector may produce a more durable alternative source of financial support, thereby enabling the NRRC to achieve greater autonomy. Furthermore, business sector membership promises to improve links to this heretofore excluded yet critical segment of the community. Including representatives from large- and medium-sized firms is therefore an important element in this proposal, as it is intended to simulate the diversity of interests and capabilities within the local business community. Because the original NRRC mandate is set to expire in 2015, ongoing efforts to renegotiate and extend this mandate to 2020 or beyond may constitute a unique opportunity to include private sector participation as part of the post-2015 strategy.

6 Conclusion

When a community confronts the threat of water-induced disaster, governments are almost always made to shoulder the burden of risk mitigation and adaptation efforts. But when a government either cannot or will not take on this responsibility, what options remain? To answer this question, this chapter analyses key findings from the disaster risk literature, identifies the endogenous and exogenous drivers of water-induced disaster risk, summarises the international response to this challenge, and examines the special case of Nepal, where ineffective governance has made the community highly vulnerable to water-induced disasters. The argument in this chapter builds on the recent UNISDR Global Assessment Report, which calls for greater private sector participation in disaster mitigation efforts. Whereas UNISDR emphasised the need for investments in risk mitigation and adaptation, this chapter argues in favour of channelling private sector investments towards disaster risk assessment and communication efforts. It notes the opportunities for framing supplier and consumer education as part of a comprehensive corporate risk mitigation strategy. It also offers a proposal for expanding the size of the NRRC to include representation from large- and medium-sized firms in Nepal.

There is a need for additional research in this area. Additional case study research may illuminate the possibilities and limitations of private sector participation in risk mitigation and adaptation efforts. There is also a need for research on the incentive structures that prove effective in prompting the private sector to act. Specifically, there is a need for greater insight into the roles that NGOs play in this context. Furthermore, additional research on the interface between government and business may shed light on the challenge of triggering private sector investments in communities with ineffective governance.

As a parting observation, this interest in private sector involvement in risk management may also signal the need for a corresponding shift in the literature on risk mitigation and adaptation. In recent years, a growing number of authors have investigated risk in the context of discourse analysis. These discourse analyses endeavoured to illuminate and problematise the power configurations behind risk management decision-making. These investigations have produced important insights into the social construction of risk and the multitude of injustices associated with risk reduction and adaptation efforts. However, more recent interest in private sector participation in risk governance may signal a fundamental transformation in the balance of decision-making power. To offer a comprehensive analysis of these contemporary developments, scholars will need to expand their focus beyond discourse to include a concern for the institutional dimensions of these developments.

References

Abramovitz JN, Starke L (2001) Unnatural disasters. Worldwatch Institute, Washington DC

Adger NW, Huq S, Brown K, Conway S, Hulme M (2003) Adaptation to climate change in the developing world. Progress Dev Stud 3(3):179–195

Allan JAT (2009) Virtual water in the real world. Conserv Biol 23(5):1331–1332

Bankoff G, Frerks G, Hilhorst D (2004). Mapping vulnerability: disasters, development, and people. Earthscan, USA

Beck U, Levy D (2013) Cosmopolitanized nations: re-imagining collectivity in world risk society. Theor Cult Soc 30(2):3–31

Benson C (2012) Economic development policy and disaster risk. In: Wisner B, Gaillard JC, Kelman I (eds) Handbook of hazards and disaster risk reduction. Routledge, New York, pp 264–275

Bogardi JJ (2006) Introduction. In: Birkmann J (ed) Measuring vulnerability to natural hazards: towards disaster resilient societies. TERI Press, New Delhi

Bohle H-G (2001) Vulnerability and criticality: perspectives from social geography. IHDP update 2(01):3–5

Bolch T, Kulkarni A, Kääb A, Huggel C, Paul F, Cogley JG, Stoffel M (2012) The state and fate of Himalayan glaciers. Science 336(6079):310–314. doi:10.1126/science.1215828

Buckle P (2012) Preparedness, warning and evacuation. In: Wisner B, Gaillard JC, Kelman I (eds) Handbook of hazards and disaster risk reduction. Routledge, New York

Chapagain AK, Tickner D (2012) Water footprint: help or hindrance? Water Altern 5(3):563–581

COE-DMHA (Center for Excellence in Disaster Management and Humanitarian Assistance) (2012) Country disaster response handbook: Nepal. Reliefweb

Collier P (2007) The bottom billion: why the poorest countries are failing and what can be done about it. Oxford University Press, New York

Comfort L, Boin A, Demchak CC (eds) (2010) Designing resilience: preparing for extreme events. University of Pittsburgh, Pittsburgh

FAO (UN Food and Agriculture Organization) (2009) Nepal forestry outlook study. In: A.-P. F. S. O. S. II (ed) Working paper series. Bangkok

Few R, Franziska M, Ahern M, Kovats S (2006) Flood hazards and health: implications for action and research. In: Few R (ed) Flood hazards and health: responding to present and future risks. Routledge, London

Gautam MR, Timilsina GR, Acharya K (2013) Climate change in the Himalayas: current state of knowledge. Policy Research working paper, no. WPS 6516. World Bank, Washington DC. http://documents.worldbank.org/curated/en/2013/06/17935389/climate-change-himalayas-current-state-knowledge

Haase TW (2010) International disaster resilience: preparing for transnational disaster. In: Comfort L, Boin A, Demchak CC (eds) Designing resilience: preparing for extreme events. University of Pittsburgh Press, Pittsburgh, pp 220–243

Habiba G, Bettencourt S, Kull D, Reid R, McCall K, Simpson A, Krausing J, Ebinger, J, Ghesquiere, F, Fay, M (2013). Main report. In: Building resilience: integrating climate and disaster risk into development—the World Bank Group experience, vol 1. World Bank, Washington DC

HNS (Himalayan News Service) (2013, 15 August 2013). Country trade deficit widens. Retrieved from http://www.thehimalayantimes.com/fullNews.php?headline=Country's+trade+deficit+widens&NewsID=387475

Hoekstra AY (2003) Virtual water trade. In: Proceedings of the international expert meeting of virtual water trade value of water research report series no. 12. UNESCO-IHE, Delft

IFRC (International Federation of Red Cross and Red Crescent Societies) (2011) Analysis of legislation related to disaster risk reduction in Nepal. Geneva

Inman M (2010) Settling the science on Himalayan Glaciers. Nat Rep Clim Change:28–30

IPCC (Intergovernmental Panel on Climate Change) (2007) Climate change 2007: the physical science basis. In: Contribution of working group I to the 4th assessment report of the intergovernmental panel on climate change, Cambridge

IPCC (2013) Climate change 2013: the physical science basis. In: Contribution of working group I to the 5th assessment report of the intergovernmental panel on climate change, Cambridge

IRIN (2013) Analysis: tensions mount as Nepal contentious election nears. Retrieved 31 October, 2013, from http://www.irinnews.org/report/99031/tensions-mount-as-nepal-poll-looms. Accessed on 31 Oct 2013

Ives JD, Shrestha RB, Mool PK (2010) Formation of glacial lakes in the Hindu Kush-Himalayas and GLOF risk assessment. International Strategy for Disaster Reduction, Kathmandu

Kates RW (2009) Cautionary tales: adaptation and the global poor. In: Schipper LF, Burton I (eds) The earthscan reader on adaptation to climate change. Earthscan, London, pp 283–294

Kellett J, Caravani A (2013) Financing disaster risk reduction. Global Facility for Disaster Reduction and Recovery. http://www.odi.org/sites/odi.org.uk/files/odi-assets/publications-opinion-files/8574.pdf

Linnerooth-Bayer J (2012) Financial mechanisms for disaster risk. In: Wisner B, Gaillard JC, Kelman I (eds) Handbook of hazards and disaster risk reduction. Routledge, New York

Ministry of Home Affairs (MoHA) (2009) National strategy for disaster risk management. MoHA-Nepal: with support from UNDP and the European Commission, Kathmandu
NRC (National Research Council) (2006) Facing hazards and disasters: understanding human dimensions. National Academies Press, Washington DC
Orr S (2009) From footprint to public policy: the business future for addressing water issues. The CEO Water Mandate: Pacific Institute and WWF
Postel S (1997) Last oasis: facing water scarcity, 1st edn. W.W. Norton, New York
Renaud FG (2006) Environmental Components of Vulnerability. In: Birkmann J (ed) Measuring vulnerability to natural hazards: towards disaster resilient societies. United Nations University Press, Tokyo, pp 117–127
Saleth RM, Dinar A (2004) The institutional economics of water: a cross-country analysis of institutions and performance. E. Elgar Publication, Cheltenham
Schellnhuber HJ, Hare W, Serdeczny O, Adams S, Coumou D, Frieler K, Martin M, Otto I, Perrette M, Robinson A (2012) Turn down the heat—why a 4 C warmer world must be avoided. World Bank
Shiva V (2002) Water wars: privatization, pollution and profit. South End Press, Cambridge
Steinbruner JD, Stern PC, Husbands JL (2013) Climate and social stress: implications for security analysis. National Academies Press, Washington DC
Taylor G, Vatsa K, Gurung M, Couture E (2013) Review of the Nepal risk reduction consortium, UN
UN WATER (2012) Survey on UN-WATER publications with specific attention to the future of the World Water Development Report process: report of survey findings. UN WATER. http://www.unwater.org/downloads/Survey_Report_UNWater_publications.pdf
UNDP (United Nations Development Programme) (2013) Human development report 2013. Retrieved 15 Nov 2013, from http://hdr.undp.org/en/Countries/Profiles/NPL.Html
UNISDR (United Nations Office for Disaster Risk Reduction) (2013a) From shared risk to shared value –the business case for disaster risk reduction. Global assessment report on disaster risk reduction. Geneva
UNISDR (2013b) Implementation of the hyogo framework for action: summary of reports 2007–2013. Geneva
UNISDR (2013c) Private sector for resilient societies: ingredients for success. Retrieved 21 Oct 2013, from https://www.youtube.com/watch?v=ePe3ugno8Gg&sns=em
Vuichard D, Zimmermann M (1987) The 1985 catastrophic drainage of a moraine-dammed lake, Khumbu Himal, Nepal: cause and consequences. Mt Res Dev 91–110
WFN (Water Footprint Network) (2013) Water footprint network launches innovative online tool to drive sustainable water use (Press release). Retrieved 13 Nov 2013, from http://www.worldwaterweek.org/documents/Press_Releases/2013-WWW/Monday/Press-Release-WFN-Launch.pdf
WFP (World Food Programme) (2013) Nepal: villagers shift away from traditional farming. Retrieved 18 Nov 2013, from http://www.wfp.org/Countries/Nepal
Wisner B, Gaillard JC, Kelman I (eds) (2012) Handbook of hazards and disaster risk reduction. Routledge, New York
World Bank (2013a) Data: Nepal. Retrieved 15 Nov 2013, from http://data.worldbank.org/country/nepal
World Bank (2013b) Data: poverty headcount ratio at $2 a day (PPP) (% of population). Retrieved 14 Nov 2013, from http://data.worldbank.org/Indicator/SI.POV.2DAY
World Bank (2013c) Gini index, from http://data.worldbank.org/indicator/SI.POV.GINI
World Bank (2013d) Migration and remittance. News and broadcast. 2013, from http://web.worldbank.org/WBSITE/EXTERNAL/NEWS/0,,contentMDK:20648762~pagePK:64257043~piPK:437376~theSitePK:4607,00.html
Worldwatch (2013) Worst flooding in a decade hits western nepal. Vital signs online. Retrieved 2 Nov 2013, from http://www.worldwatch.org/Node/4544
WWAP (World Water Assessment Programme) (2012) Managing water under uncertainty and risk. The United Nations World water development report 4, Paris

Role of Sustainability Policy Entrepreneurs in Building Water-Sensitive Cities to Respond to Climate Change: A Case Study in Adelaide, Australia

Ganesh Keremane

Abstract Climate change effects are already being felt as water shortages, storms, flooding, heat waves, and rising sea levels are all affecting the quality of life. Therefore, there is an urgency to include future climate conditions in all long-term decisions so that communities can better adapt to the impacts of climate change. Within the water sector, particularly urban water systems, adaptation is very important because current water management infrastructure and practices are inadequate in reducing the impacts of climate change on water resources. This calls for an integrated approach that seeks to better link how we plan and manage water supply, wastewater, sewerage, and storm water services and make provision for environmental flows. However, transitioning to a water-sensitive city is challenging, but through appropriate policy and governance shifts, it is possible to tackle most of the challenges. Local governments will have a major role to play in the area of climate change policy and so do 'policy entrepreneurs'—individuals who hold formal positions in the government departments and who have a strong passion for sustainable water management. This chapter focuses on such individuals, and it is based on a case study of a local council in Adelaide. The chapter discusses the context of the transition and the role of policy entrepreneurs in implementing the low-regret strategies to build a climate-resilient water-sensitive city.

Keywords Climate change · Water-sensitive city · Policy entrepreneurs · IUWM · Adelaide

G. Keremane (✉)
Centre for Comparative Water Policies and Laws, School of Law,
University of South Australia, Adelaide, Australia
e-mail: ganesh.keremane@unisa.edu.au

S. Shrestha et al. (eds.), *Managing Water Resources under Climate Uncertainty*,
Springer Water, DOI 10.1007/978-3-319-10467-6_17

1 Introduction

Climate change could significantly affect society through impacts upon a number of different social, cultural, and natural resources. These effects are already being felt as water shortages, storms, flooding, heat waves, and rising sea levels are all affecting the quality of life (Wamsler et al. 2012). According to the Intergovernmental Panel on Climate Change's (IPCC) Fourth Assessment Report (IPCC 2007a), climate change is likely to worsen current stresses on water resources from population growth and land use change, including urbanisation. A later special report (IPCC 2012) stated that there is high confidence that changes in climate have the potential to seriously affect water management systems. However, the main challenge may not be the change in climate itself; instead, it may be the uncertainties associated with future climatic conditions. Therefore, it is urgent to include future climate conditions in all long-term decisions so that the communities will be better adapted to the impacts of climate change (Hallegatte 2007).

Adaptation to climate change is 'adjustment in natural or human systems in response to actual or expected climatic stimuli or their effects, which moderates harm or exploits beneficial opportunities' (IPCC 2007b, p. 6). But adaptation to the impacts of climate change was long considered a taboo topic since policy response to climate change was initially dominated by mitigation measures (Martin-Ortega 2011). Moreover, adaptation was discussed mainly in the context of developing countries (Schipper 2006). However, perspectives have changed and adaptation is now seen as an essential part of climate policy. Pielke et al. (2007, p. 597) point out three reasons for this change in perception: (1) there is a timescale mismatch—irrespective of the mitigation measures, it will be many decades before they have a discernible effect on the climate; (2) vulnerability to climate-related impacts on society is increasing for reasons that have nothing to do with greenhouse gas emissions (e.g. rapid population growth along coasts and in areas with limited water supplies); and (3) those who will suffer the brunt of climate impacts are now demanding that the international response to climate change focuses on increasing the resilience of the vulnerable. There has also been a significant rise in awareness worldwide about climate change, and many decision-makers (water managers in this case) are now concerned about the impact of climate change on their activities and are developing strategies to adapt.

Adaptation in the water sector is important because current water management infrastructure and practices are very likely to be inadequate in reducing the impacts of climate change on water resources. Particularly, the urban water systems that include water supply, wastewater, and storm water are at risk because climate change will mainly manifest itself through alterations in the water cycle (Loftus et al. 2011). Therefore, reducing the vulnerability of society to changing water regimes due to climate variability will require policy shifts and significant investment (UN Water 2010). This means that current water management regimes, which have traditionally relied on technical solutions, should make way for more adaptive regimes that take into account environmental, technological, economical, institutional, and cultural characteristics of river basins (Pahl-Wostl 2007). It calls for an integrated approach: integrated water cycle management (IWCM) that seeks to better link how we plan and manage water supply, wastewater, sewerage, and storm water services and make provision for environmental flows. In an urban context this is commonly known as integrated urban water management (IUWM). There is an urgent need to build cities that are simultaneously resilient to future shocks, sustainable and also liveable. Examples of practical strategies to adapt to climate change impacts are found in literature; for example, Hallegatte (2009) discusses five strategies: no-regret strategies, reversible strategies, safety margin strategies, soft strategies, and strategies that reduce decision-making time horizons. The IPCC's Special Report (IPCC 2012) emphasises 'low-regret strategies' for managing the risks of climate extremes because these strategies produce cobenefits and help to address other development goals, such as improvements in livelihoods, human well-being, and biodiversity conservation (IPCC 2012; Hallegate 2009).

However, transitioning to a water-sensitive city is often challenging as it requires integrating water management with urban design, energy use, and major infrastructure systems (Loftus et al. 2011). At the same time, implementing low-regret strategies is always a challenge, and as Hallegatte (2009) identifies, it encounters three obstacles: (1) financial and technology constraints; (2) lack of information and transaction costs at the microlevel; and (3) institutional and legal constraints. In the Australian context, impediments to including 'new sources' into the urban water supply mix include lack of clarity about roles and responsibilities and accountabilities within the agencies managing the resources (Productivity Commission 2011). Other studies on transitioning to water-sensitive cities in Australia (Ison et al. 2009) have found that institutional capacity is an important issue to be considered while implementing IUWM. Institutional capacity according to Pegram et al. (2013) is multifaceted and includes at least seven dimensions: policy and legal, planning and managerial, human and infrastructural, organisational and procedural, financial, network, and stakeholder support. Our own research (McKay 2003; Keremane and McKay 2008; Keremane et al. 2011; Wu et al. 2012a) and other research (ARMCANZ 1997; Mitchell et al. 2008; Productivity Commission 2008; Ward and Dillon 2011) have identified that the property rights and ownership of the captured storm water and reclaimed water are not necessarily defined in a precise manner, thereby hindering their inclusion into the urban water supply mix.

These issues can be addressed through appropriate policy and governance shifts, which include changing the way water concerns are addressed and clearly defining the roles of different stakeholder groups involved in framing current water management policies. Accordingly, local governments will have to play a major role in the area of climate change policy. In Australian context, a shift from the traditional approach to an integrated water management approach is largely attributed to a group of key individuals (policy entrepreneurs/champions) in Western Australia, who in the early 1990s were calling for a new approach to urban planning and design (Mitchell 2006, p. 590). Furthermore, the integrated approach is a major instrument, contributing to the broader ecologically sustainable development (ESD) agenda and increasingly practised in new urban developments. Therefore, the role of ‘policy entrepreneurs’ or ‘champions’ in facilitating the transition process is important. Accordingly, the chapter focuses on these individuals who hold formal positions in the local councils or state government departments and due to their strong passion for sustainable water management go beyond the scope of their job specifications to achieve the desired results. This chapter aims to elicit the policy entrepreneurs’ role in implementing the low-regret strategies to build a city that is resilient to climate change uncertainties through a case study.

This chapter is part of an ongoing study to document case studies of individuals or a group that has influenced and/or resisted a policy change in the way water is managed in Australia. It is based on a particular case study in Adelaide and aims to address the question: what is/was the transition or policy change, and what strategies were used by the policy entrepreneurs to build a climate-resilient water-sensitive city?

2 Literature Review

A brief review of literature on climate change and urban water management in Australia followed by some discussions on policy entrepreneurs and their role in the policy process is presented to set the scene.

2.1 Urban Water Management and Climate Change Adaptation in Australia

Australian climate is both diverse and variable, and climates define the way communities live and the attitudes to water in the different regions. The urban water sector in Australia has traditionally been heavily reliant on climate-dependent sources of water such as rivers, dams, and reservoirs, which have been dominated by large government monopoly service providers and characterised by central planning and regulation (LECG 2011). However, in recent times, climate change and the risk it presents have become one of the most confronting issues and the climate change projections for Australia suggest a hotter, drier climate, wetter tropics, rising seas, and more intense fires and floods (BoM 2010). Furthermore, rapid urbanisation and increasing population have challenged the limits of traditional water sources, which are vulnerable to climate change, as was evident during the 2006 drought (White et al. 2008). Therefore, the water industry in Australia is facing an unprecedented challenge, with implications for all facets of the urban water cycle from water supply, sewerage transfer, treatment, and infrastructure, to river health, drainage, and flood management (WSAA 2012).

The early responses of the water industry in Australia were simply an incremental extension of risk mitigation applied to traditional approaches. However, in recent times, more mature and broader system-based approaches are emerging to build resilience and enhance the adaptive capacity of our urban water systems (WSAA 2012). The Commonwealth Government has made a commitment to 'support desalination projects, water recycling, and major storm water capturing projects nationwide without adding to climate change and without increasing greenhouse gas emissions' (Australian Labor Party 2007, p. 17). The Australian government's position statement on adapting to the impacts of climate change states (DCCEE 2010, p. 12),

> Climate change is already affecting water availability and security across Australia. Managing Australia's water resources will remain an important national priority into the future, and it will be important that water policies and programmes continue to factor in changes in climate.

Accordingly, many reforms and initiatives such as the Water for the Future programme (DSEWPC 2010) are already underway to consider the impact of climate change. The programme delivers a number of major streams of work to help Australians adapt to a future with less water including among others the development of effective science-based water plans to determine how valuable water resources can be shared among competing uses. All state governments in Australia are now adopting an integrated approach (IUWM) that incorporates non-conventional sources of water such as desalinated water, storm water, and wastewater into the supply mix, which had previously only included conventional sources such as rivers, dams, and groundwater (Productivity Commission 2011; DFW 2012). The IUWM approach also includes demand management strategies such as water restrictions, water use efficiency, and water conservation as other additional sources of control of the water supply. Given that climate change will have significant impacts upon the natural water resources, the IUWM approach offers a more diverse and versatile set of options for dealing with larger and more complex urban water challenges (Closas et al. 2012). However, adapting to the changes is more than the integration of different sources of water supply; it requires an understanding of the interrelations between the natural factors and the links with and within the societal factors (Oswald Spring and Brauch 2011). Brauch (2005) explains these interactions using a 'survival hexagon' of three resource challenges (air, land, and water) and three societal challenges (human population, urban systems, and rural systems) that may interact in different ways, and having a sustainability vision would help understand these interrelationships better (Oswald Spring and Brauch 2011).

2.2 *Policy Entrepreneurs and Their Role in the Policy Process*

The terms 'entrepreneur' and 'entrepreneurship' have been appearing in public policy and management literature since the 1960s and have been described in many different ways. Roberts and King (1989) use various terms to describe entrepreneurs based on their behaviour: Political entrepreneurs hold elected leadership positions in government; executive entrepreneurs hold appointed leadership positions in government; bureaucratic entrepreneurs hold formal positions in government, although not leadership positions; and policy entrepreneurs work from outside the formal government system to introduce, translate, and implement innovative ideas into public sector practice. We also find a number of scholars calling them change agents, policy advocates, and visionary leaders (Huitema and Meijerink 2009; Huitema et al. 2011). Some others describe such individuals as 'champions' (Howell and Higgins 1990; Andersson and Bateman 2000; Howell et al. 2005), characterise them as people who 'create and communicate strategic meaning around the innovation, persistently promote the innovation, sell the idea to top management in order to secure resources, and involve and motivate others to support the innovation' (Howell et al. 2005).

In the present context, the focus is on water managers who, as Teodoro (2009) describes, are public agency administrators and/or high-profile appointed officials promoting a particular public policy (Howard 2001). The most common term used to describe these individuals in the public policy and management literature is bureaucratic entrepreneurs. However, in this study, we call them 'sustainability policy entrepreneurs' because they are people who have adopted the sustainable development philosophy found in all Australian water and other natural resource laws since 1992 (NWC 2004; McKay 2005). Teske and Schneider (1994, p. 3) describe bureaucratic entrepreneurs as 'actors who help propel dynamic policy change in their community'. Huitema et al. (2011) note that bureaucratic policy entrepreneurs instigate, implement, and sometimes block transitions; they describe them as individuals from within government seeking change, who may be either politicians or bureaucrats (p. 718). Crow (2010) describes them as individuals who promote policy change and influence the policy process effectively. These individuals can affect shifts in water resource management through a set of strategies, such as advocating new ideas, defining and reframing problems, building coalitions, mobilising public opinion, managing networks, and specifying policy alternatives (Roberts and King 1991; Huitema and Meijerink 2009). They invest their time, knowledge, and skills in instigating and implementing policy change (Huitema et al. 2011) and have the energy and talent to influence alternate spheres of political activity (Schneider and Teske 1992).

Generally speaking, the role that policy entrepreneurs play in the policy process constitutes identifying/generating an issue or idea, packaging it attractively, and selling it to organisational decision-makers (Andersson and Bateman 2000). According to Roberts and King (1991), policy entrepreneurs participate in three stages: (1) developing new ideas; (2) translating the idea into a more formal statement (such as a proposal, bill, or law); and (3) help implement them (p. 151). As Brouwer and Biermann (2011) observe in their study, policy entrepreneurs employ numerous strategies to create a window of opportunity and therefore to direct policy change. Similarly, Huitema and Meijerink (2010) point out four generic strategies employed by policy entrepreneurs in affecting water transitions: (1) development of new ideas; (2) building of coalitions and marketing/promoting the ideas; (3) recognising and using windows of opportunity; (4) recognising, exploiting, creating, and/or manipulating multiple venues in modern societies; and (5) orchestrating and managing networks.

In Australian water management context, this is known as the 'champion phenomenon' (Commonwealth of Australia 2002). While examining the champion-driven processes related to sustainable urban water management in Australia, Taylor et al. (2011) identified three leadership models as being relevant; these are the transformational, distributed, and complexity leadership models (pp. 414–415). According to McKay (2010), sustainability water policy entrepreneurs in Australia can be clustered into two groups, who go beyond the scope of their job specifications to effectively unite the local region and its communities of water users: (1) informal entrepreneurs (including public servants at the local regional level, state-based public servants, Commonwealth-based public servants, members of industry, commodity, or water user groups, and members of environmental groups); and (2) formal statutory-based officers. This chapter focuses on the local-level and state-based public servants who have an influence on water policy transitions.

It is widely acknowledged in the policy science literature that policy entrepreneurs play a critical role in the policy transition process. At the same time, it is also true that the success of importing and formulating new policies ultimately depends on how well they are carried out (Teske and Schneider 1994). Furthermore, not all bureaucrats invest their time, knowledge, and skills in influencing policy transition. According to Teodoro (2009), career mobility affects the probability of administrators/bureaucrats emerging as policy entrepreneurs. It can further be argued that the likelihood of them initiating innovations in their agencies is directly related to their professional involvement. According to Howell and Higgins (1990), these individuals are able to influence the change process mostly because of intrinsic motivation and commitment rather than their formal role description. Furthermore, these individuals cannot influence the policy change process on their own (Roberts and King 1991); they need to build policy networks with the other members of the policy-making community to 'build their credibility and determine what arguments will persuade others to support their policy ideas' (Mintrom 1997).

While the emphasis here is on individuals supporting change, it is important to recognise that individuals or policy entrepreneurs may also resist change. This is true because resistance to change is a natural phenomenon and needs to be taken seriously (Coghlan 1993; Bovey and Hede 2001). According to Kotter and Schlesinger (1979), the four common reasons people resist change include a desire not to lose something of value, misunderstanding and lack of trust, a belief that the change does not make sense, and a low tolerance for change.

As described earlier, the focus of this chapter is to elicit policy entrepreneurs' role in implementing the low-regret strategies to build a city that is resilient to climate change uncertainties. Accordingly, this chapter discusses the role played by policy entrepreneurs in building a water-sensitive city in Adelaide, South Australia, including the strategies used by them to facilitate the change process and achieve the desired result of creating an environmentally sustainable city.

3 Building a Water-Sensitive City to Respond to Climate Change—A Case Study

As already mentioned, this chapter is part of an ongoing study to document case studies of individuals or a group that has influenced and/or resisted a policy change in the way water is managed in Australia. An Internet survey of water planners conducted as part of a different project helped identify such individuals across Australia. The survey asked the planners to identify individuals and/or groups who have influenced policy transition in their respective regions. The thought behind this approach was that since water planners work closely with the communities in developing the water plans, they are better placed to know whether any such individual exists in their regions. In addition, a professional network also helped in identifying some individuals. Most of the people identified hold formal positions within government bureaucracies, and due to their passion for achieving 'sustainability' in the broader sense, they went beyond their job to influence a policy change and contribute to sustainable water management in Australia. The strategies used and the actions taken by these individuals have resulted in either adoption of a water allocation plan to resolve regional water conflict, thus creating a sustainable city. The data were collected using semi-structured interviews with individuals who have contributed to Australia's plan to secure sustainable water supplies for all users, including the environment, both now and in the future.

In particular, this chapter focuses on the journey of a Local Council in South Australia: the Salisbury City Council, towards creating an environmentally sustainable city. It demonstrates how a window of opportunity (e.g. population growth and urban sprawl) combined with the challenges of climate change initiated a change process to build a water-sensitive and liveable city. It also highlights the challenges faced, including from those who resist change and the strategies used by the individual to overcome them.

3.1 Implementing Integrated Urban Water Management Strategy in the City of Salisbury, South Australia

Literature review on the one hand establishes that the IUWM approach offers a more diverse and versatile set of options for dealing with larger and more complex urban water challenges that arise due to climate uncertainty; on the other hand, it also identified the obstacles to implementing this approach, which largely include financial and technological constraints, lack of information and transaction costs at the microlevel, and institutional and legal constraints. Hallegatte (2009) argues, 'while the first two issues are well identified, more research is needed to understand the latter' (p. 244). This study (the larger study on policy entrepreneurs) fills this gap in that the case studies will be able to propose 'best practices' that could be generalised and could be a very efficient first step in a long-term adaptation strategy.

Moreover, local governments will have to play a major role in the area of climate change policy, and in assisting local communities to understand and adapt to the long term physical impacts of climate change. Equally important is the role of 'policy entrepreneurs' or 'champions' in facilitating the transition process.

Over the past 20 years, local governments have become major players in the area of climate change policy. In an urban water management context, this means that their responsibilities regarding urban water management have increased over time, particularly in developed countries including Australia. This is because in developed countries, councils are responsible for the provision of a wide range of services in their municipalities and make decisions based on a range of economic, financial, ideological, and political factors (Biswas 2006; Bel and Fageda 2007; 2009, Varis et al. 2006). This is the case with the City Council under study—the Salisbury City Council.

3.1.1 Study Area Description

The Salisbury LGA is situated 25 km north of Adelaide and covers an area of 161 km^2 stretching from the beaches of the Gulf St Vincent to the Adelaide Hills. The terrain is mostly flat with the Little Para River winding its way through the district to the sea. The City Council has pioneered many urban water management initiatives and leads the nation in implementing recycling projects. It has gained international recognition for the way it harvests urban storm water and stores it in wetlands, using it for irrigation and industrial use, or storing it in underground aquifers for later use in a process known as aquifer storage and recharge (ASR). Furthermore, the Council has created an innovative business model allowing it to separate its core responsibilities as a Local Government Authority (LGA) from its responsibilities to manage the commercial imperatives relating to the water produced. It has also developed its own climate action plans and strategies to tackle climate change.

3.1.2 The Context

The Salisbury City case study demonstrates how a window of opportunity (e.g. population growth, urbanisation, and intensity of water use) was utilised by sustainability policy entrepreneurs to initiate a change process to encourage progress towards sustainable water management. It also highlights the challenges faced, including from those who resist change and the means that they used to overcome these challenges.

The urban settlements of South Australia are primarily reliant on two key water sources: the Murray River and the Mount Lofty Ranges reservoirs. However, factors including widespread drought throughout the Murray-Darling Basin and the ever-increasing emphasis being placed on increasing environmental flows mean that less water is available from surface and groundwater sources as the frequency and severity of drought are likely to increase due to climate change. Every year, around 230,000 ML of storm water and treated effluent is pumped out to sea, which is more than Adelaide's total consumption each year (216,000 ML). Furthermore, South Australia's population is currently growing and is likely to reach 2 million by 2,050. If this occurs, then the amount of water South Australia needs will increase if water is not reused. With this in mind, the city of Salisbury has backed a 'Waterwise' strategy to harvest storm water across greater Adelaide to reduce reliance on the River Murray, to reduce pressure on groundwater resources, and to significantly reduce storm water run-off to the marine environment. As a result, various projects have been developed by the City Council as a component of the Integrated Water Cycle Management Plan (IWCMP), and the person interviewed has played a crucial role in developing and implementing this plan. To quote his words,

> The work undertaken by the City of Salisbury Council demonstrates the potential of LGAs to contribute significantly to a major public policy issue.

Furthermore, in the wake of climate change, the city has put forward a framework for action to ensure that the city continues to be a leader in sustainability practices and respond to climate change called *'Salisbury, Sustaining Our Environment: An Environmental and Climate Change Strategy'* (www.salisbury.sa.gov.au). According to the interviewee,

> This Strategy does not limit to stormwater recycling and wetlands but it includes the way we maintain our roads, collect and recycle waste, through to urban development and transport, and even the way we, as individual employees, behave in performing our everyday roles.

3.1.3 The Challenges and Strategies

Below are a couple of challenges and the strategies employed by the person in question to address them. One important thing to note here is the official role of the individual involved with the implementation of IUWM. An engineer, whose responsibility was primarily storm water management with a particular emphasis on flood (minor and major) management, went beyond the job description to achieve the desired results. This was due to sheer passion for innovative and sustainable water management and achieving sustainability in the broader sense.

Challenge 1: Community resistance to water reuse

The Salisbury Council in 1989 started working on a 'constructed wetland project' with the primary objective of cleansing storm water run-off from urban areas and reducing pollutant impact on the marine ecosystem. But later, when the council wanted to water nearby ovals, they developed an interest in using the storm water that was collected in the wetlands. However, like any reuse scheme, there was opposition to the use of recycled storm water.

Strategy employed: 'Breeding community advocates'

In the literature, we find that community participation is crucial to the process of including the 'new sources' into the urban water supply mix. Working with a community that does not have alternative water sources as its highest priority requires building participation through a combination of discussions about community outcomes and more detailed action steps of technology identification, design work, and management (Jones 2005). The author further suggests that a lack of community participation results in a wide gap between what is desired from these approaches and what is necessary to get there, and an inability to bridge this gap is the primary reason for the failure of freshwater augmentation projects. Robinson et al. (2005) argue that since it is the public who will be served by and will pay for, these water and wastewater services, the policies on its use, and management must include the human dimension. Our interviewee totally agrees with these observations and mentioned that in the case of Salisbury Council, the strategy of 'breeding community advocates' took care of the public opposition to storm water reuse and also added the human dimension to the reuse project. When asked to elaborate on the idea of community advocates, the interviewee described,

> Community advocates are a group of people within the community who support his vision and work and *include* technicians, scientists, and retired public servants, who are excited by the work this interviewee and their team are doing and want to be part of it.

Further, when asked about the methods used to recruit or 'breed' these community advocates, the response was,

> These advocates are the result of my public speaking engagements and community consultations over the years. Breeding them created resilience in relation to what we do politically because these people became our advocates in the community through the presses.

In addition, the interviewee worked closely with the elected council members to seek community support to the project and make it a success. The project that was initially trialled as a research project expanded further as it received support from other Councils, the NRM Board, the Local Government Association, and the State Office of Water Security and also received funding from the Federal Government by experts and NGOs. This was all largely due to community advocates, opines our interviewee.

As a result, within the city of Salisbury today, there are more than 50 constructed wetlands covering more than 300 ha. These wetlands allow storm water to be treated and cleansed prior to use in Council operations, as well as enhancing the landscape and providing a range of habitats for animals and plants. In addition to using in Council operations, the cleansed storm water from the wetlands is pumped into, and later recovered from, aquifers by aquifer storage and recovery (ASR) to supply customers (households, businesses, schools, other councils) via a purple pipe-recycled water reticulation network (Philip et al. 2008).

Challenge 2: Creating a market for and managing the 'new water sources'

Users' willingness to pay for the resource in question (particularly alternative or new water sources) is largely influenced by the tariff structure adopted in a particular scheme/project. However, the general tendency observed in the case of such freshwater augmentation schemes is that users might not be willing to pay more for this resource because it is considered as waste. Therefore, the tariff structure should be such that the community being served should perceive it to be appropriate, as well as taking into account the long-term viability of the service provider. In case of Salisbury Council, this challenge was addressed successfully by the formation of a subsidiary company under the SA Local Government Act, 1999, called the Salisbury Water Company. As our interviewee explained, 'this is a unique and innovative structure, and the first of its kind in Australia'.

Strategy: Creating a separate business model to manage new water sources

The Council that started storm water harvesting and reuse as a research project over the years took it to the next level. It had developed and expanded the distribution network to provide high-quality recycled storm water branded as 'Salisbury Water by the Council' throughout the Council area and beyond (DFW 2012). As a result, they developed a local market for recycled storm water. Due to an increase in the number of customers, and the management of the recycled water, it was felt necessary to separate water from the Council's other business. This was when a new governance structure was created to manage water resources in the Salisbury area.

The Salisbury Water Company is governed by the Salisbury Water Management Advisory Board. The Board is an independent body, and even though it reports to the Council, it can make policy decisions on certain issues independently, such as pricing policies and deciding on applications for water connections. More importantly, significant policy decisions can be made without referring to local government (Wu et al. 2012b). In addition to these functions, our interviewee explained the additional advantages of having a separate model,

> One, by creating a separate body: the Salisbury Water Management Advisory Board we were able to deal with the political interference by governments. And second, it has set a practical working model for the introduction of competition into the water market in South Australia.

This company has also adopted an appropriate tariff structure, which includes usage charge and residential supply charge. Compared to the current rates, the usage charge is cheaper than the mains water price, while the residential supply charge is a nominal fixed rate per quarter. While residential users had a mixed view about the tariff structure (Wu et al. 2012a), industries with high water dependency for their operations are now able to access treated storm water at cheaper rates than the mains water (City of Salisbury 2012). As a result, the dependency on mains water has declined with most residents and the industries in the city now using recycled storm water for non-potable purposes. These initiatives have resulted in environmental benefits such as protecting marine environment by reducing storm water discharge to the Gulf, and the wetlands are important habitats for flora and fauna.

4 Conclusion

Freshwater resources are vulnerable and have the potential to be strongly impacted upon by climate change, with wide-ranging consequences for human societies and ecosystems. While climate change is likely to worsen current stresses on water resources, the main challenge is to deal with the uncertainties associated with future climatic conditions. Consequently, many decision-makers are now concerned about the impact of climate change on their activities and are developing long-term strategies that include future climate conditions so that the communities will be better adapted to the impacts of climate change.

Adaptation to the water sector, particularly the urban water systems, is important because current water management infrastructure and practices are very likely to be inadequate in reducing the impacts of climate change on water resources. There is an urgent need to build cities that are simultaneously resilient to future shocks, sustainable, and also liveable. The IUWM approach offers a more diverse and versatile set of options for dealing with the complex urban water challenges; however, transitioning to a water-sensitive city is often challenging. But these issues can be addressed through appropriate policy and governance shifts including changes in the roles of different stakeholder groups in framing current water management policies. As a result, local governments will have a major role to play in the area of climate change policy. In this context, the present chapter focuses on the role of local governments as major policy players and discusses the efforts of individuals, who as sustainability policy entrepreneurs have played a key role in achieving sustainable urban water management. The case study in particular discusses the context of the transition and the strategies used by the policy entrepreneurs to build a climate-resilient water-sensitive city.

The study establishes that a strong driver for IUWM implementation, particularly in Australia, is the role of 'policy entrepreneur' or 'champions' in local councils who have a strong passion for innovative and sustainable water management. The case study emphasises the challenges and strategies used by policy entrepreneurs to encourage sustainable water management by designing innovative water management initiatives. The case study also illustrates how policy transition is an evolving process and that sustainability policy entrepreneurs have a key role in shaping policy outcomes in that they have to do more than what their job or position demands. It is evident from the case study that building trust and strong relationships with the community and thinking outside the box to come up with innovative solutions (e.g. developing a new governance structure to manage Salisbury water) are important in managing the water resources sustainably.

Nevertheless, policy entrepreneurs alone cannot steer the transition to achieving a desirable outcome. It is often the combination of personal attributes (e.g. leadership qualities) and strategies adopted by the key policy actors and contextual factors (e.g. window of opportunity, legislative environment, and the policy and social networks) that are key to influencing any policy change process. While this chapter, particularly the case study, focuses on the South Australian experience, the insights developed could provide some lessons for sustainable water management in other regions of Australia and across the globe.

Acknowledgments The funding and support from the Goyder Institute for Water Research is greatly appreciated. Thanks to Jennifer McKay and Zhifang Wu for their invaluable inputs and helpful discussions. Thanks must also go to all participants who spent time to participate in the voluntary interview process supporting this work.

References

Andersson L, Bateman T (2000) Individual environmental initiative: championing natural environmental issues in US business organizations. Acad Manag J 43(4):548–570

Agriculture and Resource Management Council of Australia and New Zealand (1997) A national framework for improved wastewater reuse and stormwater management in Australia, occasional paper number 3. Task Force on COAG Water Reform, Canberra

Australian Labour Party (2007) Labour's national plan to tackle the water crisis, Canberra

Bel G, Fageda X (2007) Why do local governments privatise public services? A survey of empirical studies. Local Gov Stud 33(4):517–534

Bel G, Fageda X (2009) Factors explaining local privatization: a meta-regression analysis. Publ Choice 139(1):105–119

Biswas AK (2006) Water management for major urban centres. Int J Water Resour Dev 22 (2):183–197

Bovey WH, Hede A (2001) Resistance to organizational change: the role of defence mechanisms. J Manag Psychol 16(7):534–548

Brauch H (2005) Environment and human security-towards freedom from hazard impacts. InterSecTions, No. 2. UNU Institute for Environment and Human Security (UNU-EHS), Germany

Brouwer S, Biermann F (2011) Towards adaptive management: examining the strategies of policy entrepreneurs in Dutch water management. Ecol Soc 16(4):5. (online) url:http://dx.doi.org/10.5751/ES-04315-160405

Bureau of Meteorology (BoM) (2010) State of the climate, CSIRO and BOM

City of Salisbury (2012) Sustainability that saves you money. (online) url:http://www.makesgoodbusinesssense.com.au/advantages/sustainability/

Closas A, Schuring M, Rodriguez D (2012) Integrated urban water management: lessons and recommendations from regional experiences in Latin America, Central Asia, and Africa. Water partnership program case profile; no. WPP 1. World Bank, Washington, DC

Coghlan D (1993) A person-centred approach to dealing with resistance to change. Leadersh Organ Dev J 14(4):10–14

Commonwealth of Australia (2002) The value of water-inquiry into Australia's management of urban water. Report of the Senate Environment, Communications, Information Technology and the Arts References Committee. Commonwealth of Australia, Canberra

Crow DA (2010) Policy entrepreneurs, issue experts, and water rights policy change in Colorado. Rev Policy Res 27(3):299–315

Department for Climate Change and Energy Efficiency (DCCEE) (2010) Adapting to climate change in Australia-an Australian government position paper. Commonwealth of Australia, Canberra

Department of Sustainability, Environment, Water, Population and Communities (DSEWPC) (2010) Water for the future-factsheet. Commonwealth of Australia, Canberra

Department for Water (2012) Water Sensitive Urban Design—consultation statement. (online) url: http://www.waterforgood.sa.gov.au/homepage/consultation/current-consultation/
Hallegatte S (2007) Do current assessments underestimate future damages from climate change? World Econ 8(3):131–146
Hallegatte S (2009) Strategies to adapt to an uncertain climate change. Glob Environ Change 19 (2):240–247
Howard C (2001) Bureaucrats in the social policy process: administrative policy entrepreneurs and the case of working nation. Aust J Public Adm 60(3):56–65
Howell J, Higgins C (1990) Champions of technological innovation. Adm Sci Q 35(2):317–341
Howell J, Shea C, Higgins C (2005) Champions of product innovations: defining, developing, and validating a measure of champion behaviour. J Bus Ventur 20(5):641–661
Huitema D, Meijerink S (eds) (2009) Water policy entrepreneurs: a research companion to water transitions around the globe. Edward Elgar Publishing, Cheltenham
Huitema D, Meijerink S (2010) Realizing water transitions–The role of policy entrepreneurs in water policy change. Ecol Soc 15(2):26. (online) url:http://www.ecologyandsociety.org/vol15/iss2/art26/
Huitema D, Lebel L, Meijerink S (2011) The strategies of policy entrepreneurs in water transitions around the world. Water Policy 13(5):717–733
Intergovernmental Panel on Climate Change (IPCC) (2007a) Climate change 2007-synthesis report. In: Pachauri RK and Reisinger A (eds) Contribution of working groups I, II and III to the fourth assessment report of the intergovernmental panel on climate change. IPCC, Geneva, Switzerland, 104 pp
Intergovernmental Panel on Climate Change (IPCC) (2007b) Climate change 2007—impacts, adaptation and vulnerability. In: Parry ML, Canziani OF, Palutikof JP, van der Linden PJ and Hanson CE (eds) Contribution of working group II to the fourth assessment report of the intergovernmental panel on climate change. Cambridge University Press, Cambridge, UK, 976 pp
Intergovernmental Panel on Climate Change (IPCC) (2012) Summary for policymakers. In: Field CB, Barros V, Stocker TF, Qin D, Dokken DJ, Ebi KL, Mastrandrea MD, Mach KJ, Plattner G-K, Allen SK, Tignor M and Midgley PM (eds.) Managing the Risks of extreme events and disasters to advance climate change adaptation. A special report of working groups I and II of the intergovernmental panel on climate change. Cambridge University Press, Cambridge, UK, and New York, NY, USA, pp 1–19
Ison RL, Collins KB, Bos JJ, Iaquinto B (2009) Transitioning to water sensitive cities in Australia: a summary of the key findings, issues and actions arising from five national capacity building and leadership workshops, NUWGP/IWC. Monash University, Clayton
Jones K (2005) Engaging community members in wastewater discussions. EcoEng Newsletter Number 11, Oct 2005
Keremane GB, McKay JM (2008) Water reuse schemes—the role of community social infrastructure. Water 35(1):35–39
Keremane GB, McKay JM, Wu Z (2011) No stormwater in my tea cup—an internet survey of residents in three Australian cities. Water 38(2):118–124
Kotter JP, Schlesinger LA (1979) Choosing strategies for change. Harvard Bus Rev 57(2):106–114
LECG Limited Asia Pacific (2011) Competition in the Australian urban water sector. Waterlines report series No. 42, National Water Commission, Canberra
Loftus A, Anton B, Philip R (eds) (2011) Adapting urban water systems to climate change-a handbook for decision makers at local level. ICLEI European Secretariat Publishing, Germany
Martin-Ortega J (2011) Costs of adaptation to climate change impacts on freshwater systems: existing estimates and research gaps. Economía Agraria y Recursos Naturales 11(1):5–28
McKay JM (2003) Who owns Australia's water?—elements of an effective regulatory model. Water Sci Technol 48(7):165–172
McKay JM (2005) Water institutional reforms in Australia. Water Policy 7(1):35–52
McKay JM (2010) Some Australian examples of the integration of environmental, economic and social considerations into decision making- the jurisprudence of facts and context. Pages 327-339 in D French (ed.) Global Justice and Sustainable Development, BRILL

Mintrom M (1997) Policy entrepreneurs and the diffusion of innovation. Am J Polit Sci 41 (3):738–770
Mitchell V (2006) Applying integrated urban water management concepts: a review of Australian experience. Environ Manag 37(5):589–604
Mitchell C, Retamal M, Fane S, Willets J Davis C (2008) Decentralised water systems—creating conducive institutional arrangements. Paper presented at Enviro 08, Melbourne 5–7 May 2008
National Water Commission (NWC) (2004) Intergovernmental agreement on a national water initiative. (online) url:www.nwc.gov.au/www/html/117-national-water-initiative.asp?intSiteID=1
Oswald Spring U, Brauch HG (2011) Coping with global environmental change—sustainability revolution and sustainable peace. In: Brauch HG et al (eds) Coping with global environmental change, disasters and security, hexagon series on human and environmental security and peace 5. Spring, Berlin Heidelberg
Pahl-Wostl C (2007) Transitions towards adaptive management of water facing climate and global change. Water Resour Manag 21(1):49–62
Pegram G, Li Y, Le Quesne T, Speed R, Li J, Shen F (2013) River basin planning: Principles, procedures and approaches for strategic basin planning. UNESCO, Paris
Philp M, McMahon J, Heyenga S, Marinoni O, Jenkins G, Maheepala S, Greenway M (2008) Review of stormwater harvesting practices. Urban Water Security Research Alliance Technical Report No. 9
Pielke R Jr, Prins G, Rayner S, Sarewitz D (2007) Lifting the taboo on adaptation. Nature 445:597–598. doi:10.1038/445597a
Productivity Commission (2008) Towards urban water reform: a discussion paper. productivity commission research paper, Melbourne, Mar 2008
Productivity commission (2011) Australia's urban water sector. report no. 55, Final Inquiry Report, Canberra
Roberts NC, King PJ (1989) Public entrepreneurship: a typology. paper presented to the academy of management meetings, Public Sector Division, Washington D.C
Roberts NC, King PJ (1991) Policy entrepreneurs: their activity structure and function in the policy process. J Public Adm Res Theor 1(2):147–175
Robinson KG, Robinson CH, Hawkins SA (2005) Assessment of public perception regarding wastewater reuse. Water Sci Technol: Water Supply 5(1):59–65
Schipper L (2006) Conceptual history of adaptation in the UNFCCC process. Rev Eur Community Int Environ Law 15(1):82–92
Schneider M, Teske P (1992) Toward A Theory of the Political Entrepreneur: evidence from Local Government. Am Polit Sci Rev 86(3):737–747
Taylor A, Cocklin C, Brown R, Wilson-Evered E (2011) An investigation of champion-driven leadership processes. Leadersh Q 22(2):412–433
Teske P, Schneider M (1994) The bureaucratic entrepreneur: the case of city managers. Public Adm Rev 54(4):331–340
Teodoro MP (2009) Bureaucratic job mobility and the diffusion of innovations. Am J Polit Sci 53 (1):175–189
UN-Water (2010) Climate change adaptation: the pivotal role of Water—UN-Water. Policy Brief. New York. (online) url:http://www.unwater.org/downloads/unw_ccpol_web.pdf
Varis O, Biswas AK, Tortajada C, Lundqvist J (2006) Megacities and water management. Int J Water Resour Dev 22(2):377–394
Wamsler C, Brink E, Rantala O (2012) Climate change, adaptation, and formal education: the role of schooling for increasing societies' adaptive capacities in El Salvador and Brazil. Ecol Soc 17 (2):2. (online) url:http://dx.doi.org/10.5751/ES-04645-170202
Ward J, Dillon P (2011) Robust policy design for managed aquifer recharge. Waterlines report no. 38, National Water Commission, Canberra

Water Services Association of Australia (WSAA) (2012) Climate change adaptation and the Australian urban water industry. Occasional Paper 27, WSAA Ltd

White S, Noble K, Chong J (2008) Policy forum: urban water pricing and supply. Reform, Risk and Reality: Challenges and Opportunities for Australian Urban Water Management, The Australian Economic Review 41(4):428–434

Wu Z, McKay JM, Keremane GB (2012a) Governance of urban freshwater: some views of three urban communities in Australia. Water 39(1):88–92

Wu Z, McKay JM, Keremane GB (2012b) Issues affecting community attitudes and intended behaviours in stormwater reuse: a case study of salisbury. S Aust. Open Access J Water 4 (4):835–847. doi:10.3390/w4040835

Managing Water Resources in Mexico in the Context of Climate Change

Úrsula Oswald Spring

Abstract This chapter analyses water management in Mexico in the context of global environmental change (GEC) and dual environmental and social vulnerability. The research questions are as follows: How can Mexico overcome the present unequal access of water without further destroying the precarious water and food security, and how could small-scale farmers depending now on rainfed agriculture support the recovery of the food sovereignty in the country? To answer this question this chapter studies the development model of integrated water management to explore the nexus between water and food. Mexico has an unequal supply of water: Only 31 % of water is available for 77 % of the population and produces 87 % of the GDP. Furthermore, 77 % of water is used in agriculture often with low efficiency in the arid northern region by agribusiness for exporting vegetables to the USA. The present situation of food insecurity in Mexico is also related to increasing food imports (virtual water) within the framework of the North American Free Trade Agreement (NAFTA). To a large extent, climate change (CC) induces floods and droughts, which are exacerbated by unsustainable urban development, where the Metropolitan Valley of Mexico City (MVMC) overuses existing aquifers. Therefore, only an environmentally sustainable management of water, recycling and reuse of treated water, will offer this densely populated country food and water security in the future.

Keywords Global environmental change (GEC) · Climate change (CC) · Dual vulnerability · Mexico · Water management · Water security · Food security

Ú. Oswald Spring (✉)
National Autonomous University of Mexico (UNAM), Centre for Regional Multidisciplinary Research (CRIM), Av. Universidad s/n, Circuito 2, Cuernavaca 62210, Morelos, Mexico
e-mail: uoswald@gmail.com

S. Shrestha et al. (eds.), *Managing Water Resources under Climate Uncertainty*, Springer Water, DOI 10.1007/978-3-319-10467-6_18

1 Introduction, Research Questions, and Conceptual Reflections

1.1 Research Questions

The increase of extreme hydrometeorological events during the last decade has obligated scientists to analyse their causes and impacts in Mexico. This chapter addresses the following research question: How can Mexico, severely affected by global environmental and climate change (CC), cope with the present unequal access of water without destroying the already precarious water and food security? Furthermore, how can small-scale farmers in rainfed agriculture contribute to Mexico's food security, thus also improving the livelihood of highly marginal people?

1.2 Organisation of the Chapter

Firstly, this chapter presents some conceptual elements on global environmental change (GEC) and dual vulnerability. The next part explores a model of integrated water management in Mexico due to more frequent and intense hydrometeorological events, resulting from GEC and CC. The third section analyses the unequal distribution of water resources in Mexico, given their regional and temporal constraints. An alternative water management system is suggested to protect the natural available resources for a growing population, where droughts, floods, and chaotic urban development are also threatening the biodiversity of the country. This part reviews also quality of water and groundwater overuse in the Metropolitan Valley of Mexico City (MVMC) (Oswald Spring 2011a) and suggests a sustainable water management system in the megalopolis. Part four revises water management in the framework of GEC and CC, where Mexico is highly exposed to hurricanes, droughts, and landslides. Part four discusses the nexus of water and food security and analyses the present situation of food insecurity related to increasing food imports (virtual water) in the context of the North American Free Trade Agreement (NAFTA). This part examines the policy of loss of food security and proposes an alternative model for dealing with the lack of maize. Climate conditions in the south and southeast of Mexico allow two harvests per annum. In this region, land is mostly used unproductively for extensive livestock by some landlords. On the other side, there are numerous peasants and indigenous people who have developed a millenarian culture of maize production, called milpa. Public investment in small-scale irrigation, credits, market mechanisms, and technical support for these peasants could overcome the present food insecurity in Mexico. This would also reduce the dependency on maize imports, improve the foreign trade balance, and overcome the existing social vulnerability of the most marginal people in Mexico. In the conclusion an integrated water and food security system is suggested able to mitigate and adapt to GEC and CC.

1.3 Orographic and Ecological Comments

Culturally, Mexico belongs to Latin America, but geographically it is located in North America. In 2013, the estimated population of the country was 120 million (estimations based on the National Census, INEGI 2010). It has a northern border of 3,145 km with the USA and a southern border with two more countries: 871 km with Guatemala and 251 km with Belize. It is culturally and environmentally highly diverse.

Mexico spans a total area of 1,964,375 km^2 and its coastlines extend almost to 11,000 km (including the islands). The country is located in the Tropic of the Cancer. Therefore, Mexico is highly exposed to warmer water conditions in the oceans due to CC. It faces cyclones from both oceans: the Pacific and the Atlantic. In 2013, tropical storms Manuel and Ingrid coincided. Further, a chain of volcanoes crosses transversally the central part of Mexico from the Pacific to the Atlantic. On both coastal lines, there are parallel mountain chains called the Sierra Madre. In Mexico, both the North American and the South American ecospheres are overlapping. Thus, in the context of ecological vocation and land use cover relating to forests, Mexico is the fourth most biodiverse country on Earth with a wide range of endemic species (Conabio 2008) and multiple ecosystems (Fig. 1). The pleasant climate in most areas of the nation has attracted many tourists due to its different altitudes, beaches, beautiful landscapes, and cultural diversity.

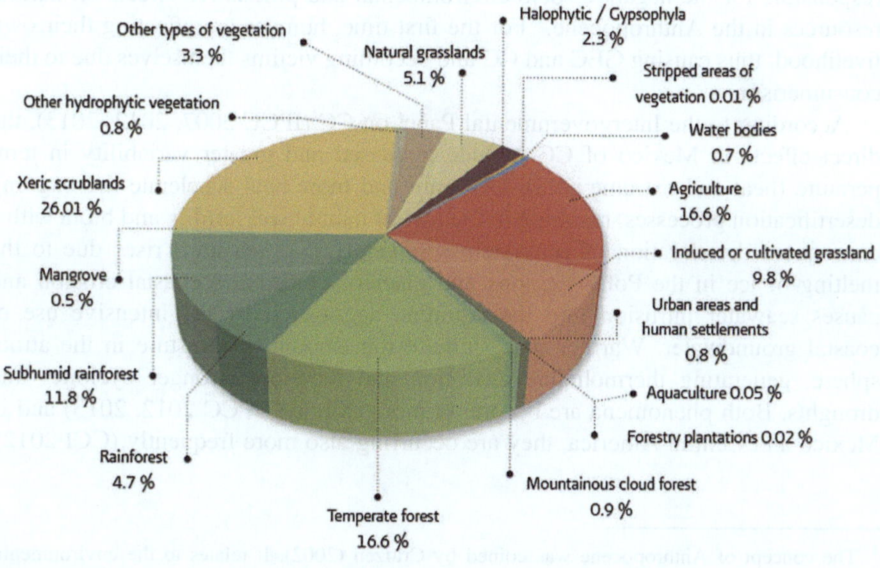

Fig. 1 Land use cover in Mexico in 2007 (*Source* CCI 2012)

The monsoon produces two clear distinguished seasons, and the rainfed agriculture relies on regular precipitation. During the last decade, climate variations have altered these traditional raining patterns, including interestival drought (Taboada 2005), which is crucial for maize production

1.4 Some Conceptual Reflections on Global Environmental Change and Dual Vulnerability

GEC (Brauch et al. 2008, 2009, 2011;) is more than CC. GEC includes the natural and human factors that produce changes in the Earth system. GEC is naturally understood to be the deterioration of soil fertility, water availability, quality of air and climate variability, loss of biodiversity, ecosystems, and services due to land use change and urbanisation. GEC is negatively interrelated and reinforced by anthropogenic factors such as population growth, chaotic urbanisation, pollution productive activities and transport systems, consumerism, and waste production. Natural and human interrelated factors produce environmental and social vulnerability, also called dual vulnerability (Bohle 2002; Oswald Spring 2013, Oswald Spring et al. 2014). This dual vulnerability is the result of deterioration in the atmosphere (physic-chemical changes, CC), the hydrosphere (rivers, lakes, seas, groundwater, and wetlands), the lithosphere (soil, minerals, sea floor, and mountains), and the biosphere (vegetation, wildlife, and ecosystems). Human beings are responsible for the negative socio-environmental and productive effects on natural resources in the Anthropocene.[1] For the first time, humans are affecting their own livelihood, thus causing GEC and CC and becoming victims themselves due to their consumerism.

According to the Intergovernmental Panel on CC (IPCC 2007, 2012, 2013), the direct effects in Mexico of CC include increased and greater variability in temperature (heat and extreme cold). Less rain and more heat accelerate the ongoing desertification processes, resulting in the loss of natural soil fertility and biota with a consequent deterioration of ecosystem services (ESS). Sea level rises due to the melting of ice in the Polar Regions and glaciers exacerbates coastal erosion and causes seawater intrusion into the aquifers, aggravated by the intensive use of coastal groundwater. Warmer seas increase the amount of moisture in the atmosphere, generating thermohaline alteration and therefore stronger cyclones and droughts. Both phenomena are becoming more extreme (IPCC 2012, 2013) and in Mexico and Central America, they are occurring also more frequently (CCI 2012).

[1] The concept of Anthropocene was coined by Crutzen (2002). It relates to the environmental changes produced predominantly by human intervention in the earth system, since the industrial revolution but especially the last five decades due to the deforestation, mining extraction, the intensive use of fossil energy, the rapid increase of greenhouse gas emissions into the atmosphere, and the pollution and warming of the seas.

2 Model of Integrated Water Management

This complex climate and human-induced development panorama affects the existing water resources reducing the supply, while at the same time more inhabitants are demanding more water. To understand this complex interrelation, a model of integrated water analysis is proposed (Fig. 2). The pressure on water is not only related to demand, but also to the changing GEC and CC conditions, which are reducing water availability. Both GEC and CC deepen the existing environmental and social contradictions and often contribute to water conflicts. Water supply is also reduced due to climate variability and inadequate water management at the local level (Martín et al. 2011). High levels of deforestation and more than half of the soils eroded (CCI 2012) reduce the infiltration of water into the aquifers, produce landslides, increase ongoing erosion, and create new risks for human settlements. As a result, water availability in groundwater, lakes, dams, soils, and wetlands is decreasing. On the demand side, the chaotic urbanisation process, cash

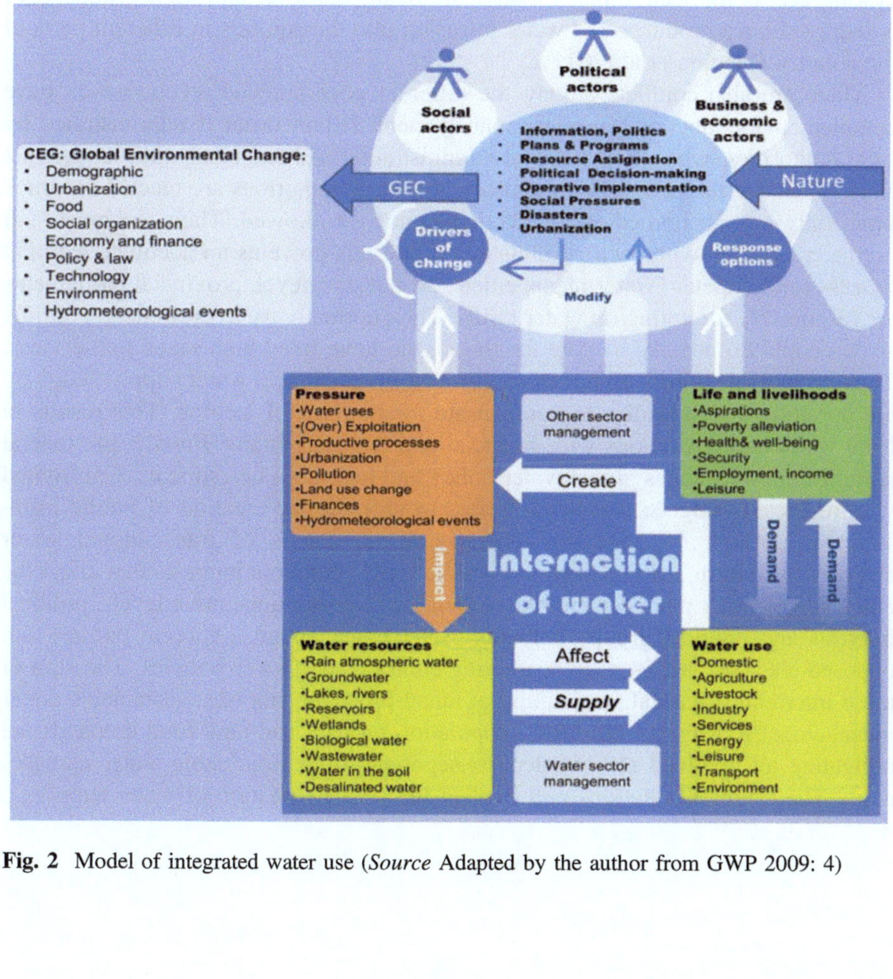

Fig. 2 Model of integrated water use (*Source* Adapted by the author from GWP 2009: 4)

crop agriculture in the northern drylands, population growth, and new hygienic conditions are demanding more water but are confronted with regional differences in precipitation. The lack of water is affecting particularly social vulnerable people, while the government is missing compensatory policies to overcome the economic differences and social inequity. On the contrary, the official policy has prioritised investment in urban facilities and irrigation districts in the northern and central drylands, responding to the political pressures from a growing urban population and the agribusiness.

There are additional social and environmental factors, such as pollution, change of land use, air contamination, urban growth, social inequity, an unsustainable model of development, and the loss of crucial ESS that further reduce water supply. Such short-term management often results in overexploitation of this natural resource and in the long-term, all social groups loose crucial ESS. Mexico is also severely affected by CC. Cyclones, floods, flash floods, falling levels of precipitation, and droughts are creating new challenges for systemic water management. More frequent and longer droughts especially in the drylands (IPCC 2012; CCI 2012) already affect irrigation agriculture in the north; Metropolitan areas are also lacking access to clean water. Finally, the rising sea level and high groundwater withdrawal have produced sea water intrusion into the aquifers in different parts of the country (Medina et al. 2011).

There are also conflicts among the different governmental levels, which have complementary functions in water management. Urban water is administrated by local authorities, who often lack the administrative and technical skills to provide clean water for the people. Furthermore, the local authorities are elected for three years only and an immediate re-election is not yet allowed. Their technical staff normally changes with each administration and this prevents an adequate learning process for competent water management. At the state level, provincial parliaments are responsible for voting on water tariffs. They normally try to promote low tariffs, since people do not like to vote for those who have fixed high taxes for services. Thus, most of the tariffs do not cover the cost of an efficient water supply and there are no additional subsidies to compensate for the lack of income. This results in local water administrations with financial and technical limits. Finally, the federal administration manages global water policy and long-term investments. Confronted with restricted budgets, this level prefers to promote the creation of public infrastructure through concessions, generally in the hands of transnational water enterprises (Barkin 2011). However, it is not only economic interests that affect the efficiency of water management. At each of the three administrative levels, political interests and political groups influence water management. Different parties with opposed electoral interests are promoting contradictory water policies. The sum of these interrelated natural, social, and political processes impedes sustainable water management in Mexico; but GEC, population growth, and new food demands are obligating all involved stakeholders to negotiate more sustainable water management. The following sections will explain the critical situation of water supply.

3 Water Use and Unequal Regional and Temporal Distribution of Water Resources in Mexico

The proposed model of water management analyses the impact on distribution of water in Mexico with negative and positive feedbacks related to GEC, such as urbanisation, population growth, globalisation, transnationalisation of the water sector, and CC. Mexico has experienced not only a growing struggle on the access and control of water, but also on failed privatisation processes, due to unjustified increases in tariffs by private firms without improving the quality and quantity of water supply. Furthermore, the increasing demand for water by different social groups, productive activities, services, regions, and political interests are aggravating the unsustainable water management in the country.

3.1 Water Availability in Mexico

Mexico receives a total of 1,522 km^3 of rainwater per year and 72 % (1,084 km^3) of this water evaporates (Conagua 2012). The average precipitation is about 711 mm each year; the aquifer recharge is estimated as 78 km^3, while 28 km^3 is extracted for human needs and productive activities. The average runoff is approximately 400 km^3 and the surface extraction (basically for agriculture) is 47 km^3. Thus, the average availability for each Mexican has declined from 17,742 m^3 in 1950 to 4,040 m^3 in 2013 due to population growth, and water is projected to fall to 3,783 m^3 for the year 2030 (Arreguín et al. 2011). The federal government has estimated the surface water availability, taking into account the existing commitments of water supply. Agriculture still uses 77 %, urban and domestic services 14 % and industry 9 %. In the official governmental communication (Diario Oficial de la Federación 2010), data for 722 basins were published (Fig. 3). The basins in the region of the Bravo River (bordering with the USA), Lerma-Chapala Lake, the MVMC and the Balsas basin (central and metropolitan region), together with the Peninsula of Baja California are short of water, densely populated, with important irrigation districts or highly industrialised. All these regions are lacking a safe and permanent water supply because they are located in the arid and semi-arid regions. The Yucatan Peninsula depends on groundwater and lacks surface currents due to karst geological conditions. Yucatan is therefore highly exposed to pollution from human sewage and industrial waste. Finally, Mexico as a whole faces temporal water constraints due to the monsoon. The rainy season is between June and September and the rest of the year little precipitation occurs (Oswald Spring 2011a).

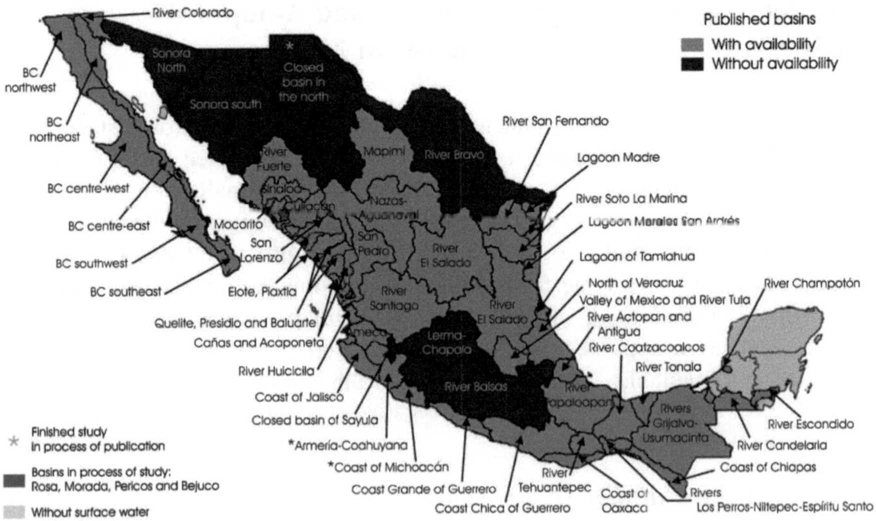

Fig. 3 Surface water availability in Mexico (*Source* Conagua 2008)

3.2 Regional Differences in Access to Water

Climate conditions play a crucial role in water availability. The distribution of rainfall is not only limited by the monsoon, but also precipitation is very unequal in Mexico. The north gets only 25 % of precipitation and the south and southeast 49.6 %. Taking into account the semi-arid and arid regions of Mexico in the north and central areas, the natural availability of precipitation is 31 % of water, where 77 % of the people live and produce 87 % of GDP. The south and southeast receive 69 % of water for 23 % of the people, who produce 13 % of GDP. This natural imbalance of precipitation was omitted in the past when industrialisation and irrigation policies were directed by the government to water scarce regions. Today, better knowledge of water availability offers new development perspectives for the south and southeast, being the region with the uppermost levels of marginality and underdevelopment. These are also the areas which are historically abandoned by governmental investment, education, services, and development possibilities. Finally, these extremely poor people are also highly exposed to hurricanes and landslides, but benefit from important freshwater reserves, high biodiversity, and crucial natural resources (oil, uranium, gas, etc.).

The lack of water is partially compensated by pumping 37 % of water from the ground (Fig. 4). Most urban, domestic, and industrial water is taken from aquifers, but increasing agricultural demand is also pumping the liquid from the ground. The most crucial imbalance of water is found in megacities and agricultural sectors in the drylands. In relation to urban and industrial development, unsustainable policies have increased the natural imbalance of water in the MVMC, Guadalajara and of

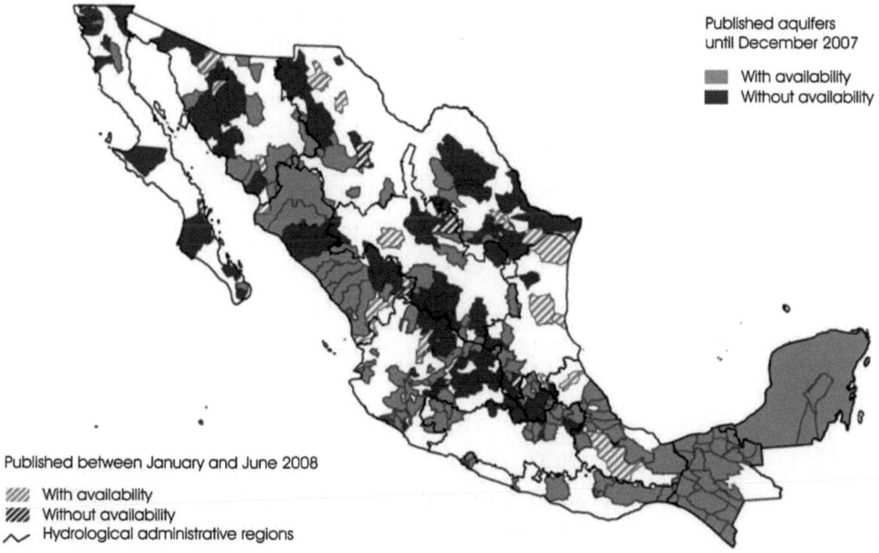

Fig. 4 Availability of aquifers (*Source* Conagua 2008)

Monterrey. In these urban areas, water consumption is high due to population density, immigration, chaotic urban development, inadequate infrastructure, and old pipes leaking up to 45 % of water. In the past, high subsidies promoted centralised urban development in this metropolitan area and created a perverse system of water consumption. On the one side, low tariffs do not cover the costs of water supply, but neither are they promoting a water saving culture. Furthermore, general subsidies have transferred necessary resources to rich colonies, industries, and the service sector. These policies have aggravated the anachronisms for unsustainable water management and avoided a permanent supply to marginal urban colonies and slums.

The chaotic urbanisation process in Mexico has taken off since the 1950s (Table 1), and in 2013, 78.3 % or 93.96 million of Mexicans are living in urban areas (INEGI 2013). The MVMC is the most critical region in terms of population growth and water supply. In Mexico City, the growth rate in 1990 was 1.64 %, but it has declined to 0.86 % between 2000 and 2005. However, the growth rates of the whole MVMC are still very high, especially in the surrounding states: Hidalgo 2.61, Tlaxcala 1.75, and Mexico 1.56 from 2000 to 2005. Due to the lack of water, the MVCM is importing one-fifth of water for human consumption and productive activities from the Cutzamala[2] system from the states of Mexico and Michoacán (northwest). Water consumption in the MVMC was 622 hm^3 per year (based on consumption in 2004). According to official projections, by 2025, the supply in the MVMC is estimated 771 hm^3 and in 2030, 830 hm^3. Water reuse is very low,

[2] The Cutzamala system provides one-fifth of the drinking water to the MVMC.

Table 1 Urbanisation processes in Mexico

Year	Per cent urban	Rural population (in 1,000)	Urban population (in 1,000)
1900	10.5	12.17	1.44
1930	12.5	13.66	2.89
1940	20.0	15.72	3.93
1950	28.0	18.57	7.21
1960	50.7	17.22	17.71
1970	58.7	19.92	28.31
1980	66.3	22.55	44.30
1990	71.3	23.29	57.96
2000	74.6	24.72	72.76
2010	77.1	24.93	87.37
2020[a]	87.8	26.79	93.35

Source Garza and Rivera 1993 and INEGI (1990, 2000, 2010)
[a] Estimations from census data in 2010

estimated at only 12 % for 2004. The MVCM is thus withdrawing most of the water from the ground, estimated 751 hm^3 in 2004. Continuing with the same model, the estimated withdrawal may increase to 1,360 hm^3 in 2025 and to 1,471 hm^3 in 2030 (Morales Novelo and Rodríguez Tapia 2011: 402–403). Today, the six aquifers of the MVMC are considered to be the most exploited on Earth (Oswald Spring 2011a) with a negative impact of subsidence in the eastern part of the MVMC of 50 cm per year. It will be impossible to continue to withdraw this amount of groundwater from existing aquifers due to the deterioration of water quality and subsidence, which is affecting the infrastructure such as the national airport, drainage system, water supply system, roads, and houses. Therefore, the MVMC must develop alternative sustainable management of its own resources (rainwater harvest, sewage, and recycling) or it will run out of water for 25 million inhabitants with serious political consequences. Short-term interests are trying to import water from surrounding regions (south and east) with high environmental and economic costs to both areas, since the MVMC is located at 2,300 m. Politically, it will be increasingly more difficult for the proposed supply regions to give their water to the metropolitan area, and there is rising political opposition in these regions against such projects.

The second and most important water use is for agriculture. The intensive use of water has salinised important areas in the irrigation districts and the extraction of groundwater (Fig. 4) has increased this process. A critical water supply situation exists in the coastal region of the Valley of Hermosillo in Sonora and Baja California, where the intensive pumping of groundwater for agriculture and urban areas in an arid region has produced sea water intrusion into the aquifers (Medina et al. 2011). Overexploitation of aquifers has resulted also in intrusion of sea water in Campeche and Yucatan.

Figure 4 indicates also that there is limited availability of groundwater where the most important demands for urban, industrial, and irrigation developments exist. By comparing the availability of surface and groundwater (Figs. 3 and 4), it is precisely in the regions with the highest water demand, where supply is seriously limited. There are no reserves for further expansion in these drylands without a drastic change in the existing model of water management. But, there is not only the physical problem of water use, but also a concentration of water demand where the limited supply has also created severe water conflicts between regions with high water availability and those with insufficient water (Oswald Spring 2011b). Before exploring other alternatives, there is a second constraint relating to the temporal availability of water in Mexico.

3.3 Temporal Constraints for Water Availability

Mexico, like most other tropical countries, depends on the monsoon for water supply. During June and September, the country receives 67 % of its annual precipitation (Conagua 2008). This creates additional pressure on water availability during the dry season, precisely because humans, agriculture, and most productive activities require water during the whole year. Agriculture is still the key user of water. Most of the irrigation districts are located in the northern arid and semi-arid regions, with little precipitation, extreme climate conditions, and thus depend on irrigation from storage water or groundwater during the dry season. Agriculture relies on the important infrastructure of water storage in the north; used to produce vegetables during the whole year. There are approximately 15,000 reservoirs with a capacity of more than 5,000 m^3; each with an overall estimated capacity of more than 150 million m^3. In tourist areas and highly populated regions, the desalination of sea and brackish water is promoted, together with the reuse of treated water in gardens and parks.

3.4 Water Use in Mexico

Agriculture and livestock consume 77 % of freshwater, urban use accounts for 14 %, industries 4 %, and the remaining 5 % is used for cooling systems in thermoelectric plants (Conagua 2012). Another 122,800 million m^3 of water is used for hydropower and is considered as non-consumptive, because this water is available for other productive activities or human needs. Mexico is also a very diverse country, where traditional systems of water management and irrigation were developed thousands of years ago. The Spaniards learnt from the indigenous people and introduced new cash crops such as sugar cane (Maldonado 1989). The diverse climate and ecosystems have created a wide range of agricultural productive systems relating to different environments and cultures but have also produced social stratification. Particularly in the south and southeast, different indigenous groups have consolidated a sustainable system of maize production with a policultivation called milpa, where beans permit nitrogen fixation from the air into the soil and then to the maize, and the maize supports the growth of beans and pumpkins. The diversity in adaptation of maize crops enabled the development of 58 species, with colour variations from white, blue, red, yellow to mixed corn species. This adaptation of regional water and climate conditions has permitted these indigenous societies to develop important civilisations such as the Maya and the Aztecs, and Mexico has created more than 600 food items in a very pluricultural settlement. Thus, Mexico and Central America are not only the regions of origin but also foremost in the adaptation of maize.

Finally, access to water is also related to social stratification and poverty. While the prosperous colonies in Mexico City get up to 800 litres per person per day (Lomas), in the poor colonies (Ixtapalapa), access is less than 35 l per day and during the dry season, these people get tap water only once per month. They are obliged to buy the vital liquid from water tanks, often at high prices. The rural areas of Oaxaca, Guerrero, and Chiapas suffer worse conditions, where officially more than 40 % of people lack any water supply (Oswald Spring 2011a). The results are periodic epidemics of water-borne illnesses.

3.5 *Quality of Water*

Increased water demand, population growth in cities without efficient drainage and treatment plants, declining supply, dryer climates, and the intensive use of groundwater have affected the quality of drinking water. Safe water does not contain damaging organisms such as bacteria, viruses, protozoa (unicellular eukaryotic organisms), and helminths (nematode worms). Safe tap water must also be free from inorganic pollutants such as arsenic, fluoride, lead, nitrates, nitrites, or asbestos. Among emergent pollutants, we find organic toxic substances (VOC: volatile organic compounds)[3] (Muñoz and Mólgora 2011), which affect the health and life quality of people. Finally, drinking water must also be without flavour, odour or colour, transparent, and free from total dissolved solids and turbidity. According to Biswas (2013), about 3,000 million people lack safe drinking water worldwide, and in Mexico, people do not trust tap water. Those who can afford it buy bottled water, a new business basically managed by multinational enterprises.

Water quality is affected by four different causes: (a) the old infrastructure of tap water whose uncountable micro-perforations allow the entry of harmful micro-organisms into the water; (b) the quality of water in the aquifers is also polluted

[3] Hydrocarbons such as benzene and toluene; chloroform used in producing drinking water; bisphenol from plastic containers and coatings; siloxanes from personal care products; brominated organics from fire-retardant clothes; and many other materials are called emergent pollutant.

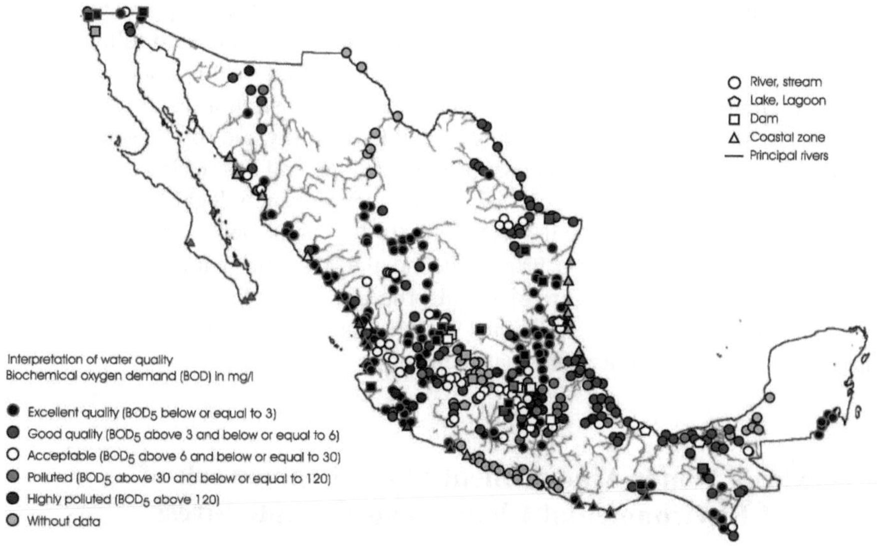

Fig. 5 Quality of water in Mexico (*Source* Conagua 2012)

naturally by arsenic and fluoride (Ávila et al. 2011), which causes brain and kidney diseases (González et al. 2011); (c) around the capital and metropolitan valleys of Mexico City, Guadalajara (Lerma-Santiago River), Monterrey, the northern Gulf, and the petrochemical production in the Central Gulf region, where the pollution relates to population concentration and industrial activities. High levels of total suspended solids (TSS), biochemical demand of oxygen (BOD), and chemical oxygen demand (COD) essentially affect coastal zones in the Pacific from Colima to Guerrero; on the Atlantic, the south of Veracruz and Tabasco, and also the rivers of Lerma-Santiago, Bravo, and Soto La Marina (Arreguín et al. 2011). The lack of industrial and domestic treatment facilities is also related to this pollution. Urban areas in Mexico are also lacking adequate treatment plants. Treatment plants use only 66 % of their installed capacity and often lack professionals for their operation (Martín et al. 2011). Finally, population growth and urbanisation have changed the hygienic pattern of people and rising temperatures, especially before the rainy season, are increasing the demand for water when the supply is lowest. In terms of quality of water, these four factors have polluted most of the surface and groundwater in the country (Fig. 5). Only in isolated mountain areas with low population density is water quality still excellent.

As an initial synthesis, the arena of water supply is insufficient and the quality of water is poor, thus creating water-borne illnesses. This brings extra costs for households, which are obliged to buy bottled water. Confronted with these bad results of integrated water use, authorities at the three levels of government (national, state, and municipality), the organised society, and the business community must change the present system of water management. Firstly, they must understand and then negotiate the changing conditions of water supply and demand related to GEC and CC. In addition, they must adapt water demand to the existing supply in different regions of Mexico. Secondly, through this concealed agreements, a common effort should orient the water administration to improve quality and overcome the temporal constraints related to the rainy and dry seasons. Thirdly, in 2012, the Senate promoted a constitutional change and has declared safe water a basic human right for any citizen. As a consequence, a new legal framework must be developed. This new Global Water Law (GWL) is an opportunity for the three sectors of Mexican society to bring order to the existing anarchy in water management. It may provide clear rules and procedures to protect the people and at the same time improve the ESS, thus granting present and future generations an integrated and sustainable water management.

4 Mexico's Water Management in the Framework of Global Environmental Change and Climate Stress

4.1 Effects of Global Environmental Change in Mexico

GEC has caused severe and complex dual vulnerabilities (environmental and social), due to the interaction of urbanisation, industrialisation, cash crop agriculture, erosion, deforestation, land use change, and extreme hydrometeorological events. The urbanisation process in the Yucatan Peninsula started over one thousand years ago with the Mayan ceremonial centres, which were considered the first urban areas in Mexico. In the fifteenth century, the Aztecs lived in the high plateau of the MVMC with more than a million inhabitants, but it is especially since 1950 that the city has grown due to immigration (Table 1). Cities depend on food, fuel, water, timber, building materials, and industrial goods, which were basically brought from outside. They form part of an intensive trade in goods, where raw materials were brought in, manufactured, and sold on. Today, a growing population, industries, and service sectors are not only consuming resources but also producing waste (some of them toxic), sewage, gases into the atmosphere, and changes in land use. The sum of these impacts is affecting crucial ESS. Also rural-urban migrants often settled in slums around the centre of cities, when their survival is threatened in the home region. In these suburbs, life quality is often low and people have no other choice than to settle in risky locations.[4] Far from their working places, the inhabitants suffer from dual vulnerabilities such as low salaries (with often precarious working conditions), threatened environmentally by landslides, and floods. Nevertheless, during the last

[4] Most of the high-quality landscape in urban areas is owned by the urban bourgeoisie and Estate holders. Therefore, for poor rural immigrants only land in the suburbs and ravines are available with risks of landslides, public insecurity, and lack of basic services.

three decades, the urban population has tripled in Mexico and urban areas have increased sixfold (INEGI 2010). These land use changes are causing further damage to natural and agricultural areas.

4.2 Impacts of Climate Change and Hazards as a Challenge to Integrated Water Management

Mexico is vulnerable to cyclones, floods, flash floods, and landslides but is also highly exposed to drought (Munich Re 2008). Given this environmental vulnerability, the government of Mexico has systematically analysed the threats and submitted five reports to the UNFCCC (CCI 1997, 2001, 2006, 2009, 2012). As a result of anthropogenic activities during the recent decades, oceans have increased in temperature. As a consequence, disaster impacts have shifted from geophysical (earthquakes, tsunamis, volcano eruptions) to hydrometeorological extreme events with a high cost to human lives, destruction of infrastructure, and economic impacts (Figs. 6 and 7).

In terms of drought, water scarcity is basically affecting the drylands in the north and centre of the country. In the urban areas, water saving technologies, changes to infrastructure and old pipes, reuse and recycling, infiltration of rainwater into the aquifers, and tariffs that promote a rational use of water could reduce this scarcity. In agriculture, the substitution of irrigation by gravity through water saving technologies (micro-aspersion) could improve crop yields. Changes from water intensive crops to those with less water requirements and the importation of food crops in

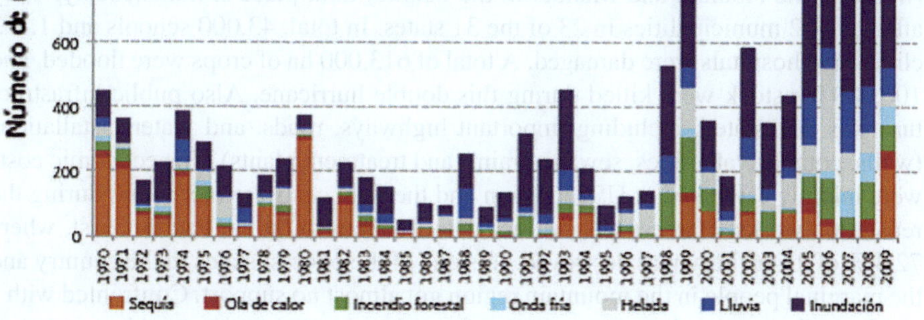

Fig. 6 Number of disasters in Mexico (*Source* Inventar, La Red, 2012 in CCI 2012)

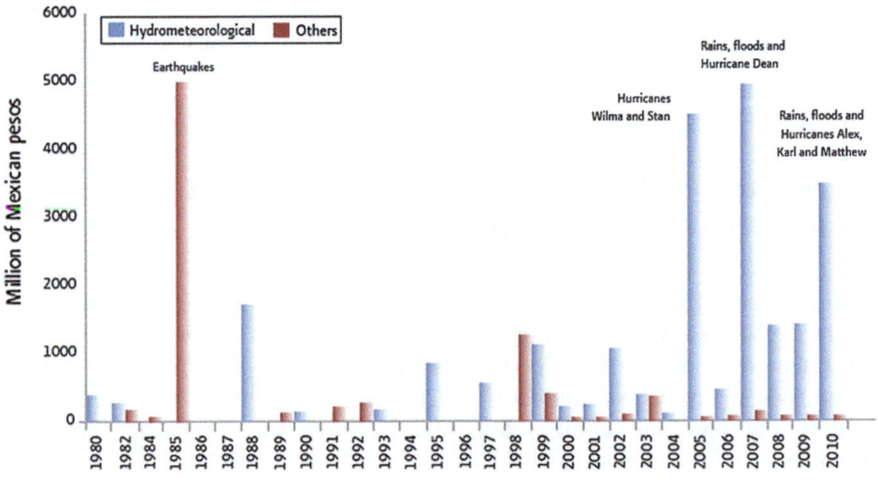

Fig. 7 Costs of disasters (Mexican pesos) (*Source* CCI 2012)

water scarce regions (virtual water) would free water for human consumption and limit aquifer overuse. Recovering destroyed ecosystems and a systematic reforestation will improve the infiltration of rain into the soil, reducing erosion and recharging aquifers. Sea water intrusion obligates coastal communities to obtain an alternative water supply or desalinate water for human consumption. Tropical storms force people and governments to intensify disaster risk management, give early warning, build refuge, and protect infrastructure from storm surges. By improving building codes for wind resistant houses and public facilities, the risks related to hurricanes, death, and damage to homes and the economy could be reduced. Communities highly exposed to storm surges must be relocated to safer areas by the government. A crucial theme in relocation is the involvement of the affected people (Bronen 2012), because in the past resettlements have been mostly unsuccessful (Oliver-Smith 2011).

Stronger, more frequent and unknown hydrometeorological events are occurring in Mexico and can be related to CC. For instance, in 2013, two tropical storms (Ingrid in the Atlantic and Manuel in the Pacific) took place simultaneously. They affected 312 municipalities in 23 of the 31 states. In total, 43,000 schools and 1,153 clinics and hospitals were damaged. A total of 613,000 ha of crops were flooded, and 100,000 livestock were killed during this double hurricane. Also public infrastructure was devastated, including important highways, roads, and water installations (wells, potable water lines, sewage drains, and treatment plants). The economic costs were initially estimated at US$5 billion and the final costs are increasing during the reparation works. Hardest hit was the state of Guerrero on the Pacific coast, where 72.4 % of its territory was destroyed. It is one of the poorest states in the country and the marginal people in the mountain region got almost no support. Confronted with a survival dilemma, an important group of people migrated to the neighbouring state of Morelos in search of safer conditions (INEGI 1990a).

5 Agriculture Confronting the Nexus of Water and Food Security

5.1 Water (Mis-)Use in Agriculture

The infrastructure of irrigation districts is located in the northern drylands where water demand and evaporation are high (Fig. 8). In 2006, the Ministry of Agriculture stated that 60.7 km^3 of water was assigned to agriculture, of which 40.6 km^3 came from surface water and 20.1 km^3 was pumped from groundwater. From 1988 to 2008, the water used in agriculture relied mostly on gravidity, which is still very inefficient. In the northern drylands, the evaporation rates are also high and transportation through open channels increases the loss of water.

Palacios Vélez and Mejía Sáez (2011) specified that the efficiency of irrigation is around 48 %, and in particular, irrigation by gravity has low efficiency. In 2010, because of a yearly reduction of 6 % in irrigated areas due to lack of water and an ongoing drought, farmers were enrolled into a government training programme in water efficiency. They increased the amount of water by improving the capacity of pumps in the wells, flow through pipes, reduction of evaporation in channels and pipes, and the use of micro-aspersion technology in the fields. Nevertheless, half of the water in agriculture could still be saved with further technological improvements such as satellite managed irrigation, pumping the water though pipes instead of open channels, levelling agricultural fields, changing from water intensive crops, and renovating old irrigation facilities with new pumps and technological systems. The most important constraints for these changes are the lack of credit and investment in these modern irrigation systems, due to low or unstable prices of cash

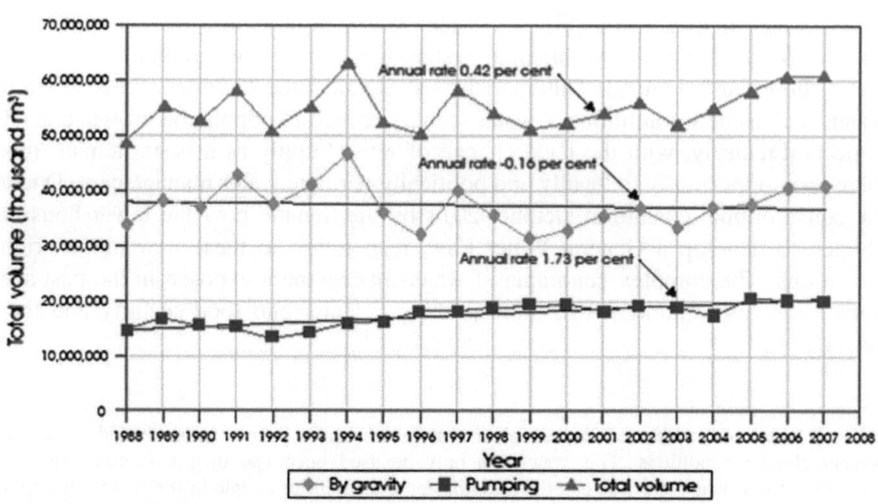

Fig. 8 Variation in volumes of water used for irrigation according to sources (*Source* Palacios Vélez and Mejía Sáez 2011: 136)

crops and periodic changes to official agricultural policies. An important issue relates to the present system of perverse subsidies, where diesel and electricity are subsidised in water-stressed regions, taking away the motivation for improving irrigation technology. Thus, export farmers and vegetable agribusinesses are unwilling to invest in irrigation efficiency with their own capital and the government has limited resources for this purpose.

As a consequence, the high amount of water in the irrigation districts of the northern states of Sonora, Baja California, Chihuahua, and Sinaloa is creating serious problems and conflicts with urban water demand. Both are using surface and groundwater, but agribusiness uses six times more water for irrigation. As a result, several aquifers in Sonora and Baja California have experienced sea water intrusion due to excessive water withdrawal and the water supply for the population is threatened. Instead of forcing agribusinesses to increase the efficiency in irrigation, the federal government in accordance with the state and the city of Hermosillo (capital of Sonora) decided to build an aqueduct from the El Novillo Dam to Hermosillo. They did not consult or negotiate with the traditional water owners, the Yaqui indigenous tribe. As a political response to this water robbery, the indigenous group blocked the highway to Hermosillo for months. Finally, the Supreme Court requested an environmental impact study assessment from the state and local authorities and negotiations to take place with the original owners of the water.

In Chihuahua, several small farmers were killed when they tried to stop the withdrawal of fossil water[5] from a deep aquifer by international agribusiness. In total, the government estimates that there are approximately 400,000 water conflicts, ranging from local to national levels such as mentioned in Sonora, Chihuahua, or in the MVMC, where indigenous women of the Mazahua group shut down the water supply to the megacity (Oswald Spring 2011a). These examples highlight the need for the government, society, and business community in Mexico to promote integrated water management (Fig. 2). Technology, policy, negotiation, and water culture together with environmental restoration can create equilibrium between water supply and demand. This model must also include regional access to freshwater, such as new conditions of water scarcity related to population growth and CC impacts. Precisely, with the legal change of water supply as a basic human right, there is an opportunity to legally and politically regulate water management. Due to the constitutional change in October 2011 by the Senate, now the lower house is obliged to develop a General Water Law, responding to these new human right conditions. The complex panorama of water management exposed in the past sections forces us to address two more problems: water and food security and their nexus.

---

[5] Fossil water was produced in geological remote times and cannot be recovered within the present climate conditions. This water can only be used once and then it is gone for ever. Therefore, fossil water must be carefully administered to grant to people in the future the supply and agricultural use cannot take away from people the possibility to leave in this region. Therefore, fossil water must be carefully administered in order to grant people water supply today and in the future. Fossil waters are not available for agriculture when the aquifer has limited resources.

5.2 Water Security: A Contested Concept

Water security has only become a scientific discussion since 2000 (Bogardi et al. 2014). It refers to complex interrelations between people, human activities, energy, ESS, and natural requirements for maintaining biodiversity in changing conditions of GEC and CC. There are also new demands relating to improved hygienic living conditions, the alleviation of poverty, urbanisation, and human aspirations of well-being. The Ministerial Declaration of the World Water Forum in The Hague in 2000 proposed a definition of water security. It includes the necessity to grant every human being their basic needs (societal security), a safe, permanent and nutritious food supply (food security), basic public health (health security), and a dignified livelihood (human security). To achieve these securities, ecosystems must be better protected (environmental security), risks managed (HUGE security[6]), shared and governed water resources (political security), and realistic water costs to grant the finances for infrastructure and administration (economic security; Oswald Spring and Brauch 2009). Figure 2 indicates that any water management has positive and negative feedbacks, affecting the water supply in quantitative and qualitative terms. Technological tools alone are not sufficient to grant water for people, industries, and agriculture, but saving techniques (reuse and recycling of water) are steps towards integrated water management together with negotiation processes, water governance, and policies. These activities include tariffs, laws, assignation of different water qualities for industries and agriculture, and safe water for human supply. Only when supply and demand produce a hydrological equilibrium, the required water quality is achieved through negotiation processes between all stakeholders and the basic human rights and food requirements are granted in any region and social class, fulfilling the Declaration of Dublin (1992). The 'tandem' system (where water is distributed for a limited number of hours per day or days per week on a rota basis operated by the municipality) allows better distribution among different users, but at the same time, it reduces the quality of water by changing the pressure within the conduits, allowing micro-organisms to penetrate through the old leaking pipes.

5.3 Food Insecurity: NAFTA and Virtual Water

Most water is used in agriculture in Mexico, since there is a direct link to food production and food security. Food security includes not only innocuous, safe, and permanent food (FAO 2012) but comprises also social and cultural factors of food and nutrition. Food security relates to land rights, seeds, credits, family ties, and

[6] HUGE security: human, gender, and environmental security concept introduced by Oswald Spring Ú (2009).

social and gender[7] relationships of a productive and consumption pattern often improving communitarian cohesion. Thus, food security depends not only on commercial agriculture, but on food produced in orchards and a diversity of non-agricultural activities, such as fishing, recollection, storing food, local markets, and hunting.

For Mexico, the projections of CC indicate harsher conditions in drylands and more rain and floods in the humid tropical regions, affecting both water and food security. In the drylands of Mexico, more severe and longer-lasting droughts have existed for several years now. There is also an increase in evapotranspiration related to higher temperatures and more water vapour in the low and high clouds are changing existing climate patterns, together with a probable alteration in global thermohaline circulation, where an excess of freshwater is now discharged into the oceans (Rosengaus 2007). As a result, 'droughts have serious social, economic, and environmental effects. Since the second half of 2010, a significant lack of rain in 19 states of Mexico became a severe drought causing losses over 15,000 million pesos' (CCI 2012: 15), which amounts to 6.39 % of GDP in the agricultural and livestock sector. Harvests of corn, beans, and vegetables severely declined and livestock perished and 2,350 communities with almost 2 million people were affected. In 2011, the drought caused losses of 1.8 million hectares; almost 5 % of the arable land (CCI 2012).

Food security is directly related to the basic food crop production of maize. The projections for CC impacts on maize production in Mexico are especially affecting the rainfed agriculture of small farmers. Between 13 and 27 % of the actual surface dedicated to maize may be lost (CCI 2012), and this will oblige subsistence farmers to immigrate to cities or abroad in order to survive. Today 2.7 million of productive unities or 66 % belong to small farmers, who cultivate less than 5 ha. Given the negative climate conditions, the yield has doubled from 1990 to 2007 achieving an average of 2.82 t/ha (Berlanga 2010).

Instead of supporting maize production in the country, the government has favoured the import of highly subsidised maize from the USA. Figure 9 indicates that Mexico increased the volume of maize imports between 1990 and 2010, especially after the signature of the NAFTA in 1994. The USA with an annual production of 250 million tons exports about 50 million tons. China is the second producer, Brazil the third, and Mexico the fourth with an average of 23 million tons, increasing to 24.12 million tons in the cycle of 2012 (SIAP 2013). Nevertheless, Mexico imports yearly 5 million tons of maize annually, basically from the USA (USDA 2013).[8] The costs are more than US$2.6 billion. This money could be productively invested in the country to promote an integrated agricultural development in the marginal

[7] Without doubt, worldwide women still produce at least half of the food for their families basically in orchards around their houses and in poor countries up to 90 %. Cash crops are basically produced for commercial means and cattle feeding, recently also for biofuel. Argentina is now the first producer of biodiesel from soya beans, basically exported to the European Union (IEA 2013).

[8] Japan is the first importer of corn with 16 million tons and the Republic of Korea the second with 8.5 million tons.

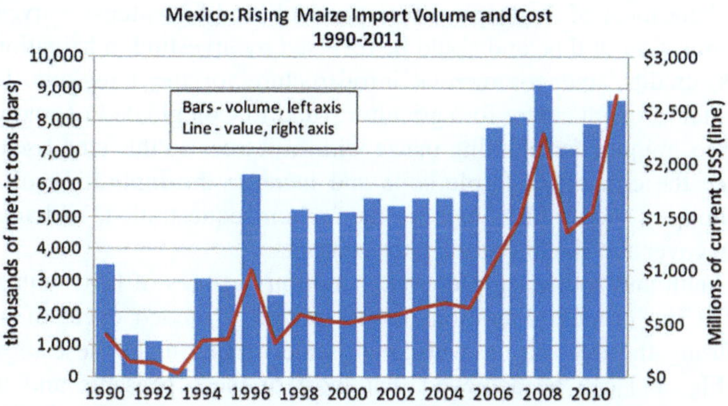

Fig. 9 Rising imports of maize in Mexico (*Source* SIAP 2013 (download, December 2013)

regions of Mexico. But, the collapse of the internal maize price without governmental compensation and the reduction in agricultural subsidies has affected the productivity of many small-scale farmers. Currently, they only produce for their subsistence.

5.4 Adaptation Under Difficult Climate Conditions and Resilience for Peasants in Rainfed Agriculture

Today Mexico produces around 23 million tons of maize in 6 million rainfed and 1.5 million irrigated hectares (Fig. 10). CC is especially threatening to this rainfed production, including the livelihood of small-scale peasants. To adapt to adverse climate conditions, more irrigated lands must be developed in regions where freshwater reserves exist. This region is located in the south and the southeast of

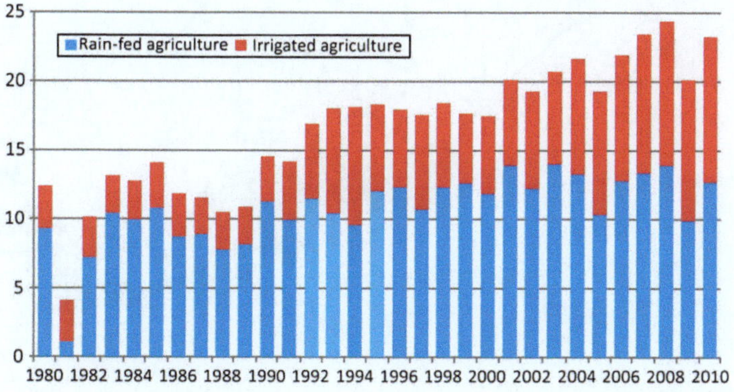

Fig. 10 Rainfed and irrigated maize production in Mexico (tons) (*Source* SIAP 2013)

Mexico, where most of the tropical forest was cleared for extensive livestock. The under-optimisation of this land could be reversed by investing in irrigation systems, extensions, credits, and commercial infrastructure for these regions. INIFAP (a governmental research centre for agriculture) and El Colegio de Posgraduados have developed techniques for planting maize with fruit trees so that small-scale farmers may reduce the erosion of fertile soils and increase the infiltration of rainwater. These techniques would also improve the carbon sequestration, recharge of aquifers, and recover the livelihood of poor peasants.

In the south and southeast, there are 9 million hectares of land with freshwater availability. The climate conditions would permit a second cycle of maize production during the dry season with small-scale irrigation. In these regions, food security (Fig. 11) can be achieved, but most of these peasants and indigenous groups are highly marginalised and have low levels of education. With credits, a small irrigation system for a second harvest, extensions and market facilities for producing, and storing and trading maize in additional tons could be added to the national maize production. With a further one million hectares of irrigated land in the south and southeast, an additional 8 million tons of maize could be harvested. Also 2 million hectares of livestock could be transformed immediately into agricultural lands to produce 16 million tons of corn (Turrent Fernández et al. 2013). Both measures would improve not only food security in Mexico but also the livelihoods of the most marginal peasants. There is a second reason to promote food security. The international prices of maize have changed dramatically during the last years. In 1990, one ton of maize costs less than US$100/t, in 2013, the price tripled to US$340/t. The agricultural trade balance of Mexico was negative in 2012

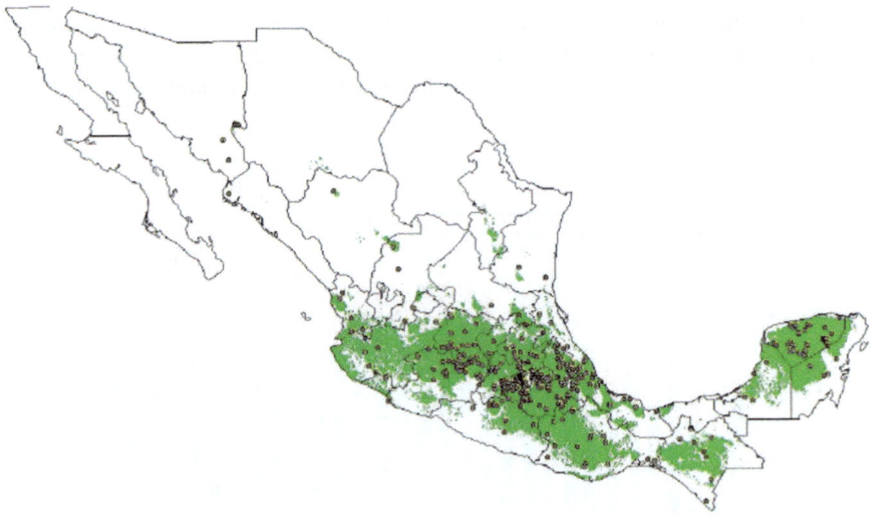

Fig. 11 Model of ecological niches for potential increase of maize production within the present conditions of climate (*Source* Ballesteros-Barrera et al. 2011)

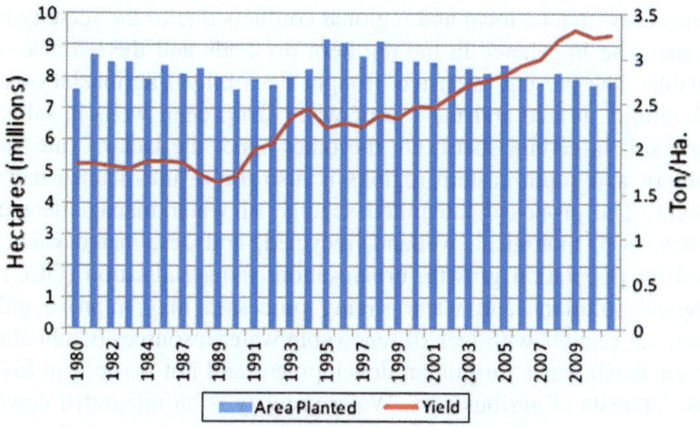

Fig. 12 Planted areas of corn and yield (*Source* Turrent Fernández et al. 2013)

with a deficit of US$2.8 billion and the costs of corn imports are almost the same amount as this trade imbalance.

A third reason relates to reduce the pressure for maize production from the northern drylands. This would enable the recovery of aquifers, reorienting water use from agriculture to human and industrial supply, and the production of less water intensive crops. Finally, the traditional knowledge of milpa is an integrated management of maize with beans and pumpkins together with agaves or fruit trees which could enable Mexico to recover its food security in the next decade. This would create resilience for the small farmers currently exposed to CC impacts and migration and would at the same time improve their livelihoods. Figure 12 indicates a constant increase of yield/ha with a clear reduction of planted hectares. The reason for the reduction of cultivated lands could be attributed to CC and to the lack of credits for production, but also to dryer conditions and difficult market conditions. Thus, most of the small-scale holders produce basically for their subsistence and could be involved into the market production with a different agricultural policy.

6 Conclusion

An integrated water and food security system where supply and demand are in equilibrium obligates Mexico to explore alternatives for better adaptation to the impacts of CC and import less maize. In relation to the two research questions posed at the beginning of this chapter, a new legal framework may grant every Mexican enough safe water, including the most marginal rural and urban groups. To achieve this goal, the organised society, agribusinesses, and governments are obliged to negotiate a water supply capable of resolving the existing and future demands without transferring water from one basin to another. The construction of aquifers has only created local and regional conflicts due to the scarcity of water in the drylands. The imbalance in the northern drylands and the overuse of water is unsustainable. The present intensive drought has obliged agribusinesses to reduce their agricultural surface and the use of water from deep wells is salinating their agricultural soil in certain states. On the other hand, the lack of safe water in the poor southern and south eastern regions is also the result of unsustainable water management. The proposed integrated scheme of water management of surface, groundwater, soil, biological, wetland, recycled, and desalinated water is further challenged by population growth, urbanisation, industrialisation (GEC), and CC. Technological solutions and water saving processes may improve efficiency in agriculture, but regions with lack of renewable water resources (Chihuahua) should reserve their fossil water for human development and not waste it in favour of the short-term interests of agribusiness. Water security is an integrated concept where negotiation processes on scarce resources are crucial to avoid dangerous water scarcity and present and future conflicts.

The diversity of climates and ecosystems enables Mexico to achieve simultaneous water and food security by focussing maize production on the south and southeast, where abundant freshwater reserves exist. These regions account for 275 km^3 of water each year. This creates the potential for irrigation in a socially neglected region. Taking into account that maize requires between 1.5 and 1.7 m of irrigation depth (with an efficiency between 60 and 70 %), between 5.7 and 6.6 million hectares could be integrated into the production process during the dry season. If Mexico promotes maize production in these irrigated lands, additionally 13.7–52.8 million tons of maize could be harvested (Turrent Fernández et al. 2013). This would not only grant sufficient maize for the country, but also produce food for livestock (cows, pigs, and chickens) or for export. This would grant food security together with water security. Mexico does not need genetic modified organisms to achieve its food security. With the existing germ-plasma and the capacity of research institutes, the peasants in the country are able to produce enough safe maize and achieve almost immediate food security. This different agricultural policy would drive millions of extremely poor people out of marginality in the south and southeast and create a true cycle of dignified livelihood security. Today, the required economic resources are wasted in the importation of maize. They could be invested in the development of infrastructure for safe water and irrigation, extensionism, credits, and market systems in the poorest and most marginal regions of Mexico. In synthesis, an integrated system of water management and food production opens new perspectives for Mexico to deal in a sustainable way with the nexus of water and food security. Currently, the only missing ingredient is the political will to act.

References

Arreguín F, López Pérez M, Marengo Mogollón H (2011) Los Retos del agua en México. In: Oswald Spring (ed) Retos de la investigación del agua en México, CRIM-UNAM, Conacyt, México, D.F., pp 21–39

Ávila JP, Sansores AC, Quintal MB, Can LA, Perera MP (2011) Environmental study on cadmium in groundwater in Yucatan. In: Oswald Spring (ed) Water resources in Mexico. Scarcity, degradation, stress, conflicts, management, and policy, Springer, Berlin, pp 239–250

Ballesteros-Barrera C, Jimémez-García D, Hernández-Cárdenas G (2011) El impacto potencial del cambio climático sobre los agroecosistemas. El caso del cultivo del maíz, proyecciones al futuro. In: Aragón GA, Jiménez D, Huerta ML (eds) Manejo Agroecológico de Sistemas, vol 2. BUAP, Puebla, Mexico, pp 1–14

Barkin D (2011) La ingobernabilidad en la gestión del agua urbana en México. In: Oswald Spring Ú (ed) Retos de la investigación del agua en México, CRIM–UNAM, Conacyt, Cuernavaca, pp 539–552

Berlanga HR (2010) The long-term view: comparing the result of Mexico's 1991 and 2007 agricultural censuses. In: Fox J, Haight L (eds) Subsidizing inequality: Mexican corn policy since NAFTA. Mexico, Woodrow Wilson International Center for Scholars; Centro de Investigacion y Docencia Economicas; University of California, Santa Cruz

Biswas (2013) Conference given at Qatar Foundation. Annual Conference, 4th October, Qatar

Bogardi J, Oswald Spring Ú, Brauch HG (2014) Water security (in press)

Bohle H-G (2002) Land degradation and human security. In: Plate E (ed) Human Security and Environment. Report on a Workshop held in Preparation for the Creation of a Research and Training Center for the UN, University in Bonn, Karlsruhe

Brauch HG et al (eds) (2008) Globalization and environmental challenges. Reconceptualizing Security in the 21st Century. Springer, Berlin

Brauch HG, Oswald Spring ÚO (2009) Securitizing the ground, grounding the security. UNCCD and Gobierno de España, Bonn

Brauch HG et al (eds) (2009) Facing global environmental change. Environmental human, energy, food, health and water security concepts. Springer, Berlin

Brauch HG et al (eds) (2011) Coping with global environmental change, disasters and security—threats, challenges, vulnerabilities and risks. Springer, Berlin

Bronen R (2012) Climate-induced community relocations: resilience and adaptation of Alaska native communities. Northern Research Forum/EPSON Akureyri, Iceland. http://www.rha.is/static/files/Espon_Enecon_project/Conference_Akureyri/Presentations/Robin_Bronin-Northern_Territories_Final.pdf

Cámara Argentina de Biocombustibles (2014), at: http://www.carbio.com.ar/es/?con=bio_estadisticas (download 3-1-2014)

CCI (1997, 2001, 2006, 2009, 2012) México Primera, Segunda, Tercera, Cuarta y Quinta Comunicación Nacional ante la Convención Marco de las Naciones Unidas sobre el Cambio Climático (CMNUC), México, D.F., Comité Intersectorial sobre el Cambio Climático, SEMARNAT-INE, Mexico, D.F.

CCI (2012) Fifth national communication to the United Nations Framework Convention on Climate Change (UNFCCC). Executive summary, Comité Intersectorial sobre el Cambio Climático, SEMARNAT-INE, Mexico, D.F.

Conabio (2008) Capital Natural y Bienestar Social, Conabio, Mexico, D.F.

Conagua (2008) Estadísticas del agua en México 2008, D.F., Conagua, Mexico, D.F.

Conagua (2012) Atlas del agua en México 2012. Conagua, Mexico, D.F.

Coneval (2013) Análisis y medición de la pobreza, Coneval, México, D.F

Crutzen P (2002) Geology of Mankind. Nature 415(3 January):23

Declaration of Dublin (1992) Declaración de Dublín sobre el Agua y el Desarrollo Sostenible, Dublin

Diario Oficial de la Federación (2010). Acuerdo por el que se da a conocer el resultado de los estudios de disponibilidad media anual de las aguas subterráneas de 30 acuíferos de los Estados Unidos Mexicanos, mismos que forman parte de las regiones hidrológicas que se indican (Agreement by which the results of average yearly water availability in 30 aquifers in Mexico come to be known. These aquifers form the hydrological regions they designate), January, Conagua, Mexico, D.F.

FAO (2012) El Estado Mundial de la Agricultura y la Alimentación, 2012. FAO, Roma

Garza G, Rivera S (1993) Censos y Población en México, Revista Mexicana de Sociología, vol 55(1). 1990, pp 177–212 (Jan–March)

Global Water Partnership (GWP) (2009) Towards water security: a framework for action. http://www.gwpsudamerica.org/docs/publicacoes/doc_78_en.pdf

González FJA, López EMR, Saldaña MCM, Barrera ALG, Juárez FJ, Sánchez JLR (2011) Potable water pollution with heavy metals, arsenic, and fluorides and chronic kidney disease in infant population of Aguascalientes. In: Oswald Spring Ú (ed) Water resources in Mexico. Scarcity, degradation, stress, conflicts, management, and policy, Springer, Berlin, pp 207–230

IEA (2013) World energy outlook special report 2013: redrawing the energy climate map. OECD/IEA, Paris

INEGI (1990) XI Censo General de Población y Vivienda, 1990, INEGI, Aguascalientes

INEGI (2000) XII Censo General de Población y Vivienda, 2000, INEGI, Aguascalientes

INEGI (2010) Censo de Población y Vivienda, 2010. INEGI, Aguascalientes

INEGI (1990a) XI Censo General de Población y Vivienda, 1990. Estado de Morelos, INEGI, Aguascalientes

INEGI (2013) Mexico en cifras, INEGI. http://www3.inegi.org.mx/sistemas/mexicocifras

IPCC (2007) Climate change 2007: synthesis report. Contribution of working groups I, II and III to the fourth assessment report of the Intergovernmental Panel on Climate Change. http://www.ipcc.ch/publications_and_data/publications_ipcc_fourth_assessment_report_synthesis_report.htm

IPCC (2012) Special report on extreme events (SREX), IPCC Geneva

IPCC (2013) Fifth assessment report: climate change 2013. The physical science basis. https://www.ipcc.ch/report/ar5/wg1/docs/WGIAR5_SPM_brochure_en.pdf

Martín A, Ortiz VJB, Gutiérrez FVC, González MÁM, Yamanaka VHA, Silvestre JM, Rendón GAO, Mejía PC, Herrera AG, Soberanis MP, de Huerta MLR, Palacios LM, Romero CEM, Tzatchkov VG (2011) Assessment of a water utility agency: a multidisciplinary approach. In: Oswald Spring Ú (ed) Water resources in Mexico. Scarcity, degradation, stress, conflicts, management, and policy. Springer, Berlin, pp 421–434

Maldonado Jiménez D (1984) Cuauhnáhuac y huaxtepec (Tlahuicas y Xochimilcas en el Morelos Prehispánico), CRIM–UNAM, Cuernavaca

Medina MR, Saavedra RM, Watts C (2011) Coastal aquifers of Sonora: hydrogeological analysis maintaining a sustainable equilibrium. In: Oswald Spring Ú (ed) Water resources in Mexico. Scarcity, degradation, stress, conflicts, management, and policy. Springer, Berlin, pp 73–86

Mendoza LA, Del Razo LM, Barbier O, Saldaña MCM, González FJA, Juárez FJ, Sánchez JLR (2011) Water quality in the state of Aguascalientes and its effects on the population's health. In: Oswald Spring Ú (ed) Water resources in Mexico. Scarcity, degradation, stress, conflicts, management, and policy, Springer, Berlin, pp 231–238

Morales Novelo JA, Rodríguez Tapia L (2011) In: Oswald Spring Ú (ed) Water resources in Mexico. Scarcity, degradation, stress, conflicts, management, and policy. Springer, Berlin, pp 395–406

Munich Re (2008) Map on geophysical and climate disasters. MunichRe, Munich

Muñoz JEC, Mólgora CGC (2011) Potable water use from aquifers connected to irrigation of residual water. In: Oswald Spring Ú (ed) Water resources in Mexico. Scarcity, degradation, stress, conflicts, management, and policy. Springer, Berlin, pp 189–201

Oliver-Smith A (2011) Sea level rise, local vulnerability and involuntary migration. In: Piguet E, Pécoud A, Guchteneire Pd (eds) Migration and climate change. Cambridge University Press, Cambridge

Oswald Spring Ú (2009) A HUGE gender security approach: towards human, gender, and environmental security. In: Brauch HG, Oswald Spring Ú, Grin J, Mesjasz C, Kameri-Mbote P, Behera NC, Chourou B, Krummenacher H (eds) Facing global environmental change. Environmental, human, energy, food, health and water security concepts. Springer, Berlin, pp 1165-1190

Oswald Spring Ú (2011a) Aquatic systems and water security in the Metropolitan Valley of Mexico City. Curr Opin Environ Sustain 3:497–505

Oswald Spring Ú (2011b) Water Security, Conflicts and Hydrodiplomacy. In: Oswald Spring Ú (ed) Water resources in Mexico. Scarcity, degradation, stress, conflicts, management, and policy. Springer, Berlin, pp 319–338

Oswald Spring Ú (2013) Dual vulnerability among female household heads. Acta Colombiana de Psicología 16(2):19–30

Oswald Spring Ú, Brauch HG (2009) Securitizing water. In: Brauch HG, Oswald Spring Ú, Grin J, Mesjasz C, Kameri-Mbote P, Behera NC, Chourou B, Krummenacher H (eds) Facing global environmental change. Environmental, human, energy, food, health and water security concepts. Springer, Berlin, pp 175–202

Oswald Spring Ú, Serrano Oswald ES, Estrada Álvarez A, Flores Palacios F, Everardo MR, Brauch HG, Ruíz Pantoja TE, Lemus Ramírez C, Estrada Villareal A, Cruz M (2014) Vulnerabilidad Social y Género entre Migrantes Ambientales, CRIM, DGAPA-UNAM, Cuernavaca

Palacios Vélez E, Mejía Sáez E (2011) Water use for agriculture in Mexico. In: Oswald Spring (ed) Water resources in Mexico. Springer, Berlin, pp 129–144

Rosengaus M (2007) Informe interno, Procedimientos para estimar tendencias del análisis parcial de datos históricos, 40 años de datos diarios, Tmax, Tmin y precitación, tendencias, promedio nacional, todos los meses y anuales de 1961 a 2000 a nivel nacional, regiones y estatal, México, Coordinación General del Servicio Meteorológico Nacional (CSMN)

SIAP (2013) Servicio de Información Agroalimentaria y Pesquera, SAGARPA, Mexico, D.F., at: www.siap.gob.mx (download, December 2013)

Taboada M (2005) Precipitación y sequía intraestival en el estado de Morelos. In: Oswald Spring Ú (ed) El Recurso Agua en el Alto Balsas, IGF/ CRIM/ UNAM, Böll, CGE, Coltlax, México, D.F., pp 155–164

Turrent Fernández A, Wise TA, Garvey E (2013) Achieving Mexico's maize potential. In: Food sovereignty: a critical dialogue, international conference, Yale University, 14–15 Sept 2013

USDA (2013). www.fsa.usda.gov

Rainwater Harvesting as an Effective Climate Change Adaptation Strategy in Rural and Urban Settings

Ezgi Akpinar Ferrand

Abstract Human societies are known for their resilience and ability to adapt to short- and long-term environmental change. Despite their pragmatism and adaptability, humans could be forced to move and/or seek better conditions for survival, especially when climate change and water availability are at issue. This chapter provides case studies on easy-to-adopt rainwater harvesting applications as effective climate change adaptation strategies in rural and urban settings to increase human resiliency.

Keywords Rainwater harvesting · Climate change adaptation · Urban and rural settings

1 Introduction

Human societies are known for their resilience and capability to adapt to short- and long-term environmental change. Despite their pragmatism, humans could be forced to move and/or seek better conditions for survival, especially when climate change and water availability are at issue (deMenocal 2001; Pandey et al. 2003; Femia and Werrell 2013). Given this reality and when water resources are in question, it becomes essential to adapt to climate change and pursue easy-to-adopt technologies to increase human resilience. In this chapter, rainwater harvesting is discussed as an important but mostly overlooked climate change adaptation tool for climatically vulnerable regions.

In the past, harvesting rainwater was widely practised (UNEP 1983; Laureano 2001; Maliva and Missimer 2012; Akpinar Ferrand and Cecunjanin 2014). Today, rainwater harvesting systems are mostly abandoned as a result of centralisation of

E.A. Ferrand (✉)
Department of Geography, Southern Connecticut State University,
New Haven, CT 06511, USA
e-mail: akpinarfere1@southernct.edu

S. Shrestha et al. (eds.), *Managing Water Resources under Climate Uncertainty*,
Springer Water, DOI 10.1007/978-3-319-10467-6_19

water resources, including increased groundwater extraction capabilities and the diversion of surface water over large distances (Mwenge Kahinda et al. 2010; Akpinar Ferrand and Cecunjanin 2014). In this chapter, a number of studies are highlighted to show that numerous documented ancient and traditional rainwater harvesting technologies were in fact developed in response to the past climate change events during the Late Holocene (Pandey et al. 2003; Lucero et al. 2011; Kennett et al. 2012).

In terms of climate change, trends from 1900 to 2005 have shown that precipitation increased considerably in the eastern parts of North and South America, Northern Europe and Northern and Central Asia (sometimes leading to severe flooding), and the levels of precipitation decreased in the Sahel, the Mediterranean, Southern Africa, and parts of Southern Asia. Climate scientists have predicted changes in rainfall and temperature regimes to reduce run-off and water availability at between 10 and 30 % in certain parts of the subtropical land regions (IPCC 2007). Given these findings and as part of climate adaptation strategies, Intergovernmental Panel on Climate Change (IPCC) recommended increasing the practice of rainwater harvesting (IPCC 2007).

Following IPCC recommendations and the findings of numerous studies, it is the premise of this chapter that easy-to-adopt rainwater harvesting technologies can act as appropriate water conservation and climate change adaptation strategies in climatically vulnerable regions. Some countries have already started paying careful attention to rainwater harvesting methods to increase their water security and as an effective buffer to climate change (UNEP 2002; Kim et al. 2012; Jiang et al. 2012).

2 Rainwater Harvesting Explained

In the past, harvesting rainwater was extensively used to secure water for drinking, household needs, and agricultural irrigation (UNEP 1983; Agarwal and Narain 1999; Gould and Nissen-Petersen 1999; Laureano 2001; Maliva and Missimer 2012; Akpinar Ferrand and Cecunjanin 2014). Frequently used rainwater harvesting practices are either no longer used or commonly carried out in the regions that they were found in (Evenari et al. 1971; UNEP 1983; Richards 1989; Doolittle 2000; Doolittle and Neely 2004). India, perhaps, provides us with some of the most telling reasons why the past pervasive sustainable water management practice is largely abandoned today. Recent studies indicate that extensive use of rainwater harvesting is absent in the country due to local communities not having the same past socioeconomic incentives, increased urbanisation and groundwater extraction abilities, previous destructive colonial policies, and also because of large-scale irrigation projects that are in place for agricultural production (Agarwal and Narain 1999; Swain 2004; Hoekstra and Mekonnen 2012).

In a recent study, fifteen major types of ancient and traditional rainwater harvesting methods were identified to be alike in practice and function regardless of the different geographic regions and cultures in which they were found (Akpinar

Fig. 1 An illustration of the selected practices from Table 1: **a.** pitting; **b.** in Situ RWH, ridging; **c.** micro-basins; **d.** rooftop RWH; **e.** traditional pond; **f.** shallow wells, example from ancient Maya; **g.** a run-off diversion system; **h.** micro-dam, an example of a khadin in association with a well; **i.** underground well (*Source* Akpinar Ferrand and Cecunjanin 2014)

Ferrand and Cecunjanin 2014). In order to harvest rainwater, rain has to flow over an impermeable catchment surface and into a storage area. The storage area can be an open pond, cistern, or soil surface. From the storage area, water can then be extracted for various needs. The study noted that the basic hydraulic principles behind rainwater harvesting have remained the same over time and that modern rainwater harvesting systems are largely built on these founding principles (Akpinar Ferrand and Cecunjanin 2014).

The following table provides a list of the fifteen identified common rainwater harvesting methods and summarises their function (Akpinar Ferrand and Cecunjanin 2014).

Figure 1 shows an illustration of the selected practices from Table 1

3 Rainwater Harvesting for Climate Change Adaption: Potential and Challenges

Binford et al. (1997) indicates that human cultures can adapt to changing environmental conditions within “a range of normal variations”. Normal variations signify short-time scale changes rather than long-term events. Paleoenvironmental

Table 1 List of commonly practised rainwater harvesting methods across dry, wet, semi-arid, and arid climatic regions (*Source* Akpinar Ferrand and Cecunjanin 2014)

Pitting:	Pits are dug and planted with cultivars on flat or sloping surfaces to conserve soil moisture and increase groundwater recharge
Contouring:	Contouring involves stone or earthen banks along a contour in a cultivated hill-slope. Contouring helps retain soil moisture, reduce soil erosion, and shorten slope length
Terracing:	Bunds in association with a trench along a contour on a sloping surface. Terracing helps reduce velocity of run-off during precipitation events, conserve soil moisture, and reduce erosion
Micro-basins:	Different shapes of small basins, surrounded by stone or earth bunds to infiltrate precipitation-related run-off
Pit courtyards:	Pit courtyards act as an impluvium to capture rainwater, surrounded by walls and impervious surfaces
In situ RWH:	Designed for increasing rainfall infiltration and reducing soil evaporation through practices such as ridging, mulching, broad bed and furrowing, hoeing, and conservation tillage
Rooftop RWH:	From rooftops, rainwater is collected and stored in storage tanks built in courtyards of houses
Traditional open ponds:	Precipitation and run-off is collected in open ponds through a catchment surface system
Cisterns:	Rainwater run-off collected and stored in underground storage reservoirs
Micro-dams:	Stored and regulated flow of rainwater for infiltration behind stone/earthen banks in a landscape with gradient
Shallow wells:	Shallow wells dug in low depressions or ponds to collect surface run-off after percolation to extract water during the dry season
Underground wells:	A proximate horizontal channel network or gallery excavated into an alluvial fan aquifer at the base of a mountain or foothill, recharged by precipitation
Run-off diversion and spate irrigation:	Diversion and spread of seasonal floods to agricultural plots from discrete rainfall events. These kinds of systems can be connected to terraces, reservoir systems, and dams of different sizes
Dams:	Stored and regulated flow of rainwater, run-off, and ephemeral streams behind large storage systems
Large reservoirs/lakes:	Precipitation and run-off collection in large-scale man-made basins

evidence strongly suggests that several subsistence-based ancient civilisations, such as the Moche IV-V, ancient Maya, Tiwanaku, Khmer Empire, agricultural dynasties of Han, Tang, and Song in China, and the native populations from the American Southwest most likely disintegrated as a result of long-term climate change, mostly because of extended aridity events (deMenocal 2001; Haug et al. 2003; Zhang et al. 2007; Buckley et al. 2010; Kennett et al. 2012). In the context of expected and currently experienced irregularities in rainfall due to ongoing global climate change, this chapter argues that the reintroduction and technological improvement of tried and tested rainwater harvesting practices could make current societies living in the climatically vulnerable regions more resilient. In the following sections, a number of case studies are presented, revealing the potential of integrating rainwater harvesting into rural and urban water management systems. Today, as aforementioned, a number of regions around the world are facing climatic challenges, including spells of increased drought and severe flooding events. For example in the USA, the Governor of California recently declared a state of emergency as a result of severe ongoing drought. According to the federal US Drought Monitor Center, almost 99 % of California is now considered abnormally dry with two-thirds of the state registered as experiencing an extreme drought. In fact, 2013 became the driest year on record, and similar aridity levels are currently being experienced in northern Nevada. The Department of Agriculture additionally announced that counties in nine other states also qualify as natural disaster areas due to water stress (Myers 2014).

The ongoing severe aridity events experienced in the USA are expected to have significant economic ramifications for the impacted populations. For the USA, it is also important to note that the current water management systems in its arid and semi-arid regions have proven unsustainable. The model of centralised water management, dam building, surface water diversion, and the use of large-scale aqueduct systems to haul water long distances to feed sprawling urban populations in arid regions and unsustainable irrigation practices of large crop fields in otherwise semi-arid areas (e.g. California) is the reality of the country (Gleick 2010).

In this chapter, given the common water resource use and management in the twenty-first century (sometimes proving unsustainable and inadequate), rainwater harvesting practices are proposed, which can act as valuable supplemental water resources. However, this basic technology needs to be integrated successfully into the existing urban and rural water management systems in the context of climate change. Yet, as mostly happens with changes to existing systems, integrating rainwater harvesting will also present a number of challenges to modern water management practitioners. Some of these challenges are discussed in Sect. 6.

4 Rural Settings

Due to their geographic characteristics, rural settings may benefit from rainwater harvesting tremendously. This is because rural settlements may lack connection to centralised water distribution infrastructures and because rainwater harvesting systems offer decentralised solutions (Mwenge Kahinda et al. 2010). In terms of global water footprint, Hoekstra and Mekonnen reported that agricultural production takes the largest share at 92 % (2012). Naturally, the majority of the agricultural production takes place in rural settings. Various studies report that supplemental irrigation with harvested rainwater during "crucial dry spells" can greatly improve rainfed cereal yields (Oweis et al. 1999; Li et al. 2001; Biazin et al. 2012).

Studies from China have further shown that if rainwater harvesting systems are combined with modern and efficient irrigation methods, significant agricultural irrigation can be achieved in crop fields, orchards and greenhouses of the arid and semi-arid regions. (Jiang et al. 2013). In rural regions, investing in agricultural water via supplementary rainwater can also help reduce poverty through higher productivity, employment, higher income, better nutrition, and better health (Biazin et al. 2012). Ultimately, these improvements will also make rural populations more resilient to various external stresses, such as climate change.

In the last few decades, there has been an increased effort to revive ancient and traditional rainwater harvesting practices in the Middle East, Africa, the Indian Subcontinent, Southeast Asia, and China (Evenari et al. 1971; Agarwal and Narain 1999; Gould and Nissen-Petersen 1999; UNEP 2002; Akpinar Ferrand and Cecunjanin 2014). In the following sections, several case studies are presented to demonstrate the potential of rainwater harvesting to achieve water and food security and reduced poverty in rural settings in the context of climate change.

4.1 China

In China, the history of rainwater harvesting dates back to 4,000 years. Some of the early techniques that were used in the country involved in situ rainwater harvesting, cisterns, pitting, terracing, and micro-dams. Since the 1980s, widespread drought and increased population pressure on water resources have led to the reintroduction of rainwater harvesting with the intention of integrating the ancient practice into modern technologies (Jiang et al. 2013). In recent decades, a large-scale governmental rainwater harvesting research programme was implemented in the dry Gansu Province in Northwest China. As a result of the programme, the province witnessed the establishment of 40,000 household rainwater harvesting cistern systems in rural households. By 2001, 1.97 million rural people had solved their drinking and household water needs. By 2007, large areas of crop fields, 10,829 ha of greenhouses, and 270,730 ha of orchards received irrigation from rainwater harvesting practices that integrated modern irrigation techniques, such as drip and sprinkler irrigation (Li et al. 2001; Zhu 2008; Jiang et al. 2013).

The implementation of rainwater harvesting systems in Gansu ultimately proved to be an effective development measure by lifting rural populations out of poverty through increased income. Also importantly, the widespread adoption of rainwater harvesting proved to be an effective climate adaptation tool by making farmers more resilient to dry spells and drought events during the water-stressed periods of crop production (Li et al. 2001; Zhu 2008). Currently in China, other water-insecure arid and semi-arid provinces in the Shanxi, Ningxia, and Inner Mongolia regions have started implementing rainwater harvesting. Additionally, semi-humid and humid provinces of the Guizhou and Guangxi Autonomous Region began utilising rainwater harvesting leading to a more sustainable and efficient water management policy in the country (Jiang et al. 2013).

4.2 India

In the case of India, the re-establishment of traditional rainwater collection and storage practices, such as ponds/reservoirs and micro-dams, has led to the revival of certain communities and made their inhabitants more resilient to drought (Agarwal and Narain 1999; Suutari et al. 2005; Tran 2013). Particularly interesting is the restoration of johad micro-dams in the Alwar district and a traditional reservoir in a village in the state of Rajasthan.

In Rajasthan, rainfall is erratic and the state overall is prone to drought events. Most of the rains occur between July and August during the monsoon season (Tran 2013). Johads are traditional earthen or rock micro-dams that catch precipitation engendered run-off with the intent of recharging groundwater through filtration. Another function of the micro-dam johad is to provide water for the livestock. In 1986, an abandoned johad in water-scarce Gopalpura was dredged with the hopes of harvesting rainwater during the upcoming monsoon season. By the following year, a well that had gone dry downstream started producing water again as a result of rainwaters collected in the johad infiltrating through the surface of the dam. By 1996, Gopalpura residents had built nine johads covering 2,381 acres and harvested 162 million gallons of rainwater. The villagers also found that the underground water level rose from an average 45 feet below surface to 22 feet. As a result, less fuel was needed to pump water closer to the surface, an important expense in rural regions. By 2005, 5000 micro-dams were built in 750 villages, covering 3,000 square miles in India leading to increased water security in the implemented areas (Suutari et al. 2005).

The village of Galandhar in Rajasthan was used in another case study of the revitalisation of an entire community after a simple rainwater harvesting initiative was reintroduced. Typically, Galandhar's work force consisted of unskilled workers who migrated to the neighbouring state of Gujarat in search of seasonal jobs. Starting in May 2012, tractors deepened an existing pond in the village, so when the monsoon rains arrived the pond would act as a storage surface. When the rains came, the rainwater collected in the enlarged pond and its waters were immediately channelled to surrounding fields for agricultural irrigation. In Galandhar, the improvement of one simple rainwater pond as a newly available water supply led to the introduction of new crops, extended the growing season to include second crops, and changed the socioeconomic structure of the village. More than half of the farmers stayed in the village instead of seasonally migrating in search of jobs (Tran 2013).

The traditional rainwater harvesting ponds have a long history of use in India as micro-dams. Besides these techniques, there are many other rainwater harvesting methods that have been successfully used in India in the past but are no longer in use today (Agarwal and Narain 1999). In Galandhar and in Gopalpura, villagers reported the lack of water and water insecurity as a key concern impacting their livelihoods (Suutari et al. 2005; Tran 2013). As an example, in Galandhar, the farmers had land but could not cultivate it due to lack of water. If small-scale ponds, micro-dam johads and other easy-to-adopt rainwater harvesting methods are reintroduced and replicated throughout India, it is likely that considerable rural areas can be transformed and become more resilient to drought spells and poverty. India, in fact, offers many valuable lessons in the use of rainwater harvesting to combat climate change. Its Thar Desert, known as the most populated desert in the world, has a continuous and dynamic record of using rainwater harvesting as a climate change adaptation tool, going back 5,500 years (Pandey et al. 2003).

4.3 Maya Lowlands, Central America

For the semi-tropic climatic regions, ancient Maya is a good example of past extensive use of rainwater harvesting. Today, the rural Central American Lowlands, where the ancient Maya thrived, may be missing out on an opportunity to use rainwater harvesting to reduce water scarcity commonly experienced in the dry–wet climate of the region. The 4–6 month dry season experienced in large areas of Central America has been likened to a seasonal desert by some experts who study the region (Haug et al. 2003).

The ancient Maya, as a civilisation, was able to flourish and sustain itself in an environment characterised by karst geology that limits the availability of surface water. Under these conditions, the Maya people proved resilient by collecting and storing rainwater in impermeable basins for human use and agricultural production in urban and rural settings between 1000 BC and Terminal Classic in the tenth century AD (Scarborough and Gallopin 1991; Dunning et al. 2002; Akpinar Ferrand and Scarborough 2012; Scarborough et al. 2012). Interestingly, the ancient Maya also experienced a number of climate aridification events throughout the civilisation's long history with the last one proving to be the most severe between 800 and 1000 AD (Hodell et al. 2001; Kennett et al. 2012).

Today, the density of modern populations in rural Central America is only a fraction of what was once populated by the ancient Maya (Rice and Rice 1990). In the Maya Lowlands, for instance, there are large numbers of abandoned ancient Maya rainwater harvesting ponds. A number of the investigated ponds display considerable volumes ranging between 2,500 and 10,000 m^3, indicating their serious potential to increase water security in Central America (Akpinar Ferrand 2011; Akpinar Ferrand et al. 2012). Modern experiments also reveal the possibility for aquaculture in these ponds, adding to their promise to increase food security (Flores-Nava 1994; Zambrano et al. 1999; Akpinar Ferrand and Scarborough 2012). Given the agricultural water usage and shortage statistics from ancient Maya settings like Mexico, integration of past rainwater harvesting technologies may offer tremendous potential for rural populations plagued by water insecurity and erratic rainfall (National water commission of Mexico (NWC) 2010; Akpinar Ferrand and Scarborough 2012; Faust et al. 2012; Akpinar Ferrand and Cecunjanin 2014).

5 Urban Settings

UNESCO (2012) reports that there is serious water demand cast by increasing urban populations and industrial activities. Urban development, population increase, and the surging water demand have begun to put stress on existing water resources (Villarreal and Dixon 2005). This chapter proposes that rainwater collection can make urban settings more water secure and resilient to climate change. In terms of climate change, the use of rainwater harvesting for storm water management to prevent flooding (i.e. given the impervious fabric of urban settings) and the collection of rainwater for domestic use in drought-prone regions are among the most cited uses (Villarreal and Dixon 2005; Abdulla and Al-Shareef 2009; Stump et al. 2012; Kim et al. 2012). A recent case study from the city of Suwon in South Korea has shown that with government support and incentives for rainwater harvesting, it can be successfully integrated into urban contexts and prove very useful in increasing water capacities, reducing water bills, and preventing flooding (IRHA 2013). In water-scarce semi-arid and arid urban regions, rainwater harvesting has similarly shown considerable potential to increase water security in residential buildings (Abdulla and Al-Shareef 2009).

5.1 Europe and USA

In the populated cities of Europe, attention has shifted to looking for alternative supplementary water resources. A number of rainwater harvesting studies point to urban roofs as important and ubiquitous catchment surfaces offering a significant possibility for rainwater harvesting (Villarreal and Dixon 2005; Ward et al. 2012). In Sweden, widespread urbanisation and the related creation of centralised water infrastructures have led to water shortages and quality issues (Villarreal and Dixon 2005). In the country, it is estimated that 20 % of household water is used for flushing toilets, 15 % for laundry, and 10 % for car washing and cleaning. Villarreal and Dixon (2005) proposed that water collected from rainfall may help offset some of this water use, leading to many economic and environmental benefits. Environmental benefits in particular include storm and waste water management. Their study recommended the installation of rainwater tanks as part of a household dual water supply and incorporation of water efficient appliances. If low flush toilets are integrated into households, the study additionally found that rainwater tanks can be used to save almost 40 % of the water demand. Moreover, in a neighbourhood, the study also found the harvested rainwater from rooftops could supply 60 % of the water needed for irrigation during the summer months (Villarreal and Dixon 2005).

In connection to the large water demand that toilet flushing involves, another study from Exeter in England, focusing on a single office building reported that harvested rainwater provided enough water annually to fulfil the full WC flushing demand of the building. The study encouraged the application of rainwater harvesting to office buildings due to reduced water demand and cost savings (Ward et al. 2012). In another example in London, a large-scale rainwater harvesting system is reported to collect rainfall from the roof of the Millenium Dome, a surface with an approximate area of 100,000 m^2. The famous facility discharges the collected water either into the River Thames via a storm water culvert or to a treatment plant used for the city based on the amount of rainfall (Villarreal and Dixon 2005).

Moving on to Germany, in Berlin, at Daimler Chrysler Potzdamer Platz, roof run-off collected from 19 buildings (with a total area of 32,000 m^2) is used for flushing toilets, watering gardens, and to replenish a pond. Another example is the Belss-Luedecke-Strasse building and its surroundings (i.e. streets, pathways, and parking areas). The collected rainwater is routinely used for toilet flushing and garden watering. A study found that 58 % of the rainfall was captured locally using the building's system, leading to reduced run-off. A ten-year simulation also showed that 2,430 m^3 of potable water savings per year can be achieved via rainwater collection (Villarreal and Dixon 2005).

Lastly, in New York City, Hurricane Sandy in 2012 once again demonstrated how vulnerable New York City is to flooding and combined sewage overflow (CSO). CSO is essentially a common problem in many of the US cities, where shared piping is used for waste water and curbside rain run-off collection. Many times, the system cannot handle the volume of storm water and waste water when it rains hard and ends up releasing untreated sewage and storm water into adjacent waterways. In New York City, Brooklyn's Gowanus is testing a countermeasure. In 2014, construction begins for a series of 10-by-5-foot plant filled pits trenched into the edge of the Gowanus Canal. These pits are designed to capture rainwater before it makes its way into the sewage pipes. The city sees additional benefits to this system, such as evaporation from the planted areas creating a cooling effect and the plants being watered regularly without having to use the city water (Treat 2013). In urban settings plagued by CSO problems with predicted rainfall irregularities and flooding, these kinds of solutions should be considered carefully.

5.2 *Asia*

In Japan, three multipurpose stadiums located in Tokyo, Nagoya, and Fukoka harvest rainwater for flushing and irrigation of plants. The catchment surface areas for the stadiums are 16,000, 25,900 and 35,000 m^2, respectively, with rainwater storage tank volumes of 1,000, 1,800, and 1,500 m^3. In the Fukoka stadium, rainwater is reported to provide 65 % of the volume of water used in the stadium, representing a significant economic saving for the facility. At the stadium in Tokyo, collected rainwater is also used to help cool the building, leading to energy savings. Other rainwater harvesting systems exist in Japan with the intent of reducing flooding problems, combat droughts, and decrease dependency on main water supplies, to reduce water costs and to provide a backup water supply (Villarreal and Dixon 2005; Kim et al. 2012).

In South Korea, the first law to promote rainwater harvesting was passed in 2001, known as the "Water Law." In the country, the rainwater is managed with the goal of mitigating urban water flooding, to conserve water, control non-point source pollution, and recharge groundwater and to alleviate urban heat island effect, yielding clear climate change adaptation benefits. Since the Water Law, hundreds of small-to-larger scale rainwater harvesting projects have been implemented around the country. One project involves the implementation of a real estate development complex with approximately 58 buildings in a 5 ha site. The catchment area for the complex comprised of 51,200 m^2 surface area with a rainwater storage capacity of 3,000 m^3. The system was ultimately found to save 26,000 m^3 of rainwater per year, consisting of 47 % of the annual rainfall the complex received, leading to an effective storm and water management. Additionally, between the study dates of June 2007 and May 2008, a considerable 8.9 MWh of electricity was saved using the harvested rainwater (Kim et al. 2012). Part of the reason why rainwater harvesting can lead to energy savings is because the conventional centralised water supply systems are generally inefficient and energy intensive (Grady and Younos 2008).

5.3 Dry Regions of the USA and Jordan

In the semi-arid and arid parts of USA, rainwater harvesting is making a comeback in states such as Texas due to severe droughts and the state reservoirs and aquifers reaching to distressingly low levels. Some of the state's residents have started showing interest in rooftop rainwater harvesting as it is less costly than opening wells that now need to go deeper due to the dropping water tables. Residents who use rainwater harvesting often indicate that they prefer the taste and softness of rainwater to well water as a potable water alternative. The majority of the rooftop rainwater harvesting systems are fitted with filtration (sand and/or carbon filters) and disinfection systems (ultraviolet or chlorination) to make sure the water collected in the tanks meets the national water standards (Stump et al. 2012). Studies have shown that pollutants can be introduced into collected water from rooftops from organic matter, inert solids, faecal deposits from animals and birds, and trace amounts of certain metals from housing parts used in construction (Villarreal and Dixon 2005). Stump et al. (2012) study investigated 36 households in Texas that installed rooftop rainwater harvesting systems with capacities that ranged from 9 to 189 m^3. The study found that with the existing filtration and disinfecting systems, collected rainwater generally met the USEPA drinking water standards. However, the study also found lead contamination to be a potential hazard from piping and atmospheric origins and recommended the frequent testing of rainwater if collected as a potable water source (Stump et al. 2012). For this problem, EPA states that filtering devices certified to remove lead and other metals could be used to ensure potability of the collected water (1993).

Moving on, Jordan, a country characterised by arid and semi-arid climates, has faced water deficits since the 1960s. The country also ranks within the world's top ten water-stressed countries. Urban development and population increase are some of the important factors that put stress on already scarce water resources in the country (Abdulla and Al-Shareef 2009). Jordan, like China and India, has a long rainwater harvesting history. Given the water scarcity in the country, since 1995, rainwater harvesting has become part of government policy in most governorates of Jordan. According to the Population and Housing 2004 Census, about 33,229 rainwater cisterns with average volumes of 20 m^3 have been built in the country. The Abdulla and Al-Shareef study (2009) strongly encourages the increase of rainwater harvesting in residential spaces to supplement centralised water systems. This could be achieved through government incentives and will ultimately reduce the overall public water costs. The Jordanian Government has requested rainwater harvesting systems for all newly constructed homes. Abdulla and Al-Shareef recommend an extension of the policy for the existing buildings as well (2009).

6 Challenges

The above-presented case studies show that rainwater harvesting strategies have significant potential in rural and urban settings to combat extreme weather events, such as droughts and flooding. The importance of rainwater harvesting as a buffer against extreme events has been mostly overlooked in water planning until recently. In a study concerning climate change and increased water security in South Africa, it was found that there can be a 5–20 % increase in water security in dry sub-humid, humid, and semi-arid parts of the country with the inclusion of rainwater harvesting (Mwenge Kahinda et al. 2010). Although the presented case studies highlight opportunities, a number of challenges exist regarding rainwater harvesting. A survey of the literature on the subject reveals that these challenges relate to socioeconomic, cultural, health, and technological areas and should be addressed if rainwater harvesting is to be incorporated into existing water systems successfully.

Rural rainwater harvesting studies in developing regions pose a list of advantages and disadvantages of collecting rainwater for household and agricultural use. Advantages include water collection for small-scale productive activities, agricultural and livestock purposes, creation of drinking water, less time spent in fetching water for women, elderly and the children by creating a water source close to the household, less waterborne diseases if the necessary care is taken, improved sanitation, more time for education, improved food security, and a reduction in malnutrition. Disadvantages may include danger of water and vector-borne diseases, cost of implementation, need of training for construction, operation and maintenance, and loss of revenue for water utility companies (Sturm et al. 2009; Mwenge Kahinda et al. 2010). Studies also mention the perception of rainwater as a "second class" water supply system that diminishes its popularity among users (Sturm et al. 2009; Mwenge Kahinda et al. 2010). A relevant point is that rainwater does not contain any minerals. As a result, it does not carry any taste and may not be desirable for some (Abdulla and Al-Shareef 2009).

To elaborate on the disadvantages of rainwater harvesting, it is a fact that due to its randomness, rainfall makes an intermittent water source, constrained by catchment surfaces and storage areas. In addition, any collected, stagnant water may pose health risks. Users of rainwater harvesting need to be trained in health-related risk management measures. It is known that rainwater quality may not always meet general health guideline values due to possible chemical and microbiological contamination. Yet, in a study in a rural Namibian village, Sturm et al. (2009) found that a well-constructed rainwater harvesting system may indeed prove to be safer than unprotected water sources that the village typically received for their consumption, such as local rivers or earth dams. Once the rainwater is collected, its quality can and should be improved by methods such as filtering, boiling, exposure to nature UV radiation (e.g. solar water disinfection-SODIS), and chlorination (Environmental Protection Agency (EPA) 1993; Sturm et al. 2009).

In urban settings, government policies need to incorporate rainwater harvesting in legislature and provide incentives in order to encourage its wide-scale adoption. China and Korea set a good example of this. In countries such as India, where rainwater harvesting shows great potential, government policies are still lagging behind for both rural and urban settings (Kim et al. 2012). This chapter presented case studies that illustrate the potential usefulness of rainwater harvesting to combat flooding and droughts in urban settings. Yet, a survey of the literature shows that the challenges remain, including contamination, backflow, lack of clarity in various rainwater management methods, uncertainty in calculating the benefits of the system, possible inflexibility of design criteria in ordinances, and the general lack of government incentives and knowledge in the installation and decision-making process for overall rainwater harvesting systems (Abdulla and Al-Shareef 2009; Kim et al. 2012; Stump et al. 2012).

7 Conclusion

Rainwater harvesting has a long past as an effective climate change adaptation tool and an easy-to-adopt water management method. As indicated in this chapter, the basic hydraulic principles behind rainwater harvesting have remained the same over time, and modern rainwater harvesting systems are largely built on these founding principles. Rainwater harvesting systems are, as a result, relatively low-technology solutions. Sect. 2 of this chapter shows the fifteen most common ancient and traditional time-tested rainwater harvesting methods identified in many parts of the world. Today's technologies will only improve these simple methods. Many case studies in rural and urban settings have proven this point.

In the end, rainwater harvesting can act as a significant supplemental water resource if integrated successfully into the existing water systems. There are, however, a number of challenges. With careful planning, government incentives, and technological improvements of the existing methods, there is a great possibility that rainwater harvesting can act as an important supplementary water source, making rural and urban settings more water secure and ultimately less vulnerable to climate change.

References

Abdulla FA, Al-Shareef AW (2009) Roof rainwater harvesting systems for household water supply in Jordan. Desalination 243(1–3):195–207

Agarwal A, Narain S (1999) Making water management everybody's business: water harvesting and rural development in India. The Gatekeepers Series No. SA87. London: International Institute for Environment and Development

Akpinar Ferrand E, Cecunjanin F (2014) Potential of rainwater harvesting in a thirsty world: a survey of ancient and traditional rainwater harvesting applications. Geogr Compass 8 (6):395–413

Akpinar Ferrand E, Dunning NP, Lentz DL, Jones JG (2012) Use of aguadas as water management sources in two southern maya lowland sites. J Ancient Mesoamerica 23:85–101

Akpinar Ferrand E, Scarborough VL (2012) Ancient maya solutions to water and food insecurity: low technology lessons for contemporary development. Global Water Forum Discussion Paper 1215. Article available online. http://www.globalwaterforum.org/2012/04/23/ancient-maya-solutions-to-water-and-food-insecurity-low-technology-lessons-for-contemporary-development/Biazin et al. 2012

Akpinar Ferrand E (2011) Aguadas: a significant aspect of the southern maya lowlands water management systems, PhD Dissertation: University of Cincinnati. Available online. http://etd.ohiolink.edu/send-pdf.cgi/Akpinar%20Ezgi.pdf?ucin1307320694

Biazin B, Sterk G, Temesgen M, Abdulkedir A, Stroosnijder L (2012) Rainwater harvesting and management in rainfed agricultural systems in sub-Saharan Africa-a review. Phys Chem Earth 47–48:139–151

Binford MW, Kolata AL, Brenner M, Janusek JW, Seddon MT, Abbott M et al (1997) Climate variation and the rise and fall of an Andean civilization. Quatern Res 47:235–248

Black R, Bennett SRG, Thomas SM, Beddington JR (2011) Climate change: migration as adaptation. Nature 478(7370):447–449

Bruins HJ, Evenari M, Nessler U (1986) Rainwater-harvesting agriculture for food production in arid zones: the challenge of the African famine. Appl Geogr 6:13–32

Buckley BM, Anchukaitis KJ, Penny D, Fletcher R, Cook ER, Sano M, Nam LC, Wichienkeeo A, Minh TT, Hong TM (2010) Climate as a contributing factor in the demise of Angkor Cambodia. Proc Nat Acad Sci 107(15):6748–6752

deMenocal PB (2001) Cultural responses to climate change during the late holocene. Science 292:667–673

Doolittle WE, Neely JA (2004) A checkered landscape. In: (eds.) The Safford Valley grids: prehistoric cultivation in the southern Arizona desert. University of Arizona Press, Tucson

Doolittle WE (2000) Cultivated landscapes of native North America. Oxford geographical and environmental studies series. Oxford University Press, Oxford

Dunning NP, Beach TP, Luzadder-Beach S, Jones JG, Scarborough VL, Culbert P (2002) Arising from the bajos: the evolution of a neotropical landscape and the rise of maya civilization. Ann Assoc Am Geogr 92(2):267–283

Environmental Protection Agency (EPA) (1993) Actions you can take to reduce lead in drinking water. EPA 810-F-93-001. Article Available Online. http://water.epa.gov/drink/info/lead/lead1.cfm

Evenari M, Shanan M, Tadmor N (1971) The Negev: the challenge of a desert. Harvard University Press, Cambridge

Faust BB, Anaya Hernández A, Geovannini H (2012) Reclaiming the past to respond to climate change: mayan farmers and ancient agricultural techniques in the Yucatan Peninsula of Mexico. In: Castro AP, Taylor D, Brokensha DW (eds) Climate change and threatened communities: vulnerability, capacity, and action. Practical Action Publications, Warwickshire, pp 139–151

Femia F, Werrell C (2013) Climate change before and after the Arab awakening: the cases of Syria and Libya. In: Werrell CE, Femia F (eds) The Arab spring and climate change. The Center for Climate and Security, Washington DC, pp 23–33

Flores-Nava A (1994) Some limnological data from five water bodies of Yucatan as a basis for agriculture development. Anales del Instituto de Ciencias del Mar y Limnologia, 1–2 (21)

Gleick PH (2010) Roadmap for sustainable water resources in southwestern North America. Proc Nat Acad Sci 107(50):21300–21305

Gould J, Nissen-Petersen E (1999) Rainwater catchment systems for domestic supply: designing, construction and implementation. ITDG Publishing, London

Grady C, Younos T (2008) Analysis of water and energy conservation of rainwater capture system on a single family home. Virginia Water Resources Research Center No. SR39-2008. Study available online. http://www.cabellbrandcenter.org/Downloads/sr392008.pdf

Haug GH, Gunther D, Peterson LC, Sigman DM, Hughen KA, Aeschlimann B (2003) Climate and the collapse of maya civilization. Science 299:1731–1735

Hodell DA, Brenner M, Curtis JH, Guilderson T (2001) Solar forcing of drought frequency in the maya lowlands. Science 292(5520):1367–1370

Hoekstra AY, Mekonnen MM (2012) The water footprint of humanity. Proc Nat Acad Sci 109 (9):3232–3237

IPCC AR4 (2007) Synthesis report contribution of working groups I, II and III to the fourth assessment report of the intergovernmental panel on climate change. Available online. http://www.ipcc.ch/pdf/assessment-report/ar4/syr/ar4_syr.pdf

IRHA (2013) Rain City Suwon, City of Suwon Presentation to IRHA. Available online. http://www.youtube.com/watch?v=nxk-EQsuiJc

Jiang Z, Li X, MA Y (2012) Water and energy conservation of rainwater harvesting system in the Loess Plateau of China. J Integr Agric 12(8):1389–1395

Kennett DJ et al (2012) Development and disintegration of maya political systems in response to climate change. Science 338(6108):788–791

Kim R, Mun, J, Park J (2012) Water resource management: policy recommendations for the development of eco-efficient infrastructure. UNESCAP and KOICA Policy Paper. Available Online. http://www.unescap.org/esd/environment/lcgg/documents/roadmap/Policy-Papers/Water-Resource-Management.pdf

Laureano P (2001) Water atlas, traditional knowledge to combat desertification. Bollati Boringhieri, Torino

Li F, Cook S, Geballe GT, Burch WR (2001) Rainwater harvesting agriculture: an integrated system for water management on Rainfed Land in China's semiarid areas. Ambio 29 (8):477–483

Lucero LJ, Gunn JD, Scarborough VL (2011) Climate change and classic maya water management. Special issue, "managing water resources and development in a changing climate." Water, 3(2):479–494

Maliva R, Missimer TM (2012) Arid lands water evaluation and management. Springer, New York

Mwenge Kahinda J, Taigbenu AE, Boroto RJ (2010) Domestic rainwater harvesting as an adaptation measure to climate change in South Africa. Phys Chem Earth 35:742–751

Myers J (2014) California Governor declares drought emergency. USA Today. Article Published January 17, 2014. Article available online. http://www.usatoday.com/story/weather/2014/01/17/california-drought-emergency/4581761/

National water commission of Mexico (NWC) (2010) Statistics of water in Mexico, 2010 Edition

Oweis TY, Bruggeman A (2004) The role of indigenous knowledge in improving present water-harvesting practices. In: Oweis TH, Bruggeman A (eds) Indigenous water harvesting systems in West Asia and North Africa. International Center for Agricultural Research in the Dry Areas, Aleppo, pp 1–20

Oweis TY, Hachum A, Kijne J (1999) Water harvesting and supplemental irrigation for improved water use efficiency in dry areas. SWIM Paper 7. ICARDA Institute, Colombo

Pandey DN, Gupta AN, Anderson DM (2003) Rainwater harvesting as an adaptation to climate change. Curr Sci 85(1):46–59

Rice DS, Rice PM (1990) Population size and population change in the Central peten lake region, Guatemala. In: Culbert TP, Rice DS (eds) Precolumbian population history in the Maya Lowlands. University of New Mexico Press, Albuquerque, pp 123–148

Richards T (1989) Evidence of ancient rainwater concentrating structures in northern Egypt as seen on Landsat MSS imagery. Int J Remote Sens 10(6):1135–1140

Scarborough et al (2012) Water and sustainable land use at the ancient tropical city of Tikal, Guatemala. Proc Nat Acad Sci 109(31):12408–12413

Scarborough VL, Gallopin GG (1991) A water storage adaptation in the Maya lowlands. Science 251(4994):658–662

Stump M, McBroom M, Darville R (2012) Demographics, practices and water quality from domestic potable rainwater harvesting systems. J Water Supply: Res & Technol-AQUA 61 (5):261–271

Sturm M, Zimmermann M, Schutz K, Urban W, Hartung H (2009) Rainwater harvesting as an alternative water resource in rural sites in central northern Namibia. Phys Chem Earth 34 (13–16):776–785

Suutari A, Marten G, Brooks S (2005) Environmental tipping points: a new slant on strategic environmentalism. World Watch 18(6):10–14

Swain A (2004) Managing water conflict: Asia, Africa and the Middle East. Routledge, London

Thomas T (1998) Domestic water supply using rainwater harvesting. Building Res & Inf 26 (2):94–101

Tran M (2013) Water-hungry Indian villagers find new reservoirs of solidarity. The Guardian, February 26, 2013. Article Available Online. http://www.theguardian.com/global-development/poverty-matters/2013/feb/26/water-indian-villagers-reservoirs-solidarity?CMP=twt_gu

Treat J (2013) Global view: waste not. Nat Geogr 224(5):20–21

UNEP (2002) Rainwater harvesting and utilization: an environmentally sound approach for sustainable urban water management: an introductory guide for decision-makers, Vol 2. IETC urban environment series. Available online. http://www.unep.or.jp/ietc/publications/urban/urbanenv-2/index.asp

UNEP (1983) Rain and storm water harvesting in rural areas. Tycoly International, Dublin

UNESCO (2012) The fourth world water development report. Marseilles, France

Villarreal EL, Dixon A (2005) Analysis of a rainwater collection system for domestic water supply in Ringdansen, Norrköping, Sweden. Build Environ 40:1174–1184

Ward S, Memon FA, Butler D (2012) Performance of a large building rainwater harvesting system. Water Res 46:5127–5134

Zambrano L, Perrow MR, Marcias-Garcia C, Aguirre-Hidalgo V (1999) Impact of introduced carp (Cyprinus carpio) in subtropical shallow ponds in Central Mexico. J Aquat Ecosyst Stress Recovery 6:281–288

Zhang DD, Brecke P, Lee HF, He YQ, Zhang J (2007) Global climate change, war, and population decline in recent human history. Proc Nat Acad Sci 104(49):19214–19219

Zhu Q (2008) Rainwater harvesting in dry areas: the case of rural Gansu in China. UNEP Tech Monitor Special Feature: Water Harvesting Technologies. Sep–Oct 2008

Coevolving Water Infrastructures for Adaptation to Climate Change

Tse-Hui Teh

Abstract London's water infrastructure has coevolved over centuries and has been used as a model for water infrastructure systems in modern cities around the world. However, it is a system that is stretched to the limit as evidenced by the need to build a desalination plant in the Thames estuary in order to secure supplies of freshwater and ongoing problems with sewerage overflows into aquatic environments. The system will only come under further strain in the future as the population of London increases, and climate change alters rainfall patterns. It is also a system that inherently degrades aquatic environments, preventing it from being sustainable. A methodology using an actor-network theory (ANT) coevolutionary theoretical approach was used to understand how people would adapt their water infrastructures and practices in the extreme water conditions of excess and shortage as projected by climate change models. From this, coevolutionary pathways were found to suggest how water infrastructures in London could be adapted for climate change and a more sustainable water resource future.

Keywords Actor-network theory · Coevolution · Flood · London · Water scarcity

1 Introduction

London is a city with an unfounded reputation of having plentiful supplies of water due to its commonly grey skies, fog, drizzle, and rain. However, this does not take into account its per capita consumption and its growing population with its incumbent urbanisation of areas causing deleterious effects on groundwater recharge and water quality. It has an annual rainfall of less than 650 mm/year; therefore, it currently has less water available per capita than places with a reputation for dryness, such as Sudan or Syria (Waterwise 2009). All this makes London

T.-H. Teh (✉)
Bartlett School of Planning, University College London, London, UK
e-mail: t.teh@ucl.ac.uk

S. Shrestha et al. (eds.), *Managing Water Resources under Climate Uncertainty*,
Springer Water, DOI 10.1007/978-3-319-10467-6_20

a water-stressed city, with already over abstracted water supplies (Environment Agency 2006). Climate change models project that this water stress will be exacerbated by an increasing propensity for intense rainfall which will lead to greater surface water run-off, less water infiltration, and therefore a less constant supply of water, which will cause problems for both water supply and wastewater discharge (IPCC 2007; World Wildlife Fund-UK 2009).

This research investigates how the urban water cycle could coevolve in London in response to both climate change and to alleviate water difficulties inherent to this form of water supply and drainage. The theoretical basis and methods of this research could be applied to other places of human settlement facing water difficulties to develop an infinite array of innovative water cycles tuned to local water practices and sources.

This chapter first outlines the changing visibility of water infrastructure in the urban environment and how this affects the relationship between people's water use and aquatic ecologies. It then describes the theoretical stance from which the methodology was developed. It follows this with the description of the methodology employed for a research case study conducted in the lower Lea River Basin in east London from 2009 to 2011 and its results, before drawing conclusions about the wider applicability of this approach.

2 Background: Hidden and Celebrated Water Infrastructure

Presently, London is served by large centralised water supply and wastewater infrastructures. Water is abstracted from rivers and aquifers, filtered, treated, and distributed via a myriad of pressurised pipes that are mostly underground and hidden from view. It is conveniently piped to and drained from the locations where it is needed: taps in kitchens and bathrooms, showers, toilet cisterns, and gardens. Once the water has exited the water supply, the majority of it will then enter the wastewater system of non-pressurised gravity pipes that transfer the water and waste materials to one of several wastewater treatment plants where solid and many chemical wastes are removed. From here, the water is discharged into a natural watercourse such as a brook or a river. The solid waste has several paths depending on the wastewater treatment plant. It can go to a biodigester before being landfilled or incinerated; it can go straight to landfill, or the sludge can be spread on farmland.

This centralised infrastructure encourages the unconscious wasteful use of water because it continually flows from the source points to places of habitation, all the while concealing the consequent environmental impact on aquatic environments from water abstraction and receiving waters for discharge. This has led to a trend of increasing consumption (Butler and Memon 2006; Environment Agency 2010), which has led to a never-ending chase to find more water resources (Butler and Memon 2006). Even without climate change projections, this is an unsustainable infrastructure system because water demand in London is higher than water availability, which means that it kills aquatic environments.

The use of this form of centralised water supply and wastewater infrastructure has multiple problems which are derived from its encouragement towards the unthinking use of water. Aquatic food stocks are affected by both the over abstraction of freshwater from the environment and the discharge of nutrient-laden water into the environment, both of which cause the degradation of aquatic ecologies that are essential for the ongoing viability of habitats for fish and aquatic invertebrates (Kosmala et al. 1999), which are prerequisites to our food stocks (Newson 2007). The use of water to transport waste also leads to pollution from people dumping toxic wastes into the sewer system. The use of waterborne waste transport also leads to a dilution and mixing of fertiliser sources with other waste materials, making it difficult to harvest the fertiliser effectively (Speers 2007). The continued over abstraction of water from aquifers leads to salt water intrusion, rendering them useless as a water source. An over abstraction of water from the environment also causes liminal land to dry out, which makes settlement on these low lands possible (Acreman 2000), and these settlements are then susceptible to flooding because are they built not only on lowlands but also on the former marshlands. Therefore, they lack the protection of liminal land that retains and absorbs volumes of water from a deluge of rain or a high tide (Acreman 2000).

Despite the inherent problems that the hidden nature of the existing infrastructure system leads to, the solutions in London to climate change projections have continued the same vein of engineering solutions. A new energy-intensive desalination plant has been built to provide another water source, and a super sewer is proposed to intercept wastewater overflows in times of heavy rainfall (Thames Water 2007). These solutions remain hidden to the large population of people using the drinking water and wastewater infrastructures; hence, they remain ignorant of their environmental impact due to their water use at times of water scarcity or abundance.

Water has not always been so hidden in the urban environment. Many ancient cities celebrated the water that helped the citizens of the cities thrive with fanfare and mythology. Rome had magnificent aqueducts that fed water to resplendent fountains that provided water for its populace (Chant 1999). Hangzhou and Suzhou are two famed water towns with canals and lakes that continue to be regaled as paradise on earth by the common saying 'Just as there is *paradise* in heaven, there are *Suzhou* and *Hangzhou* on earth'. Mexico City was built in the middle of a great lake (Chant 1999). Angkor's temple complexes were built around canals and water reservoirs (Rigg 1992).

In London, 'The vital importance of water in the everyday life of the City was reflected in the form and siting of the Conduits constructed as the public supply points on the various pipelines bringing water in from the north and west. These were impressive structures, large and highly decorated, often sited in the middle of major thoroughfares' (Flaxman and Jackson 2004). Each of these waters was a necessary infrastructure, a connection to water availability, and a source of pleasure for the populace. These infrastructures were a celebration of water that encompassed both the natural water features on which the city was founded and the human ingenuity that transformed natural water sources into a resource for the citizens. It made obvious the connections between the water source and water consumption. It also gave information on water availability, allowing people to alter their water use depending on its abundance. This infrastructure innately displayed environmental information which allowed people to react and modify their behaviours accordingly, unlike present-day infrastructure.

3 Approach: Actor-Network Theory Coevolution

In order to explore the current and changing relationships between people and their water infrastructure, this research used an actor-network theory (ANT) coevolutionary framework. This framework extends the insights from two existing frameworks which originate from Science and Technology Studies. The theoretical framework is important because unlike other studies about water, it does not focus on natural water resources, or technological solutions, or social answers. Instead, it focuses on how all these different actants relate to each other to form the existing water infrastructure and probe the directions in which it has likelihood to change, given people's current unique practices and the imagined changes if water was in excess or scarcity. These relationships are not just understood as either historic or contemporary relationships, but also projected to understand how they may alter in the future. This is different to other approaches to understanding water infrastructure because the actants, such as people, taps, and water, which are involved in the relational network of water infrastructure, are viewed as having an equal effect on the network and cross traditional disciplinary boundaries.

Actor-network theory has been developed for over 30 years and originated in Science and Technology Studies, where it was first used to understand how scientific knowledge was formed and technological innovations invented (Callon 1986; Latour 1987; Law and Callon 1988). It is a relational theory, which has an ontological stance in which all things and phenomena in the world are composed symmetrically of human and non-human relations. In other words, our perception of things in the world is limited to those that we have found ways to recognise and that this recognition requires both a world out there to perceive and a human to perceive it. This ontology results in an epistemological stance that concentrates on investigating the relationships between humans and non-humans that bring phenomena into being. This means that no knowledge can be free from the cultural and political biases of the human, but equally, these cultural and political biases are formed from the material world that surrounds the human (Latour 1993, 2005; Law and Hassard 1999).

The ANT approach has been used to understand many different phenomena including water infrastructures from the large-scale systems of water provision in Paris and Istanbul (Dinckal 2008; Latour and Hermant 2006), to the bush pumps of Zimbabwe (de Laet and Mol 2000) and the taming of the Rhine River for navigation and trade (Disco 2008). The ANT analysis of these phenomena are snapshots of the relationships between people and things that make up the water infrastructure as it exists or existed. Sometimes, it simply describes the relations of the system as it exists, such as the Paris drinking water system (Latour and Hermant 2006), and in other instances, it describes a change in technology, which is an example of an intense change in network relations, such as the spreading of water pump technology (de Laet and Mol 2000) and the control of urbanising water navigation systems (Disco 2008).

The intense change in network relations can be characterised by five distinctive stages: problematisation, interessement, enrolment, mobilisation, and stabilisation (Callon 1986; Latour 2004). Problematisation is when an actant decides that they want to effect a change in the world. Interessement is when an actant starts to probe the relationships between things to find out whether other actants would be willing to make this type of a change in the world. Enrolment is when the actant gathers other relevant actants to make this change. Mobilisation is when the actant effects all the other actants and makes the desired change in the world. Finally, the stabilisation occurs when all the actant relations are stable, are continually mobilised, and no longer questioned. This way of looking at change in network relations is useful, but does not probe the question of how new problematisations arise.

However, other Science and Technology Studies that use a socio-technical approach to understand water infrastructures have shown that the change in network relations appears from problematisations that exist from tensions in the existing network relations between people and technologies. This means that network relations are not just subject to short periods of intense change, but that these changes in network relations can occur through gradual change over long durations of time. In other words, seismic shifts in relations between things can develop through many tiny changes over a long time frame: a coevolution of relations (Geels 2005; Shove 2004). Coevolution shows that new relations arise from pre-existing relationships between things. There is a trajectory to the change that occurs which is formed by existing relations (Geels 2010).

By developing a new framework, ANT coevolution, it is possible to take from ANT the idea of symmetry of influences between human and non-human relationships in creating actants and networks. It also uses the ANT concept for the process of new network formation and stabilisation. It joins this with the socio-technical understanding of incremental network change over time, so that the relations between humans and non-humans are always in flux, always being remade, and thus continually open to change. But also that this change is not random, and it has a particular trajectory that is based upon pre-existing relations. This moves both theoretical frameworks from describing existing situations to being able to show trajectories of coevolutionary pathways that exist within the tensions of current network relations.

4 Method: Developing New Water Infrastructures in London

The ANT coevolutionary theoretical basis drives research that is focused on finding the existing relationships between humans and non-humans, probing their relative stabilities, and finding probable new relationships. The framework was used to develop methods to specifically probe the relationships that form the urban water cycle in London in order to find what coevolutionary pathways exist in response to the extreme weather projected from climate change models. The combination of methods that were used for this research included interviews, group discussions, photographic diaries and notebooks, design illustrations, and models.

In order to understand the existing relations between actants that form the existing stabilised water infrastructures, three methods were employed: interviews, group discussions, and photographic diaries and notebooks of water interactions. Fifty-three people took part in this initial stage. Thirty-five of these people returned water diaries describing their everyday interactions with water, such as toilet flushing and drinking, as well as things that they did on an intermittent, but regular basis, such as laundry and rowing on the river, and some people also recorded unusual things they did with water such as picnics by the reservoir and a boat trip. The combination of the group discussions, interviews, and diaries helped to trace the relationships between the humans and non-humans that create the extant water infrastructure in London.

These initial interviews and group discussions also investigated how people speculate that their water interactions would alter, if they were faced with the problematisations of extreme shortage or an extreme excess of water. Floods and water scarcity are both projected to have a high likelihood of occurring in climate change scenarios (Jenkins et al. 2009). In the case of extreme scarcity, the scenario posed was one where direct water supply to the point of use was stopped, such that supply to showers, taps, and toilets had ceased; the only available water source was a standpipe about 750 m/820 yd away from the front door of the home. In the case of extreme excess, the scenario considered was where there was seasonal flooding to the property of approximately 30 cm/1 foot deep, which occurred for about a fortnight annually. The conjectures about the imagined behaviour from the respondents in these situations revealed which relations would be most likely to change. It also indicated the likelihood of such a strategy being adopted by the number of people who suggested the same behaviour.

By understanding these existing, stable, and fragile relationships that create the water infrastructure in London, it was then possible to speculate on probable ANT coevolutionary pathways that could form in the future to adapt to climate change. These coevolutionary pathways were explored through an interessement. Firstly, this interessement was tested by using the imagined changed relationship between people and water to formulate design propositions which responded to the suggestions of people in the interviews and group discussions. These design propositions were expressed in illustrations, models, and verbal descriptions. Secondly, this interessement was tested by employing these coevolutionary design propositions as the basis for further interviews and group discussions. These were conducted in a further sets of interviews and group discussions with forty of the original 53 people and then finally in group discussions with 15 of the initial participants and the introduction of 23 new participants. The new participants were invited in order to test whether the coevolutionary pathways would remain valid to people not included in the process of their development.

All the participants understood that the scope of the research was to investigate the likelihood of different reconfigurations of the urban water cycle for adaptation towards greater sustainability and to cope with climate change. All the participants remain anonymous and are referred to in the research via pseudonyms. Furthermore, all participants reserved the right to withdraw from the research without penalty. There were no ethical issues for the participants in providing their private viewpoints for this research.

Together the design propositions, and the interviews and group discussions exposed the human and non-human relationships that were most likely to be mobilised in the future due to a high degree of interessement from the participants and those that were least likely to occur due to a lack of interessement. This developed the ANT coevolutionary pathways of change.

5 Results: Coevolutionary Pathways for New Water Cycles

This method found that within the 53 participants, most people had the same relationships to obtain and use water. There were some variations which included ways to reuse and conserve water some of which involved altering the way technology was used such as using the bath or a washing up bowl as a reservoir to reuse water, or not flushing the toilet after urination; others introduced new material configurations such as installing a grey-water recycler, or a water diverter to the shower and basin water. These variations formed the basis for coevolutionary design suggestions.

It was found that there was no coevolutionary pathway for adaptation to regular floods; however, there were two distinct coevolutionary pathways that would create new water cycles in times of water scarcity. One coevolutionary pathway was the development of polyculture water reuse communities. The second coevolutionary pathway was the development of an alternative sanitation infrastructure that would reuse by-products of human waste and treatment for fertiliser and energy.

5.1 Flood Landscapes

In the first interviews and group discussions, there was no interessement from the human participants to prepare themselves for a scenario of regular flooding. The overwhelmingly common response to flooding was that people would move house rather than adapt to regular flooding. Many people would not engage with the idea of flooding as they believed that it would not affect them.

The few people that had experience of regular flooding either personally or through friends or family did not suggest moving; instead, they mentioned modifying their home to prevent water damage. This shows that experience and knowledge limit imagined reconfigurations.

However, when these suggestions were put forward in a set of design interessements, the overwhelming majority maintained their initial response that they would rather move home than adapt to regular flooding. Therefore, there was no coevolutionary pathway for this climate change circumstance.

5.2 Water Reuse

When asked how people thought they would alter their water practices in times of water scarcity, the largest response was to attempt to reduce existing water use. One of the ways to reduce drinking water usage was to reuse water from one use to another; another way was to collect rainwater and finally to find other alternative water sources. Grey-water reuse was the alternative most often suggested (30), followed by rainwater harvesting (17). These potential ANT coevolutionary pathways were tested through a series of four design interessements.

5.2.1 Kitchen Water Reuse

The first design interessement was to modify the existing washing up bowl actant. This was a discrete, widely used actant that was already being used by a few people to reuse water today. However, the convenience of this water reuse was hampered by the awkwardness of handling the bowl. Therefore, the ANT coevolution was to provide lids and handles to the washing up bowl so that it would be easier to lift out of the sink and move around without splashing water around the house. The cleanliness of the water after its initial use of washing vegetables and washing or rinsing dishes would determine its next use. The suggested uses included flushing the toilet, watering the garden, cleaning bikes, and the house. More sophisticated lids could also be developed that also incorporated spouts, and water roses to enable the watering of pot plants, or sprinkling the garden.

The majority of people stated that this was a way of reusing water that they would be willing to do. However, a further ANT coevolutionary pathway was indicated as people speculated that others may not be willing to undertake the heavy lifting required for this form of water reuse. This led to alternative proposals of ANT coevolutions by these human actants to make water reuse even easier. This included dual waste pipes for the kitchen sink which could divert water to either an irrigation tank or the sewer and using a hand-powered pump to drain water to a tank outside. Another suggestion was that water from showers would be a more effective source of water reuse than the washing up bowl. This was the basis for the development of the next design interessement.

5.2.2 Bathroom Water Reuse

The second design interessement was to modify the water flow from shower trays and baths from discharging into the centralised wastewater system. Instead, it would be held in a tank for reuse primarily for toilet flushing. The grey water in the reuse tank could also be connected to drip irrigation in the garden and a garden tank for other water uses. The tank would overflow into the centralised sewer system once it was full or if it needed to be drained for maintenance. On the tank itself would be an indicator, much like those on the side of an electric kettle that would show the water volume inside the tank.

This system was gravity-fed and relied on a common configuration of housing in London, whereupon most houses were two-storeyed attached single dwellings with the wet areas of the house stacked on top of each other to the rear of the building. These houses were on their own individual lots with both a front and back garden. This means that a grey-water tank installed between the two levels would collect water from the shower and basin on the upper floor to flush the toilet on the ground floor and irrigate the garden without requiring the use of any additional energy from a pump. The grey-water tank would have a screen to make sure solids, such as stray hair could be removed before reuse. If the ground floor toilet was the only one used, on average (Butler and Memon 2006), the shower and basin water would be used within a 24-h period for toilet flushing negating the need for water treatment.

The majority of people (27) thought that this was a system that they would be willing to use. People also suggested further ANT coevolutions which were mostly centred on adding more water sources to the tank. This included rainwater, the grey-water from the washing machine, and the grey water from the hand basin. One person suggested that if the water was treated, it could be used for drinking. One person thought that heat could be extracted from the grey-water to lower the energy use of heating in the house.

5.2.3 Shared Household Grey-Water Recycler

The third design interessement was based on a unique participant who had installed a grey-water recycler in his home. In order for the grey-water recycler to be more effective in terms of initial cost, energy efficiency, water treatment, and continuing operating costs, the grey-water recycler was proposed to be shared among a few (3–4) neighbouring households. This relied on the repetitive forms of the building stock common in London. The repetitive forms meant that house lots were approximately the same size, as were the buildings, which means that household sizes and water use would also be approximately alike. Thus, the sharing of this technology would be relatively equitable.

The grey-water recycler system would have all water from baths, shower trays, hand basins, and washing machines diverted for treatment and reuse. This design uses the existing grey-water recycler technology to gather and treat the water before it is pumped to a reservoir tank, from which water is then gravity-fed and distributed into the house for reuse in the washing machine, for toilet flushing, for irrigating the garden, and for house cleaning. The tank would also overflow to a wetland that would provide additional polishing to the water and recharge to the soil water and groundwater, before it is either pumped to the reservoir tank or discharged to the centralised wastewater system. This wetland system enabled people to observe how their water use affected natural systems. It would only be wet when the grey-water recycler overflowed and would give visual feedback as to the water availability from the water balance between grey-water creation and grey-water use.

Most people (31) did not think that this was a possible ANT coevolutionary pathway because they did not believe that neighbouring properties would be willing to share the cost and management of the equipment needed for this form of water reuse. The reasons given were the rapid turnover of tenant occupants, absentee landlords, difficult neighbours, and a high turnover of property ownership.

The ANT coevolutionary pathway suggested by the participants to overcome the lack of commitment between neighbouring properties was for the centralised management of grey-water treatment and reticulation. Some people suggested that this could be done in existing centralised managed buildings such as flats, housing estates, and social housing.

5.2.4 Polyculture Water Reuse Community

Following the idea of a centralised management system for grey-water recycling was the development of an ANT coevolutionary design interessement which would involve an entire neighbourhood block, rather than a few neighbours. Again, it was based on the reuse of grey-water, but it primarily used a set of constructed ponds and wetlands to treat the water rather than a grey-water recycler. This allowed the nutrients in the grey water to be used to feed plants and fish that could then be used as a food stock for livestock, for humans, or for biofuel. The uptake of nutrients by plants, fish, and other biota also cleans and filters the water for reuse by people. This system is based on traditional models of integrated aquaculture and agriculture systems that have been practised in different forms around the world prior to industrial aquaculture and agriculture techniques (FAO 2001). In more recent times, the Vietnamese have promulgated this as VAC (Vuon: garden; Ao: fish-pond; Chuong: pigsty or poultry shed) (Hop 2003; Luu 2001; Ogle et al. 2003). The modification of the grey-water treatment to provide food stocks responds to the additional problematisations of lowering resource use for the importation of food and increasing the appreciation of the beneficial effects between people and biodiverse ecologies. This design actant also adds electricity via microgeneration from the running water between ponds to this established system of water treatment; this also aerates the water increasing its oxygen content. Additionally, to balance the production of treated grey water with water use, it proposes to export grey water to offices, schools, and other industrial uses, where this water could be used for cleaning, irrigation, and toilet flushing.

The majority of people thought that this was a water system that they would be willing to participate in. The ANT coevolutionary pathway of centralised water reuse infrastructure and management was a successful interessement.

The further ANT coevolutions that were suggested from this design interessement included adding rainwater to the system and the reuse of water overflow as an alternative drinking water supply. Another person identified that heat energy could be harvested from the grey-water to heat housing before it was reused. Finally, renewable energy supplies such as wind and solar power were suggested as additional ways in which this design interessement could provide mitigation not just adaptation to climate change.

5.2.5 Elaborated Polyculture Water Reuse Community

The final water reuse design interessement responds to the ANT coevolutionary pathways suggested by adding more energy-harvesting functions, water sources, and water uses to the previous polyculture reuse design actant. This design interessement was both a coevolution and a validation of this coevolutionary pathway because it was only discussed with the 15 initial and 23 new participants in a group discussion format. The polyculture water reuse community is plumbed to collect grey-water from washing machines, basins, baths, and showers in all buildings in the surrounding area and rainwater from roofs. It uses a heat exchanger to harvest the heat from the grey-water, which can be used as a source of heat for the buildings. This grey-water first fills a grey-water tank that is used for any flushing toilet. When this is full, the water is then used in a series of ponds and wetlands where its nutrients are used to grow food and the energy of flowing water is used for electricity microgeneration. These ponds and wetlands also collect surface water from rainfall. Overflow water from the ponds and wetlands is treated by a grey-water recycler. The water that has been treated either by the wetlands and ponds or by the grey-water recycler is pumped up to communal reservoir towers. These reservoir towers are located in the street and combine photovoltaics and wind turbines for electrical energy generation. The towers ensure that the photovoltaics gain maximum sunlight; the wind turbines maximum airflow; and the reservoirs adequate water head for pressurised water distribution. At the base of each tower is a small rain garden, which indicates the availability of water in the local environment. These towers are also a tangible material signal of a polyculture community. The treated water is reused in all building types in the local area as well as irrigation for street trees and parklands. In addition, the water can be boiled and used as a source of drinking water. However, this is not a closed system and assumes that centralised piped drinking water will still be supplied and wastewater sewer systems are used for toilet flushes, surface water and grey-water overflow.

Despite all the additional ANT coevolutions added to this design interessement, there was no additional degree of interessement from the human actants it was tested with, nor were any new ANT coevolutionary pathways suggested. This shows that the ANT coevolutions had reached a saturation point using this methodology.

This ANT coevolutionary pathway in response to water scarcity arose from people who were already conducting their own ad hoc water recycling by using diversion techniques from bathrooms and kitchens to water the garden. By making these actions easier and introducing the idea to more people, coevolutionary pathways developed that led to ever more sophisticated systems of water reuse.

The polyculture reuse was a simplification of their labour-intensive techniques, but also greater sophistication in the methods the water could be reused and the numbers of households that could be enrolled within the same infrastructure system. It repurposed part of the private back garden to create the space for this infrastructure and connected it to nearby public and commercial buildings. This implies new types of social relationships between neighbouring properties. The management of the infrastructure would be a new type of job, and the product that could be harvested from aquaculture is a new market for food crops, animal feed, or biofuels. This ANT coevolutionary pathway shows the potential ways our lives, infrastructures, and places in which we live need to alter in order to adapt to climate change projections of water scarcity.

6 Alternative Sanitation

In times of water scarcity, people also stated that they would reconsider their practices and technologies of sanitation. People firstly suggested that they would begin by practicing 'yellow mellow', flushing only after defecating. This was a technique already in use by some participants who were committed to conserving their use of water. However, in times of water shortage, this can only be a temporary measure because the toilet will eventually need to be flushed. As a more permanent response, seven people suggested that they would change their toilet technology altogether so that it no longer needed to use water to flush waste. The majority of the technologies mentioned were composting toilets, but night soil collection and a chemical toilet were also put forth as options. The suggestions of replacing the flushing toilet show that in times of water scarcity, this technology will begin to be considered ineffective as a means of sanitation and new technologies would be coevolved to replace it. This potential coevolutionary pathway was tested through two design interessements.

6.1 Remove and Compost

This design interessement follows the ANT coevolutionary pathway suggested by some people during times of water scarcity whereupon several people suggested that they would consider changing their toilet infrastructure to a composting toilet in order to save water.

It replaces the flush of the toilet actant with a dry urine-separating sanitation system. The removal of the flush takes away the decision of what is an appropriate time to flush, thereby stabilising non-flushing behaviour. The replacement of the flushing toilet with dry sanitation occurs because the flushing toilet is a technology with a scope of reconfiguration that is limited to altering the volume of water used for flushing and finding alternative water sources used for flushing. Replacing the flushing toilet technology with dry sanitation is an ANT coevolutionary trajectory that maximises the water conservation of the flushing toilet. Using a dry sanitation system removes human waste from the wastewater infrastructure, thus removing many pollutants from entering waterways during sewer overflow events. It also reduces the number of sewer overflow events because wastewater drains would not contain the volume of water from the toilet flush. Moreover, dry sanitation responds to the supplementary rationale of repurposing human waste was a resource by concentrating the nutrients from human waste, making it simpler to extract. Using human waste to create fuel, compost, and fertiliser was common in London's past (Ackroyd 2008), but it has not been used as such within the living memory of the participants.

The human waste of urine and faeces is chemically different. Urine is sterile and can be used as a fertiliser immediately, while faeces contain pathogens that need to be broken down through a composting process to become sterile. Hence, separate collection means that urine can be repurposed immediately and faeces can be decomposed to treat the pathogens. This is done by using aurine-separating toilet bowl. This is similar to current toilet bowls, but has a division between the front and the back, so that the urine is separated from the faeces. To use this effectively, all people will need to sit to pee. This type of toilet bowl is available on the market and has been installed in an office in Switzerland (Tilley et al. 2008). The two types of human waste are collected in two removable canisters beneath the toilet bowl.

These canisters would be collected every 2–3 days, similar to the collection of recycling or the milk delivery, whereby the filled containers are collected and cleaned empty ones are delivered simultaneously. The nutrients collected can then be applied as fertilisers to the land for agriculture or gardens, thereby increasing the sources of nutrients.

The remove-and-compost system responds to the existing built form of most dwellings in London, where there is no space to store the waste from a composting toilet on site. The separation of the two waste materials and its collection demonstrate its value as a source of fertiliser and fuel.

Overall, the replacement of the existing toilet for a new infrastructure was thought to be a good idea by most people. Thirteen people said that they would be willing to change their infrastructure given certain caveats of smell, collection, and hygiene. Five people would be happy to do so if they were forced by external circumstances. Another 13 people thought that the idea was laudable, but they did not yet feel comfortable enough to change their toilet. Only nine people were outright negative about changing their toilet infrastructure.

The ANT coevolutionary pathways put forth were for a more convenient system of waste removal such as the use of a chute or a vacuum. A participant also suggested that the biogas from the decomposition of the solid waste could be used to run the vehicle rather than a petrol combustion engine. A few other people also suggested that the collection of the waste could be combined with existing waste and recycling collection services.

6.2 Remove, Gas, and Compost

Using these ideas, the second design interessement modifies the previous remove-and-compost dry sanitation system by adding uses for the gas produced by the decomposing faecal waste. It also uses a prototype of a remove-and-compost toilet that is currently being developed at Imperial College (Gardiner, 2010) for more hygienic and less smelly waste handling, with no divided toilet bowl. This design interessement was both the coevolution and validation of this coevolutionary pathway because it was only deliberated with the 15 initial and 23 new participants in a group discussion format.

The prototype dry sanitation toilet uses a starch liner within the toilet bowl to seal and package the human waste. It separates the urine and faeces using patented manufacturing separation techniques; thus, the bowl does not have a separating division for the urine and the faeces. The packaging system is operated by a hand crank that winds the waste down into the removable storage chamber below. The urine is separated from the faeces at this point in time. Once the hand crank has been wound, all that remains in the toilet bowl is the clean starch bag ready for the next toilet use. This contains the waste, making it cleaner, more hygienic, and less smelly to handle. The removable storage chamber below the toilet is about the size of a wheeled carry-on piece of luggage for a passenger aircraft and can contain up to 15 kg of waste. One person produces approximately 1 kg of waste a day; therefore, this toilet needs to be emptied at least once a week for a two-person household. The chamber is sealed on removal from the toilet bowl and can be taken to an anaerobic digester for decomposition. The starch bag is decomposed at the same time as the human waste; thus, there is no need to open the starch storage bag prior to its deposit in the digester. The storage chamber is reused.

The decomposing of faecal waste produces two gases that are useful to human life. One is biogas (methane produced from a contemporary decomposition of organic matter) that can be used to drive turbines to produce electricity, used immediately for cooking or lighting, or compressed, and used as a fuel to power vehicles. The use of biogas as a cooking fuel is a technology that has been applied globally, but usually in rural areas of developing countries (Tilley et al. 2008). The decomposition of faecal wastes also produces carbon dioxide, and this could also be siphoned off and used to fertilise the air in greenhouses for growing food. This is a technology being used in a combined heat and power plant in Ontario (EBR Staff Writer 2009).

This new design actant did not show any significant additional human interessement in comparison with the remove-and-compost system, despite the addition of better uses for all the products produced from the anaerobic digestion process and a cleaner more convenient method of removing the waste. Similar sorts of ANT coevolutionary pathways were suggested for this second iteration of the design interessement as the first, such as having a chute, vacuum, or conveyor belt to collect the waste rather than a manual system. This ANT coevolutionary pathway had reached its saturation point using this methodology.

This coevolutionary pathway arose from the ineffectiveness of 'yellow mellowing' in response to prolonged water scarcity. A toilet that required no flushing would overcome this problem and recycle human waste efficiently. A new sanitation system entails that physical infrastructure changes people's bathrooms as well as requires new infrastructure for the collection, treatment, and reuse of human waste, products, jobs, and management. The treatment could be done on existing sites of wastewater treatment, or new sites could be developed in locations close to places of urban agriculture. The products from this system would be fertilisers and biofuels for which there are existing markets. There would be new jobs to operate and manage this system. It also implies new types of relationships between fertiliser and food production.

Like polyculture water reuse, this ANT coevolutionary pathway also shows another potential way our lives, infrastructures, and places we live needs to alter in order to adapt to climate change projections of water scarcity. These paths are nascent, and their implementation depends on many contingent factors such as the duration and severity of water shortages, political will, and resource availability, which cannot be predicted. However, these are the two strongest ANT coevolutionary pathways for London's adaptation to water scarcity.

7 Conclusion

Using ANT coevolution as a theoretical framework to develop this set of methods enabled a detailed investigation into how people might change their water use to adapt to climate change and expand upon these ideas to show ways in which these would coevolve into changing infrastructure, urban form, governance, market, and management in the London context. It first uses ethnographic techniques to document and make self-conscious to the participants of the research their water use. It then uses this self-consciousness to speculate on how this could change in the future, should climate change projections of floods and droughts occur. This process found two different thematic strands within current water uses and speculations which were then expanded on using design propositions to find whether this would form interessement with the research participants. It also found that there was no ANT coevolutionary response to flooding because nobody believed that this would affect them, and if it did, they would move.

This set of methods resulted in two ANT coevolutionary pathways to adapt to climate change-projected water scarcity in London: polyculture water reuse communities and dry sanitation. Both of these ideas can be seen as a radical change to the existing regime, yet they are both based on the way people currently use their water and their ideas of how they imagine they would adapt to a water-scarce future.

ANT coevolution shifts ANT and socio-technical coevolution from historic studies and contemporary ethnographies about the creation of knowledge and technologies. It uses these understandings of the network relations between people and things to develop a set of methods that takes the insights from both perspectives to create explorations of possible new futures that address contemporary problematisations, in this case, adapting to water scarcity in London.

Furthermore, the ANT-coevolved reconfigurations of people and things from this process made more obvious the link between the ecological quantities of resources and the people using it. By making this relationship obvious gives information to the people who are part of the system. This creates the possibility that further ANT coevolutions can occur to continually tune the infrastructural system to the resources available to prevent increasing ecological crises.

Water excess and scarcity are presently a problem for people all over the world. Climate change models project more extreme weather patterns, exacerbating this problem (IPCC 2007). Moreover, different places have different water problems (Gleick et al. 2006). One type of water infrastructure solution cannot possibly suit all situations. By using an ANT coevolutionary approach, it is possible to firstly understand the existing local water cycles and their relationships to infrastructures with its innate tensions and then probe these nascent ANT coevolutionary pathways by testing different problematisations and interessements with other actants. This makes it possible to develop larger infrastructure moves that are tuned to both the unique water situations of the locality and the water practices of the people who live there because it equally examines the implications of both in relation to the other.

The combination of the ANT coevolutionary framework and these methods forms ways in which people's everyday changing material practices and desires can be linked to changes in infrastructure, urban form, culture, and social relations. This framework has shown to be of value in probing responses of how water infrastructures could be adapted to the climate change projection of water scarcity in London and also revealed a lack of capacity for adaptation to flooding. This methodology would be valuable to apply to other places also facing water shortages or excess to see what ANT coevolutionary pathways exist in other locales with different water problems, infrastructures, urban forms, and practices. Equally, this is a framework that joins people and things and could also be used to explore other climate change adaptations beyond water.

References

Ackroyd P (2008) Thames: the biography. Chatto and Windus, London

Acreman M (ed) (2000) The hydrology of the UK: a study of change. Routledge, London; New York

Butler D, Memon FA (2006) Water demand management. IWA Publishing, London

Callon M (1986) Some elements of a sociology of translation: domestication of the scallops and the fishermen of St. Brieuc Bay. In: Law J (ed) Power, action and belief: a new sociology of knowledge?. Routledge, London, pp 196–223

Chant C (1999) The pre-industrial cities and technology reader. Routledge, London, New York

De Laet M, Mol A (2000) The Zimbabwe bush pump. Soc Stud Sci 30(2):225–263

Dinckal N (2008) Arenas of experimentation: modernizing Istanbul in the late Ottoman Empire. In: Hard M, Misa TJ (eds) Urban machinery. The MIT Press, Cambridge, Massachusetts; London, England, pp 49–69

Disco C (2008) Taming the Rhine: economic connection and urban competition. In: Hard M, Misa TJ (eds) Urban machinery. The MIT Press, Cambridge, Massachusetts; London, England, pp 23–47

Environment Agency (2006) The London Catchment Abstraction Management Strategy, Environment Agency. Available at: http://www.environment-agency.gov.uk/research/planning/33432.aspx. Accessed 24 July 2009

Environment Agency (2010) Household water use in London—indicator two. Available at: http://www.environment-agency.gov.uk/research/library/publications/41051.aspx. Accessed 29 May 2011

FAO (2001) Integrated agriculture-aquaculture. United Nations Food and Agriculture Organisation, Rome. Available at: www.fao.org/DOCREP/005/Y1187E/y1187e00.htm. Accessed 28 May 2014

Flaxman EW, Jackson EW (2004) Sweet and wholesome water: five centuries of history of water-bearers in the City of London. E.W. Flaxman, Cottisford

Gardiner V (2010) Loowatt waterless toilet system. Available at: http://www.loowatt.com/. Accessed 13 Sept 2011

Geels F (2005) Co-evolution of technology and society: the transition in water supply and personal hygiene in the Netherlands (1850–1930)—a case study in multi-level perspective. Technol Soc 27(3):363–397

Geels F (2010) Ontologies, socio-technical transitions (to sustainability), and the multi-level perspective. Res Policy 39(4):495–510

Gleick PH et al (2006) The World's water: the biennial report on freshwater resources. Island Press, Washington, Covelo, London

Hop LT (2003) Programs to improve production and consumption of animal source foods and malnutrition in Vietnam. J Nutr 133:4006S–2009S

IPCC (2007) Intergovernmental panel on climate change. Fourth assessment report

Jenkins G, Perry M, Prior J (2009) The climate of the UK and recent trends. Met Office Hadley Centre. Available at: http://ukclimateprojections.defra.gov.uk/content/view/816/9/. Accessed 26 June 2009

Kosmala A et al (1999) Impact assessment of a wastewater treatment plant effluent using in stream invertebrates and the Ceriodaphnia dubia chronic toxicity test. Water Res 33(1):266–278

Latour B (1987) Science in action: how to follow scientists and engineers through society. Harvard University Press, Cambridge, Massachusetts

Latour B (1993) We have never been modern. Harvard University Press, Cambridge, Massachusetts

Latour B (2004) Politics of nature: how to bring the sciences into democracy. Harvard University Press, Cambridge, Massachusetts, London, England

Latour B (2005) Reassembling the social: an introduction to actor-network-theory. Oxford University Press, Oxford

Latour B, Hermant E (2006) Paris: invisible city. Available at: http://www.bruno-latour.fr/virtual/index.html. Accessed 8 Aug 2009
Law J, Callon M (1988) Engineering and sociology in a military aircraft project: a network analysis of technological change. Soc Probl 35(3):284–297
Law J, Hassard J (1999) Actor network theory and after. Wiley-Blackwell, Oxford
Luu LT (2001) The VAC system in Northern Viet Nam. In: Integrated agriculture-aquaculture: a primer. FAO fisheries technical paper. Food and Agriculture Organization of the United Nations, Rome, p 149
Newson M (2007) Land, Water and Development: Sustainable Management of River Basin Systems, 2nd edn. Routledge, London; New York
Ogle BM et al (2003) Food, feed or medicine: the multiple functions of edible wild plants in Vietnam. Econ Bot 57(1):103–117
Rigg J (1992) The gift of water: water management, cosmology and the state in south east Asia. School of Oriental & African Studies, London
Shove E (2004) Comfort, cleanliness and convenience: The Social Organization of Normality, illustrated edn. Berg Publishers, Oxford; New York
Speers A (2007) Water and cities—overcoming inertia and achieving a sustainable future. Cities of the Future. IWA Publishing, London, pp 18–31
Thames Water (2007) Taking care of water: the next 25 years our plan for a sustainable future
Tilley E et al (2008) Compendium of sanitation systems and technologies. Swiss Federal Institute of Aquatic Science and Technology, Dubendorf, Switzerland
Waterwise (2009) Reducing water wastage in the UK: the facts. Waterwise. Available at: http://www.waterwise.org.uk/reducing_water_wastage_in_the_uk/the_facts/the_facts_about_saving_water.html. Accessed 13 July 2009
World Wildlife Fund-UK (2009) Rivers on the edge. WWF-UK, Surrey
EBR Staff Writer (2009) GE energy inaugurates first North American greenhouse cogeneration facility. Energy Bus Rev. Available at: http://fossilfuel.energy-business-review.com/news/ge_energy_inaugurates_first_north_american_greenhouse_cogeneration_facility_090710. Accessed 13 Sept 2011

MIX
Papier aus verantwortungsvollen Quellen
Paper from responsible sources
FSC® C105338

If you have any concerns about our products,
you can contact us on
ProductSafety@springernature.com

In case Publisher is established outside the EU,
the EU authorized representative is:
Springer Nature Customer Service Center GmbH
Europaplatz 3, 69115 Heidelberg, Germany

Printed by Libri Plureos GmbH
in Hamburg, Germany